AF324065

Antioxidants and Stem Cells *for* Coronary Heart Disease

Philip B. Oliva, M.D.

Colorado Heart Research and Education Association, USA

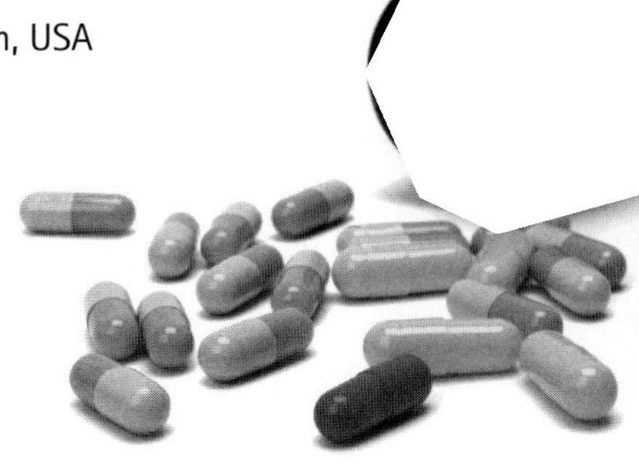

Antioxidants and Stem Cells
for
Coronary Heart Disease

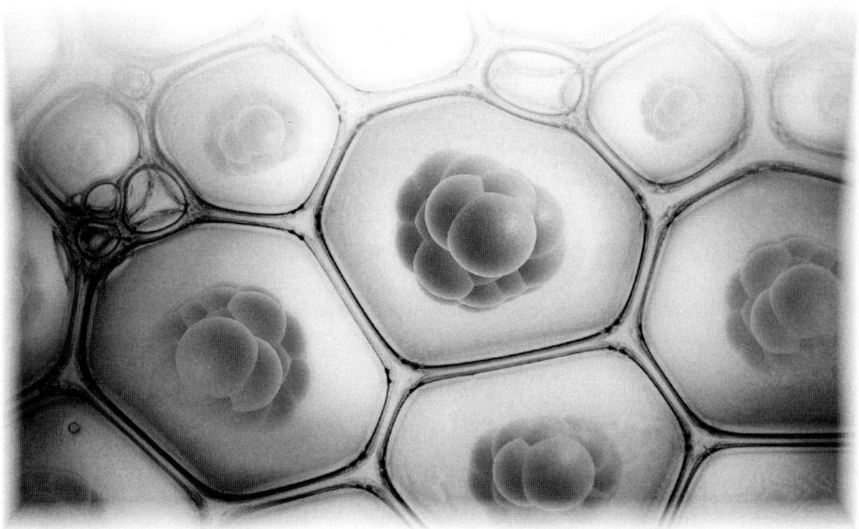

World Scientific

NEW JERSEY · LONDON · SINGAPORE · BEIJING · SHANGHAI · HONG KONG · TAIPEI · CHENNAI

Published by

World Scientific Publishing Co. Pte. Ltd.

5 Toh Tuck Link, Singapore 596224

USA office: 27 Warren Street, Suite 401-402, Hackensack, NJ 07601

UK office: 57 Shelton Street, Covent Garden, London WC2H 9HE

Library of Congress Control Number: **2013956293**

British Library Cataloguing-in-Publication Data
A catalogue record for this book is available from the British Library.

ANTIOXIDANTS AND STEM CELLS FOR CORONARY HEART DISEASE

ISBN 978-981-4293-44-0

Typeset by Stallion Press
Email: enquiries@stallionpress.com

Printed in Singapore by Fuisland Offset Printing (S) Pte Ltd

CONTENTS

PREFACE

Science and medicine need not be arcane. Both can be expressed in nonscientific terms. That is the aim of this book—to present in *plain English* information regarding vitamin antioxidants and stem cells *vis-à-vis* coronary heart disease, the leading cause of death in America, Canada, Europe, Southeast Asia, and Australia.

The title of this book is more apt to catch the eye of folks over 35 — the target audience — those under 35 being the invincible ones. That is because while older adults may experience the terrible outcomes of this devilish disease, its roots track back to infancy and all we ate during our childhood, adolescence and our younger adult years. It takes egg upon egg, hot dog upon hot dog, year upon year for plagues to insidiously narrow down coronary — and carotid — arteries, but just a few seconds of time for a blood clot to form leading to a heart attack or stroke. All those eat-whatever-we-want years are suddenly paid for as sirens wail on the way to the ER.

You may wonder, *What's the connection between antioxidants and stem cells?* The answer is that they're both hot topics, be it in print, television commentaries, or happy hour conversations. There's also a disjunctive relationship that will become apparent as time goes by. Anticipate the unexpected — or at least recognize the unexpected when it happens and figure out why it did.

No work such as this book is accomplished without the strong support of others. Many people have chipped in, in a variety of ways — particularly people with advanced computer skills. I am especially grateful to several individuals affiliated with the University of Colorado Medical School in Denver: *Ryan Laterzo,* then an undergraduate at the University of Colorado, now himself a cardiologist; Sergey Markovitch, a very smart graduate student from eastern Europe; and David H. Spodick, M.D., professor emeritus at the

University of Massachusetts in Worcester for his inspirational writing style over the years. Thanks are also due to copy editors Marta Victoria Colon, Shelley Chow, and Jihan Abdat of World Scientific Publishing Company for their useful comments and suggestions...and patience; a number of librarians in the Metro Denver region, the Mayo Clinic, and the NIH; and my son, John, and my daughter, Julie (now in their mid-30s and early-40s, respectively), for putting up with a dad who has holed up in science libraries for a big chunk of the last five years or so.

Part I

1

THE FRENCH PARADOX

Three culturally distinct races — the French, the Japanese, and the North American (as well as Greenlandic) Eskimos — share a medical blessing: they have fewer heart attacks than the natives of many other countries. Their coronary resistance resides in what they eat and/or drink. In the case of the French, although they eat a lot of rich food, they also drink a lot of (red) wine. In the case of the Japanese, they eat a lot of low-cholesterol seafood and drink a lot of sake and beer. In the case of the Eskimos (or the Inuit[a]), they eat a lot of marine life — fish, whale, and seal[1] — and drink a lot of homemade or distilled spirits.

These dietary, cultural or medical facts spawned curiosity about compounds in food and/or alcoholic beverages that might protect against coronary heart disease. Advancement of the concept that low-density lipoprotein (LDL), the principal carrier of cholesterol in the blood, must be oxidized before it becomes dangerous[2] has led to intensive research on ways to slow or prevent its oxidation. And after wine was discovered to have antioxidant properties in 1993,[3] scientists conjectured that alcohol might link the three dissimilar cultures and their shared resistance to heart disease.

A trickle of publications regarding antioxidants and coronary heart disease appeared in the medical literature in the mid-1990s, and has now grown into a sea of papers throughout the scientific literature. The published material records the findings of a considerable body of research concerning the clinical use of antioxidants to prevent or treat coronary heart disease, and, to an even greater extent, it registers the results of an extraordinary volume of research by basic scientists pertaining to the biochemistry of how LDL is oxidized. The

---

[a]The term "Eskimo" means "he who eats (meat) raw." "Inuit" translates best as "real people." Most North American and Greenland Arctic Circle dwellers prefer the latter term.

last chapter of the antioxidant story has yet to be written. When it finally is put into print, it will be a memorable one. But, even today, the murky waters that encircled antioxidants just a few years ago have become much clearer. It's a tale worth telling even though its ending is still up in the air.

In the Beginning. . .

It all began in 1979, when a French physician published a paper in England's leading medical journal, showing that the mortality rate due to coronary heart disease (CHD) was lower in France — where wine consumption was higher — than in 17 other countries, including the U.S. (Fig. 1.1).[4]

At about the same time as the "French paradox" was being recognized, a group of researchers in the Department of Molecular Genetics at the University of Texas (in Dallas) came upon another seeming contradiction.

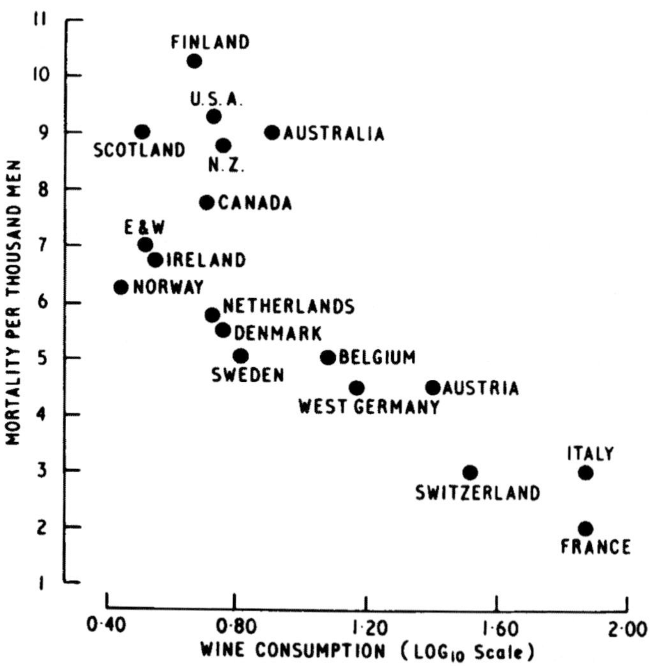

Fig. 1.1. The inverse relationship between mortality from coronary heart disease and wine consumption among men aged 55–64 in 17 countries.

(*Source:* Reproduced with permission from the publisher.) Reference 4.

They discovered that "bad cholesterol" (i.e. low-density lipoprotein)[b] was taken up by cells lining the walls of blood vessels at a rate too slow to result in any significant accumulation of cholesterol within their cytoplasm.[c] Yet blood vessels critically narrowed by cholesterol deposits had accumulated a huge amount of the lipid — a fat that, by definition, is not soluble in water. The scientists also learned that patients with familial hypercholesterolemia (elevated blood cholesterol level),[d] an inherited disorder of lipid metabolism leading to blood cholesterol levels often in the range of 400–1,000 mg/dL, similarly had vascular cells that did not take up any bad cholesterol. . .at least when they were incubated in a laboratory with their own blood containing very high levels of cholesterol. They were essentially invulnerable to harmful cholesterol. Yet the same type of cells clearly took up cholesterol when they were within the human body in direct contact with circulating blood, leading to severe atherosclerosis[e] even in the absence of any other risk factors. To

---

[b] Low-density lipoprotein is a large compound made up of 22,000 cholesterol molecules and 2616 fatty acid molecules, which together constitute 79% of the compound's weight, attached to one protein molecule accounting for the remaining 21% of its weight. The lipoprotein is the principal carrier of cholesterol in the bloodstream. Cholesterol, being a lipid, is not soluble in the watery plasma, which constitutes 55% of the blood volume. Therefore, a carrier is needed to transport cholesterol from its origin in the liver or digestive tract to its destination at cells throughout the body. The modifier "low-density" refers to the compactness of the lipoprotein, expressed as the ratio of mass to volume. In the case of cholesterol-carrying lipoproteins, such as LDL and HDL, the units of density are g/cm^{-1}, with 1.063 g/cm^{-1} separating low- from high-density lipoproteins. The range is 1.019–1.210 g/cm^{-1}.

[c] Such cells are known collectively as the intima — the innermost coat of arteries and veins — though only the intima of arteries takes up cholesterol. The *single* layer of intimal cells in direct contact with blood is called the endothelium.

[d] Individuals with this rare condition affecting one in one million persons inherit a defective gene governing LDL receptors (to be discussed in Chap. 3) from each parent. The defective gene usually leads to a very high cholesterol level (600–1000 mg/dL), which, in turn, results in a heart attack during childhood.

[e] The term "atherosclerosis" refers to the disease process that causes the deposition of LDL in arteries. It describes the mushy consistency of cholesterol in the wall of arteries during the early and intermediate stages of the disease. The term comes from the Greek words "*athero*" (meaning "porridge") and "*sclerosis*" (meaning "hardening"). The term "arteriosclerosis" refers to the more advanced stage of the disease, when the arteries become brittle. That stage of the disease process is often not reversible. "Arteriosclerosis" comes from the Greek words "*arterio*" and "*sclerosis*" (meaning "artery" and "hardening," respectively). The phrase "hardening of the arteries" is often applied to such vessels. In this book, "Atherosclerosis" is used almost exclusively to discuss (for the most part) the early and mid-phases of the disease.

explain the paradox of how fat-filled cells in the lining of such cholesterol-narrowed blood vessels became overloaded with the lipid in spite of taking it up too slowly in the laboratory to result in a buildup, and to account for why LDL was also taken up too slowly to cause atherosclerosis — even in individuals *without* a genetic predisposition to develop the disease — it was suggested that LDL must be changed into a form that could both penetrate cells and be taken up more rapidly than "native" LDL.[f] Only then could massive deposits of LDL occur throughout the body. In 1984, that related form was shown to be "oxidized" LDL.[5]

References

1. Bang HO, Dyerberg J, Hjorne N. (1976) The composition of foods consumed by Eskimos. *Acta Med Scand* **200**: 69–73.
2. Steinberg D, Parthasarathy S, Carew TE, Khoo JC, Witzum JL. (1989) Beyond cholesterol: modifications of low-density lipoprotein that increase its atherogenicity. *N Engl J Med* **320**: 915–924.
3. Frankel EN, Kanner J, German E, Parks E, Kinsella JE. (1993) Inhibition of oxidation of low density lipoprotein by phenolic substances in red wine. *Lancet* **341**: 454–457.
4. St Leger AS, Cochrane AL, Moore F. (1979) Factors associated with cardiac mortality in developed countries with particular reference to the consumption of wine. *Lancet* **313**: 1017–1020.
5. Steinbrecher UP, Parthasarathy S, Leake DS, Witztum JL, Steinberg D. (1984) Modification of low density lipoprotein by endothelial cells involves lipid peroxidation and degradation of low density phospholipids. *Proc Natl Acad Sci USA* **81**: 3883–3887.

---

[f]The term "native LDL," rather than "natural LDL," is traditionally used in science because LDL does not occur "naturally," i.e, it does not occur in nature. The adjective "native" is used to indicate that the LDL exists in its original (i.e, its unoxidized) form. Native (unoxidized) LDL is harmless. It does not damage cells until it undergoes oxidation.

2

LDL OXIDATION — BRIEFLY

But just what is "oxidized LDL"? How does native (unoxidized) LDL become oxidized? How does its oxidation accelerate the deposition of cholesterol in the intima of arteries? How do antioxidants affect those deposits, particularly those in the coronary arteries? And, perhaps most importantly, can taking an antioxidant prevent a buildup — and perhaps, one day, a heart attack?

We bandy the term "oxidized cholesterol" around as if it were a well-known compound, a common dietary substance — like flour or rice. We use the term "oxidation" (with reference to LDL) as if it were a straightforward process — like boiling an egg. We even throw around phrases related to oxidized LDL — such as "free radicals," "omega-3 fatty acids," "cytokines," "superoxide ions," and "trans fats" — at cocktail parties and during dinner table conversations as if we were talking about last Sunday's NFL games or today's stock market activities. But our ease with these technical terms belies the complexity of the LDL oxidation process.[a] That process will be put in a nutshell to encourage

---

[a]Understanding the process of LDL oxidation gets confusing right off the bat, because the process doesn't necessarily require oxygen.

The term "oxidation" was used originally to describe a process during which a substance, such as iron, chemically combined with oxygen to form an altered compound, — in the case of iron, rust. However, the definition of "oxidation" was broadened when it was learned that at the atomic level it was the loss of an (negative) electron from the substance being oxidized that "oxidized" it, irrespective of whether oxygen was available or not. In the case of iron changing to rust, oxygen is present and it converts reduced, ferrous iron (Fe^{++}) into its oxidized ferric (Fe^{+++}) form. But at an even more fundamental level, it's the loss of one electron from the ferrous (Fe^{++}) ion that converts reduced iron into its more positive state, the oxidized Fe^{+++} ion, which becomes rust — ferric oxide. Thus, oxidation is now defined as a loss of one or more electrons, while reduction is defined as a gain of one or more electrons, whether or not oxygen is involved in the reaction. The Mnemonic "oil rig" is a convenient way to remember this oxidation/reduction, electron-number relationship, "oil rig" standing for "oxidation is loss, reduction is gain."

continued reading, before going into more detail and losing whatever reader interest exists at this point. Readers bored by biochemistry, even if condensed, should turn to Chapter 10.

LDL Oxidation. . .Briefly

Unoxidized LDL is harmless as it circulates in the bloodstream. It would not lead to any accumulation of cholesterol in the walls of the blood vessels through which it flows if it remained *within* the bloodstream. It would not lead to an artery being blocked by plaque if it stayed in solution within the channel of the artery, just as iron in water would not lead to a pipe being clogged by rust if the iron remained in solution i.e. if it did not react with oxygen in the air above/around the water or dissolved in it). But (unoxidized) LDL *does* cross from blood within the lumen of a blood vessel into cells making up the blood vessel's wall, simply because the concentration of LDL in the blood is higher than its concentration in the wall.

At first, the unoxidized lipoprotein is confined to a narrow space within the wall of arteries where it remains innocuous, until it is acted on by certain waste products — known as free radicals — made by cells in the arteries' wall. The byproducts are, in essence, biological (potentially) toxic wastes generated by arterial cells as they burn oxygen to use nutrient food molecules. . .just as potentially toxic automotive wastes are generated by cylinders in a car engine as they burn oxygen to use gasoline. These (human) oxidative waste products now oxidize LDL (even though the cells' oxygen has been burned and none is left to oxidize LDL) by removing electrons from the LDL molecule, thereby

---

The majority of oxidative processes in the human body are not dependent on the immediate presence of oxygen. They are accomplished by the enzymatic removal of a hydrogen atom or atoms (each hydrogen atom having one electron) from a molecule, such as the successive removal of several hydrogen atoms from adenosine triphosphate (ATP) during the enzymatic conversion of glucose to pyruvate or lactic acid during exercise.[1] The highly efficient Krebs cycle, the hub of the metabolic system, is another example of oxidation without oxygen. Three of the eight sequential steps in the cycle involve oxidation achieved by the transfer of hydrogen atoms from one compound to another, with energy released at each step. As a result, 11 times more energy is produced by utilizing the Krebs cycle than would have been produced by burning glucose directly. Oxygen is finally converted to water by nicotinamide dinucleotide (NAD) in its reduced form, NADH, but only after much energy for vital functions has been released or stored through the metabolism of fats, carbohydrates, and proteins — the raw materials for the cycle.[2] NAD itself is formed by cells from niacin, also known as vitamin B3.

meeting the modern definition of oxidation. Oxidizing LDL is the physiologic equivalent of rusting.

A Prelude to a More Detailed Description of LDL Oxidation in Chap. 9

That's the nutshell version of how LDL is oxidized. The expanded version picks up the LDL molecule as it is diffusing passively — simply due to its higher concentration in the bloodstream — across the exceedingly thin lining of an artery into the space immediately beyond; that is to say, immediately underneath the innermost lining — the subendothelial space. But, unlike outer space, which is filled with asteroids, comets, and tens of millions of stars — not to mention objects we can't see, such as ultraviolet rays and quasars — the subendothelial space is (normally) devoid of biological bodies, such as cells. Instead, it is filled with a gelatin-like substance that appears clear under a conventional microscope, although strands of connective tissue (collagen) become visible under an electron microscope. The tangled collagen strands make the subendothelial space look more like the crawlspace under a house in the Louisiana bayou filled with (plumbing) pipes and (electrical) wires. And, akin to a crawlspace, the subendothelial space is largely ignored until something goes wrong.

The Endothelium and the Subendothelial Space

The term "endothelium" describes a delicate tissue lining the inside of the heart, arteries, and veins, as well as the abdominal and thoracic cavities. The prefix "endo-" means "inside," while the suffix "-thelium" refers to a tissue that covers the surface of or lines a body cavity. The endothelium, then, lines the extensive lining of the entire cardiovascular system, an estimated distance — including capillaries — of 62,000 miles, or more than twice the circumference of Earth.[3]

Before the 1960s, scientists tended to ignore the endothelium and the subendothelial space (Fig. 2.1). The vascular endothelium was viewed as simply a "sheet of cellophane"[5] between the blood and the more peripheral layers of a blood vessel's wall, while the subendothelial space was looked on as if it were an extraterrestrial "black hole" that might be interesting to explore. . .someday. But the technology wasn't on hand for subendothelial-space exploration in the 1960s, though it was available for outer-space exploration, as many

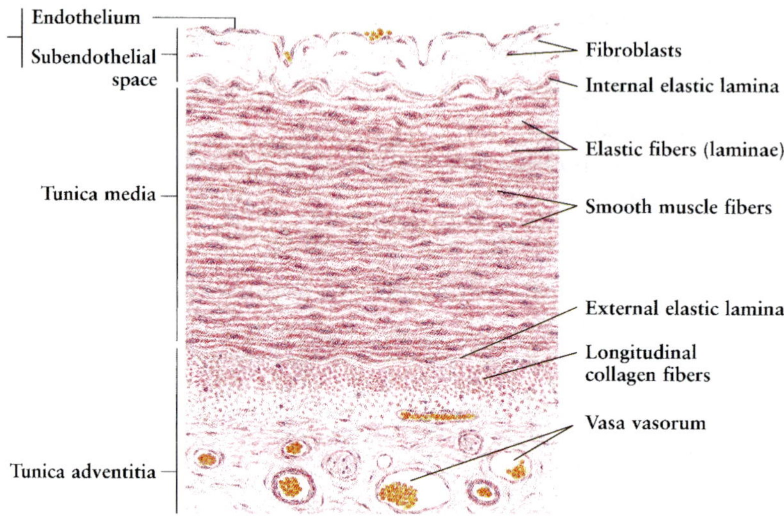

Fig. 2.1.   A photomicrograph of a human artery. The endothelium is only one cell thick. The subendothelial space is the acellular (free of cells) region between the endothelium and the internal elastic membrane. The tunica media is the thickest layer of an artery. Its constituents are listed on the right hand side of the figure. The tunica adventitia is the outermost layer and is composed of fat cells, nerve fibers, and tiny blood vessels (vasa vasora).

(*Source:* Reproduced with permission from the publisher.) Reference 4.

witnessed in July 1968, when Neil Armstrong took "one small step for a man, one giant step for mankind."

Back on planet Earth in the 1970s, however, the endothelium and its underlying space began to emerge as targets for scientific investigation and research dollars. A new breed of scientists was born who shared an intense interest in the biochemical changes at the cellular and molecular levels that precede (clinical) coronary heart disease and cancer, and that accompany the aging process. The disciplines of "cellular biology," "molecular biology," and "vascular biology" came into being. The development of the electron microscope and the introduction of cell culture techniques, allowing the growth of cells in a laboratory, were important technological advances.[6,7] Soon, it was learned that the single-celled endothelium lining every blood vessel in every human organ regulates the manufacture and/or distribution of important, diverse molecules such as blood thinners, blood-clotting factors, and certain hormones, as well as the distribution of red blood cells, white blood cells, and platelets. It is a veritable factory manufacturing thousands of enzymes — an enzyme being a protein that changes the speed of a chemical reaction without its own structure being changed — which are then transported into the subendothelial space, where an even larger number of biologic reactions take place. More recently, it has been further established that some endothelial reactions stimulate other enzyme pathways in the vessel wall, while yet other endothelial reactions inhibit vascular cellular responses. In short, the endothelium is now recognized as an extremely dynamic tissue — not an inert one — *and* one that is crucial to the prevention and the treatment of vascular diseases.[8]

Endothelial research is a fascinating scientific frontier.

References

1. Robergs RA, Ghiasvand F, Parker D. (2004) Biochemistry of exercise-induced metabolic acidosis. *Am J Physiol Regul Integr Comp Physiol* **287**: R501–R506.
2. La Noue K. (2001) Citric acid cycle. In: *Encyclopedia of Life Sciences.* Macmillan, London, New York, Tokyo, Vol. 4, pp. 552–558.
3. http://en.wikipedia.org/wiki/blood_vessel
4. Cormack DH. (1987) The circulatory system. In: *Ham's Histology,* 9th ed. JB Lippincott, Philadelphia, London, Mexico City, New York, St Louis, Sao Paulo, Sydney, pp. 423–449.
5. Florey L. (1966) The endothelial cell. *Br Med J* **2**: 487–490.
6. Jaffe EA, Nachman RL, Becker CG, Minick CR. (1973) Culture of human endothelial cells from umbilical veins. Identification of morphologic and immunologic criteria. *J Clin Invest* **52**: 2745–2756.
7. Goldstein JL, Dana SE, Brunschada Y, Brown MS. (1975) Genetic heterogeneity in familial hypercholesterolemia: evidence for two different mutations affecting functions of low-density lipoprotein receptors. *Proc Natl Acad Sci* USA **72**: 1092–1096.
8. Cines DB, Pollack ES, Buck CA, Loscalzo J, Zimmerman GA, McEver RP, Pobec JS, Wick TM, Konkie BP, Schwantz BS, Barnathan ES, McCrae KP, Hug BA, Schmidt AM, Stern DM. (1998) Endothelial cells in physiology and pathophysiology of vascular diseases *Blood* **91**: 3527–3561.

3

ACETYLATION OF LDL

That was the stage in 1972 when two early molecular biologists, Michael Brown and Joseph Goldstein (University of Texas, Dallas), began their research careers. Puzzled by why vascular cells from patients with familial hypercholesterolemia — an inherited disorder in which vascular cells contain excessive cholesterol — seemingly did not take up "bad" cholesterol when they were *incubated* with blood having an extremely high cholesterol level, Brown and Goldstein postulated that *circulating* cholesterol must somehow be altered before it could be taken up by vascular cells to form so-called "foam cells." Foam cells are cholesterol-filled cells that create the appearance of frothy bubbles in the foam on an ocean beach. (Their formation and function will be discussed in Chap. 12.) Accordingly, they chemically modified LDL "simply" by acetylating[a] it in the laboratory and found that the altered LDL was, indeed, taken up by cells 20 times faster than before it had been modified, leading to a 38fold increase in the cellular content of harmful LDL[b,1]. Besides that, such chemically modified LDL was taken up exclusively by specific receptors on the surface of those cells that had ingested it and it was not taken up by other receptors on the same cells that took up unmodified (unacetylated) LDL. The structural change of the LDL particle induced by acetylation rendered it sufficiently different from the unacetylated particle[c] that the

------

[a]Acetylation refers to the introduction of a very short two-carbon compound — acetic acid — into the molecular structure of another molecule. Acetic acid is the same as store-bought vinegar, the latter being a weak (approximately 6%) solution of the former.

[b]During the years between the initiation of their studies on LDL and their discovery that acetylation converted LDL into a form that was taken up rapidly by cells, Brown and Goldstein treated the lipoprotein with many other chemicals before one chemical reaction — acetylation — was successful.[2]

[c]During everyday conversations, we generally use the word "particle" to describe a very small amount of a substance that we can see, such as a particle of dust. The term may also be used

modified version would not "fit" into the LDL receptor, just as a three-pronged plug will not fit into a two-pronged receptacle. In reality, acetylation neutralized the positive charges on the terminal amino groups of lysine molecules in LDL, thus changing the electrical charge of the LDL particle from negative to slightly more negative, making it less attractive to the negatively charged LDL receptor and quite attractive to a different, slightly positive receptor not regulated by the intracellular concentration of cholesterol.[3]

LDL Receptors and "Scavenger Cells"

Since (acetylated) LDL was taken up specifically by cells in the vascular wall that literally ingest foreign material, the biologists who first described their function referred to them as "scavenger cells,"[2] rather than as "macrophages,"[d] their technical name (establishing that scientists occasionally can be whimsical). Now the paradox of why vascular cells from patients with familial hypercholesterolemia (high blood cholesterol) were seemingly invulnerable to cholesterol — at least in the laboratory — was resolved: chemically modified (but not native) LDL attached to specific receptors on the surface of scavenger cells, whereupon it was incorporated into the cells' cytoplasm (Fig. 3.1). (Macrophages cultured from the blood of patients with familial hypercholesterolemia turned out, by the way, to be endowed with a normal number of acetyl-LDL receptors despite having a genetic deficiency of receptors for

---

metaphorically: for example, "That doesn't make a particle of sense," or "He doesn't have a particle of doubt."

In science, the term "particle" technically refers to one of the components of molecules, such as electrons, protons, or neutrons. However, it is often used in the nonscientific sense of the word as a substance that may not be visible with the "naked" (unmagnified) eye. Consequently, LDL is variably characterized as a "particle" or a "molecule" in scientific papers. It will be referred to as a particle in this chapter, implying that it is composed of molecules.

[d]The term "macrophage" comes from the Greek words "*macros*" (meaning "large") and "*phagos*" (meaning "to eat"). A macrophage, then, is a "big eater." Or, more scientifically speaking, a macrophage is a large cell that ingests (any) foreign material, be it in the blood or in other tissues throughout the body. (A macrophage might be envisioned as if it were a Pac-Man gobbling up everything in its path.) A "bacteriophage," on the other hand — while logically it should be a more discriminating type of macrophage that ingests. . .that's right, bacteria — is not. A bacteriophage, or "phage," as it is commonly referred to in scientific and medical circles, is a virus that kills — believe it or not — bacteria. Phages and bacteriophages will be dealt with much later, in Chap. 30.

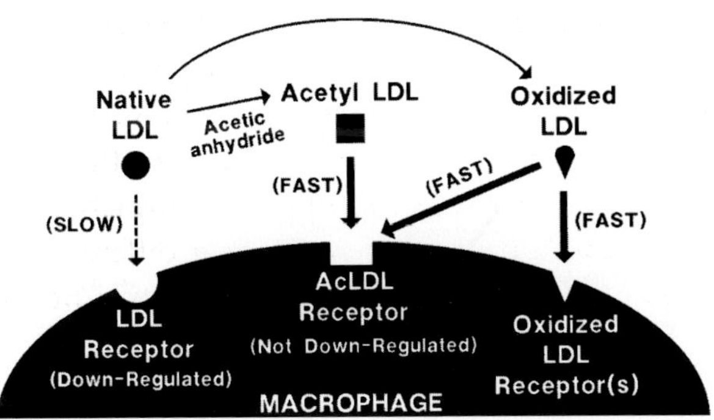

Fig. 3.1.   A schematic representation of the unoxidized (native) LDL receptor — depicted as a half-moon — and the oxidized (scavenger) LDL receptor — depicted as a niche. The former is regulated by the cholesterol content of the cell, while the latter is not regulated by cholesterol. Consequently, native LDL is harmless, while oxidized LDL may lead to foam cell formation.

(*Source:* Reproduced with permission from the publisher.) Reference 4.

unacetylated LDL). Unlike native (unacetylated) LDL receptors, whose activity decreased as cholesterol accumulated within cells (a common biological feedback mechanism limiting the extent of a reaction), the uptake of acetylated (i.e. oxidized) LDL by scavenger cell receptors was unregulated.

The important relationships among the blood cholesterol level, the number of oxidized LDL receptors, and the typical age at which a heart attack occurs are displayed in Fig. 3.2. The relationships appear "illogical" because, *a priori*, it seems clear that a greater number of LDL receptors should lead to more cholesterol being available for uptake by cells throughout the body. In fact, just the opposite happens because the number of LDL receptors is limited by a built-in feedback mechanism: as the intracellular cholesterol level goes up, cells "recognize" the rise and quickly reduce their synthesis of LDL receptors to avoid a potentially harmful cholesterol overload. Consequently, in the Texas scientists' patients with familial hypercholesterolemia who lacked LDL receptors, cholesterol uptake continued unchecked in their vascular cells. The end result was frequently a heart attack before age 30.

Applying a Laboratory Discovery to a Patient

Once in a while in medicine, a single case report illustrates a vague biochemical point or makes a more lasting clinical impression on a physician than a

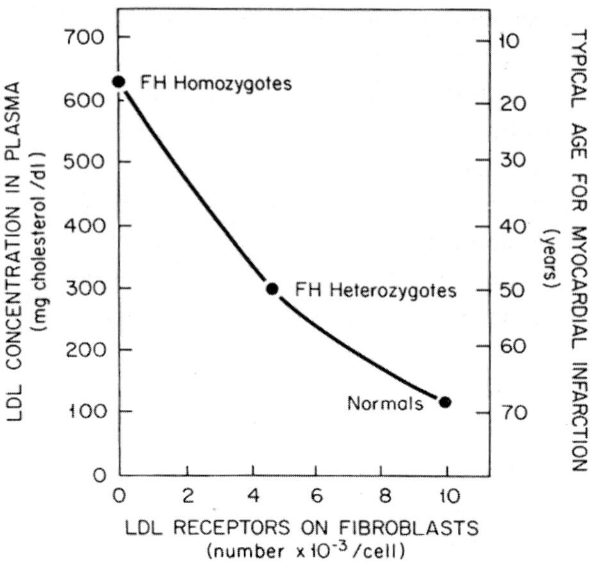

Fig. 3.2. The number of LDL receptors per cell is plotted against the plasma LDL concentration and the typical age at which an acute myocardial infarction occurs in persons with a normal number of LDL receptors and in subjects with the heterozygous and homozygous forms of familial hypercholesterolemia.

(*Source:* Reproduced with permission from the publisher.) Reference 5.

thousand similar — but not exactly the same — cases. In 1984, a six-year-old girl with an adult disease accomplished that just five years after the two types of cell surface receptors were discovered. The child had inherited a gene for familial hypercholesterolemia from each parent and hence had the pure (homozygous) form of the disease. As a result, she had a serum cholesterol concentration of over 1000 mg/dL. By February 1984, she had had an acute myocardial infarction — a heart attack. Continued angina after the heart attack required two coronary bypass operations. But her angina went on.

The grim outlook afforded an opportunity to apply the concept that LDL receptors were necessary to take up physiological amounts of cholesterol (necessary for cell membrane and adrenal hormone synthesis), whereas scavenger receptors (along with their unregulated uptake of cholesterol) stepped into the picture only when the number of (regulated) LDL receptors was insufficient to take up enough native LDL to meet the body's basic metabolic requirements (as above).

Based on the expectation that *more* LDL receptors would *lower* the blood concentration of cholesterol *and* nullify the opportunistic uptake of cholesterol by scavenger cells, the child underwent combined transplantation of her liver and her severely damaged heart. Since three-quarters of the total number of LDL receptors in the body are located in the liver, it was anticipated that by providing a healthy liver with a normal number of LDL receptors the serum cholesterol level would fall. And it did. Dramatically. Within three days, it dropped from over 1000 to 302 mg/dL (Fig. 3.3). Within two months, fatty cholesterol deposits in the patient's skin regressed notably. In December 1984, when her case was reported,[6] her new heart and liver were performing splendidly. (The heart transplant was done because her birth heart had been extensively damaged by a heart attack.) The serum cholesterol remained below 300 mg/dL during the next four years when a followup report was published.[8]

The remarkable drop of her blood cholesterol level after liver transplantation underscored the validity of the LDL receptor concept. We *need* these regulated receptors to take in the cholesterol necessary for normal, physiologic purposes. In recognition of their research establishing that a relatively minor structural change of LDL (induced by acetylation) resulted in

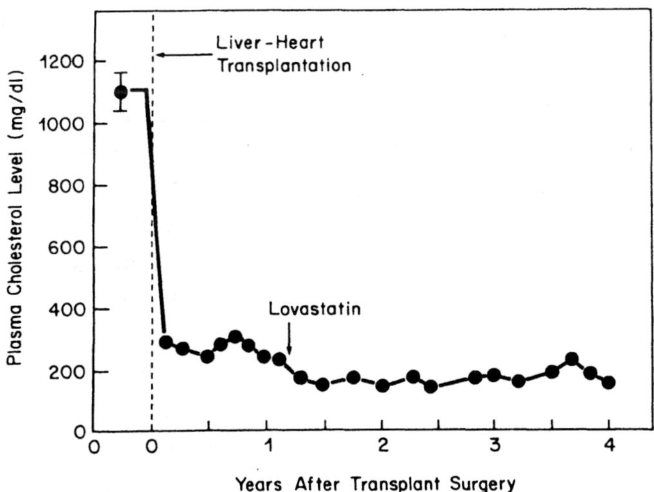

Fig. 3.3. The plasma cholesterol concentration of a child during the first four years following combined liver–heart transplantation. Within a week of surgery, the concentration fell from about 1100 mg/dL to nearly 300 mg/dL. Following the introduction of lovastatin (Mevacor) it dropped to the near-normal 200 mg/dL range.

(*Source:* Reproduced with permission from the publisher.) Reference 6.

an accumulation of excessive cholesterol by macrophages — mediated by specific, previously unrecognized receptors on the cells' surfaces — Brown and Goldstein received the 1985 Nobel Prize in Physiology or Medicine.[e]

The Link Between Acetylated LDL in the Laboratory and Oxidized LDL in the Human Body

Although their discovery generated enormous excitement in basic science circles, the biologic relevance of their laboratory observations, if any, wasn't clear at first because there was no evidence that LDL could be similarly modified *in vivo*, i.e. within the human body. While the acetylation of proteins occurs in a number of human enzyme systems, the recognition of acetylated LDL by macrophage receptors would require the acetylation of an "implausible" number of lysine molecules within the LDL particle.[8] (More than 80% of the 360 lysine groups in the protein component of LDL would need to be acetylated for LDL uptake by macrophage scavenger receptors.[9] There was simply no evidence that acetylation of LDL occurred to any extent, *in vivo*.

Baffled, Dr. Daniel Steinberg at the University of California, San Diego, designed an experiment to see if human endothelial cells could generate a form of LDL similar to — but not the same as — chemically modified (acetylated) LDL. If they could *and* if such a biologically altered LDL used the acetyl-LDL (scavenger) receptor, then. . .then, an important link between the laboratory and human atherosclerosis would be established.

The clear-thinking molecular biologist sought an answer through a simple experiment. . .simple because he appreciated the extraordinary complexity of cholesterol/lipoprotein metabolism, allowing him to know exactly where to begin the search and what to look for along the way. In his N.I.H.-supported laboratory, he incubated endothelial cells extracted from readily available and cheap human umbilical veins with LDL for 24 hours (ever wondered if an umbilical vein might be good for *something* if the nutrient cord in which it had lain for months weren't allowed to dry up and fall off within a week of birth?) and then collected the endothelial-cell-modified LDL to assess its

---

[e]Research regarding the role of cholesterol in causing coronary heart disease has led to 13 Nobel Prizes in Chemistry or in Physiology and/or Medicine. The first was in 1928, while Brown and Goldstein's award is the most recent one. Yet, in spite of the incalculable number of research hours and dollars spent on cholesterol and coronary heart disease, it remains the leading cause of death in the Western world.

effect on macrophage cholesterol metabolism. The study indicated an intricate process during which endothelial cells modified LDL by a mechanism involving the generation of "free radicals" (defined in Chap. 4) by the endothelial cells, themselves. As if the process wasn't already complex enough, he then removed the cells and replaced them with a specific amount of oxidized copper ions (Cu^{++}) on the hunch that when they were reduced to unoxidized copper ions (Cu^{+}), the reduced ions, now being an oxidizing agent, might (alone) also modify LDL. And, they did. This established that the only essential contribution of endothelial cells to the LDL modification (i.e. the LDL oxidation) process was the generation of free radicals — the reduced copper ion (Cu^{+}) being the free radical in this case. And it illustrated that neither cells nor oxygen is necessary to oxidize LDL! In a landmark paper[10] that complemented Brown and Goldstein's 1979 acclaimed publication, Steinberg and his colleagues at UCSD established that (human) vascular cells *could* and *did* oxidize LDL, thereby converting it into a form that was recognizable by the same macrophage (scavenger) receptors that took up acetylated LDL.[f] And, just as importantly, they found that antioxidants completely prevented these changes.

References

1. Goldstein JL, Ho YK, Basu SK, Brown MS. (1979) Binding site on macrophages that mediates uptake and degradation of acetylated low density lipoprotein producing massive cholesterol deposition. *Proc Natl Acad Sci USA* **76**: 333–337.
2. Brown MS, Basu SK, Falck JR, Goldstein JL. (1980) The scavenger cell pathway for lipoprotein degradation. Specificity of the binding site that mediates the uptake of negatively-charged LDL by macrophages. *J Supramol Struct* **13**: 67–81.
3. Basu SK, Goldstein JL, Anderson RG, Brown MS. (1976) Degradation of cationic low density lipoprotein and regulation of cholesterol metabolism in homozygous familial hypercholesterolemia fibroblasts. *Proc Natl Acad Sci USA* **73**: 3178–3182.

---

[f]A report from the Cleveland Clinic published in the July/August 1984 issue of *Arteriosclerosis* verified that LDL could be oxidized by free radicals.[11] The Clinic's report referred to LDL that had been altered so that it was taken up by "scavenger" receptors as "oxidized LDL." Steinberg's concurrent paper called it "modified LDL." In this book, the two terms will be used interchangeably. Sometimes a third term will be employed — "oxidatively modified LDL." While an occasional investigator differentiates among the three terms, they will be used equivalently in this book for simplicity.

4. Steinberg D. Lewis A. (1997) Connor Memorial Lecture: Oxidative modification of LDL and atherogenesis. *Circulation* **95**: 1062–1171.

5. Goldstein LJ, Brown SM. (1977) The low-density lipoprotein pathway and its relation to atherosclerosis. *Annu Rev Biochem* **46**: 897–930.

6. Bilheimer DW, Goldstein JL, Grundy SM, Starzl TE, Brown MS. (1984) Liver transplantation to provide low-density receptors and lower plasma cholesterol in a child with homozygous familial hypercholesterolemia. *N Engl J Med* **311**: 1658–1664.

7. Goldstein JL, Brown MS. (1989) Familial hypercholesterolemia. In Scriver CR, Beaudet AL, Sly WS, Valle D (eds.), *The Metabolic Basis of Inherited Diseases*, 6th ed. McGraw-Hill, pp. 1215–1250.

8. Brown MS, Goldstein JL. (1983) Lipoprotein metabolism in the macrophage: implications for cholesterol deposition in atherosclerosis. *Ann Rev Biochem* **52**: 223–261.

9. The Modified LDL. In: Parthasarathy S (ed.), *Modified Lipoproteins in the Pathogenesis of Atherosclerosis*. RC Landes Austin, 1994, pp. 1–10.

10. Steinbrecher UP, Parthasarathy S, Leake DS, Witzum JL, Steinberg D. (1984) Modification of low density lipoprotein by endothelial cells involves lipid peroxidation and degradation of low density phospholipids. *Proc Natl Acad Sci USA* **81**: 3883–3887.

11. Morel DW, Di Corleto PE, Chisolm GM. (1984) Endothelial and smooth muscle cells alter low density lipoprotein by free radical oxidation. *Arterioscler Thromb Vasc Biol* **4**: 357–364.

4

FREE RADICALS

Now, the race was on. The pace of research regarding the mechanism of LDL oxidation quickened. The intensity of the research on how LDL is oxidized was heightened by the simultaneous — though coincidental — publication by the NIH of a consensus statement, reflecting a December 1984 conference, saying that "it has been established beyond a reasonable doubt that lowering blood cholesterol levels (specifically, blood levels of low-density lipoprotein cholesterol) will reduce the risk of heart attacks caused by coronary heart disease."[1] Although Steinberg had shown that free radicals oxidized LDL (in a laboratory setting, using umbilical veins), he had not determined if the oxidation process occurred in the circulation or in the wall of the blood vessel… or even if oxidation was an all-or-none process, as opposed to a stepwise one. Also, while Steinberg had shown that human endothelial cells could oxidize LDL, he hadn't shown whether human plaques contained oxidized LDL.[a] It couldn't be assumed that one plus one equaled two. It had to be proven. (It wouldn't be until 1989 that human plaques would be proven by the presence of antibodies — at first synthetic and later natural antibodies — to LDL, to contain oxidized, but not native, LDL.) Plenty of uncertainties remained. As so often occurs in scientific research, "answers" begot more questions. Results from laboratory experiments or clinical studies can be synthesized in unforeseen ways to create fascinating "conclusions," which require more probing. It is an endless activity fueled by two three-letter words — "why" and

---

[a]There is no individual LDL particle representing "oxidized LDL." Instead, there is a family of oxidized LDLs and one family member doesn't look much like another. Since anywhere from one to all of the 2616 fatty acids and the 4536 amino acids in native LDL could be oxidized, the number of permutations and combinations of oxidized LDLs is almost infinite.[2,3] Even *unoxidized* (native) LDL is not unique; its fatty acid composition reflects the fatty acids consumed in the diet.[4]

"how" — and an open mind. In the final paragraph of his 1984 landmark paper,[5] Steinberg used two comparable words — "whether" and "if" — making his open-mindedness clear. He acknowledged that although "these [laboratory] findings implicate free radicals…as key elements in the generation of endothelial cell-modified LDL…whether a similar process occurs *in vivo* remains to be determined"; "…if the processes described here do play a role *in vivo,* it would in theory be possible to intervene and alter the atherogenic process by using antioxidants…."

Free Radicals and Antioxidants

Those concluding sentences included two scientific terms — "free radicals" and "antioxidants" — that would soon become buzzwords across America. "Free radicals" would become the equivalent of the sinister Darth Vader (*circa* 1977), while "antioxidants" would act as the hero, Luke Skywalker. But, in 1984, even within many medical circles, discussions on free radicals and antioxidants were regarded as being on the "fringe" of science. However, the field of cardiovascular research embraced the concepts of free radical biology and antioxidants to help explain some phenomena that had been perplexing in the past. The excitement generated by free radicals as oxidizing agents and antioxidants as their adversary was reflected by an exponential increase in the number of papers on each topic[b] and the appearance of specialized journals on the two themes: *Free Radicals in Biology and Medicine* (1985) and *Free Radical Research Communication* (1985) are two free-radical journals, while

Table 4.1   Number of papers published on antioxidants and coronary angioplasty between 1979 and 1990.

	1979–1984	1985–1990
Antioxidants	755	1832
Coronary angioplasty	1184	4709

Source: Pub Med.

---

[b]Antioxidants were the focus of 755 peer-reviewed papers published between 1979 and 1984, while 1832 more were published between 1985 and 1990, By way of comparison with another new (at the time) treatment for coronary heart disease, coronary angioplasty was the subject of 1184 papers between 1979 and 1984, whereas 4709 were published from 1985 to 1990 (**Table 4.1**).

Current Opinions in Lipidology (1990) and *Eicosanoids* (1990) are two journals specializing in antioxidants. National and international meetings on both topics followed quickly. Within five years, free radicals became fashionable in many clinical medicine circles. Five more years would pass before they gained respectability in more general circles.

The History of Free Radicals

Free radicals are not a consequence of living in the late-20th and early-21st century. In fact, the term "radical" — though not "free radical" — was first used in the 18th century by the renowned French scientist Antoine Lavoisier to explain the structure of organic acids. (An organic acid is an organic compound with acidic properties.) After the notion of a radical gained a foothold during the first half of the 19th century, a theory emerged regarding why radicals existed. Basically, they were thought simply to be "organic analogs" of inorganic elements, such as copper and iron, that were known to exist in a free (unbound) state. The next logical thought was that if *elements* can exist in a free state, so must *radicals*. Pursuing that hypothesis, several mid-19th century chemists did, indeed, isolate a few free radicals.

However, efforts to replicate their findings using different experimental techniques soon failed. As nonconfirmatory studies appeared during the second half of the 19th century, organic chemists came to believe in the "fundamental impossibility of isolating free radicals....."[6] The idea that free radicals existed seemed "buried," according to Rostanstev and Loshadkn, two Russian chemists writing a review of free radical history in 2001.[6]

In 1900, Russian-born chemist Moses Gomberg accomplished the "impossible" by first isolating a (organic) free radical while working at the University of Michigan.[7] His achievement was viewed so skeptically that it was not put to use until the 1930s. At that time, Imperial Chemical Industries in London began to use free radicals to generate polyethylene polymers, essential for manufacturing counter tops, telephones, toys, and many other products. Yet, 30 more years would go by before free radicals were appreciated to play a pivotal role in human biology. Then, in 1969, two biochemists at Duke discovered an enzyme that inactivates a biological free radical called "superoxide."[8] (Its function will be discussed later in this chapter.) That was the turning point. Today, many industrial and biological reactions depend on free radical reactions.

Definition of a Free Radical … Versus a Radical

Just as the pendulum has swung back and forth with regard to the existence of free radicals, so has their definition changed over time. At first, in the early and mid-19th century, when only "bound" radicals were recognized, a "free" radical was surmised to be an unbound radical — if it existed at all. Since Gomberg's discovery in 1900 that free radicals exist — quite apart from bound radicals — the distinction between the two types of radicals, curiously, has diminished — in printed text, though not in atomic structure. Today, the terms are often used interchangeably, although a distinction between them will be made in this chapter.

A radical is now defined as "a group of atoms that act as a unit," but it usually does not exist in the "free" state. For example, the familiar household solvent ammonia, with its pungent odor, is a "radical" designated as NH_3, itself composed of one nitrogen (N) and three hydrogen (H) atoms, acting as a unit. The N atom carries an electrical charge of positive three, and each H atom carries a negative charge of one, yielding a stable, electrically balanced product — ammonia. However, if one hydrogen atom is added, the result is a positively charged group of atoms — a cation — designated as NH_4^+ and called ammonium. Both ammonia and ammonium are *radicals* — not free radicals — because neither compound has an unpaired electron. If one nitrogen atom is replaced by three hydrogen atoms, each with a negative charge (valence) of one, the result is again the natural, stable gas ammonia. Definitions of radicals, free radicals, atoms, and molecules are provided in Table 4.2.

Table 4.2   Definitions of atoms, molecules, radicals, and free radicals.

Atom: The smallest particle of an *element* that can exist. It consists of a nucleus with a positive charge and a variable number of negatively charged electrons revolving or oscillating around the nucleus in a vast empty space. An "element," in turn, is a substance all of whose atoms have an identical nuclear charge. There are now 118 known elements, ranging from hydrogen, with an atomic number of 1, to ununoctival, with an atomic number of 118, synthesized in October 2006.

Molecule: The smallest particle of a *substance*. A substance, in turn, is any material of a specific nature but of no particular shape or dimensions. The term "substance" is far more inclusive than "element."

Radical: A group of atoms that act as a unit, such as the ammonium radical, NH_4^+.

Free radical: An atom with at least one unpaired — i.e. "free" — electron in its outermost shell; it is capable of independent existence.

Atomic Bonds and Atomic Clouds

The "electrical forces" that link — i.e. bond — atoms together to form molecules are represented by solid lines (*e.g.* H–O–H, representing H_2O) in linear, chemical formulae, for simplicity. Electrons are the glue. In reality, bonds vary in number, length, and angularity, producing three-dimensional structures rather than linear ones. As a result of differences in the strength of bonds arising from these three variables, there are differences in molecular geometry among various compounds. These geometric differences are better displayed in two- or three-dimensional diagrams of molecules. Water has a boomerang shape (Fig. 4.1), because only two of the oxygen atom's six electrons are used to bond with hydrogen atoms. This leaves four of its electrons unbound. The unbonded electrons repel each other and try to get as far away from each other as possible. The four electrons left unbonded push the bonding pairs closer together, resulting in a boomerang configuration. The usefulness of water as a solvent is attributable to the molecule's shape and the difference in charge between the side of the molecule attached to the two hydrogen atoms (positive) and the opposite side occupied by the oxygen atom (negative).

Two- and three-dimensional diagrams imply that electrons move in space, as they do in reality. Such electronic motion may be depicted in a two-dimensional drawing by two atomic orbits overlapping in a limited region. Typically, electrons seek the company of another of their kind, so that they can pair up in a system of lower energy. It takes less energy for a pair of electrons to spin together than it does for either to spin alone. At one time, electron pairs were thought to move within a defined space — an *orbit* — around the nucleus, akin to a single plane making a transoceanic flight within a certain altitude range. Today, electrons are believed to circle the nucleus "haphazardly" in a three-dimensional space — an *orbital* — more

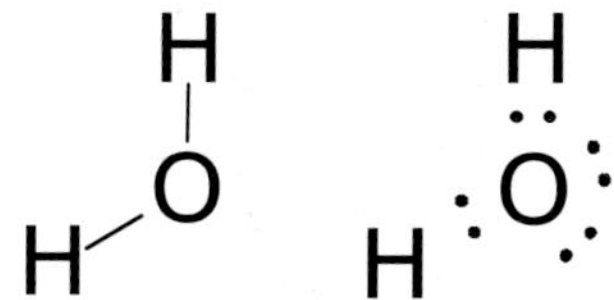

Fig. 4.1.   Electromagnetic forces within the simple water molecule lead to repulsion of its two hydrogen atoms by oxygen's unbound electrons. This results in a higher electron density around water's single oxygen atom driving the two hydrogen atoms apart to form a 107.5° angle — in the shape of a boomerang.

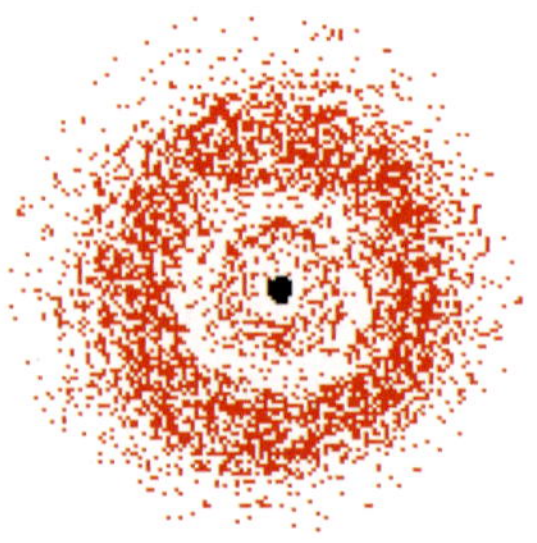

Fig. 4.2. Two clouds of electrons randomly circling the nucleus of an atom. The nucleus is much, much smaller than depicted, in reality.[11] If a nucleus were the size of a ping-pong ball, an electron would be 1 km (0.6 miles) away. Most of an atom is empty space.

like thousands of planes flying across/around the globe at any one time in different orbits than unidirectional planes in a single orbit. Because electrons move in unpredictable directions within an orbital, they are more realistically depicted as clouds of different densities, circling the nucleus (Fig. 4.2). Within an orbital, every electron in the *outermost* shell has a complementary electron that spins in the *opposite* direction. Opposites attract. A free radical, however, contains one or more unpaired electrons, an unpaired electron being one that is alone in an orbit. By convention, a superscript bold dot denotes a free radical, indicating the presence of a "free," i.e. unmatched, electron. Thus, the free radical superoxide (discussed in detail later) is designated as SO$^\bullet$, while the hydroxyl radical is designated as OH$^\bullet$.

A free radical searches for an additional electron to spin with its unpaired electron and thereby to achieve electrical stability. In so doing, a free radical "steals" an electron from a stable atom or molecule, converting it into a free radical, while stabilizing itself. The new free radical adopts the same MO and itself quickly steals an electron from another atom or molecule. A chain reaction follows. It can continue *ad infinitum* if not interrupted by an antioxidant. Specific antioxidants will be discussed in Part II of this book.

Species of Free Radicals

The generic term "free radicals" includes four different molecules: the so-called superoxide anion, hydrogen peroxide, nitric oxide (not to be confused with its cousin, nitrous oxide, laughing gas) and the most potent free

$$OH- \qquad\qquad OH^{\bullet}$$

Hydroxyl ion   Hydroxyl radical

Fig. 4.3. The hydroxyl ion and the hydroxyl radical.

radical, the hydroxyl radical,[c] [also not to be confused with its cousin, the hydroxyl ion (Fig. 4.3)]. Any reader uncertain about the difference should be comforted by the knowledge that the two are "repeatedly confused (even) in the biomedical literature."[12]

Neither superoxide nor hydrogen peroxide is particularly dangerous alone; however, they can conspire to generate the mighty hydroxyl radical — the bully of the subendothelial space — through a variety of reactions.[13] Technically, *atomic* oxygen also is a free radical, because two of its six outer-shell electrons are unpaired, qualifying it as a free radical. *Molecular* oxygen also has two unpaired electrons, making it a free radical as well (Fig. 4.4). Molecular oxygen is much stronger than atomic oxygen, because yoked electrons are simply more powerful than single electrons, much like a pair of horses pulling a plow, rather than a single horse.

Eager to find mates for its two unpaired electrons, even oxygen "steals" electrons from another source, seeking electrical stability. It is because of oxygen's two unpaired electrons that the molecule is such a powerful oxidizing agent. However, its two unpaired electrons are in separate orbits with *parallel* rotations (spins), whereas most molecules with which oxygen reacts have *opposite* rotations. Since reactions between molecular oxygen and other molecules occur more easily if their electrons rotate in the same direction as oxygen's electrons, oxygen is not very reactive, until it is "activated." Iron (in its reduced ferrous form) — common in cells — is the "activator." The conversion of oxygen into superoxide requires the presence of iron as a catalyst — a substance which causes or accelerates a chemical reaction without itself being consumed or changed in any way whatsoever. While oxygen alone burns nutrients such as carbohydrates and fats slowly, iron, functioning as a catalyst, speeds the reaction up. The iron requirement is a protective mechanism to allow the cell to control the rate at which it consumes O_2 and to help

---

[c]Technically, neither hydrogen peroxide nor nitric oxide is a "free radical," but atomic oxygen is because one of its six outer-shell electrons is unpaired.

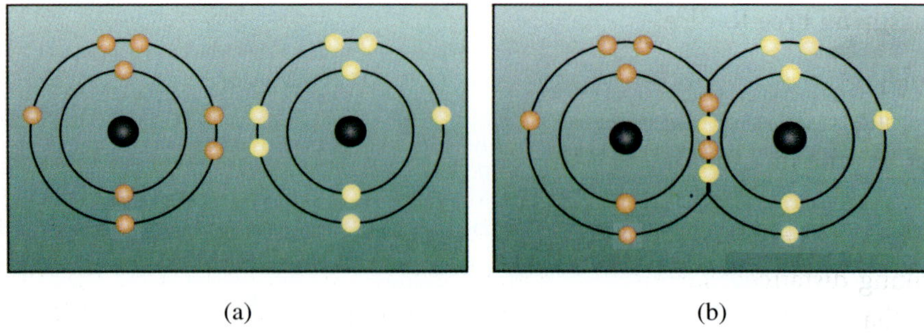

(a)   (b)

Fig. 4.4.   A diagram of atomic oxygen (a) and molecular oxygen (b). Atomic oxygen has eight electrons arranged in two shells. The inner shell holds two electrons and the outer shell holds six more, with a capacity of eight. Equally important, two of the six outer-shell electrons are unpaired, making atomic oxygen a free radical. Because atoms are not stable unless they have eight electrons in their outer-shell (the octet rule), atomic oxygen seeks two more electrons, both to satisfy the quota and to achieve stability.

Molecular oxygen (effectively) acquires eight outer-shell electrons by sharing four electrons between its two oxygen atoms, satisfying the octet rule. Six of the outer-shell electrons rotate in pairs with opposite spins (opposites attract), while two electrons rotate in an orbit with parallel spins, and consequently remain unpaired. The reason that two electrons remain unpaired — even after sharing a pair — is that when electrons are added to orbitals they do so in succession, in a way that saves energy. The lowest total energy expenditure is reached when the eight outer-shell electrons are arranged in four subshells. Those electrons in the three innermost subshells are paired and have opposite rotations (which require more energy, because opposite forces demand more power), while those in the outermost subshell have parallel spins and remain unpaired. The two unpaired electrons — which are the "oxidizing" electrons — need more energy for oxidizing purposes, but in the interest of conserving the molecule's total energy expenditure, they rotate in a parallel direction.

The four outer-shell electrons are spaced equally in one outer shell in this diagram. In reality, they are distributed in four distinct subshells, akin to halos around the moon.

(*Source*: Modified and reproduced with permission from the publisher.) Reference 14.

prevent haphazard reactions involving O_2 that could destroy the cell's interior.

When iron donates one of its electrons to molecular oxygen during the metabolism of carbohydrates and fats, the extra electron converts oxygen to superoxide, thereby activating oxygen. However, most biologic oxidations involve the transfer of two electrons, both of which must spin in the *opposite* rotation to oxygen's. This requirement imposes a restriction that forces oxygen to accept its electrons one at a time. Were it not for this spin restriction slowing combustion down, the reaction of oxygen with organic compounds would theoretically lead to "spontaneous combustion," followed by disintegration of the cell![15]

Measuring Free Radicals in the Laboratory

A major practical problem facing free radical research is the brief existence of free radicals, due to their inherent instability. How do you measure something that is gone in a billionth of a second? And even if they could be measured, how can you deal with a radical (in particular, the hydroxyl radical) capable of reacting with — and thereby destroying — every other molecule within striking distance, that distance being the space immediately adjacent to the cell that produced it? The hydroxyl radical may be the ultimate weapon of biological warfare — one we all hope will never be used.

Free Radical Hazards

Initially, free radicals were viewed as invisible enemies. For years they were linked to environmental hazards, including smog, herbicides, pesticides, and ionizing radiation (i.e. ultraviolet light). And without a doubt they can and do contribute to the breakdown of Earth's protective ozone layer by industrial chlorofluorocarbons (CFCs).

In 1914, Marie Curie cleverly put a twist on the destructive reputation of free radicals when she introduced the use of X-rays for the treatment of cancer.[d,16] The success of radiation therapy depends on the release of free radicals by damaged malignant cells. During World War II, scientists sought to exploit the toxic effects of an enormous number of free radicals on human cells by generating the radicals through the "splitting" of plutonium atoms — atomic fission — leading to the development of the atomic bomb. The devastating consequences of thermonuclear explosions at Hiroshima and Nagasaki counterpointed the success of treating malignant tumors with radiation delivered *in a controlled way*. The human damage caused by radiation, ultraviolet light, and even air pollutants depends on a common element — free radicals released by cells exposed to those hazards.

In addition to environmental free radicals, which can *indirectly* harm the human body, free radicals generated during normal cellular metabolism can *directly* harm the human body. That sounds masochistic and unnatural, since harmful processes and little-used "parts" of the body of primates (like tails)

---

[d]Madame Curie was the first woman to receive a Nobel Prize. She is also the only woman to be awarded two Nobel Prizes, the first in physics in 1903, and the second in chemistry in 1911. Both were in recognition of her extraordinary work in radiation physics and radioactivity.

were weeded out by evolution. Logic aside, human diseases such as atherosclerosis, rheumatoid arthritis, cataracts, Parkinson's disease, muscular dystrophy, and Alzheimer's disease now lead a long list of common conditions that have been shown to be related to the production of biological free radicals. Motivated by scientific curiosity, many investigations have been carried out since the mid-1980s aimed at evaluating the pathologic role of free radicals in human disorders.

Atherosclerosis tops the lineup.

Hazard of Free Radical Deficiency

While an excess of free radicals can be deadly, so can a deficiency. This is made clear by a rare, inherited disorder called chronic granulomatous disease.[e,17] Its victims are unable to make superoxide. Insufficient superoxide leads to insufficient H_2O_2 and the powerful hydroxyl radical — the principal bacteria-killing free radical. Given victims' inability to fight infections, even minor skin infections can be devastating, let alone deep-organ infections, such as a liver abscess, pneumonia, and osteomyelitis, which can be lethal. Thus, although free radicals are a byproduct of cellular metabolism, they also serve to maintain our health. "We need free radicals to live," *The New York Times* reported in 1993.[18] They can kill bacteria and cripple cancer cells.[19] Whether free radicals hurt us or help us depends on their concentration, which in turn depends on the *balance* between the generation of free radicals by cells and their elimination by intrinsic defense systems. (Recall that free radicals are the biological waste materials of other cells and they do not wear halos on the job. The other cells might do well to wear protective helmets, however.)

The intrinsic defense systems take the form of enzymes (notably superoxide dismutase) that "neutralize" free radicals and of free radical scavengers, particularly vitamin E. Free radicals in small amounts are beneficial. In large quantities, they can be dangerous. High levels can contribute to cancer; low levels can slow its growth. It's the balance that counts. Too much of anything — even money and food — can lead to problems.

---

[e]Granulomatosis refers to the presence of many small nodules — granulomas — composed largely of macrophages responding to an infection.

Discovery of Superoxide Dismutase

Nevertheless, free radicals were still considered to be fundamentally destructive and of no human biological importance (other than when they were the result of a nuclear explosion), since free radicals were not known to exist in living organisms. In 1969, a turning point in the history of free radicals occurred when a crucial paper was published establishing that an enzyme, termed "superoxide dismutase," was present in [bovine] red blood cells[f] and that its function was to inactivate superoxide.[g,18] Twenty years later, the two men who found the enzyme acknowledged that the discovery was made by "serendipity."[21] On top of that, superoxide, the free radical that superoxide dismutase inactivates, was "not known to be a normal product of biological oxidation prior to the discovery of superoxide dismutase activity."[22] In other words, superoxide dismutase was discovered before superoxide was known to exist in humans, although superoxide and the hydroxyl radical had been proposed to be the cause of "oxygen poisoning" in mice, in 1954, based on the biochemical similarities of oxygen poisoning and radiation injury: both processes led to shortened survival, most likely due to an excess of free radicals.[23] Years later, biological free radicals were likened by a Johns Hopkins physician to little "blowtorches" burning everything in their paths as antioxidant "fire extinguishers" raced after them attempting to put out the fire.[24]

Superoxide — a Matter of Semantics?

Curiously, although the term "superoxide" implies that it is stronger than oxygen, it really isn't. It has only one unpaired electron in its outer shell, while

---

[f]Although the enzyme was first found in bovine red blood cells, it was immediately shown to be identical to a known human enzyme present in circulating red blood cells.

[g]The term "superoxide" refers to a free radical formed by all human cells when oxygen is "burned." Oxygen is reduced to superoxide. The prefix "super" means "beyond" or "after", in this context "not greater in quality or degree," such as in "supersaturated" or "Superman." Having only one unpaired electron, superoxide is actually less "super" than oxygen as an oxidizing agent.[20]

The term "dismutase" refers to a reaction in which atoms or electrons are transferred between two identical molecules to form a different product. The prefix "dis" "means to reverse or do the opposite of," while "mutate" means "to change." Thus, superoxide dismutase is an enzyme that simultaneously oxidizes and reduces superoxide, thereby protecting cells against damage by superoxide.

the oxygen molecule having two such unmatched electrons (qualifying *it* as a free radical!) makes oxygen a more powerful oxidant than "superoxide." Oxygen is a better oxidizing agent *because* it is more electron-deficient than superoxide. Superoxide is a rather weak oxidant compared with either oxygen before it or the formidable hydroxyl radical that follows it. Thus, although the term "superoxide" is grammatically correct — it is "beyond" or "above" oxygen as implied by the prefix "super" — it is weaker than oxygen. (At the molecular level, the addition of an extra electron into oxygen's outermost orbit — its bonding orbit — stretches the bond between the two oxygen atoms ever so slightly, weakening it and making superoxide less potent than oxygen itself.)

The superoxide radical forms after oxygen has oxidized its substrate, leaving behind $SO^{\bullet}$. Oxygen is converted to water by the addition of four electrons in the stepwise fashion illustrated in Fig. 4.5. The $SO^{\bullet}$ radical is largely responsible for the lethal effect of oxygen on bacteria (such as the bacterium *Gonococcus,* which causes gonorrhea) that grow only in the absence of air. Many such anaerobic bacteria not only simply fail to grow in the presence of air, but also are killed by oxygen because they lack superoxide dismutase (to be discussed soon). Superoxide is formed in only small amounts during normal cellular metabolism. It is *estimated* that just 1%–3% of the oxygen consumed daily by an adult at rest is converted into superoxide.[20] Or, restated, that 97%–99% of superoxide generated by the escape of electrons from the electron transport chain are either inactivated by superoxide dismutase or scavenged by the antioxidant network, particularly vitamin E. It is a very efficient system. After an insult to a cell or tissue, such as by trauma or infection, superoxide production increases because scavenger cells responding to the site of injury/infection release their free radicals to counteract the injury or to kill invading bacteria.[14] Superoxide's strength lies in its ability to quickly react

$$O_2 + e^- \longrightarrow O_2^- \quad \text{Superoxide}$$
$$O_2^- + e^- + 2H^+ \longrightarrow H_2O_2 \quad \text{Hydrogen peroxide}$$
$$H_2O_2 + e^- + H^+ \longrightarrow H_2O + OH^{\bullet} \quad \text{Hydroxyl radical}$$
$$OH^{\bullet} + e^- + H^+ \longrightarrow H_2O \quad \text{Water}$$

$$\text{Overall: } O_2 + 4e^- + 4H^+ \longrightarrow 2H_2O$$

Fig. 4.5. The four-electron reduction of oxygen (O_2) to water (H_2O) by stepwise addition of electrons. All of the intermediaries formed are reactive and toxic to cells.

Source: Reproduced with permission from the publisher.) Reference 26.

with whatever molecule is adjacent to it, be that molecule a carbohydrate, a fatty acid, or a protein … though not DNA. By adding a second electron from H_2, hydrogen peroxide (H_2O_2) is formed, which is also weakly reactive. However, if a third electron is then added to H_2O_2, the powerful hydroxyl radical ($OH^\bullet$) is generated[h] (Fig. 4.5). It can be toxic to cells.

References

1. Consensus Conference. (1985) Lowering blood cholesterol to prevent heart disease. *JAMA* **253**: 2080–2086.
2. Chisolm GM, Steinberg D. (2000) The oxidative modification hypothesis of atherogenesis: an overview. *Free Rad Biol Med* **28**: 1815–1826.
3. Steinberg D. (1996) Oxidized low density lipoprotein — an extreme example of lipoprotein heterogenicity. *Isr J Med Sci* **32**: 469–472.
4. Reaven P, Parthasarathy S, Grasse BJ, Miller E, Almazan F, Mattson JC, Khoo JC, Steinberg D, Witzum JL. (1991) Feasibility of using an oleate-rich diet to reduce the susceptibility of low-density lipoprotein to oxidative modification in humans. *Am J Nutr* **54**: 701–706.
5. Steinbrecher UP, Parthasarathy S. Leake DS, Witzum JL, Steinberg D. (1984) Modification of low density lipoprotein by endothelial cells involves lipid peroxidation and degradation of low density phospholipds. *Proc Natl Acad Sci USA* **81**: 3883–3887.

---

[h] H_2O_2 can also react with more $SO^\bullet$ (with iron as the catalyst) to form $OH^\bullet$ in the so-called Haber–Weiss reaction. A major determinant of how much damage the OH radical causes may be how much iron is available to catalyze lipid oxidation.[14] Some iron is necessary for free radicals to oxidize LDL. Free iron (i.e. iron not bound to its protein carrier) in a high concentration can even oxidize LDL by itself.

Individuals who take iron tablets or vitamins containing iron may be concerned about this potentially harmful reaction. Any concern should be tempered, however, by the knowledge that in the body the majority of iron is secured within red blood cells to transport oxygen to the tissues. Therefore, the availability of (free) iron to catalyze the Haber–Weiss reaction is limited. Also, individuals with hemochromatosis, a genetic disorder that leads to excess iron in the blood and liver, do not develop premature atherosclerosis.[25] Their iron overload does, however, lead to severe damage of the liver, in particular.

On the other hand, recent studies have suggested that the reason women have fewer heart attacks (during their child-bearing years) than men is related not only to the protection provided by their female hormones, but also to the protection afforded "coincidentally" by the loss of iron during menstruation.

6. Rozantsev EG, Loshadkn DV. (2001) The history and modern problems of free radical chemistry. 100 years of free radical chemistry. *Design Mono Polym* **4**: 281–300.

7. Gomberg M. (1900) An instance of trivalent carbon: triphenylmethyl. *J Am Chem Soc* **20**: 757–771.

8. McCord JM, Fridovich I. (1969) Superoxide dismutase. An enzymic function for erythropreim (hemocupreim). *J Biol Chem* **244**: 6049–6055.

9. Pryor WA. (1976) The role of free radical reactions in biological systems. In *Free Radicals in Biology.* Academic, New York, San Francisco, London, Vol. 1, pp. 1–23.

10. *Blakiston's Medical Dictionary.* 4th ed. (1979). McGraw-Hill, New York, St Louis, San Francisco, Auckland, Bogotá, Düsseldorf, Johannesburg, London, Madrid, Mexico, Montreal, New Delhi, Panama, Paris, Sao Paulo, Singapore, Sydney, Tokyo, Toronto.

11. http://www.chemguide.co.uk/basicorg/bonding.orbitals.html

12. Halliwell B, Gutteridge JMK. (1999) *Free Radicals in Biology and Medicine,* 3rd ed. Oxford Univerity Press, Oxford, New York, Auckland, Bangkok, Buenos Aires, Cape Town, Chennai, Dar es Salaam, Delhi, Florence, Hong Kong, Istanbul, Karachi, Kuala Lumpur, Madrid, Melbourne, Mexico City, Mumbai, Nairobi, Sao Paulo, Shanghai, Taipei, Tokyo, Toronto, pp. 1–35.

13. Cheeseman KH, Slater TF. (1993) An introduction to free radical chemistry. In: Cheeseman KH, Slater TF (eds.) *Free Radicals in Medicine.* Churchill Livingston, Edinburgh, London, Madrid, Melbourne, New York, Tokyo. *Br Med Bull* **49**: 481–493.

14. Reference needed.

15. Halliwell B, Gutteridge JMK. (1999) *Free Radicals in Biology and Medicine, 3rd ed.* Oxford Univerity Press, Oxford, New York, Auckland, Bangkok, Buenos Aires, Cape Town, Chennai, Dar es Salmam, Delhi, Florence, Hong Kong, Istanbul, Karachi, Kuala Lumpur, Madrid, Melbourne, Mexico City, Mumbai, Nairobi, Sao Paulo, Shanghai, Taipei, Tokyo, Toronto, pp. 105–245.

16. Curie M. (1914) *Sur les ecarts de poids atomiques obtenus avec le plomb provenant. C R Hebd Séance Acad Sci* **158**: 1676–1678.

17. Klempner MS, Malech HL. (1992) Phagocytes: normal and abnormal host defenses. In: Gorbach SL, Bartlett JG, Blacklow NR (eds.), *Infectious Diseases.* W.B. Saunders, Philadelphia, London, Toronto, Montreal, Sydney, Tokyo.

18. Angier N. (1993) Free radicals: the price we pay for breathing. *The New York Times,* Apr. 25, Magazine Section.

19. Wardman P. (2001–2002) Free radicals and new prodrugs for cancer therapy. *Scientific Yearbook.* pp. 36–39.

20. Halliwell B, Gutteridge JMK. (1999) *Free Radicals in Biology and Medicine,* 3rd ed. Oxford Univerity Press, Oxford, New York, Auckland, Bangkok, Buenos Aires,

Cape Town, Chennai, Dar es Salaam, Delhi, Florence, Hong Kong, Istanbul, Karachi, Kuala Lumpur, Madrid, Melbourne, Mexico City, Mumbai, Nairobi, Sao Paulo, Shanghai, Taipei, Tokyo, Toronto, pp. 1–35.

21. McCord J, Fridovich I. (1988) Superoxide dismutases: the first twenty years (1968–1988). *Free Radic Biol Med* **5**: 363–369.

22. McCord J, Fridiovich I. (1977) Superoxide dismutases: a history. In: Michelson AM, McCord J, Fridovich I (eds.), *Superoxide and Superoxide Dismutases*. Academic, London, pp. 1–10.

23. Gerschman K, Gilbert DL, Nye SW, Dwyer P, Fenn WO. (1954) Oxygen poisoning and X-irradiation: a mechanism in common. *Science* **119**: 623–662.

24. Bulkley G. Free radicals and reactive oxygen species. http://www.cosmos-club.org/web/journals/2002/bulkley.html

25. Smith LH, Jr. (1990) Overview of hemochromatosis. *West J Med* **153**: 296–308.

26. Brock TD. (1988) *Biology of Microorganisms*. Prentice-Hall Englewood Cliffs, NJ.

5

THE OXYGEN PARADOX

Free radicals are the result of a long evolutionary process that allowed plants to produce the oxygen that man needed for survival, but without the danger of breathing *too much* oxygen, which can harm humans. Humans cannot exist without oxygen, yet oxygen endangers the very existence of humans. This occurs because when oxygen is "burned," the process generates byproducts that may damage the very tissue(s) that used the oxygen. Thus, "the superoxide anion radical, H_2O_2 and the extremely damaging hydroxyl radical ($OH^•$) are common products of life in an aerobic environment, and these agents appear to be responsible for oxygen's toxicity".[1]

A vivid example of the damage that can be caused by excess oxygen occurred in the 1940s, when premature infants were first treated with oxygen. Too much oxygen caused blindness due to retrolental fibroplasia[2] (Fig. 5.1). An epidemic of the disease occurred in the U.S. between 1943 and 1953. This soon led to the controlled administration of oxygen and by 1959 the problem was practically eliminated.[4] Ironically, with advances in neonatal care during the 1970s and 1980s, the incidence of retrolental fibroplasia has risen to 80% of premature babies who weigh less than 1000 g (2.2 pounds),[5] although it is now known as "retinopathy of prematurity." (Table 5.1 records quotations from a major U.S. textbook of pediatrics from 1948 to 2004 regarding the likelihood of retrolental fibroplasia. Changing opinions about how much oxygen a premature infant should receive are also mentioned.) This poses a dilemma, because very premature babies need supplemental oxygen if they are to survive at all. Singer Stevie Wonder, born in 1950, is a well-known victim. Later, it was learned that damage from excess oxygen is not limited to premature babies. Adults placed in a hyperbaric oxygen chamber to treat life-threatening infections (such as necrotizing fasciitis, caused by the bacterium *Clostridium perfingens*), may also suffer impaired vision, temporarily.[6]

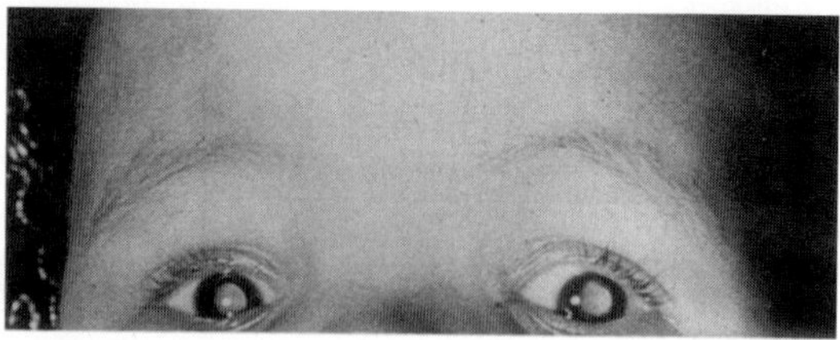

Fig. 5.1.   A blind child with retrolental fibroplasia. The retina has detached and been displaced forward, obstructing the pupil and causing blindness.

(*Source*: Reproduced with permission from the publisher.) Reference 3.

Table 5.1   Excerpts from Nelson Textbook of Pediatrics on the Incidence of Retrolental Fibroplasia From 1946 to 2004.

1946 (4th edition):	No mention of retrolental fibroplasia in this edition.
1950 (5th edition):	No mention of retrolental fibroplasia in this edition.
1954 (6th edition):	Ambivalence about the incidence of the disease and the amount of oxygen a premature baby should receive is reflected by the writings of separate contributors to this edition. One chapter writer states that "small premature infants should be placed in 40 to 60 percent oxygen." A second author states that ". . .it seems wise to limit routine use of oxygen to concentrations of 30 to 40 percent. . ." And a third pediatrician says that "currently it is the practice in most clinics to use oxygen therapy to a much less extent than formerly in the care of the premature infant."
1959 (7th edition):	Before 1956, retrolental fibroplasia occurred in "5 to 25% of. . .infants. . . below 4 pounds (1800 gm)." Now, the disease "rarely occurs."
1969 (9th edition):	"The toxicity to the retina of high concentrations of oxygen. . .has been amply demonstrated."
1983 (12th edition):	"At greatest risk is the very small premature infant. . .who has required oxygen therapy to sustain life."
2004 (17th edition):	"Generally, the lower the birth weight and the sicker the infant, the greater the risk for retinopathy of prematurity."

1943–1953: The First Epidemic of Retrolental Fibroplasia

With the conversion of the original incubator, which was basically a heated bed, into a "chamber for oxygen therapy" in the early 1940s, an epidemic of retrolental fibroplasia occurred. The term "retrolental" comes from the Latin words "*retro*" and "*lental,*" meaning "behind" and "lens," respectively. The term "fibroplasia" comes from the Latin words "*fibro*" *and* "*plasia,*" meaning "fibrous or scar" and "formation or development," respectively. Initially, no connection was made between the high-inspired oxygen concentration associated with an enclosed incubator and the loss of eyesight in a spate of premature infants. Practicing pediatricians were told in a manual of standard care for babies issued by the U.S. Government Printing Office that "oxygen is of great value. . .in the treatment of premature infants. It should be used freely.. . ."[7] Accordingly, no attempt to measure the oxygen concentration in the incubator was made since oxygen was deemed beneficial for premature babies.[a]

The idea that oxygen — the elixir of life — might be responsible for retrolental fibroplasia arose from the observations of an Australian pediatrician comparing the treatment of premature infants in America to their treatment in England. She noted that "in America, where retrolental fibroplasia is a problem and oxygen [is] used freely, [while] in England, where retrolental fibroplasia is seen rarely. . .oxygen [is] used sparingly."[10] She also noted, somewhat paradoxically, that the disease occurred more frequently in the premature babies of indigent patients than in the premature babies of more affluent patients, in whom oxygen was restricted because the expensive gas had to be paid for by the more well-to-do parents of the latter group of infants.[10,11] In the indigent group, the incidence of retrolental fibroplasia was 4 of 58 infants (6.9%), while in the babies of more affluent parents it was 23 of 123 infants (18.7%).

By 1954, independent studies from Australia, Canada, England, and the U.S. shown had that the disease was not confined to a single country and that it was occurring in "phenomenal proportions" on both sides of the

---

[a]Three significant changes in the treatment of premature babies emerged in the 1940s: the administration of multivitamins in which fat-soluble vitamins were made miscible with water, the administration of iron in greatly increased quantities, and the more frequent administration of oxygen.[8] The invention of wall outlets to deliver oxygen, rather than cumbersome oxygen tanks, was a practical reason for the widespread use of oxygen in premature nurseries. The author of this 1949 paper misattributed the increased incidence of retrolental fibroplasia between 1942 and 1948 to the increased use of iron and vitamin supplements, rather than oxygen, during those years. This is understandable, both because no one imagined that supplemental oxygen might be harmful and because the amount of iron and vitamin E in iron and vitamin preparations, respectively, was established by pharmaceutical companies and confirmed by the FDA. But the concentration of oxygen in the body could not be measured until the invention of the oxygen electrode in 1956.[9]

Atlantic.[12,13] In a quirk of the retrolental fibroplasia story, a report from Johns Hopkins appeared in the *American Journal of Ophthalmology* suggesting that the disease could be caused by "vitamin E deficiency".[b,14] Of the fat-soluble vitamins — A, D, K, and E — A, D, and K were routinely given shortly after birth, but vitamin E was not included. Because premature babies do not absorb fat well, they were given a low-fat diet with supplements of vitamins A, D, and K, but vitamin E was not given, for unknown reasons. In addition to poor absorption of vitamin E, by premature babies, the amount of vitamin E in the newborn is further reduced by the administration of vitamin A and iron. Vitamin A supplements (fish liver oils) increase the requirement for vitamin E, because the former vitamin contains a large amount of unsaturated fat that can be oxidized. Vitamin E — an antioxidant — acts to protect the unsaturated fat from oxidation. Iron can also oxidize unsaturated fat and further increase vitamin E needs.

To test the hypothesis that vitamin E deficiency might lead to retrolental fibroplasia, Owens and Owens gave vitamin E to 23 premature babies weighing less than 1360 g (3 pounds) and did not give any vitamin E to 78 other premature infants. Over 10 months, only 1 of the 23 (4.4%) vitamin-E-treated babies developed retrolental fibroplasia, while 17 of the 78 (21.8%) untreated infants suffered the disease. Although the difference in frequency of retrolental fibroplasia development was impressive in the Owens' study, the results could not be replicated in two other studies at Columbia University's Presbyterian Hospital[15] and the University of California,[16] respectively, performed in 1951. A set of identical twins, in the University of California report, strikingly emphasized the lack of a correlation between vitamin E deficiency and retrolental fibroplasia. One twin received vitamin E and developed retrolental fiboplasia. The other twin did not receive vitamin E and did not develop the disease. As a result of the conflicting reports about the efficacy of vitamin E and the contradiction highlighted by the identical twins, by 1953 vitamin E administration "had been discontinued in most nurseries throughout the world".[17] By the late 1950s, the use of vitamin E for retrolental fibroplasia had ended.

---

[b]Retrolental fibroplasia is not an inheritable disorder and it is not present at the time of birth. It becomes apparent by routine ophthalmoscopic examination of the retina about 4–6 weeks after birth.

Is Oxygen the Cause of Retrolental Fibroplasia?

The controversy over vitamin E and its role in the treatment of retrolental fibroplasia was mild compared to a concurrent debate about the role of oxygen in the disease. Campbell's 1951 paper incriminating oxygen notwithstanding, there was "reluctance to consider something as necessary as oxygen for the treatment of small infants could be harmful".[11] Indeed, at least one report in 1952 came to the opposite conclusion. It stated that retrolental fibroplasia is "produced by retinal oxygen deficiency" and is "preventable by. . .oxygen administration".[18] Couple the debate over whether oxygen was the cause or the cure of retrolental fibroplasia with the knowledge that the incidence of cerebral palsy — which was known to be due to too little oxygen reaching the brain of extremely low-birth-weight premature infants[19] — would probably rise if the administration of oxygen to premature infants were curtailed, and the stage was set for a dramatic medical showdown. As usual in medicine, important clinical questions are answered only through the performance of one or more large, prospective, randomized trials comparing one potential cause or treatment of a disease with a control group in which the cause or treatment is withheld. Between 1952 and 1956 three such studies were conducted. The results were conclusive and ended a dark era in American medicine.

The First Controlled Study

The first controlled trial of oxygen on the incidence of retrolental fibroplasia was published in September 1952.[20] Conducted at a municipal hospital in Washington, D.C., infants weighing less than 1500 g were alternately assigned to receive either 65%–70% oxygen or less than 40% oxygen.[c]

Over 65 infants were assigned to the high- or low-oxygen groups. Seventeen of the 28 (60.7%) infants in the high-oxygen group developed retrolental fibroplasia, while only 6 of the 37 (16.2%) in the low-oxygen group developed the disease (Table 5.2). Importantly, 25% of the babies in the high-oxygen group had more advanced grades (grades 3 and 4) of retrolental fibroplasia,

---

[c]The administration of less than 40% oxygen was, in fact, a concession by the research investigators to the hospital's "powers that be" concerned about not giving any supplemental oxygen to newborns, which was the initial intent of the study's designers.[17]

Table 5.2   Incidence of retrolental fibroplasia in high-and low-oxygen groups.

Oxygen concentration	Number of infants with retrolental fibroplasia (%)
High (65%–70%)	17/28 (60.7%)
Low (under 40%)	6/37 (16.2%)

Table 5.3   Five-Grade classification of retrolental fibroplasia (National Eye Institute).

Grade	Description
1	The mildest form. It usually resolves spontaneously and does not cause any visual disturbance.
2	A moderate form that also resolves spontaneously. The difference between stage 1 and stage 2 is the extent of blood vessel damage visible through an ophthalmoscope.
3	A more severe form that requires treatment to avoid retinal detachment.
4	Partial retinal detachment leading to significant visual impairment—myopia (near-sightedness), glaucoma, cataracts.
5	Complete retinal detachment leading to blindness. The most effective treatment is laser or cryotherapy. But neither is likely to avoid significant visual impairment.

Source: Reference 21.

whereas none in the low-oxygen group developed advanced disease. There are five grades of retrolental fibroplasia, as listed in Table 5.3.

The authors concluded in 1952 that "the data suggest strongly that high oxygen administration is a factor in the pathogenesis of retrolental fibroplasias".[20] Yet, looking back nearly 25 years later in a symposium published by the American Academy of Pediatrics on the history of retrolental fibroplasia, the editors said that "skeptical pediatricians [found it difficult] to accept. . .oxygen restriction and to implement such a radical change in practice".[17]

The Second Controlled Study

An additional layer of assurance came with the 1954 publication of a study from Bellevue Hospital in Manhattan, of 64 premature babies treated with supplemental oxygen.[22] As with the first controlled study, the infants were divided into high- and low-oxygen groups. In contrast to the first study, the low-oxygen group received oxygen for only a specific clinical indication, most often visible cyanosis (a bluish discoloration) of the fingertips and toes.

Among 36 infants in the high-oxygen group, eight (22.2%) developed advanced retrolental fibroplasia, whereas none of the 28 in the low-oxygen group showed severe disease. The results led the authors to conclude, "We believe that RLF is directly related to the excessive administration of oxygen and can be controlled by severely limiting oxygen therapy to premature infants. Such restriction does not appear harmful."

The Third Controlled Study

The results of the second study were published after the third and largest controlled study had been started. Therefore, many pediatricians — aware of the bigger, ongoing study — reasonably decided to await its completion before making any change in the way they administered oxygen to premature infants.

The third study consisted of a collaboration among 18 university hospitals to determine whether the incidence of retrolental fibroplasia rose significantly in premature infants treated in an "oxygen-enriched environment".[23] Because it was an academic collaboration, it was termed the "Cooperative Study of Retrolental Fibroplasia and the Use of Oxygen." It quickly became famous and subsequently was referred to as simply the "Cooperative Study." That designation is maintained in this summary of the study.

Over just one year (July 1, 1953 to June 30, 1954), 620 premature infants weighing less than 1500 g were randomized to one of two supplemental-oxygen groups. The high-oxygen group was called a "routine oxygen group" and the low-oxygen group was referred to as a "curtailed-oxygen group." The high-oxygen group received 50%, or more, oxygen for 28 days; the low-oxygen group received less than 50% oxygen and only if the infant's pediatrician thought that any oxygen was needed at all. Among those in the routine-oxygen group, the incidence of retrolental fibroplasia (all grades of severity) was 94%, while in the curtailed-oxygen group it was 40%. More severe forms of the disease developed in 17% and 4.7% of the two groups, respectively. Of equal importance, spontaneous regression from the milder forms of the disorder occurred in almost 90% of cases in both groups. Mortality during the newborn period was not different in the two groups despite the difference in the concentration of oxygen delivered. Table 5.4 records the more important findings of the Cooperative Study. It concluded that "the length of time the premature infant is kept in an oxygen-enriched environment is the important

Table 5.4   Major findings of the cooperative study.

1. The incidence of retrolental fibroplasia in the high-oxygen group was approximately twice that in the low-oxygen group.

2. The incidence of the more severe forms of retrolental fibroplasia in the high-oxygen group was approximately 3½ times that in the low-oxygen group.

3. The mortality rate in the low-oxygen group was the same as that in the high-oxygen group.

4. Spontaneous regression of milder forms of retrolental fibroplasia occurs in nearly 90% of infants.

factor in the production of RLF" and that "there is no concentration of oxygen in excess of that in air that is not associated with risk of developing RLF."[d] The Study's final recommendation to practicing pediatricians was that "the length of time a premature infant. . .is kept in an environment containing oxygen in concentrations in excess of that of air should be kept to an absolute minimum."

The disease of retrolental fibroplasia seemed to have ended with such a decisive conclusion and recommendation, and two papers from the United States and the United Kingdom supported the notion that the disease was — or soon would be — over. In a commentary from Bellevue Hospital in New York, the authors said that "RLF can be either completely or almost completely eliminated by administering oxygen at times of clinical need and then for as brief periods as possible at concentrations less than 40 percent",[24] while an editorial from London published in response to a paper from the University of Manchester showing no instance of blindness during the three years after restricted oxygen administration to low-birth-weight babies was instituted (in contrast to 21 such instances in 37 infants over the preceding 4 years, when oxygen was administered liberally),[25] asserted that "there can be no further grounds for doubting the danger of oxygen in the treatment of premature babies".[26]

---

[d]During the performance of the three controlled trials in the early 1950s, no technology existed to measure the concentration of oxygen in a baby's blood. Even after introduction of the oxygen electrode in 1956,[9] serial measurements of the oxygen concentration in a baby's blood did not become widely available because the techniques used to measure the oxygen content in circulating blood required the withdrawal of a substantial amount of blood. Micro techniques appeared in the late 1960s.

References

1. Davies KJA. (1995) Oxidative stress: the paradox of aerobic life. In Free Radicals and Oxidative Stress: Environment, Drugs and Food Additives. *Biochem. Soc. Symp.* **61**: 1–31.

2. Terry TL. (1942) Extreme prematurity and fibroblastic overgrowth of persistent vascular sheath behind each crystalline lens. *Am J Ophthalmol* **25**: 203–204.

3. Patz A. (1957) The role of oxygen in retrolental fibroplasia. *Pediatrics* **19**: 504–524.

4. Nelson WE. (1959) Textbook of Pediatrics. WB Saunders, Philadelphia, London.

5. http://www.visionchannel.net/retinopathy

6. Balentine JD. (1982) The eye. In Pathology of Oxygen Toxicity. Academic, New York, London, Paris, San Diego, Sao Paulo, Sydney, Tokyo, Toronto, p. 149.

7. Dunham EC. (1948) Premature infants: A Children's Bureau Publication No. 325. 42–43, U.S. Government Printing Office.

8. Kinsey VE, Zacharias L. (1949) Retrolental fibroplasia: incidence in different localities in recent years and a correlation of the incidence with treatment given infants. *J Am. Med.* Assoc. **139**: 572–578.

9. Clark LC, Jr. (1956) Monitoring and control of blood and tissue oxygen tensions. *Trans Am Soc Artif Intern Organs* **2**: 41.

10. Campbell K. (1955) Intensive oxygen therapy as a possible cause of retrolental fibroplasia: a clinical approach. *Med J Austr* **2**: 48–50.

11. Orellano J. (1986) Identification of retinopathy of prematurity/retrolental fibroplasia. In: *Retinopathy of Prematurity.* McPherson AR, Hittner HM, Kretzer FL (eds.), BC Dekker, Toronto, Philadelphia.

12. Gordon HH. (1954) Oxygen administration and retrolental fibroplasia. *Pediatrics* **14**: 543–546.

13. Cross VM, Evans PJ. (1952) Prevention of retrolental fibroplasia. *AMA Arch Ophthalmol* **48**: 83–87.

14. Owens WC, Owens EU. (1949) Retrolental fibroplasia in premature infants. The use of alpha tocopheryl acetate. *Am J Ophthalmol* **32**: 1–21.

15. Reese AB, Blodi FC. (1951) Retrolental fibroplasia. *Am J Ophthalmol* **34**: 1–24.

16. Laupus WE, Bousquet FP, Jr. (1951) Retrolental fibroplasia: the role of hemorrhage in its pathogenesis. *Am J Dis Child* **81**: 617–626.

17. James LS, Lanan JT (eds). (1976) History of oxygen and retrolental fibroplasia. *Pediatrics* **57 (Suppl 4)**: 591–598.

18. Szewczyk TS. (1952) Retrolental fibroplasia: etiology and prophylaxis. *Am J Ophthalmol* **35**: 301–310.

19. Asher P, Schowell FE. (1950) A survey of 400 cases of cerebral palsy in childhood. *Arch Dis Child* **24**: 360–379.

20. Patz A, Hoeck LF, De La Cruz E. (1952) Studies on the effect of high oxygen administration in retrolental fibroplasia: nursery observations. *Am J Ophthalmol* **35**: 1248–1252.
21. http://www.nei.nih.gov/health/rop/index.asp
22. Lanman JT, Guy LP, Dancis J. (1954) Retrolental fibroplasia and oxygen therapy. *J Am Med Assoc* **155**: 223–226.
23. Kinsey VE. (1956) Retrolental fibroplasia: cooperative study of retrolental fibroplasia and the use of oxygen. *Arch Ophthalmol* **56**: 481–543.
24. Guy LP, Lanman JT, Dancis J. (1956) The possibility of total elimination of retrolental fibroplasia by oxygen restriction. *Pediatrics* **17**: 247–249.
25. Forrester RM, Jefferson E, Naunton WJ. (1954) Oxygen and retrolental fibroplasia: a seven-year study. *Lancet* **2**: 258–260.
26. Editorial. (1954) Oxygen and retrolental fibroplasia. *Lancet* **2**: 275–276.

6

RETROLENTAL FIBROPLASIA FROM 1956 TO 1972: THE QUIET TIME

The incidence of retrolental fibroplasia plummeted in the late 1950s (Fig. 6.1) and remained low during the 1960s. Unfortunately, the incidence of hyaline membrane disease — a severe lung disease seen only in premature infants — rose and infant mortality increased.[1] The term "hyaline membrane disease" refers to the clear, glasslike appearance of lung air sacs (under a microscope) leading to respiratory distress during life. The disease is also known as the "infant respiratory distress syndrome," referring to the newborn's clinical condition.[3] Thus, the optimal use of oxygen in premature babies involves balancing a lesser chance of blindness (by restricting supplemental oxygen) against increasing the infant mortality rate from hyaline membrane disease — in which very small infants need supplemental oxygen, if they are to survive at all — and the risk of cerebral palsy (Fig. 6.2). Like retinopathy of prematurity, cerebral palsy is largely a disease of infants weighing less than 1,500 grams (3.3 pounds). But, unlike retinopathy of prematurity, which is related to too much oxygen reaching the immature retina, cerebral palsy is related to too little oxygen reaching the underdeveloped brain.

The Cost of Preventing Retrolental Fibroplasia

During the 1960s, a pediatrician in London sensed that more premature infants were dying in his nursery after the practice of curtailing oxygen was adopted in 1956. To determine if a rise in premature infant deaths was occurring at other institutions, he reviewed the available data on infant death rates

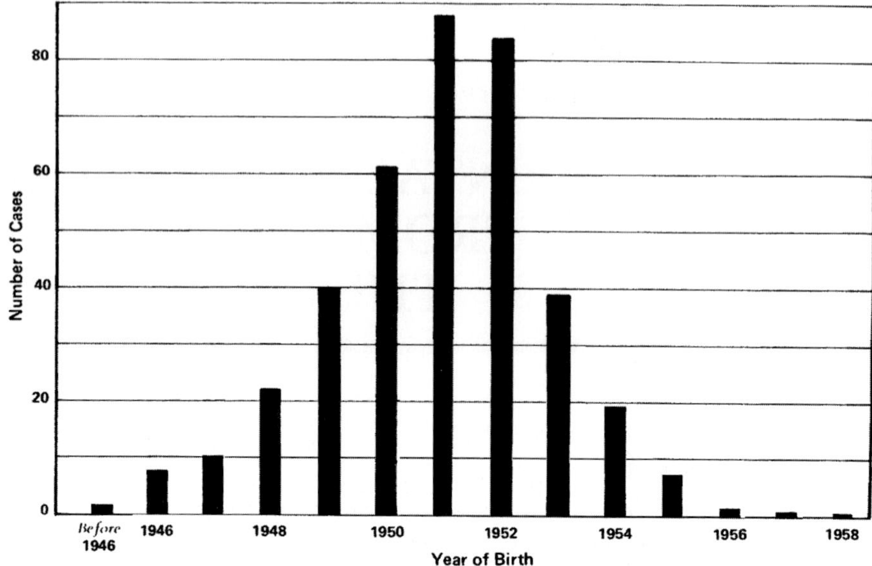

Fig. 6.1.   A bar graph showing the number of cases of retrolental fibroplasia from 1946 to 1958 in southern California. The incidence dropped dramatically after oxygen was restricted.

(*Source*: Modified and reproduced with permission from the publisher.) Reference 2.

kept by the Department of Vital Statistics (in the U.S.) and the National Health Service in England. After digesting the figures, he wrote (in 1973) that "while the policy of restricting the amount of oxygen in incubators has diminished the number of cases of retrolental fibroplasia. . .it has concurrently increased the number of deaths in the first 24 hours of life. A rough estimate suggests that for each case of blindness prevented, there is an excess of 16 deaths".[5]

He then analyzed the infant mortality records of the New York City and State Departments of Health over a 35-year period. When they were put into graph form, a distinct "bump" in the number of low-birth-weight infant deaths was seen between 1955 and 1970 (Fig. 6.3).[6] The graph on the right (Fig. 6.4) shows the *actual* infant death rate from 1935 to 1950 and the *projected* death rate from 1951 to 1970 if oxygen therapy had not been restricted.[6]

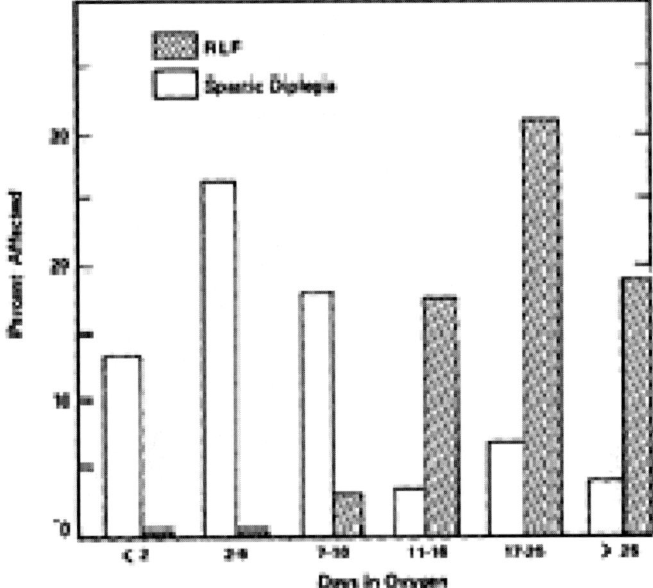

Fig. 6.2. A bar graph showing the see saw relationship between the incidence of spastic diplegia (cerebral palsy) and of retrolental fibroplasia (RLF) with respect to the duration of oxygen therapy. The shorter the time infants are exposed to a high oxygen concentration, the lower the incidence of retrolental fibroplasia and the higher the incidence of cerebral palsy, and vice versa.

(*Source*: Reproduced with permission from the publisher.) Reference 4.

Pathophysiology of Retrolental Fibroplasia

The disease begins shortly after birth, when the immature retinal capillaries are exposed to a higher oxygen concentration than they experienced *in utero*. The abrupt rise in oxygen tension and the release of associated oxygen-derived free radicals when a high concentration of oxygen is given, coupled with the extremely low level of vitamin E at birth, injure the incompletely developed retinal capillaries. The administration of oxygen paradoxically leads to *too little* oxygen reaching the retina because of constriction (caused by oxygen itself) of the blood vessels bringing oxygen to the retina (Fig. 6.5). Blood vessels throughout the body naturally dilate when the oxygen tension falls to ensure that more oxygen reaches the tissue(s) being nourished by

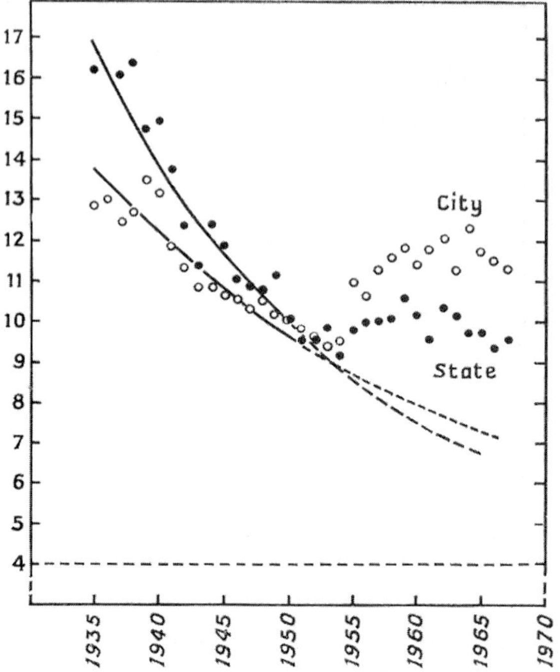

Fig. 6.3.   The number of deaths per 1000 low-birth-weight infants (ordinate) is plotted over a 35-year span in New York City and State. A distinct bump occurred after 1955.

(*Source*: Reproduced with permission from the publisher.) Reference 5.

those vessels.[a] Thus, the retinal vessels constrict when the oxygen tension rises. The autonomic nerves in the walls of the two small arteries bringing blood to the eyes do not "know" that the tissue it is nourishing is a very immature, susceptible retina. The "oxygen-induced constriction leads to sludging and eventual clotting of blood in the tiny retinal arteries and capillaries, and finally to destruction of the retina" four weeks to four months after birth.[8]

---

[a] A common expression of such vasodilatation is the occurrence of a headache when a person from sea level arrives at high altitude. Cerebral blood vessels dilate in an attempt to provide more oxygen to the brain.

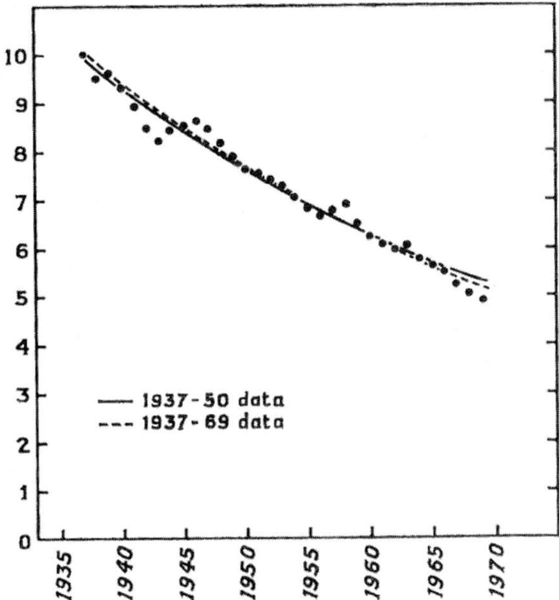

Fig. 6.4. The number of deaths per 1000 full-term infants during the first week of life over a 35-year span in the U.S. The death rate followed an exponential curve from 1935 to 1970.

(*Source*: Reproduced with permission from the publisher.) Reference 6.

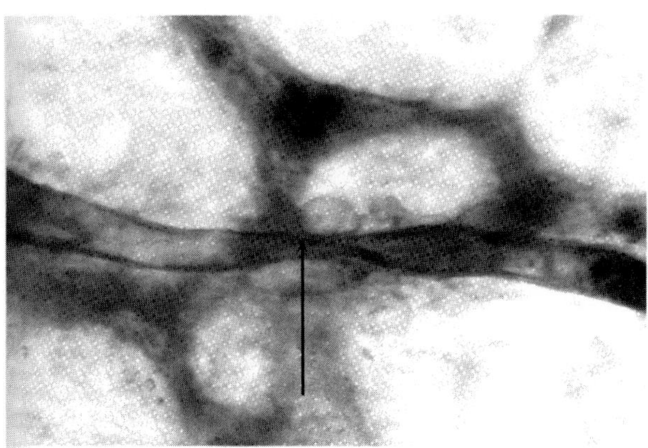

Fig. 6.5. A photomicrograph of the retina after exposure to oxygen for 48 hours. The small blood vessel is constricted at the arrow.

(*Source*: Reproduced with permission from the publisher.) Reference 7.

References

1. Avery ME, Oppenheimer EH. (1960) Recent increase in mortality from hyaline membrane disease. *J Pediatr* **57:** 553–559.
2. Silverman WA. (1980) The first decade of RLF. In *Retrolental Fibroplasia: a Modern Parable.* Grune & Stratton, New York, London, Toronto, Sydney, San Francisco, Chap. 3.
3. Rodriguez RJ, Martin RJ, Fanaroff AA. (2002) *Neonatal–Perinatal Medicine: Diseases of the Fetus and Infant.* In: Fanaroff AA, Martin RJ (eds.) 7th ed. Mosby, Elsevier, Philadelphia pp. 1001–1011.
4. Silverman WA. (1980) The consequences of oxygen restriction. In *Retrolental Fibroplasia. A Modern Parable.* Grune & Stratton, New York, London, Toronto, Sydney, San Francisco, Chap. 8.

7

THE SECOND EPIDEMIC OF RETROLENTAL FIBROPLASIA: 1972–20??

After a hiatus of nearly 20 years, the second epidemic began. In the early 1970s, newborn intensive care units were introduced. With them came an unexpected resurgence of retrolental fibroplasia (Fig. 7.1). About the same time, the term "retrolental fibroplasia" itself was replaced by "retinopathy of prematurity." (In keeping with that change, the newer term will be used from this point on.) The resurgence was not due to overuse of oxygen. It merely reflected the greater number of premature babies who survived the first year of life as a result of neonatal intensive care units.[1] Retinopathy of prematurity seems to be the price small infants pay for better neonatal care, much like Alzheimer's disease is the price adults pay for living longer at the end of life as a result of better cardiac care.

Cryotherapy for Retinopathy of Prematurity

In the mid-1980s, a new era began in the treatment of retinopathy of prematurity. Cryotherapy of the retina in premature infants weighing less than 1251 g (2¾ pounds) with retinopathy of prematurity was introduced.[a]

---

[a] Cryotherapy, also called cryoablation, is a procedure for freezing a small portion of the surface of the eye. It employs an instrument — a small probe — that briefly (2–3 s) delivers a temperature of –30°C (–22°F) — cold enough to generate small ice crystals — to the sclera overlying the retina.[5] Forty to fifty separate freezes over 20–30 min are typically delivered.[6] Each brief period of intense hypothermia destroys the underlying (small) area of the retina that is closest to the probe. During the freezing process, the retina prone to detachment is "sealed" from the rest of the retina, preventing future detachment.

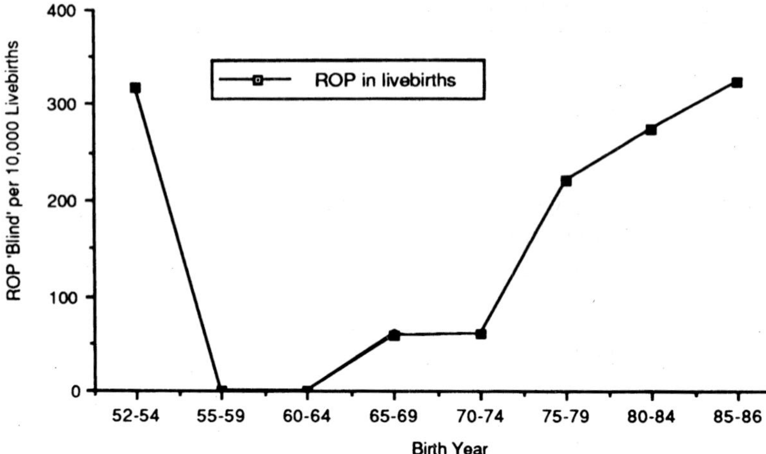

Fig. 7.1. The incidence of retinopathy of prematurity (ROP)-induced blindness per 100,000 live-born infants weighing 750 to 999 g from 1952 to 1986. The incidence in the 1980s equaled that in the 1950s. The reduced incidence in the late 1950s and early 1960s followed the careful regulation of oxygen to premature babies. The re-rise was related to an increased survival rate of the smallest newborns due to improvements in technology and the advent of neonatal intensive care units. See text for explanation of this paradox. The incidence of retinopathy of prematurity was the same (68%) in a more recent (2005) natural history study of premature infants in the U.S.[2] as the incidence was in a 1991 report (65.8%).[3]

(*Source*: Adapted and reproduced with permission from publisher.) Reference 4.

The procedure was considered only when the retinopathy reached a "threshold" indicating a risk of retinal detachment of approximately 50%. Over two years beginning in January 1986, 9751 such small babies at 23 university hospitals in the U.S. had cryotherapy done on one eye, while the other eye did not have cryosurgery performed and served as the control eye.[5] The infants were then evaluated 12–16 months after surgery. They were divided into two groups: those with an "unfavorable" *anatomic* outcome and those with an "unfavorable" *visual* outcome. An unfavorable anatomic outcome was defined as retinal detachment or a "macular fold" (a ridge of tissue formed by the doubling back of the retina) that interfered with vision. An unfavorable visual outcome was defined as eyes with poor vision (as assessed by the infant's ability or inability to consistently fix his or her vision on an object) or no vision (i.e. blind) at 12–16 months following surgery.[7] An unfavorable anatomic outcome was found in 25.7% of treated eyes and 47.4% of control eyes, with a similar degree of retinopathy of prematurity. An unfavorable visual outcome

Table 7.1   Visual acuity 15 years after cryotherapy during infancy.

Visual acuity outcome	Number of treated eyes (%)	Number of untreated eyes (%)
Favorable	(55.3%)	(35.7%)
20/20 or better	12 (6.3)	18 (9.7)
Worse than 20/20 but 20/40 or better	34 (17.9)	27 (14.6)
Worse than 20/40 but 20/60 or better	23 (12.1)	7 (3.8)
Worse than 20/60 but better than 20/200	36 (18.9)	14 (7.6)
Unfavorable	(44.7%)	(64.3%)
Equal to or worse than 20/200	16 (8.4)	17 (9.2)
Equal to or worse than 20/200	13 (6.8)	12 (6.5)
Blind	16 (8.4)	50 (27.6)
Exempt	40 (21.0)	40 (21.6)
Total	200	195

Eighty children were exempted from the 15-year study because of blindness due to retinal detachment as determined before the final examination.
(*Source*: Modified and reproduced with permission from the publisher.) Reference.[9]

was noted in 35.0% of the treated eyes and 56.3% of the control eyes. The authors concluded that the data demonstrated "the efficacy of cryotherapy in reducing by approximately one-half the risk of an unfavorable retinal outcome" in infants with moderately advanced retinopathy of prematurity.

In 2001, the 10-year followup report on these children noted that the untreated eyes had a 30% greater chance of legal blindness (vision worse than or equal to 20/200) than the treated eyes.[8] And in the final 15-year report — 2005 — of the Cryotherapy for Retinopathy of Prematurity Study Group,[9] an unfavorable visual outcome — defined as 20/200 or worse visual acuity — was found in 44.7% of treated eyes and 64.3% of untreated eyes ($p < 0.001$) (Table 7.1). Interestingly, only 6.3% of children with treated eyes had 20/20 or better vision 15 years after cryosurgery, whereas 9.7% of children with untreated eyes had normal vision 15 years later.

Laser Therapy for Retinopathy of Prematurity

In 1994, a meta-analysis of three small studies conducted between 1991 and 1993 on retinal laser ablation was published. The report also included 132 of the authors' patients, bringing the total number of eyes treated to 293.[10]

In these small, premature babies, one eye received then-standard cryoablation of the retina while the other eye had laser ablation of the retinal area likely to detach. Both cryo- and laser ablation of the periphery of the retina apt to detach destroy that area and the corresponding (future) lateral vision.[b] However, the more important central vision, which will be needed for reading and driving, is preserved.

The 1994 Laser ROP Study Group learned that eyes treated with laser ablation had an 8.4% chance of an unfavorable outcome during infancy, whereas those eyes treated with cryoablation had a 19.2% chance. A 1998 report from the Willis Eye Hospital in Philadelphia found that among 42 infants who had cryotherapy of one eye and laser therapy of the other, at an average followup time of 5.8 years the "odds an eye treated with laser had a good outcome were 6.91 times greater than for eyes treated with cryotherapy" (Table 7.2).[11]

A 2002 report from the National University of Ireland and the Willis Eye Institute (Philadelphia) backed up these numbers. Among 66 infants randomly assigned to receive either cryotherapy or laser therapy, the eyes treated with laser were 7.2 times more likely to have 20/50 or better vision 10 years later. The authors concluded that "visual acuity [was] significantly better in those eyes treated with laser photocoagulation into late childhood".[12]

The improved outcome with laser therapy coupled with the observation that "cryotherapy damages the sclera of the eye, treatments [are] painful and [the] equipment is cumbersome"[13] soon led to the adoption of laser as the preferred therapy for ROP. As reported in 1998, laser is "more effective in restoring vision and is easier to perform by the ophthalmologist and it is better tolerated by the infant".[14]

Table 7.2

Treatment modality	Good vision (20/50 or better)	Poor vision (20/60 or worse)	Total
Laser	17 (81%)	4 (19%)	21
Cryotherapy	8 (38%)	13 (62%)	21
Total	25 (60%)	17 (40%)	42

(*Source*: Reproduced with permission of the publisher.) Reference 11.

---

[b]With laser ablation, the peripheral retina is "burned away" rather than frozen. The damaged area results in scar tissue — which is strong — that seals the peripheral retina, preventing or deterring its detachment.

Guidelines for Practicing Physicians Regarding Cryotherapy Versus Laser Ablation for Retinopathy of Prematurity

Guidance for practicing physicians often comes in widely read textbooks on a particular subject. These texts are typically written by university professors. Quotations from two such textbooks follow.

In a book by Frank Oski from Johns Hopkins published in 2006, the author states that the "current treatment of ROP consists of laser surgery.... A large multicenter trial (the Early Treatment for Retinopathy of Prematurity Cooperative Study[15]) has documented the value of this therapy.... Laser treatment appears to be better tolerated by infants and achieves [a] similar favorable outcome as compared to the more invasive and less-well-tolerated cryotherapy."[16]

Nelson's 2007 edition of his *Textbook of Pediatrics* agrees. "Laser is the *treatment of choice*" (author's italics) for retinopathy of prematurity.[17] And an editorial in the December 2003 issue of the *Archives of Ophthalmology*[18] pointed out that since 1984, progress in the treatment of moderately advanced retinopathy of prematurity has been "nothing short of revolutionary." The initial headway was made through cryotherapy. However, the editorialist continued, "It has...[been] demonstrated that this treatment [cryotherapy] is no panacea, because 10 years after treatment 45.4% of the treated eyes had a visual acuity of 20/200 or worse." Encouragingly, a new report[19] in the same issue of the journal found "a benefit of earlier treatment" [at 10–12 weeks of age], largely attributable to laser therapy. The editorial concluded optimistically, "Further refining of the indications for treatment [of ROP] will undoubtedly improve the outcome of many neonates and infants with severe ROP." Good news for parents of children with this sight-robbing disease.

References

1. Phelps DL. (1982) Retinopathy of prematurity: an estimate of vision loss in the United States — 1979. *Pediatrics* **67**: 924–926.
2. Early Treatment for Retinopathy of Prematurity Study Group. (2005) The incidence and course of retinopathy of prematurity: findings from the early treatment for retinopathy of prematurity study. *Pediatrics* **116**: 15–23.
3. Palmer EA, Flynn JT, Hardy *et al.* (1991) Incidence and early course of retinopathy of prematurity. *Ophthalmology* **98**: 1628–1640.

4. Gibson DL, Shepa SB, Uh SH, Schechter MT, McCormack AO. (1990) Retinopathy of prematurity-induced blindness: birth weight-specific survival and the new epidemic. *Pediatrics* **86**: 405–412.

5. Cryotherapy for Retinopathy of Prematurity Cooperative Group. (1988) Multicenter trial of cryotherapy for retinopathy of prematurity. Preliminary results. *Arch Ophthalmol* **106**: 471–479.

6. Brown GC, Tasmar WS, Naidoff M, Schaffer DB, Quinn G, Bhutani V. (1990) Systemic complications associated with retinal cryoablation for retinopathy of prematurity. *Ophthalmology* **97**: 855–858.

7. Cryotherapy for Retinopathy of Prematurity Cooperative Group. (1990) Multicenter trial of cryotherapy for retinopathy of prematurity: one year outcome — structure and function. *Arch. Ophthalmol* **108**: 1408–1416.

8. Cryotherapy for Retinopathy of Prematurity Study Group. (2001) Ophthalmologic outcomes at 10 years. *Arch Ophthalmol* **119**: 110–118.

9. Cryotherapy for Retinopathy of Prematurity Study Group. (2005) 15-year outcomes following threshold retinopathy of prematurity. *Arch Ophthalmol* **123**: 311–318.

10. Laser ROP Study Group. (1994) Laser therapy for retinopathy of prematurity. *Arch Ophthalmol* **112**: 154–156.

11. Connolly BP, McNamara JA, Sharma S, Regillo CD, Tasman W. (1998) A comparison of laser photocoagulation with trans-scleral cryotherapy in the treatment of threshold retinopathy of prematurity. *Ophthalmology* **105**: 1628–1631.

12. Ng EYJ, Connoly BP, McNamara JA, Regillo CD, Vander JF, Tasmas W. (2002) A comparison of laser photocoagulation with cryotherapy for threshold retinopathy of prematurity at 10 years. *Ophthalmology* **109**: 928–935.

13. Hunter DG, Repka MX. (1993) Diode laser photocoagulation for threshold retinopathy of prematurity: a randomized study. *Ophthalmology* **100**: 238–244.

14. Subramaian KNS. Retinopathy of prematurity. www.emedicine.com/ped/topic1998.htm

15. Early Treatment for Retinopathy of Prematurity Cooperative Group. (2003) Revised indications for the treatment of retinopathy of prematurity: results of the early treatment for retinopathy of prematurity randomized trial. *Arch Ophthalmol* **121**: 1684–1696.

16. Traboulski EI. (2006) Pediatric ophthalmology. In: McMillian JA, Feigin RD, De Angelis CD, Jones MD, Jr (eds.), *Oski's Pediatrics: Principles and Practice,* 4th ed. Lippincott, Willams & Wilkins, pp. 801–827.

17. Olitsky SE, Hug D, Smith LP. (2007) Disorders of the eye. In: Kliegman RM, Behrman RE, Jenson HB, Stanton BF (eds.), *Nelson Textbook of Pediatrics,* 18th ed. Saunders Elsevier, Philadelphia, PA, p. 2600.

18. Fielder AR. (2003) Preliminary results of treatment of eyes with high-risk threshold retinopathy of prematurity in the early treatment for retinopathy of prematurity randomized trial. *Arch Ophthalmol* **121**: 1769–1771.

19. Hardy RJ, Palmer EA, Dobson V, Summers G, Phelps DL, Quinn GE, Good WV, Tung B for the Retinopathy of Prematurity Study Group. (2002) *Arch Ophthalmol* **121**: 1647–1701.

8

CYTOKINES AND CHEMOKINES

At about the same time as free radicals were being indicted, another class of substances was also gaining notoriety . . . for the most part, however, favorable notoriety. These substances, known as chemoattractant cytokines[a] (or chemokines, for short), were discovered in the early 1970s.[1] They quickly captured the interest of basic scientists in the US and the UK. In the 1990s, they became familiar to citizens of both countries through science-update articles in The *New York Times* and *The London Times.*[2,3]

Among researchers in the 1990s, cytokines and chemokines became more thoroughly investigated, thanks to faster access to abstracts and related database links through the Internet. Even the sometimes confusing and erratic terminology associated with cytokines has diminished over time, also due in part to Internet communication, allowing scientists to communicate with one another more effectively than by telephone or at once-a-year national meetings. Publication of full-text papers remains the gold standard, of course, since they are peer-reviewed and contain details not available either in abstract form or during brief oral presentations at symposia.

Like free radicals, cytokines are produced by most cells. And like free radicals, there are a heady variety of cytokines.[b,c] They both stimulate and inhibit

---

[a]The term "cytokine" comes from the prefix "cyto-," meaning "cell," and the suffix "-kine," meaning "action" or "motion." All cytokines, then, affect some action of a cell, the specific action affected being reflected by the preceding modifier, *e.g.* "chemoattractant". A chemoattractant cytokine literally attracts one cell to another, chemically.

[b]Cytokines are a group of small proteins that are released by most cells to cause a change in a function of the target cell. The target cell(s) is (are) most often adjacent to or near the source cell. A distinctive characteristic of cytokines is that they are generated only in response to a stimulus. Therefore, they are produced intermittently.

[c]Chemokines represent a subgroup of cytokines produced defensively by macrophages when stimulated by bacterial toxins. They thereby recruit white blood cells (polymorphonuclear cells

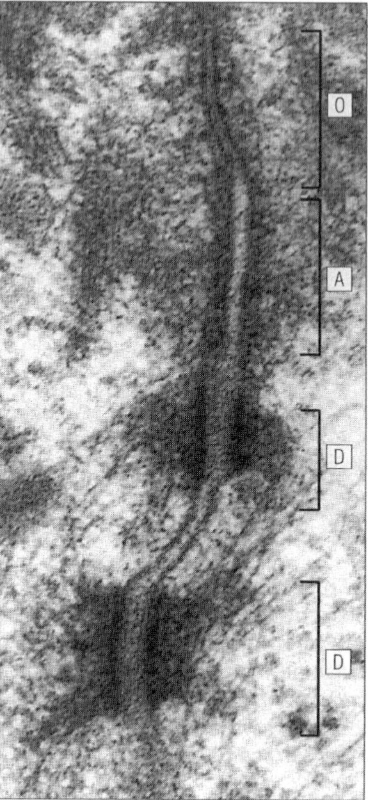

Fig. 8.1.　A gap junction is a 2–4-nm-wide channel that transmits small sugar molecules and even smaller ions from one cell to an adjacent cell. Each gap junction provides a means to even out spontaneous fluctuations in the intercellular concentration of small molecules. Macromolecules such as LDL with a mass of 3×10^6 daltons cannot use gap junctions to get from cell to cell. An electron microscope is needed to see a gap junction.

(*Source*: Reproduced with permission from the publisher.) Reference 4.

cells affecting the body's normal physiology and its defense against bacteria, aiming to interrupt the body's everyday functioning. As such, cytokines maintain and/or restore the normal balance of physiological processes, known as

in response to bacteria) to the site of an infection to combat the microorganisms. In the setting of atherosclerotic vascular disease, macrophages also produce chemokines, this time in response to the presence of LDL in the subendothelial space. Chemokines, then, mediate adhesion of other white blood cells (mononuclear cells in the case of LDL) to the endothelium and promote their migration through the endothelium.

homeostasis. These relatively small but powerful molecules are synthesized in substantial amounts by macrophages in the subendothelial space and in much smaller amounts by endothelial cells. Cytokines released by macrophages attract mononuclear cells (best known for their prevalence in infectious mononucleosis) called monocytes to the endothelial surface. They then guide the monocytes through the intact endothelium into the subendothelial space through narrow gap junctions lined by adhesion molecules on each side of the endothelial layer (Fig. 8.1). There, monocytes are transformed by free radicals, also made by macrophages, into more macrophages, which make more cytokines, and so on. As a result, free radicals and cytokines work hand in hand to oxidize LDL in a Goliath-versus-David struggle, where David has only a few antioxidants as his stones. (On the other hand, LDL might be viewed as the Goliath in this contest, because its molecular weight is 2.5 million daltons, while the combined molecular weight of the four free radicals (superoxide, nitric oxide, hydrogen peroxide, and the hydroxyl radical) plus cytokines is a mere 20,000 daltons!)

While cytokines and free radicals work hand in hand to oxidize LDL, with the former guiding monocytes into the subendothelial space where the latter oxidizes LDL, another "joint venture" aimed at oxidizing LDL is going on in the same region, this one between free radicals and copper ions. In 1984, a collusion between copper (or iron) and free radicals was established by showing that the oxidation of LDL (*in vitro*) could be prevented either by antioxidants (such as vitamin E, which scavenges free radicals) *or* by agents (such as EDTA) that keep copper ions in their reduced state (Cu^+) preventing copper from oxidizing LDL (or by simply not including metal ions in any form in the culture medium surrounding the cells). If either iron or copper ions weren't available, LDL oxidation by free radicals didn't happen.[5] Even very low concentrations of metal ions led to the oxidation of LDL by cells, through free radicals generated by the cells.

References

1. Oppenheim JJ, Feldmann M. (2001) Introduction to the role of cytokines in innate host defense and adaptive immunity. In: Durum SK, Hirano SF, Vilcek J, Nicola NA (eds.),*Cytokine Reference Book,* Vol. 1 (Ligands). Academic, San Diego, San Francisco, New York, Boston, London, Sydney, Tokyo, pp. 3–20.
2. Kolata G. (1995) The double-edged molecule. *The New York Times,* Dec. 31, p. 2E.
3. *London Times,* Apr. 10, 1997.

4. Stevens A, Lowe JS. (2005) Epithelial cells. In: Stevens A, Lowe JS (eds.), *Human Histology, 3rd ed.* Elsevier Mosby, Philadelphia, Edinburgh, London, New York, Oxford, St Louis, Sydney, Toronto, pp. 11–54.
5. Steinbrecher UP, Parthasarthy S, Leake DS, Witzum JL, Steinberg D. (1984) Modification of low density lipoprotein by endothelial cells involves lipid peroxidation and degradation of low density lipoprotein phospholipids. *Proc Natl Acad Sci USA* **81**: 3883–3887.

9

LDL OXIDATION BY FREE RADICALS — IN DETAIL

The complexity of the LDL oxidation process is predictable given the enormous size and intricacy of the LDL particle (Fig. 9.1). This is a large, spherical structure measuring 22 nanometers (nm) in diameter,[a] and it is composed of 700 molecules of phospholipids, 600 molecules of cholesterol (free), 1699 molecules of cholesterol esters, 185 molecules of triglycerides, and 1 (very large) molecule of apolipoprotein B[b] containing 4536 amino acids. For comparison purposes, three other large biomolecules (biological molecules) — insulin, hemoglobin and DNA — are shown in Fig. 9.2. An LDL particle drawn to scale is included. Cholesterol molecules (shown in pink in Fig. 9.2 and structurally displayed in Fig. 12.10) are less than 1 nm in diameter, accounting for why 600 cholesterol molecules (plus 1699 molecules of cholesterol esters) are part of a single, gigantic LDL particle.

Although all LDL particles are big, their composition varies according to a person's diet.[3] Particles from individuals who eat foods rich in saturated fatty acids are predominantly small, dense particles more susceptible to oxidation than particles from individuals who consume foods rich in monosaturated fatty acids that are larger, lighter, and more resistant to oxidation.[3,4] (The effect of diet on atherosclerosis will be addressed more fully in Chap. 17.) It is speculated that the single apolipoprotein molecule "circumnavigates" the huge LDL sphere as short segments "dive in and out" of its cholesterol core to

---

[a]One nm equals one-millionth of a mm. One LDL particle is 22 times larger than one water molecule, whose diameter is a little less than 1 mm.

[b]Apolipoproteins are a family of large biological molecules embedded in the outer shell of lipoproteins. Apolipoprotein B is the specific family member that transports cholesterol from the liver to all other organs, where it binds to LDL receptors.

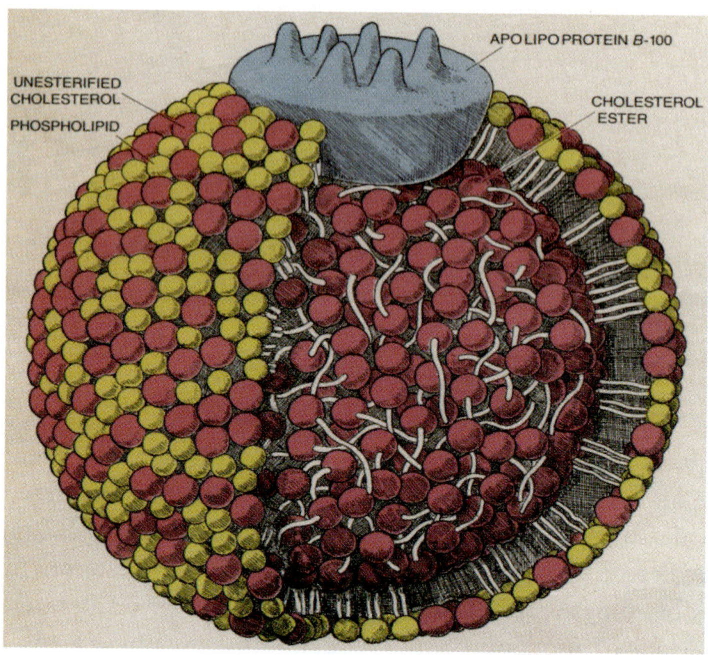

Fig. 9.1. A single LDL particle is 22 times larger than one water molecule, whose diameter is a little less than 1 nm. An LDL particle is composed of 700 molecules of phospholipids, 600 molecules of cholesterol (free) within its oily core, 1699 molecules of cholesterol esters, 185 molecules of triglycerides, and 1 molecule of apolipoprotein B containing 4536 amino acids, including 270 lysines.

A cholesterol ester simply has a long-chain fatty acid linked to a cholesterol molecule, making it easier to transport while it is in the watery plasma because it is much less polar than unesterified cholesterol.

(*Source*: Adapted with permission from Scientific American.) Reference 1.

bind with it.[5] The particle also comes equipped with its own vitamin E to protect its cholesterol from oxidation. But, because the LDL particle contains many more polyunsaturated fatty acids than it does vitamin E, its antioxidant abilities are overwhelmed quickly. As a result of the "heterogeneous nature of the LDL particle," and the "extraordinarily complex chemistry involved in LDL oxidative modification," there are a "wide variety of oxidized LDLs."[6] In theory, one to all of the 2616 fatty acids and one to all of the 4536 amino acids in LDL could be oxidized, and the product of each would qualify as "oxidized LDL." As a result, the number of permutations and combinations of oxidized LDL approaches infinity.[7,8]

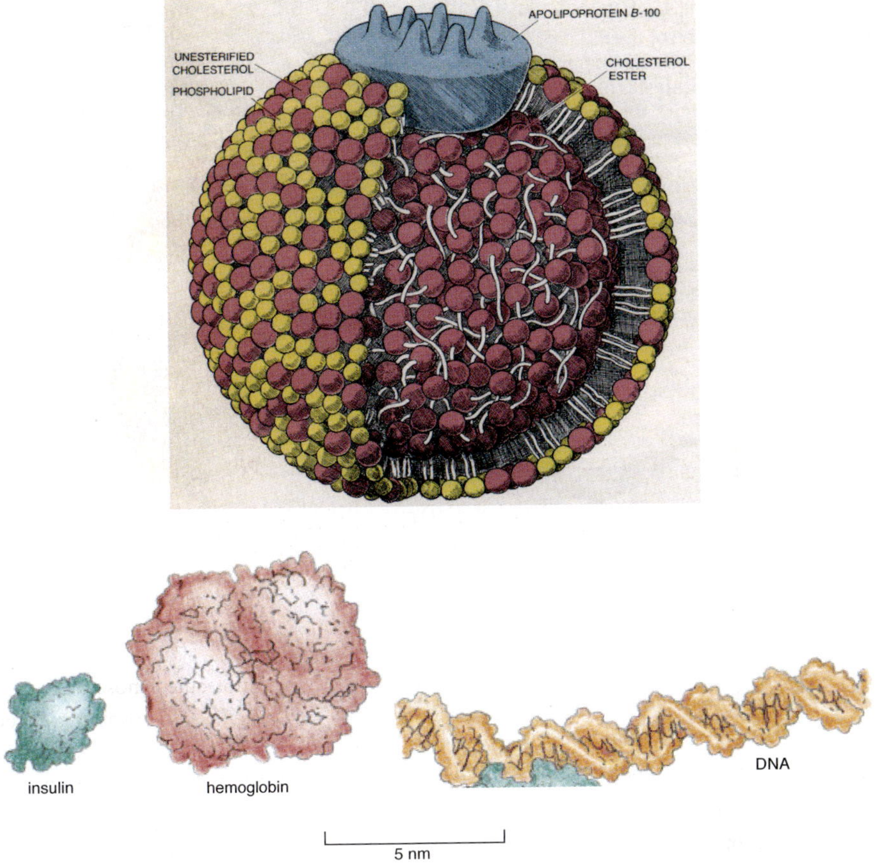

Fig. 9.2.　A collection of three biomolecules drawn to the same scale. For comparison, an LDL particle is reillustrated. The LDL particle dwarfs all others.

(*Source*: Reproduced with permission from the publisher.) Reference 2.

Good and Bad Cholesterol

The LDL cholesterol measured during a routine blood test following an overnight fast is often referred to as "bad cholesterol." The total cholesterol — which should be under 200 mg/dL — consists of LDL cholesterol and HDL cholesterol. A value of less than 100 mg/dL of LDL cholesterol is considered "optimal" in the U.S., while an HDL cholesterol of 60 mg/dL or more is recommended[9] (Table 9.1). However, LDL cholesterol is not inherently "bad."

Table 9.1 Current classification of LDL cholesterol and total cholesterol levels.

LDL cholesterol	Total cholesterol	Classification
<100 mg/dL	<160 mg/dL	Optimal
100–129 mg/dL	160–179 mg/dL	Near-optimal
130–159 mg/dL	180–199 mg/dL	Borderline-high
160–189 mg/dL	200–239 mg/dL	High
>190 mg/dL	>240 mg/dL	Very high

Source: Reference 9.

It received its unsavory reputation in the 1960s and '70s before it was established, largely through the work of Brown and Goldstein, that there are two types of cell-surface receptors for 'bad cholesterol' — LDL receptors and so-called scavenger receptors. The former bind unoxidized LDL, while the latter bind oxidized LDL. Unoxidized LDL (native LDL) is harmless and, in fact, is necessary physiologically for building cell membranes and sex hormones. Even "minimally oxidized" LDL is not harmless, but it is not measurable in the blood because oxidation of LDL is carried out in the subendothelial space — as will be explained — and because abundant antioxidants (most notably vitamin E) are present in the blood. Thus the "[LDL] cholesterol level" measured by a routine laboratory test is virtually all unoxidized. LDL becomes harmful only when its concentration in the blood exceeds 150 mg/dL, allowing it to enter the subendothelial space, where it undergoes oxidation and becomes harmful. In light of the now-proven hypothesis that LDL cholesterol leads to coronary heart disease only *after* it is oxidized, it might be more accurate to refer to unoxidized LDL cholesterol as "harmless" or "physiologic" and oxidized LDL cholesterol as potentially "harmful" or "unphysiologic," leaving HDL cholesterol as "good cholesterol." Alternatively, unoxidized LDL cholesterol might be called "good" cholesterol, since it is used for normal, physiologic purposes, while oxidized LDL is referred to as "bad" cholesterol and HDL cholesterol as "better" cholesterol, since it actually removes oxidized LDL. My goodness, if LDL cholesterol builds cell membranes and sex hormones, it's got to be "good." If it builds plaques in blood vessels, it's got to be "bad." And if HDL cholesterol *removes* plaques from blood vessels, it's got to be "better" than good!

The Problem of Measuring Oxidized LDL Cholesterol

Given that oxidized, but not native, LDL leads to coronary heart disease, it would seem, *a priori*, that measuring the oxidized portion specifically would be more predictive of coronary heart disease than measuring unoxidized LDL cholesterol. Although such reasoning is valid, the technology to quantitate oxidized LDL isn't available. The reason it isn't is simply that even if it did, there is very little oxidized LDL in the blood to measure, for four reasons. Initially, the high concentration of antioxidants in plasma prevents the oxidation of circulating LDL. Second, the majority of LDL oxidation takes place in the subendothelial space, where heavily oxidized LDL is taken up by and stored in scavenger cells, preventing its entry into the bloodstream. Third, minimally oxidized LDL, though not removed by scavenger cells, is taken up by LDL receptors throughout the body, which are exquisitely regulated by the intracellular concentration of cholesterol. And fourth, oxidized LDL is a complex array of lipids, rather than a single quantifiable substance. Physiology has, in effect, foiled technology!

The Danger of Too Little Cholesterol

While the consequences of too much cholesterol are concerning, the consequences of too little cholesterol are also worrisome. In 1964, a "new" inherited disorder of cholesterol metabolism was first described.[10] Although the syndrome was initially called the "RSH syndrome" (after the first letters of the last names of the first three children with the disorder,[11] it is now better known as Smith–Lemli–Opitz syndrome (representing the last names of the three geneticists at the University of Wisconsin who first described it). The disorder occurs in 1 of 20,000–40,000 live births, is most common in Caucasians of eastern European ancestry, and is rarely seen in African and Asian populations.[11,12]

Although the syndrome was recognized as a genetic disorder in the 1960s, its cause was not learned until 1994, when Tint and his colleagues at the New Jersey School of Medicine[13] discovered that children with the disorder lack the final enzyme (7-dehydrocholesterol-delta-reductase) in the 30-step cholesterol synthesis process.[c] (Since 1994, five more disorders due to enzyme

---

[c]The complex cholesterol biosynthesis pathway has 30 steps, utilizing 22 enzymes.[14] This is not in discord with the "one gene–one enzyme" hypothesis proposed in 1941 by Beadle and

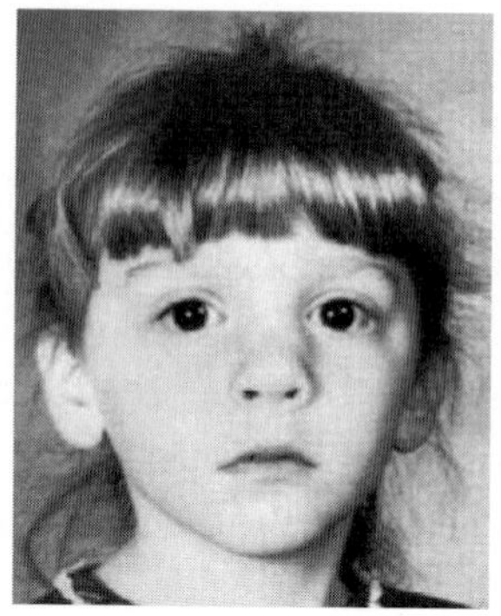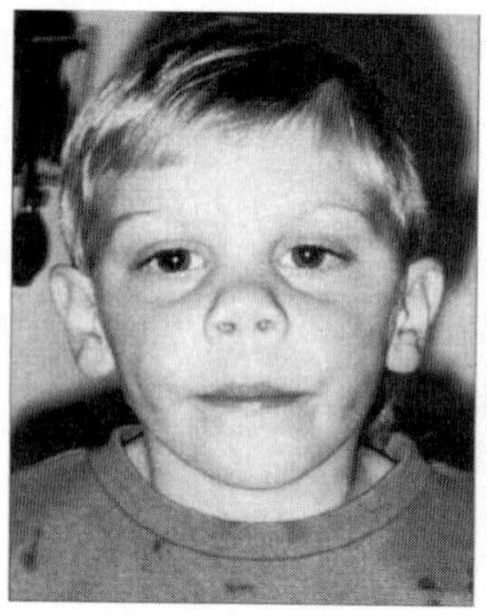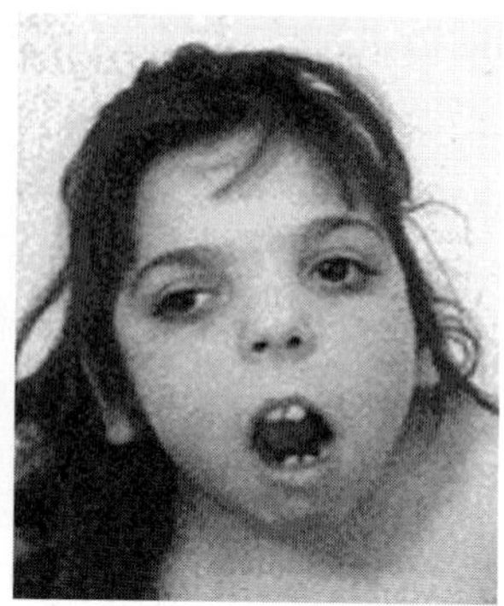

Fig. 9.3.   A 5-year-old girl (left), a 2½-year-old-boy (center), and a 10-year-old girl (right) with Smith–Lemli–Opitz syndrome. The 5-year-old girl has subtle facial features such as low-set ears, a small chin, and an upturned nose. The boy displays more apparent facial features, including droopy eyelids, epicanthal folds, a broad nasal bridge, and an upturned nose. The girl on the right shows more advanced facial features (Epicanthal folds are hereditary and common among Asians). The canthus of the eyes is where so-called "sleepers" or "sleep sand" forms at night from the accumulated secretions of eyelid glands.

(*Source*: Reproduced with permission from the publisher.) References 16 and 13.

defects in cholesterol synthesis have been found. Smith–Lemli–Opitz syndrome remains by far the most common.[14] Because cholesterol is an essential nutrient needed for cell membrane structure, the white matter of the brain, and nerve sheaths, infants lacking cholesterol have a variety of developmental defects. Most patients have distinctive facial features, including microcephaly (a small head), a high forehead, droopy eyelids, a broad nasal bridge, low-set ears, an upturned nose, a cleft palate, and micrognathia (a small jaw) (Fig. 9.3).[16,17]

Anomalies of the fingers and toes — such as short thumbs and webbing of the second and third toes (Fig. 9.4) — and congenital heart disease occur in some children.[15] Mental retardation is also common. A few children have only one or two malformations; many more have multiple defects.

As a result of the fundamental cholesterol enzyme deficiency, the plasma cholesterol in children with the disorder is usually <50 mg/dL.[17] The low

---

Tatum,[15] because several enzymes catalyze more than one step. The originators of the hypothesis — later proven — received the Nobel Prize in Physiology or Medicine in 1958.

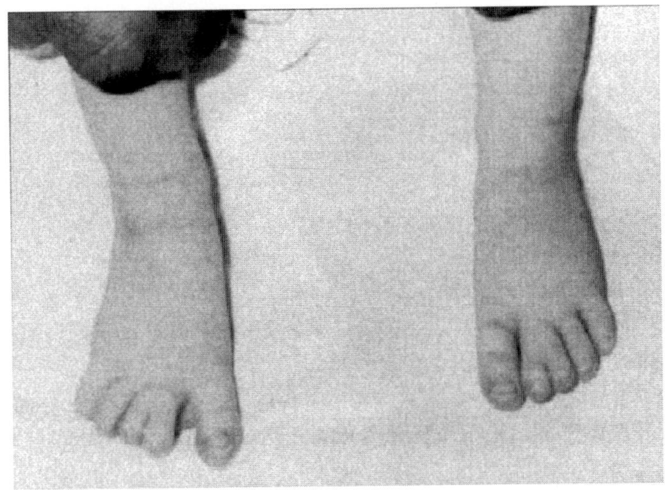

Fig. 9.4.   Webbing of the second and third toes of both feet and a "fork-toe" appearance of the first and second toes of the right foot.

(*Source*: Reproduced with permission from the publisher.) Reference 16.

cholesterol concentration is compounded in affected infants by interference with the basic sucking process needed for nourishment. Consequently, they do not grow at the expected rate.

Treatment of the disorder consists in dietary supplementation with foods rich in cholesterol — mostly eggs and cream.[d] Unfortunately, many older children find the diet "unpalatable" and require a feeding tube to get sufficient cholesterol.[17] Adequate dietary supplementation can "restore a normal growth pattern. . .alleviate behavioral abnormalities and improve general health"[13] Regrettably, cholesterol supplementation has "minimal effect on intellectual function."[14] Perhaps the best treatment for infants and children afflicted with Smith–Lemli–Opitz syndrome is in the offing. Recently two mouse models of the disease have been developed through gene targeting and homologous recombination[14] (see Chap. 13). While no therapeutic genetic breakthrough has been reported yet, the mouse model has "permitted identification of one

---

[d]The yolk of an egg is a rich source of cholesterol. One egg contains roughly 300 mg of cholesterol, the maximum recommended daily dietary intake. In contrast, peanut oil has a very low cholesterol content. One would have to drink about 9.6 (19.57 pounds) of peanut oil to ingest the amount of cholesterol in a single egg.[18,19]

of the major mechanisms that suppresses" cholesterol synthesis in children with the syndrome.[20] That offers reason to hope that gene therapy for the condition is not too far away. Understanding the mechanism(s) of a disease process (usually) precedes an effective therapy.

Apolipoprotein B — The Cholesterol Carrier

Because cholesterol — a lipid (fat) — will not mix with the water in the bloodstream, it needs a carrier while it is within the blood. Apolipoprotein B is the molecule (in LDL) that the much smaller cholesterol molecule attaches to as it is being transported from the liver to peripheral tissues, such as the heart. Apolipoprotein B itself is made up of lipids and proteins, — more lipids than proteins. It also includes the specific binding site needed to bind the bulky LDL particle to its receptors in every tissue.

Every Cell Makes its Own Cholesterol

Although dietary cholesterol is the source of about 25% of a person's cholesterol, cholesterol is also synthesized by all cells in the body.[21] By far most (85%) of the cholesterol required by cells is synthesized within each cell throughout the body (as opposed to being made in the liver and later extracted by each cell for its own metabolism). In essence, each cell takes care of its own cholesterol requirement. And, a human being — as an entity, composed of many cells — is especially stingy in its synthesis of cholesterol. For example, humans synthesize the *least* cholesterol per kilogram of body weight (9 mg/day/kg), while rats and mice synthesize the most (50 mg/day/kg)[22] (Fig. 9.5) among a variety of warm-blooded mammals.

The Subendothelial Space: The Source of Oxidized LDL

Virtually no LDL is oxidized in the blood. Abundant water-soluble antioxidants — most notably vitamin E — prevent its oxidation while the particle is within the bloodstream. To escape from circulating antioxidants, the LDL particle must reach the sanctuary of the subendothelial space, where antioxidants are scarce. Two routes are available. One is simply diffusion of the particles through large pores (20–30 nm) in the endothelium that allow the 22 nm (1 nm equals 1 millionth of a mm) diameter particle to enter the

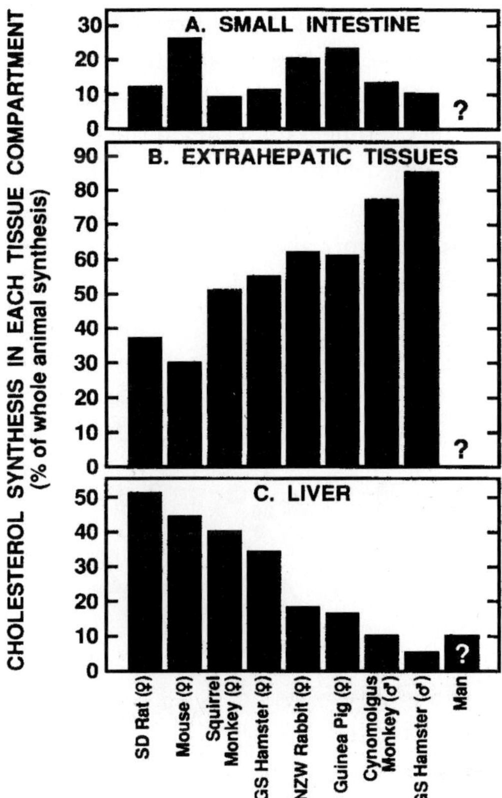

Fig. 9.5.　A graph showing the amount of cholesterol synthesized by the liver, small intestine and other extra-hepatic sites in a variety of animals. Rats and mice synthesize the most, while man makes the least.

(*Source*: Modified and reproduced with permission of the publisher.) Reference (2).

subendothelial space.[23] This is a purely passive, concentration-dependent process that does not require attachment of the particle to a receptor on the endothelial cell membrane.[6]

The second route is a more ingenious pathway. Known as "transcytosis," it is used exclusively by LDL and a few other macromolecules, such as insulin, hemoglobin and DNA (Fig. 9.2), and by viruses and bacteria trying to enter a cell. Upon reaching the endothelium, LDL is taken up in clathrin-coated[e]

---

[e]Clathrin is a protein that coats cell-surface pits destined to become vesicles. Lying external to the pits and vesicles, it provides a supportive cage for the delicate cell membrane within.

"pits," formed by "invagination of the cell membrane."[24] The pits are lined with clusters of receptors — binding sites — that recognize the macromolecule.[25] Once filled with LDL, the pits pinch off to form clathrin-coated vesicles (reminiscent of a diving chamber) underneath the endothelium. The vesicles then shuttle LDL across the cell to its opposite surface, where they evaginate, discharging their cargo into the subendothelial space (Fig. 9.6) — the estuary where free radicals and LDL mix and where LDL is oxidized.

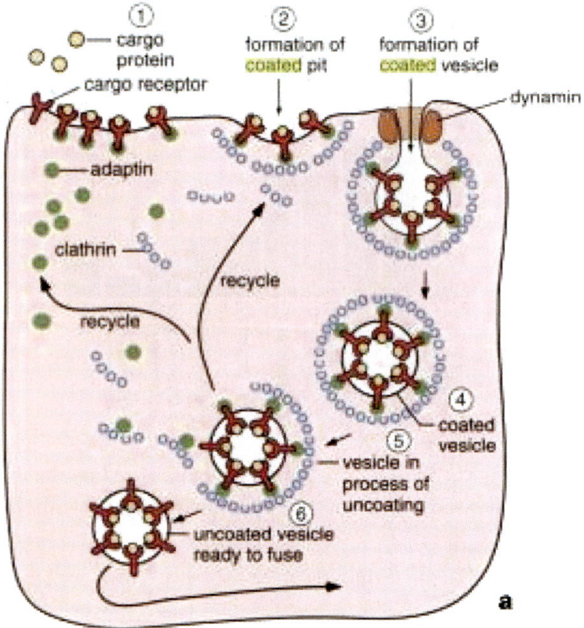

Fig. 9.6.   This diagram depicts in six steps how LDL is transferred from the blood into the subendothelial space. In step 1, so-called LDL cargo receptors on the surface of a cell recognize and bind LDL in the blood flowing by them. In step 2, the receptors (depicted in red and carrying small balls of LDL) attach to clathrin molecules (blue circles) and assemble into a shallow pit. In step 3, the pit pinches off from the cell surface to form a clathrin-coated vesicle. In step 4, the vesicle has reached the midpoint of its journey through the cell. In step 5, the vesicle begins to shed its clathrin receptors as it prepares to dock on the other side of the cell. In step 6, the uncoated vesicle prepares to discharge its cargo into the subendothelial space (not shown). After docking and discharging, the clathrin receptors are recycled.

(*Source*: Reproduced with permission of the publisher.) Reference 26.

Entrapment of LDL

Once within the subendothelial space, which occupies less than 1% of an artery's wall (Fig. 9.7) — yet is 100 times wider than the overlying endothelium — LDL is literally trapped in a three-dimensional network of collagen fibers and fibrils (Fig. 9.8).[28] Short protein segments within the single apolipoprotein B component of the LDL particle connect with proteoglycans in the matrix of the subendothelial space.[29] The strength of the chemical connection explains why a 70-fold greater concentration of "minimally" oxidized LDL (to be discussed soon) is present in the artery wall than in the plasma despite the greater concentration of *un*oxidized LDL in the plasma, favoring its diffusion into the wall.[30,31] LDL bound to collagen is not only trapped and resigned to being oxidized, it is also shielded by the endothelium from water-soluble antioxidants (such as vitamin C) in the plasma.

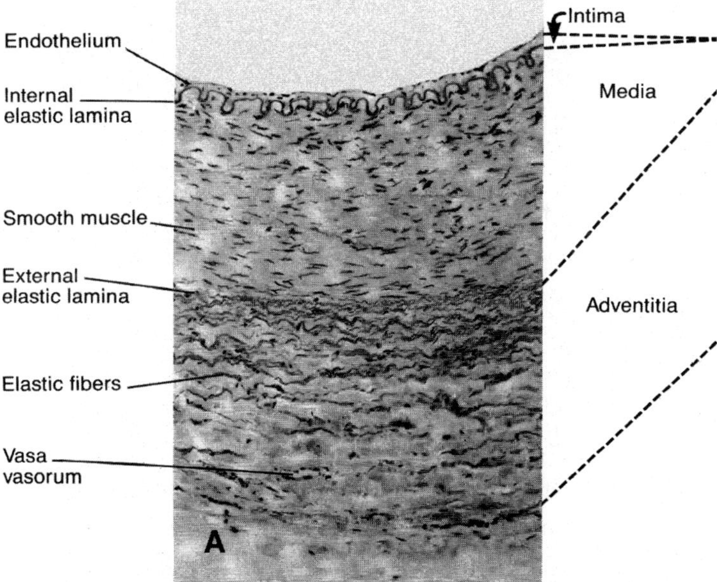

Fig. 9.7. A photomicrograph of a normal human artery. The subendothelial space lies immediately below the single-cell layer of endothelium. The endothelium, the subendothelial space, and the wavy, relatively thick internal elastic membrane together constitute the intima. The bulk of an artery's wall is composed of the media and adventitia. The intima constitutes about 5% of the wall thickness, and the subendothelial space less than that.

(*Source*: Reproduced with permission from the publisher.) Reference 27.

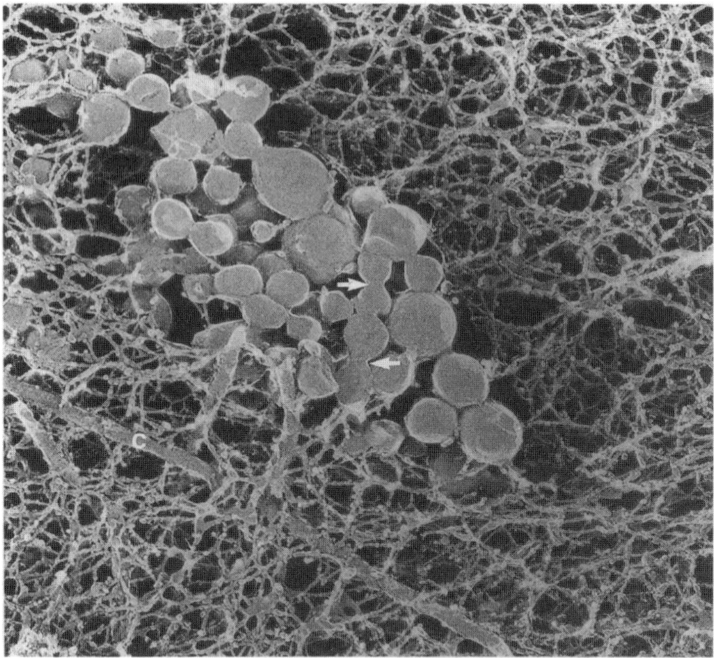

Fig. 9.8. A cluster of LDL particles entrapped by a network of filaments presumed to be proteoglycans. The filaments connect collagen fibrils, one strand of which is marked "c" (running diagonally in the lower left quadrant), together into a mesh. Several LDL particles are fusing (at arrows) to form a larger LDL particle. This photomicrograph was made 1–2 weeks after beginning a high-cholesterol diet in a laboratory animal.

(*Source*: Reproduced with permission from the publisher.) Reference 29.

The Beginning of LDL Oxidation

Now that the lipoprotein is anchored securely to cross-connecting collagen fibers in the subendothelial space, the process of LDL oxidation begins.[f] (Readers with a low or no tolerance for biochemistry are again encouraged to

---

[f]LDL oxidation happens at a woefully slow pace. Oxidation isn't like toasting bread, where the finished product pops up in 1–2 min. It's more like grilling a steak, where the finished product is cooked to varying degrees, according to taste. Yet it differs from grilling a steak, which may take an hour or so outdoors, because oxidizing LDL to the point where all of its several thousand fatty acids and amino acids are oxidized — when the huge particle is "well done" — takes 24–30 h.

turn to Chapter 10. Matters only get worse from this point through the end of the chapter.) The series of oxidative events is initiated by free radicals produced by both the overlying endothelial cells and the underlying smooth muscle cells. (The free radicals simply diffuse into the oxidative chamber.) Since at this very early stage of the transformation process the subendothelial space itself contains few or no cells directly releasing their free radicals into the region (the subendothelial space is normally devoid of cells), the concentration of pro-oxidants in the matrix is low and the resulting modified LDL is only minimally oxidized. Although only minimally modified, the revised LDL particle has two important biological properties not possessed by either native LDL or the LDL that is oxidized to a greater degree soon to follow. The first of these is its ability to stimulate endothelial cells to release a substance known originally (in 1989) by the matter-of-fact, though clumsy, term "vascular endothelial cell adhesion molecule, number 1" (VCAM-1).[g] VCAM-1 — a cytokine — initially attracts circulating monocytes[h] and then attaches them to the surface of those vascular cells in direct contact with the blood stream, *i.e.* the endothelial cells themselves.[i] (The two most important steps in the initial development of an atherosclerotic plaque are the adhesion of monocytes to the endothelium followed by the oxidation of LDL.) The second

---

[g]Simplicity yielded to scientific jargon in 1991, when VCAM-1 was renamed "selectin-E." The "E" stands for the endothelium that manufactures the compound, while "selectin" owes its name to the fact that lectin is the substance on the surface membrane of endothelial cells that specifically (selectively) binds monocytes to the membrane.[32]

[h]Monocytes are large white blood cells with a single kidney-shaped nucleus. Monocytes make up 5% of the circulating leukocyte population, whereas polymorphonuclear leukocytes constitute about 65%.

[i]Endothelial cells lining both arteries and veins are able to oxidize LDL. Thus, endothelial cells from umbilical veins can be used to study the process of LDL oxidation. However, monocytes, which ingest oxidized LDL, thereby transforming themselves into macrophages, do not accumulate in veins. As a result, atherosclerosis does not develop naturally in veins. On the other hand, if (saphenous) leg *veins* are used to bypass blocked coronary *arteries* (in humans), the re located veins can develop atherosclerosis. The high pressure within the arterial system allows monocytes to enter the subendothelial space of veins, where the monocytes begin to scavenge oxidized LDL. If the auto transplanted veins accumulate enough oxidized LDL to form plaques that narrow their lumens, then they can become blocked, just as the arteries that the veins had been used to bypass originally became blocked. Obstruction of leg veins used to bypass blocked coronary arteries occurs following one-third of heart operations done to relieve coronary artery disease. It is the principal reason that a second (or third, etc.) operation is needed within 5–10 years of the first procedure.

property of minimally modified LDL is its ability to stimulate endothelial cells to express another potent compound, the intercellular adhesion molecule (ICAM-1) — also a cytokine — which allows the now-attached monocyte to migrate in between the lining cells into the underlying region through gap junctions,[j] passageways similar to the large pores LDL particles used to reach the subendothelial space. VCAM-1 and ICAM-1 together form an "adhesion pathway" for monocytes to get from point A (the surface of endothelial cells) to point B (the subendothelial space).

Monocytes, Macrophages, and Minimally Oxidized LDL

In addition to large cell–surface adhesion molecules, such as VCAM-1 and ICAM-1, cytokines and chemokines contribute to the adhesion of monocytes to endothelial cells. After entering the subendothelial matrix, monocytes differentiate into macrophages, akin to embryonic stem cells dividing to produce lines of more complex cells. The mature macrophage looks little like its predecessor (Fig. 12.3). It behaves differently as well, as it begins to accumulate minimally oxidized LDL, acting in its new role as a "scavenger cell." (Until now, LDL has resided only *outside* of cells, both as it circulated in the blood and as it diffused through large endothelial pores into the subendothelial space, or while it was being ferried across an endothelial cell in a transport vesicle (Figure 9.6).) At this point, the LDL has not been altered sufficiently to affect its recognition by (or attachment to) the closely regulated (through the cholesterol-feedback mechanism) LDL receptors of macrophages. However, since macrophages are a rich, additional source of oxygen free radicals generated during their path-clearing activities, the LDL oxidation process within the subendothelial space picks up. As more and more oxidative wastes are released into the subendothelial matrix, more and more LDL is oxidized into a highly modified form that is gradually recognized and taken up by unregulated oxidized LDL (scavenger) receptors,[33] resulting in foam cell formation.[34]

---

[j]Both a monocyte's attraction to and its migration through the endothelial cell layer are accomplished by the binding of selectin and monocyte chemotactic protein, respectively, to their endothelial homologs. These lookalike compounds form a bond with each other, allowing the monocyte to pass through the single layer of endothelial cells — as if on a magic carpet.

$$^+NH_3$$
$$^+NH_3CH_2CH_2CH_2CH_2CHCO_2$$
Lysine

$$CH_3(CH_2)_4CH=CHCH_2CH=CH(CH_2)_7COOH$$
Linoleic acid

Fig. 9.9. Structure of lysine and linoleic acid

Investigators at UCLA have shown threshold recognition of LDL by scavenger cell receptors after neutralization of 16% of their lysine molecules.[35] Macrophages have both LDL and scavenger receptors, but only the latter slowly admit oxidized LDL into the cell. When 16% of the positive $^+NH_3$ groups within the 270 lysine molecules per LDL particle are neutralized by negative (aldehyde) fragments arising during the oxidation of *fatty* acids (predominately linoleic acid (Fig. 9.9), further uptake of LDL shifts to the macrophage's scavenger receptors.[36] Lysine (Fig. 9.9) is an essential *amino acid*[k] that contains the specific site (its terminal amino group — $^+NH_3$) where 3–9-carbon aldehyde fragments, generated during the oxidation of *fatty acids*, attach to the lysine components of apolipoprotein B, converting the lipoprotein into a potentially harmful form. Situated on the outside of the LDL particle, with its positive end facing outward, the positively charged, terminal amino ($^+NH_3$) group of the short lysine molecule is in a prime position to be neutralized by (negative) aldehydes generated during the free- radical-mediated oxidation of fatty acids contained in LDL. Isn't it interesting that an 18-carbon molecule weighing 280.5 grams/mol can dictate the rate of oxidation of the massive LDL particle[l]

Investigators at UCLA have shown threshold recognition of LDL by scavenger cell receptors after neutralization of 16% of their lysine molecules.[35] Macrophages have both LDL and scavenger receptors, but only the latter slowly admit oxidized LDL into the cell. When 16% fatty acid (linoleic acid)

---

[k]An essential amino acid is one that is required by the human body for normal metabolism, yet is not made by the body. It must be provided through the diet.

[l]The size and weight of LDL particles varys a bit — according to the percentage of triglycerides within the gigantic particle[40,41] — but all LDL particles weigh in under $2.0–2.5 \times 10^5$ g category, making them the "light–heavy weights" among the five classes of lipoproteins with HDL — the "*good* cholesterol" being the heaviest. For once, being overweight has an advantage!

and a six-carbon amino acid (lysine) with a combined molecular weight of 424 govern the conversion of the huge LDL particle (molecular weight 2,200,000) from its harmless to its harmful form? Or, put another way, isn't it odd that 1,100 linoleic acid molecules.[37,38] and 356 lysine molecules with a combined molecular weight of 356,776 — representing 16% of the weight of a single LDL particle — regulate the conversion of unoxidized to oxidized LDL?

Middle Stages of LDL Oxidation

Implicit in this gradual recognition process is the fact that LDL oxidation occurs in stages, not all at once. In the laboratory, it takes 8–16 h to oxidize LDL with copper ions.[39] In the human body, it also happens gradually, as first endothelial and smooth muscle cells and later macrophages generate increasing amounts of free radicals. It seems illogical that a collection of cells — endothelial cells — resourceful enough to synthesize more than 50 enzymes would shoot themselves in the foot by manufacturing potentially harmful free radicals. And, in reality, they aren't hurting themselves. They are simply carrying out their assigned metabolic function, fueled by oxygen, to manufacture the many important enzymes needed to guard the wall of an artery. They are counting on macrophages to do *their* job of getting rid of the waste products created during the enzyme-making process. And the macrophages in the subendothelial space do. . .up to a point. They remove potentially lethal oxidized LDL, thereby preventing damage to the overlying, vulnerable endothelium — until the scavenger cells burst.

Once their cytotoxic contents have been released into the subendothelial space, the toxins rapidly diffuse into the overlying endothelial cells, injuring them and forcing them to release their free radicals into the subendothelial space — making matters worse. Endothelial cells, by the way, produce only small amounts of free radicals and then do so only after an "insult" (such as injury), in agreement with their philosophy of not hurting themselves.[42,43] But the "big eaters" — the macrophages — do generate a lot of toxic byproducts. A busy digestive tract simply begets more waste than a relatively idle one.

Vitamin E — Both an Antioxidant and a Pro-oxidant — and Copper

Just as endothelial cells are exquisitely dependent on metal ions for their ability to oxidize LDL, so are metal ions (copper and iron) intimately dependent

Vitamin E the Antioxidant Becomes Vitamin E the Pro-oxidant

$$
\begin{array}{lll}
 & \quad\quad\quad O^{\bullet} & \quad\quad\quad\quad\quad \overset{\displaystyle H}{O} \\
 & \quad\quad\quad O^{\bullet} \quad \text{copper} & \quad\quad\quad\quad\quad O \\
\text{Vitamin E} + & =\text{CHCHC} \xrightarrow{\quad\quad\quad} & \text{Vitamin E}^{\bullet} + \quad =\text{CHCHCH}= \\
\text{(oxidized)} & \quad \text{peroxyl} & \text{(reduced)} \quad\quad \text{hydroperoxide} \\
 & & \quad\quad\quad\quad\quad\quad\quad\quad \text{radical}
\end{array}
$$

Fig. 9.10.   In this reaction, oxidized vitamin E is reduced to vitamin E• by a peroxyl radical in the presence of copper ions. In the process, vitamin E itself becomes a free radical — having gained an electron — and, as such, is a pro-oxidant.

on free radicals generated by cells for their contribution to the oxidation of LDL. The metal ions must first be reduced (by gaining an electron), before they can oxidize LDL by forcing it to lose an electron. It's that simple. To generate reduced copper ions (Cu^+), oxidized copper ions (Cu^{++}) interact with α-tocopherol (vitamin E), which is included, by nature, in the LDL particle in order to *limit* the oxidation of LDL by free radicals. But since vitamin E, by *its* very nature as an antioxidant, also fluctuates between its reduced and its oxidized states, it functions as a *pro*-oxidant when it is reduced, as shown in Fig. 9.10. It is as if vitamin E were at once a queen in her castle and a vassal in someone else's kingdom.

As a free radical, reduced vitamin E does just what other free radicals do: it oxidizes LDL. However, it can't get the job done alone. It needs help — albeit by default — from vitamin C. Vitamin E can only oxidize LDL *if* vitamin C (ascorbic acid) — also an antioxidant — is absent. Hence, *in the blood*, where water-soluble vitamin C is an effective antioxidant, ascorbic acid provides a "backup" antioxidant system if vitamin E shifts from being an antioxidant to being a pro-oxidant. But once LDL leaves the aqueous bloodstream and enters the subendothelial space, where only fat-soluble molecules exist, the supportive co-antioxidant role of vitamin C is lost. Vitamin E becomes a turncoat, promoting the oxidation of LDL. These opposing actions of vitamin E may account for its inconsistent effect in preventing death and heart attacks in clinical trials, as will be discussed in Chap. 20.

Iron

Iron, like copper, must first be reduced before it can oxidize LDL. But, unlike copper, the substance being reduced — which was vitamin E in the case of copper — is not built into LDL. The substance reduced by iron — superoxide — is, however, nearby in the subendothelial space. Although superoxide is not as potent a pro-oxidant as (reduced) vitamin E, it is more available within the subendothelial space, being released in large quantities by macrophages below and, to a lesser degree, by endothelial cells above the subendothelial space. Couple the availability of superoxide with the much higher concentration of iron than copper in the human body,[43] and a profitable LDL oxidation process by iron is easy to envision. In fact, there is no faster way to oxidize LDL than by adding iron to LDL *in the laboratory.*

That's the bad news, particularly since iron is readily available in everyday foods such as red meats, leafy vegetables, and fortified breads. . .not to mention over-the-counter multi vitamin preparations. The good news is that, *in the body,* iron is tightly bound to proteins, making the concentration of *free* iron almost nonexistent. Iron remains protein-bound in fatty streaks characteristic of early atherosclerosis, making it "unlikely that free iron is important in the oxidation of LDL early in [the] disease process."[45] However, in "advanced lesions [where] dissolution of foam cells is a prominent feature of complex atherosclerotic lesions, release of free metal ions may promote lesion development".[46] If these observations are translated into clinical practice, a case could be made for using iron supplements early in life (which, of course, coincides with menstruating years in females), but avoiding supplemental iron in later years, when iron might promote the oxidation of LDL.

Particle Size and Density

All LDL particles are not the same. They differ in their fatty acid content — due to differences in diet — and density, as mentioned earlier. The type and amount of the seven polyunsaturated fatty acids in LDL reflect the type and amount of each polyunsaturated fatty acid eaten by an individual.[47,48] LDL particles are not only dissimilar in their fatty acid content, but they also differ in their size and density, as mentioned earlier. Small, dense particles are able to move from the bloodstream into the subendothelial space more easily than large, light particles.[49] And they are more prone to oxidation.[50] Luckily, only 10%–15% of the American population has small, dense — so-called type B — LDL particles.[51] (A person with a "type A" personality and a "type B" LDL profile might be at more risk for coronary heart disease than other Americans.) While a large part of a person's LDL particle size and density is determined

Table 9.2 Similarities and differences among oleic, linoleic, and linolenic acid

Fatty acid	Spelling origin	Structure	Source
Oleic	*Oleum,* Latin for "derived from oil"	$CH_3(CH_2)_7CH=CH(CH_2)_7COOH(C_{18}H_{34}O_2)$	Olive oil and canola oil
Linoleic	*Linon,* Latin for "flax," plus *oleum*	$CH_3(CH_2)_4CH=CHCH_2CH=CH(CH_2)_7OH(C_{18}H_{32}O_2)$	Sunflower oil and safflower oil
Linolenic	*Linon,* Latin for "flax," plus the consonant "*n*," a connective in Latin, plus *oleum*	$CH_3CH_2CH=CHCH_2CH=CHCH_2CH=CH(CH_2)_7COOH\ (C_{18}H_{30}O_2)$	Sunflower oil and safflower oil

genetically, everyone can influence his or her LDL size and density, once again through diet. After only five weeks on a diet enriched with monosaturated fatty acids, LDL from even healthy adults is less susceptible to oxidation.[52] (see Chap. 20). And humans need as many oxidation-resistant fatty acids as possible, since the ratio of polyunsaturated fatty acids to antioxidants in LDL is 150:1.[53] There is plenty of fuel to fan the LDL oxidation fire, and once the (antioxidant) wall around LDL comes tumbling down, the pace of polyunsaturated fatty acid oxidation accelerates.[m]

Oleic Acid, Linoleic Acid, and Linolenic Acid

Oleic acid, linoleic acid, and linolenic acid are the three fatty acids in LDL oxidized by free radicals. The three compounds share similarities in spelling, structure, and function, as shown in Table 9.2, with their source being the principal difference. All three are "unsaturated" fatty acids, "saturation" referring to the number of hydrogen atoms attached to the carbon atoms in the carbon chain. If the carbon atoms carry as many hydrogen atoms as possible, the fatty acid is "saturated." By convention, when the carbon chain has less than a full complement of hydrogen atoms, the "missing" ones are represented by a double bond, symbolized as =. Oleic acid is a monounsaturated

---

[m]The vitamin E "built into" LDL is depleted in within 3–6 h after the oxidative process begins, leaving LDL at the mercy of free radicals.

fatty acid, having only one double bond, while linoleic acid and linolenic acid are polyunsaturated fatty acids, having two and three double bonds, respectively. At each double bond, the adjacent carbon atoms may have hydrogen atoms attached on the same side or on the opposite side, as shown in Fig. 9.11.[53] Adjacent carbon atoms with hydrogen atoms on the same side are referred to as being in a "cis-" configuration; "cis-" is a prefix meaning "on the same side," as in "cisatlantic." Similarly, adjacent carbon atoms with hydrogen atoms on the opposite side are referred to as being in a "trans-" configuration; the prefix "trans-" means "across from or opposite," as in "transatlantic." As a consequence of the cis configuration, the carbon chain bends at the double bond site due to a slight electrical imbalance created by the side-by-side hydrogen atoms. Thus, *cis*-oleic acid with one double bond is slightly bent at its double-bond site. A carbon chain with two double bonds (linoleic acid) has a more pronounced bend, and a carbon chain with three double bonds (linolenic acid) takes on a hook like shape (Fig. 9.12).

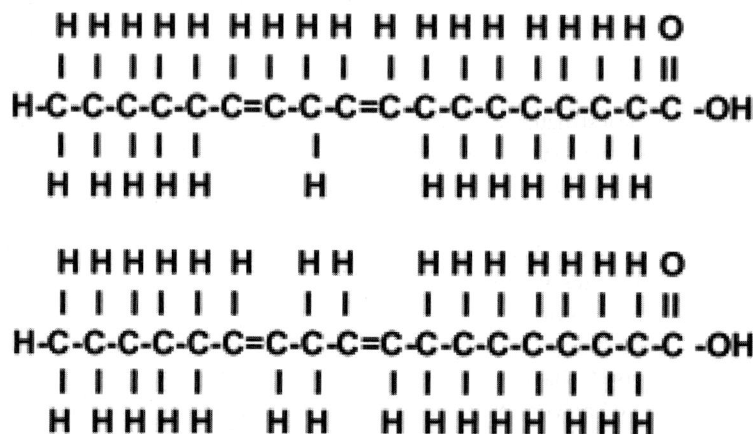

Fig. 9.11.   The upper diagram shows a fatty acid (linoleic acid) with its hydrogen atoms arranged in a cis configuration on the carbon atoms adjacent to the two double bonds. The hydrogen atoms repel one another and the carbon chain kinks to compensate for the empty space on the opposite side of the chain, as seen in Fig. 9.12. In the lower diagram, the hydrogen atoms on the carbon atoms adjacent to the double bonds are arranged in a trans configuration in an oleic acid molecule. This reduces the electrical repulsion created by the cis configuration. As a result the carbon chain is a bit straighter, as seen in Fig. 9.12.

(*Source*: Reproduced with permission from the publisher.) Reference 53.

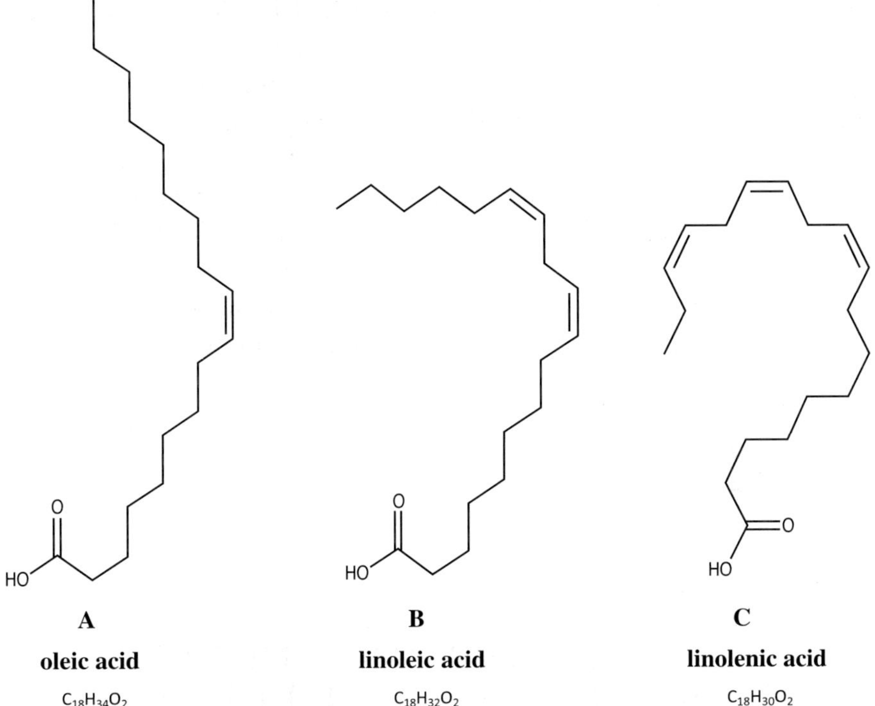

Fig. 9.12.   The chemical structures of oleic, linoleic, and linolenic acid. All carbon atoms are represented by the peaks and valleys of the chain and all hydrogen atoms are omitted except for the ones in the hydroxyl (–OH) group at the end of each chain. All three compounds contain 18 carbon atoms.

Oleic acid, a monounsaturated fatty acid, has a single double bond. The hydrogen atoms attached to the carbon atoms at both ends of the bond are on the same side, creating a minor electrical imbalance. As a result, the chain is somewhat bent.

Linoleic acid, a polyunsaturated fatty acid, has two double bonds. It has a more definite bend, assuming a reverse C shape.

Linolenic acid, also a polyunsaturated fatty acid, has three double bonds. It takes on an upside-down hook like shape.

Source: Reproduced with permission from the publisher. Reference 54, 55, 56.

Linoleic acid and linolenic acid are essential fatty acids and they are not interchangeable. An essential fatty acid is one that cannot be synthesized by a human — because humans lack the enzyme to synthesize the fatty acid — yet it is required for optimal health. Therefore, it is "essential." The only two essential fatty acids in human beings are linoleic acid and linolenic acid. (Although linoleic acid and linolenic acid are essential dietary amino acids, *all* amino acids are essential for life.) Sunflower oil and safflower oil are rich sources of linoleic acid, while marine oils and flaxseed oil have generous amounts of linolenic acid. While linoleic acid and linolenic acid are both essential polyunsaturated fatty acids needed for the synthesis of cell membranes (among other things), polyunsaturated fatty acids are highly susceptible to oxidation by free radicals.[52] In contrast, monounsaturated fatty acids, such as oleic acid, are more resistant to oxidation. This biochemistry bears on the success of the Mediterranean diet — rich in monounsaturated fatty acids present in olive oil — in protecting against coronary heart disease (see Chap. 17).

LDL Oxidation is a Seven-Step Biochemical Process[n]

Superoxide free radicals start the oxidation a rolling by abstracting (subtracting) a hydrogen atom from a polyunsaturated fatty acid. Characterized by their multiple double bonds, polyunsaturated fatty acids predominate in plant (*e.g.* sunflower, safflower, and peanut) and seafood (*e.g.* salmon, herring, and mackerel) oils. The type and amount of the several polyunsaturated fatty acids in LDL reflect the type and amount of each polyunsaturated fatty acid eaten by an individual.[47,48]

Superoxide is the principal free radical capable of initiating the oxidation of LDL in man.[57] It does not catalyze any other oxidative step. However, it gives rise to two other radicals — H_2O_2 and $OH^\bullet$ — by the four-step reduction process shown in Fig. 4.5. Both can oxidize LDL. Superoxide is generated in limited amounts by normal endothelial cells. However, a high cholesterol level boosts $SO^\bullet$ production by endothelial cells, which then become a source for more H_2O_2 and $OH^\bullet$ radicals that propagate the reaction started by superoxide.[58]

Choosing to make oxidation as easy as possible, $SO^\bullet$ selects the middle carbon atom in the short three-carbon segment ($=CHCH_2CH=$) of the linoleic or linolenic acid molecule shown in Table 9.2 to attack. So-called bisallyic

---

[n]Although LDL oxidation is presented as a straightforward, seven-step chemical reaction leading to a single substance (oxidized LDL), the reaction is not simple and the substance is far from homogeneous.

hydrogen atoms are readily extracted from any polyunsaturated fatty acid. In the biochemical vernacular, the abstraction of a hydrogen atom from the CH_2 group occurs because of "lowered bond dissociation energy."[59] In simpler terms, the strength of the two double bonds tugging on the central carbon atom with only two single bonds weakens them, allowing a hydrogen atom to be more easily abstracted — akin to the horse-pulling-a-plow analogy mentioned in Chap. 4 — as shown in Fig. 9.13, step 1. This leaves a short three-carbon (free) radical that is unstable and promptly undergoes "molecular rearrangement,"[60] as shown in step 2, to gain stability. The weakened carbon

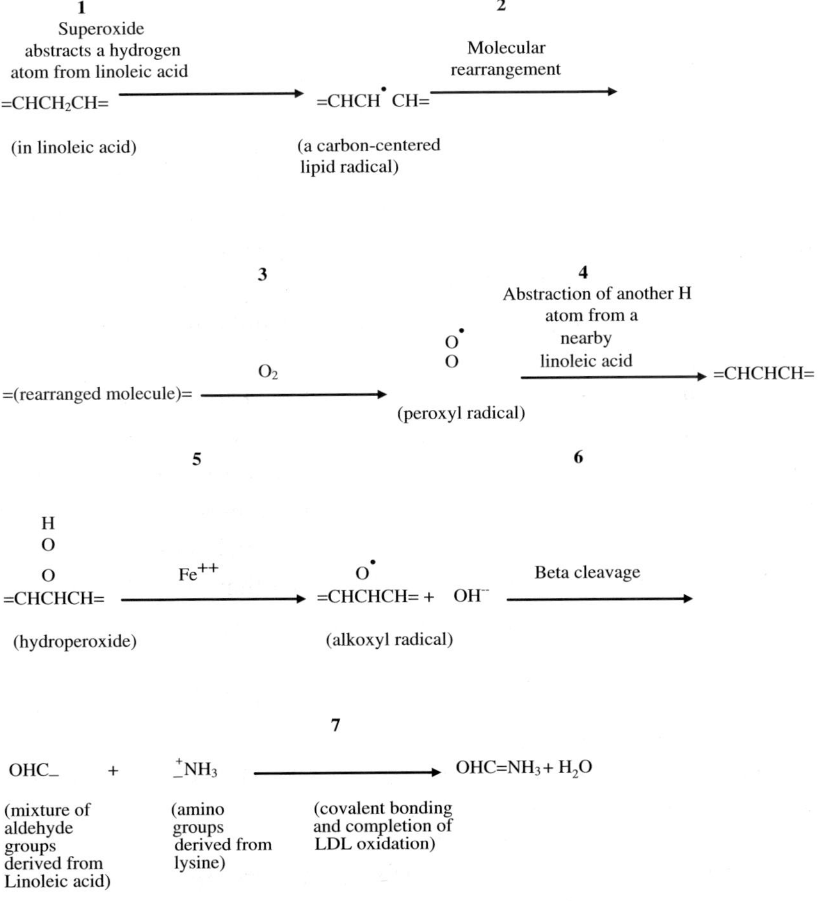

Fig. 9.13. Oxidation of LDL in a seven-step process.

atoms stabilize by forming alternating single and double bonds. The rearranged molecule is now ready for the insertion of an oxygen molecule step 3, yielding a peroxyl radical. The peroxyl (free) radical quickly extracts a hydrogen ion from an adjacent linoleic acid molecule, forming a hydroperoxide radical (step 4), which decomposes, in the presence of iron, into a "variety of breakdown products," including alkoxyl radicals (step 5).[o,61] The conversion of lipid hydroperoxides to alkoxyl radicals is pivotal to the efficient continuation of the oxidative process, for two reasons. It is the rate-limiting step (contrary to the rate-limiting step usually being the slowest step, it is the fastest one) and it requires the presence of iron ions, — the only step needing metal ions.[59]

The Buck Stops Here

Next in step 6 the alkoxyl radicals undergo beta cleavage — splitting of the double bonds on either side of the middle carbon atom in the =CHCHCH= group — yielding an array of 3–9-carbon aldehyde groups. These 3–9-carbon fragments may be the most important link in the oxidative chain of events, although since it is a "chain of events" no link is inconsequential. The short, negatively charged, fatty-acid-derived aldehyde groups (OHC$^-$–) generated by step 6 now interact with terminal, positively charged, amino-acid-derived amide groups ($-^+NH_3$) on lysine molecules in the LDL particle, neutralizing them in step 7 (Fig. 9.14).[64] The change of the electrical charge on LDL is sufficient to attract it to positively charged scavenger receptors, just as acetylation of LDL in the laboratory increased its negative charge, making it attractive to scavenger receptors. Although only a breakdown product of a polyunsaturated fatty acid oxidation, the aldehydes form critical cross-links (covalent bonds) between the fatty acids and the amino acids, completing their

---

[o]The initiation phase (steps 1–4) of LDL oxidation is slow. When human LDL is exposed to oxygen continuously by simply bubbling oxygen through plasma in the laboratory, 3–6 h of such oxygenation are required to deplete the pre-existing vitamin E and other antioxidants within the LDL particle.[63] After the antioxidants are gone and peroxides have formed rapidly,[4] the reaction slows again as the peroxides gradually break down into at least seven different aldehydes[62] that link the fatty acid (linoleic) to the amino acid (lysine) in LDL. This portion of the reaction continues for 12–24 h.[4] As a result of the glacial rate of LDL oxidation, more than 75% of linoleic acid (the major polyunsaturated fatty acid in LDL) is still present after 24 h of oxidation.[62]

The Last Step

$$OHC^- + -^+NH_3 \longrightarrow OHC=NH_3 + H_2O$$

Fig. 9.14. An aldehyde group (OHC$^-$) in a fatty acid merges with an amide group ($-^+NH_3$) from an amino acid to form a covalent bond (C=N) — the final link in the LDL oxidation process.

individual transformations! LDL *finally* is "oxidized" — albeit to different degrees, depending on how many of its 2616 fatty acids and 4536 amino acids formed covalent bonds and were recognized by scavenger receptors.

After multiple steps that began when superoxide abstracted a single hydrogen atom from a linoleic acid molecule, the buck stops here.

References

1. Brown MS, Goldstein JL. (1984) How LDL receptors influence cholesterol and atherosclerosis. *Sci Am* **251**(5): 58–66.
2. www.ncbi.nlm.nih.gov/books/bv.fcgi?rid=mboc4.figgrp.429
3. Parthasarathy S, Khoo JC, Miller E, Barnett J, Witzum JL, Steinberg D. (1990) Low density lipoprotein rich in oleic acid is protected against oxidative modification: implications for dietary prevention of atherosclerosis. *Proc Natl Acad Sci USA* **87**: 3894–3898.
4. Jilal I, Devaraj S. (1992) Low-density lipoprotein oxidation, antioxidants and atherosclerosis: a clinical biochemistry perspective. *Clin Chem.* 498–506.
5. Young, SG, Parthasahathy S. (1994) Why are low-density lipoproteins atherogenic? *West J Med* **160**: 153–164.
6. Berliner JA, Navab M, Fogelman AM, Demer LL, Edwards PA, Watson BS, Lusis AJ. (1995) Atherosclerosis: basic mechanisms. *Circulation* **91**: 2488–2496.
7. Steinberg D. (1996) Oxidized low density lipoprotein — an extreme example of lipoprotein heterogenicity. *Isr J Med Sci* **32**: 469–472.
8. Chisolm GM, Penn MS. (1996) Oxidized lipoproteins and atherosclerosis. In: Fuster V, Ross R, Topol EJ (eds.), *Atherosclerosis and Coronary Artery Disease.* Lippincott-Raven Philadelphia, pp. 129–149.
9. www.americanheart.org/presenter.jhtml?identifier=4500.
10. Smith DW, Lemli L, Opitz JM. (1964) A newly recognizd syndrome of multiple congenital anomalies. *J Pediatr* **64**: 210–217.
11. www.emedicine.com?PED/topic2117.htm
12. ghr.nim.nih.gov/condition=smithlemliopitzsyndrome

13. Tint GS, Irons M, Elias ER, Batta AK, Friden R, Chen TS, Salen G. (1994) Defective cholesterol biosynthesis associated with the Smith–Lemli–Opitz syndrome. *N Engl J Med* **330**: 107–113.

14. Herman GE. (2002) Disorders of cholesterol biosyntheis: prototypic metabolic malformation syndromes. *Hum Molec Genet.* **12**: R75–R78.

15. Beadle GW, Tatum EL. (1941) Genetic control of biochemical reactions in *Neurospora. Proc Natl Acad Sci* **27**: 499–506.

16. Malgorzata JM, Nowaczyk MD, Whelan DT, Heshka TW, Hill RC. (1999) Smith–Lemli–Opitz syndrome. a treatable inherited error of metabolism causing mental retardation. *CMAJ* **161**: 165–170.

17. www.smithlemliopitz.org.

18. Behrman EJ, Gopalan V. (2005) Cholesterol and plants. *J Chem Educ* **83**: 1779–1793.

19. http:// www.amereicanheart.org/presenter.jhtml?identifier=1516

20. Fitzky BU, Moebius FF, Asaoka H, Waage-Baudet H, Xu L, Xu G, Maeda N, Kluckman K, Hiller S, Yu H, Batta AK, Shefer S, Chen T, Salen G, Sulik K, Simoni RD, Ness GC, Glossmann H, Patel SB, Tint GS. (2001) 7-dehydrocholesterol-dependent proteolysis of HMG-CoA reductase suppresses sterol biosynthesis in a mouse model of Smith–Lemli–Opitz syndrome. *J Clin Invest* **108**: 905–915.

21. Tabas I. (2004) Cellular cholesterol metabolism in health and disease. In: Chien KR (ed.), *Molecular Basis of Cardiovascular Disease.* Saunders, 414–431.

22. Spady DK, Wollett LA, Deitschy JM. (1993) Regulation of plasma LDL cholesterol levels by dietary cholesterol and fatty acids. In: *Ann Rev Nutr* **13**: 355–358. Olson RE, Bier DM, McCormick DB (eds.), Annual Reviews Inc., Palo Alto, CA.

23. Rippe B, Haraldsson H. (1987) How are macromolecules transported across the capillary wall? *News Physiol. Sci* **2**: 135–138.

24. Marsh M. (2002) Cathrin-coated vesicles and receptor-mediated endocytosis. In *Encylopedia of Life Sciences.* Macmillan, London, New York, Tokyo, pp: 567–573.

25. Goldstein JL, Anderson RGW, Brown MJ. (1979) Coated pits, coated vesicles, and receptor-mediated endocytosis. *J Cell Biol* **279**: 679–685.

26. Ross R, Pawlina W. (2006) The cell cytoplasm. In *Histology: A Text and Atlas with Correlated Cell and Molecular Biology.* Lippincott Williams & Wilkins, Baltimore, pp. 23–70.

27. The circulatory system. In: *Ham's Histology,* 9th ed. Cormack DH (ed.), JB Lippincott, London, Mexico City, New York, St Louis, Sao Paulo, Sydney, pp. 423–449.

28. Berliner JA, Gerrity RG. (1992) Pathology of atherogenesis. In: Lusis AJ, Potter JI, Sparkes RS (eds.), *Molecular Genetics of Coronary Artery Disease.* Karger, Basel, Freiburg, Paris, New York, New Delhi, Bangkok, Singapore, Sydney.

29. Carmejo G, Olofsson S-O, Lopez F, Carlson P, Bondjers G. (1988) Identification of apo B-100 segments mediating the interaction of low density lipoproteins with arterial proteoglycans. *Arteriosclerosis* **8**: 368–377.

30. Hoff HL, Heideman CL Gotto AM, Jr, Gaubatz JW. (1989) Apolipoprotein B retention in the grossly normal and atherosclerotic human aorta. *J Biol Chem* **264**: 2599–2604.

31. Nishi K, Itabe H, Uro M, Kitazato KT, Horiguchi H, Shinno K, Nagahiro S. (2012) Oxidized LDL in carotid plaques associates with plaque instability. *Arterioscler Thromb Vasc Biol* **22**: 1649–1654.

32. Bevillacquam E, Butcher E, Furie B, Furie B, Gallatin M, Gimbrone M, Harlan J, Kishimoto K. Weissman F, Zimmerman G. (1991) Selectins: a family of adhesion receptors. *Cell* **67**: 233–234.

33. Sparrow CP, Parthasarathy S, Steinberg D. (1989) A macrophage receptor that recognizes oxidized low density lipoprotein but not acetylated low density lipoprotein. *J Biol Chem* **264**: 2599–2604.

34. Brown MS, Goldstein JL. (1990) Scavenging for receptors. *Nature* **343**: 508–509.

35. Haberland ME, Fogelman AM. (1987) The role of altered lipoprotein in the pathogenesis of atherosclerosis. *Am Heart J* **113**: 573–577.

36. Salonen JT, Korpela H, Salonen P, Nyyssonen KK, Yla-Herttuala S, Yamamoto R, Butter S, Parlinski W, Witzum JL. (1992) Autoantibody against oxidized LDL and progression of carotid atherosclerosis. *Lancet* **339**: 883–887.

37. Esterbauer H, Gebick J, Puhl H, Jurgens G. (1992) The role of lipid peroxidation and antioxidants in the oxidative modification of LDL. Free *Rad Biol Med* **13**: 341–390.

38. Yang C-Y, Pownall HJ. (1992) Structure and function of apolipoprotein B. In: Rosseneu M (ed.), *Structure and Function of Apolipoproteins.* 1992 CRC Boca Raton, Ann Arbor, Tokyo.

39. Steinberg D. (1997) Low density lipoprotein oxidation and its pathobiological significance. *J Biol Chem* **272**: 20963–06.

40. Pagano IS, Strait NB (2009) HDL and LDL Cholesterol. Physiology and Clinical Significance In: Nova Biomedical Books, New York. Pp. 113–127.

41. Myant NB (1990) Cholesterol Metabolism, LDL and the LDL Receptor. Academic Press Inc. Harcourt Bruce Jovanovich Publishers. San Diego, New York, Boston, London, Sydney, Tokyo, Toronto. pp: 99–111.

42. Halliwell B. (1994) Free radicals and antioxidants: a personal view. *Nutr Rev* **52**: 253–265.

43. Arroyo CM, Carmichael AJ, Bouscarel B, Liang JH, Weglicki WB. (1990) Endohtelial cells as a source of oxygen-free radicals: an ESR study. *Free Radic Res Comm* **9**: 287–296.

44. Aust SD. (1982) The role of iron in enzymatic lipid peroxidation In: Pryor W (ed.), *Free Radical Biology*, Vol. 5. Academic, New York, London, Paris, San Diego, San Francisco, Sao Paulo, Sydney, Tokyo, Toronto, pp. 1–28.

45. Heinecke JW. (1997) Mechanisms of oxidative damage of low density lipoprotein in human atherosclerosis. *Curr Opin Lipidol* **8**: 268–274.

46. Esterbauer HW, Wag G, Puhl H. (1993) Lipid peroxidation and its role in athero-sclerosis. *Br Med Bull* **49**(3): 566–576. In: *Free Rdicals in Medicine.* Cheesman KH, Slater TF (eds.), Churchill Livingston, Singapore pp. 566–576.

47. Berry EM, Eisenberg D, Haratz D, Freidlander Y, Norman Y, Kaujmann NA, Stein Y. (1991) Effects of diets rich in monounsaturatred fatty acids on plasma lipo-proteins — the Jerusalem Nutrition Study: high MUFAs vs PUFAs. *Am J Clin Nutr* **53**: 899–907.

48. Zilversmit DB, Nordestguard BG. (1989) Comparison of arterial intimal clear-ances of LDL from diabetic and non-diabetic rabbits. *Arteriosclerosis* **9**: 176–183.

49. Stampfer MJ, Krauss RM, Ma J, Blanche PJ, Holt LG, Sacks FM, Hennekens CH. (1996) A prospective study of triglyceride level, low-density lipoprotein particle diameter, and risk of myocardial infarction. *JAMA* **276**: 882–888.

50. Slyper AM. (1994) Low-density lipoprotein density and atherosclerosis: unravel-ing the connection. *JAMA* **272**: 305–308.

51. Reaven P, Parthasarathy S, Grasse BJ, Muller E, Almazan F, Mattson FH, Khoo JC, Steinberg D, Witzum JL. (1991) Feasibility of using an oleate-rich diet to reduce the susceptibility of low-density lipoprotein to oxidation in humans. *Am J Clin Nutr* **54**: 701–706.

52. Cheeseman KH, Slater TF. (1993) An introduction to free radical chemistry. In *Br Med Bull* **49**: 481–493. *Free Radicals in Medicine,* eds. Cheeseman KH, Slater TF. Churchill Livingston, Edinburgh, London, Madrid, Melbourne, New York, Tokyo, pp. 481–493.

53. www.thepaleodiet.com/nutritional_tools/fats.shtml

54. www.3dchem.com/moremolecules.asp?ID=384

55. www.3dchem.com/moremolecules.asp?ID=382

56. www.3dchem.com/moremolecules.asp?ID=240

57. Cheeseman KH, Slater TF. (1993) An introduction to free radical chemistry. In *Br Med Bull* **49**: 481–493. *Free Radicals in Medicine,* eds. Cheeseman KH, Slater TF. Churchill Livingston, Edinburgh, London, Madrid, Melbourne, New York, Tokyo. pp. 481–493.

58. Hiramatsu K, Rosen H, Heinecke JW, Wolfbauer G, Chait A. (1987) Superoxide initiates oxidation of low density lipoprotein by human monocytes. *Arterioscler* **7**: 55–60.

59. Ohara Y, Peterson TE, Harrison DG. (1993) Hypercholesterolemia increases endothelial superoxide anion production. *J Clin Invest* **91**: 2546–251.

60. Wayner DDM, Clark KB, Rank A, Yu D, Armstrong A. (1997) C-H bond dissociation energies of alkyl amines: radical structure and stabilization energies. *J Am Chem Soc* **119**: 8925–8932.

61. Halliwell B, Gutteridge JMC. (1998) Role of free radicals and catalytic metal ions in human disease: an overview. In: Packer L, Glaser AN (eds.), *Methods in Enzymology.* Academic, San Diego, New York, Boston, London, Sydney, Tokyo, Toronto, Vol. 186, pp. 1–85.

62. McCall MR, Frei B. (2000) Mechanisms of LDL oxidation. In: Keaney JF, Jr (ed.), *Oxidative Stress and Vascular Disease.* Kluwer, Boston, Dordrecht, London, pp. 75–98.

63. Esterbauer H, Jurgens G, Quehenberger O, Keller E. (1987) Autoxidation of human low density lipoprotein: loss of polyunsaturated fatty acids and vitamin E and generation of aldehydes. *J Lipid Res* **28**: 495–509.

64. Palinski W, Rosenfeld ME, Yla-Herttula S, Gurtner GC, Socher SS, Butler SW, Parthasarathy S, Carew TE, Steinberg D, Witzum JL. (1989) Low density lipoprotein undergoes oxidative modification *in vivo*. *Proc Natl Acad Sci* **86**: 1372–1376.

10

TWO OTHER WAYS TO OXIDIZE LDL: LIPOXYGENASE AND NITRIC OXIDE

Although free-radical-mediated LDL oxidation is one mechanism — and possibly the most important mechanism — of LDL oxidation, two other oxidative pathways are available. Certain enzymes within cells can, in the *absence* of free radicals, convert polyunsaturated fatty acids into lipid hydroperoxides, the important intermediate compounds involved in step 5 of the LDL oxidation process shown in Chap. 9. These hydroperoxides may then be transferred out of the cell into the subendothelial space, where pesky free radicals complete the process of oxidation, nonenzymatically. Lipoxygenase is the principal enzyme suggested to play such an oxidative role in the absence of free radicals.

The similarities between the free-radical- and lipoxygenase-mediated oxidation of LDL outweigh the differences. Both lipoxygenase[a] and free radicals

---

[a] Lipoxygenase, known as lipoxidase until the 1970s, is an enzyme that speeds up — catalyzes — the oxidation of linoleic acid by the insertion of oxygen, leading to the metabolism of the fatty acid. Lipoxygenase can not only oxidize linoleic acid but also incorporate oxygen into other unsaturated fatty acids located within different types of cells. Such cells may be found in either the animal or plant kingdom, which helps explain why lipoxygenases were discovered in plants before in man, in 1979.[1] There are more than 50 plant lipoxygenases, but only 6 mammalian lipoxygenases. The newest mammalian lipoxygenase is 15-lipoxygenase, the one that oxidizes LDL. It does so with high specificity. Only the linoleic acid molecules in LDL are oxidized. It does not even oxidize the cholesterol in LDL.[2]

abstract a hydrogen atom from linoleic acid to initiate the oxidative process. The principal difference between the two processes is that lipoxygenase, being an intracellular enzyme, oxidizes LDL within the cell; free radicals, being byproducts of cellular metabolism released into the extracellular (subendothelial) space by endothelial cells, macrophages, and smooth muscle cells, do so outside the cell.[2] A second difference between the two processes is that although lipoxygenase can initiate the LDL oxidative process within the cell, it can't sustain the reaction because there is no free iron in the cell. In other words, lipoxygenase can start the ball rolling, but it can't finish the job it started because step 5 requires iron, not available within the cell. Therefore, the hydroperoxide formed during step 4 must be transferred out of the cell to be further oxidized.[2,3] How this transfer occurs is unsettled, a point that lessens the credibility of lipoxygenase-mediated LDL oxidation.

The initial transformation of linoleic acid into its hydroperoxide by lipoxygenase (steps 1–4) *within cells* is often referred to in the LDL vernacular as "seeding" of LDL (with its hydroperoxide). In other words, the introduction — the seeding — of hydroperoxides into LDL leads to its further oxidation by free radicals *outside cells*.[3,4] These similarities and differences apply both to macrophages and to endothelial cells, the two principal cell types in early atherosclerotic lesions that promote the oxidation of LDL. And, curiously, foam cells — macrophages in early lesions that have become stuffed with cholesterol — make lipoxygenase, but nearby cells in the normal blood vessel wall do not.[3,5]

Lipoxygenase Needs the Help of Cholesterol to Oxidize LDL

Contrary to expectations, if the concentration of lipoxygenase in macrophages increases, that *by itself* does not lead to an increased synthesis of

---

The specificity of 15-lipoxygenase underscores both that cholesterol itself is not a fatty acid and that the term "cholesterol" is not synonymous with "LDL." LDL is the carrier of both cholesterol and fatty acids, while they are being transported in the blood. Once inside the subendothelial space, LDL is oxidized, freeing the roughly 200 molecules of cholesterol and fatty acids bound to each LDL particle. Again, at the risk of repetition, both cholesterol and fatty acids are necessary for a person's well-being, cholesterol being used for the synthesis of cell membranes, estrogens, and androgens, while fatty acids are used for energy every day. Only at blood levels above 160 mg/dL does cholesterol begin to form fatty streaks.

oxidized LDL.[6] However, if the concentration of lipoxygenase is increased in conjunction with a high plasma cholesterol level — acting as a "governor" — *then* the synthesis of oxidized LDL steps up sharply.[6] As a result, a threshold level of cholesterol must be reached before lipoxygenase begins to oxidize LDL, and then it does so briskly…much like a threshold temperature must be attained before water begins to boil, and then the full boiling point is reached quickly.

On the other hand, even when the blood cholesterol level is markedly increased by taking in too much dietary cholesterol, no accumulation of cholesterol occurs in blood vessels *if* the increased cholesterol intake is preceded by the administration of a lipoxygenase inhibitor. In other words, fatty streaks can be prevented by a drug that blocks lipoxygenase activity even when the blood cholesterol level is high. However, so far, such an impressive result from a drug has been confirmed only in rabbits.[7]

Lipoxygenase Protects Arteries Against Atherosclerosis

As often happens there are two sides to a story. In medicine, the other side of the story is frequently a troublesome side effect that offsets a desirable effect of a drug. The unexpected twist in the lipoxygenase story is that the LDL-oxidizing enzyme may *protect* blood vessels from atherosclerosis. An excess of lipoxygenase reduced the extent of atherosclerosis after 13.5 weeks of a high-cholesterol diet. However, as in the case of fatty streaks being prevented by a drug that blocks lipoxygenase, this favorable outcome has so far been confirmed only in rabbits.[8] Thus, early in the process of atherosclerosis, lipoxygenase somehow facilitates the metabolism of LDL, reducing its incorporation into a fatty streak. At a later stage, lipoxygenase is released from foam cells within the streak and thereby fosters the conversion of LDL into its oxidized form that contributes to atherosclerosis.[2] These opposing actions suggest that "lipoxygenase is the major contributor to [LDL] oxidation at some critical time during, but not throughout, atherogenesis."[7]

How Important is Lipoxygenase?

Among the several ways to oxidize LDL, lipoxygenase was the most suspect mechanism until 1997. Doubt lessened when the development of atherosclerosis in rabbits was dramatically reduced by a drug (PD 146176) that inhibits the action of lipoxygenase.[9] Prevention of a consequence through inhibition of a reaction is usually accepted as strong evidence of the importance of a reaction. Pharmacologists at Columbia University and Parke-Davis found that the compound had "a profound inhibitory effect against the development of atherosclerosis in cholesterol-fed rabbits."[9] Parke-Davis merged with Pfizer in 2000, when Pfizer bought its parent company, Werner-Lambert and development of PD 146176 seemingly ended without any clinical trials being done, according to Dr. Steven Feinmark at Columbia.[10] Sarai Spohn,[11] a representative with BIOMOL International, a company with headquarters in the U.S. (Philadelphia) and the U.K. (Exeter) which manufactures enzymes and enzyme inhibitors, also seemed puzzled, commenting that although "*in vivo* results with the compound were remarkable" there have been "no clinical trials carried out with it." The drug was "dropped from development and is now just a research tool".[12]

Another testimony to the importance of lipoxygenase came in 1999, by preventing its synthesis in the first place.[12,13] That was made possible in the molecular biology laboratory by deleting the gene that makes the lipoxygenase enzyme in the new genetically engineered mouse model (see Chap. 13). Mice, by nature, are resistant to atherosclerosis.[14] Their natural resistance can be increased by deleting the single gene that governs lipoxygenase synthesis. Such lipoxygenase-deficient mice, unable to oxidize LDL by that enzyme, had fewer atherosclerotic plaques in their blood vessels after one year of not being able to oxidize LDL through the lipoxygenase pathway (Fig. 10.1).[12]

During that time, their plasma cholesterol levels were in the 500–600 mg/dL range. Such high cholesterol levels were achieved by cross-breeding lipoxygenase-deficient mice with mice in whom a second gene that governs the synthesis of another substance — known as apolipoprotein E — that *removes* excess cholesterol from the blood was also deleted, and by then allowing the mice to eat a cholesterol-unrestricted diet. This led to an experimental situation analogous to the clinical circumstances related to the girl with a deficient number of LDL receptors that accelerated the process of atherosclerosis to the point of requiring a heart and liver transplantation at age six (see Chap. 3). A control group of mice was matched in every way to the lipoxygenase-deficient group in order to ensure that any reduction in the extent of atherosclerosis seen in the lipoxygenase-deficient group could reasonably be attributed to the absence of the enzyme in question. As a consequence of this clever study using "double knockout mice" — one deletion resulting in reduced

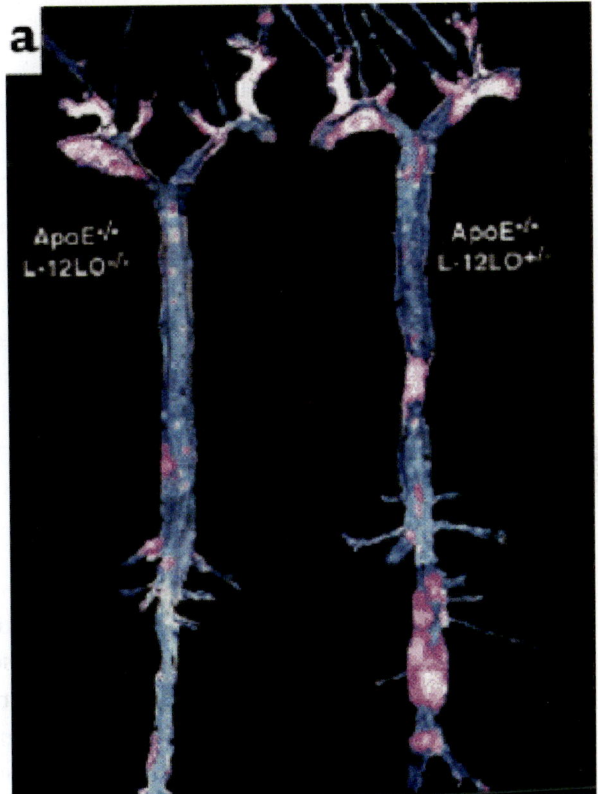

Fig. 10.1. The aortas of two mice. Both mice ate the same normal mouse-chow diet for one year. The aorta on the left came from a mouse unable to fully oxidize LDL because it lacked lipoxygenase. The aorta on the right came from a control mouse with intact lipoxygenase synthesis able to fully oxidize LDL. Atheromatous plaques (stained pink) are at a much more advanced stage in the mouse able to make lipoxygenase. The aortas have been opened longitudinally and then laid flat, with the intima containing the plaques facing the viewer.

(*Source*: Reproduced with permission from the publisher.) Reference 12.

LDL oxidation (by lipoxygenase), and the other resulting in reduced reverse-cholesterol transportation to the liver — the idea that lipoxygenase can oxidize LDL in a living creature — not just when incubated with LDL in a laboratory for 24 h — received its first direct confirmation (as opposed to indirect confirmation implied by blocking lipoxygenase activity with a drug).[13]

Lipoxygenase and Free Radicals as an Oxidative Tandem

A concerted effort between lipoxygenase and free radicals to oxidize LDL has some biological appeal, since endothelial cells as well as macrophages contain lipoxygenase and both cell types release free radicals into the subendothelial space. While lipoxygenase can initiate LDL oxidation directly by producing peroxyl and alkoxyl radicals, continuation of the reaction requires free radicals.[15] The superoxide anion free radical inhibitor, superoxide dismutase (see Chap. 4), inhibited LDL oxidation by 25% in rabbits, while a lipoxygenase inhibitor "reduced (its) oxidation by as much as 70–85%".[16] Together the duo almost completely inhibited LDL oxidation. In spite of the greater protective effect of lipoxygenase inhibitors (compared with free radical inhibitors), the role of lipoxygenase in the oxidation of LDL is more controversial than that of free radicals, but less so than that of the last LDL oxidant, nitric oxide.

Nitric Oxide

The third mechanism proposed to contribute to LDL oxidation involves nitric oxide. This smaller-than-oxygen molecule (molecular weight of nitric oxide, 30; oxygen, 32) was better known as an environmental villain in the 1970s and early 1980s. It was associated with such unsavory characters as smog, cigarette smoke, automobile exhaust, and acid rain.[17,19] But its reputation changed in the mid-1980s, when it was realized that the molecule wasn't *always* a bad boy. It was discovered that nitric oxide has biological functions in the human body. This "came as a complete surprise and caused quite a stir".[20] Basic scientists showed that the "sometime poison is a fundamental player in [maintaining] the everyday [healthy] business of the human body".[19] It is "essential to activities … [ranging] from digestion and blood pressure regulation to antimicrobial (antibacterial) defense."[19] And nitric oxide really gets busy after the sun goes down. Given the right company, candlelight, and soft music, the brain sends a message to key, male pelvic nerves. In response, nitric oxide is released and blood rushes into the male sex organ, which rises to the occasion. Nitric oxide might be thought of as the body's "blood pressure police" or its "goddess of love" — take your pick.

A Signaling Molecule

Nitric oxide is one of only a few "signaling molecules" in the form of a gas. (Most of the others are nongaseous molecules.) It's called a signaling

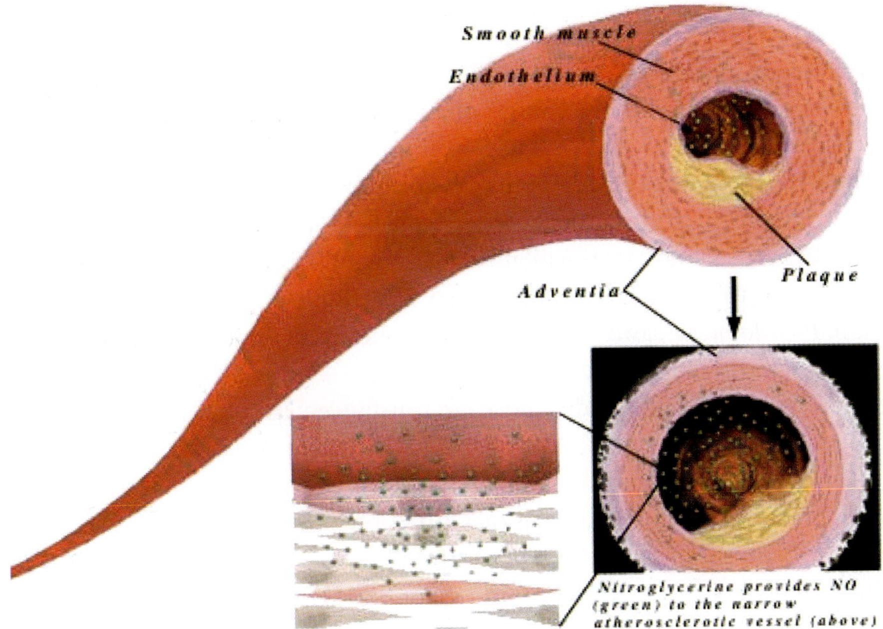

Fig. 10.2.   The thin endothelium manufactures nitric oxide for use by the thicker smooth muscle layer directly atop it. The muscle relaxes, the blood vessel dilates, and blood flow increases. Nitroglycerin tablets are converted into nitric oxide in the body with the same effect. An atherosclerotic plaque partially narrows the artery.

(*Source*: Reproduced with permission from the publisher.) Reference 21.

molecule because it "tells" cells and tissues what to do. For example, if the blood pressure is too high, the endothelium in a blood vessel's wall (its inner lining) (see Figs. 2.1 and 10.2) synthesizes nitric oxide[b] to signal smooth muscle cells in the muscular media directly atop the endothelium to relax (Fig. 10.2). As the chemical pours out of the endothelium, it has essentially zero distance to travel — that of a tiny gap — before it enters contracted smooth muscle cells, relaxing them. And as smooth muscle cells unwind, blood vessels dilate, blood pressure falls, and blood flow improves.

---

[b] The molecule is synthesized by the endothelium of blood vessels (the inner lining; see Figs 2.1 and 10.2) from two simple molecules: the amino acid arginine and oxygen. Arginine is a nonessential amino acid (nonessential because the body can make it — other than during infancy) which donates the nitrogen atom, while (molecular) oxygen donates the oxygen atom.

A Killing Molecule

The above vasodilating actions of nitric oxide might be viewed as being akin to an offensive line taking charge of matters; but the chemical is a potent defensive weapon as well. It attacks iron groups in enzymes, including those in the DNA of tumor cells, crippling them so that they grow slower or die. But while some nitric oxide can kill cancer cells — (and bacteria), too much nitric oxide can cause circulatory collapse, *i.e.* shock. *This* type of shock — which may be due to overwhelming bacterial infections, such as those caused by *E. coli, Pseudomonas,* and *Salmonella* — is true circulatory shock, not the variety of shock when the word is used colloquially — he or she is "in shock" upon hearing such and such. Under those circumstances, "stunned" might be a better descriptor. And, yeah, I'm a cardiologist, but not a totally insensitive one.

Nitric oxide's Dr. Jekyll and Mr. Hyde personality is particularly apparent in the brain. There, the chemical may function as a healthy neurotransmitter or as a killer of brain neurons. For example, following a stroke, "neurons containing [nitric oxide] synthase [the enzyme that makes nitric oxide] may actually kill their neighbors."[19] The nitric oxide synthase–containing cells "— themselves eerily resistant to stroke damage — apparently may become overstimulated by the excess neurotransmitter released during a stroke and flood nearby cells with toxic amounts of [nitric oxide]."[19] And so, nitric oxide's "Cinderella story — from... noxious gas to powerful queen of communication and defense — offers a classic example of nature's continuing power to surprise."[19]

The Appeal of Simplicity

One feature that makes nitric oxide so appealing is its simplicity. It's nothing more than a single nitrogen atom attached to a single oxygen molecule. As such, its chemical formula is NO, — no kidding. Its simple structure makes it the "smallest, lightest — and the first gas — to act as a biological messenger."[19] But because this bizarre molecule has positive attributes, as well as negative ones, and to avoid making the reader see "NO" 60 times or so within the span of four pages, the abbreviated version will be eschewed.

A second feature unique to this tiny neurotransmitter is that it is synthesized in a very different manner. It's made on the spot, when needed, and is not stored for future use. *And* when it's no longer needed — its mission accomplished — it is not terminated by degradation; rather, it is removed by contact with its target (a smooth muscle cell), kamikaze style.[22]

A third odd feature of this wacky molecule is that it's a free radical. It has a single, unpaired electron orbiting its nucleus, qualifying it as such. That's also the reason it's so evanescent. It's here one second, gone the next.

A fourth peculiar feature is that it does not need an identically configured membrane receptor — like a lock and key — to enter a cell (see Figs. 3.1 and 9.6). It simply enters a cell's inner sanctum by crossing its protective membrane like a ghost.

The final unusual feature of this unconventional biological messenger is that should excess nitric oxide be produced, it simply diffuses into the bloodstream, where it is "rapid(ly)" bound to hemoglobin, the same hemoglobin that transports oxygen to all tissues.[23] It then diffuses out of the body.

But maybe we ought not be surprised about nitric oxide's exceptional biological characteristics. They are in harmony with its wild, wild history going back to the mid-19th century.

A brief history of nitric oxide

Nitroglycerin — a synthetic compound — was developed long before nitric oxide was recognized, even though nitric oxide is a natural compound. Around 1850, liquid nitroglycerin was initially synthesized in Italy by a chemist, who immediately realized its destructive potential as an explosive and warned against its use.[24] Still, humans, being humans, tried to put it to use.

Following a series of explosive catastrophes in Europe,[c] during the 1860s, none other than the Swedish engineer Alfred Nobel ingeniously figured out a way to stabilize nitroglycerin by mixing it with the "fossilized remains" of "hard-shelled algae."[25] The earth in which these fossils were mixed was a soft form of rock easily crumbled by human fingers, dating back to the Tertiary Period of evolution, some 65 to 2 million years ago, a relatively recent parcel of time compared with the Jurassic Period, popularized in the 1993 film *Jurassic Park*. By adding fossilized algae, liquid nitroglycerin was converted

---

[c] During construction of America's First Transcontinental Railroad in the 1860s, nitroglycerin was also used to blast through the Rocky Mountains and the Sierra Nevada Range — with a number of fatalities.[29] Onehundred years later, safer sticks of dynamite (Fig. 10.3) and a yet safer "mole" — a tunnel-boring machine — were used to drill, more so than blast, two tunnels at an altitude of about 11,100 feet above sea level and 890 feet under the Continental Divide, in Colorado.[28] Construction on the challenging Eisenhower Tunnel (Fig. 10.4), 60 miles west of Denver, started on March 15, 1968, and was completed on December 21, 1979, nearly 12 years later[28] — at a cost of US$108,000,000 (equivalent to US$529,000,000 in 2010).

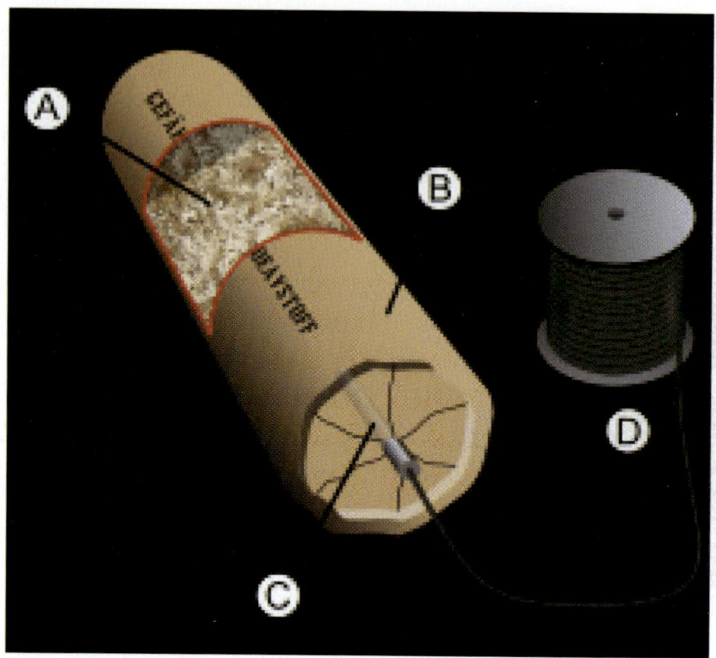

Fig. 10.3.   A stick of dynamite. Sawdust soaked in liquid nitroglycerin (A) is surrounded by a protective coating (B). From a blasting cap (C), an electrical cable (D) is used to detonate the explosive.

(*Source*: Reproduced with permission of the publisher.) Reference 26.

into a far more stable compound "much safer [to] transport than nitroglycerin in its raw form."[25] In 1867, Nobel named the explosive "dynamite" and patented it.[30]

Having one unpaired electron rotating around its nucleus, nitric oxide qualifies as a free radical. Yet, unlike its relatives, nitric oxide is relatively stable — if left alone. But, put in the company of superoxide, it becomes a bad boy, a strong pro-oxidant radical on a par with the hydroxyl radical.[33] And, like other free radicals, it lasts but one or two fleeting seconds and then it's gone as quickly as it arrived. It's a cellular assassin. It strikes out of nowhere. It's gone in a flash, leaving a cellular corpse ... with no DNA to trace.

Clinical Applications of Nitric Oxide

Nitric oxide's vasodilating properties were quickly put to use in 1998 by the Pfizer Pharmaceutical Corporation through the release of a drug patented

Fig. 10.4.   A photograph of the eastern portal of the Eisenhower Tunnel on I-70, 60 miles west of Denver, as it bores under the Continental Divide at 11,100 feet above sea level. The American Society of Civil Engineers designated the tunnel as one of the Seven Wonders of the United States in 1994. The other six include the Golden Gate Bridge, the Hoover Dam, The Empire State Building, Mount Rushmore, The Grand Canyon and the Internet.[27]

More recently, civil engineers added a second architectural wonder along I-70, 60 or so miles west of the Eisenhower Tunnel. Amidst great fears that the natural beauty of Glenwood Canyon would be forever marred by constructing a four-lane highway through it, they cleverly designed a double-decker highway along the canyon's steep, rugged walls. The new road, if anything, enhanced the canyon's beauty (Fig. 10.5). The 15-mile segment opened in 1992.

(*Source*: Reproduced with permission from the publisher.) Reference 28.

just two years earlier, called Viagra — generic sildenafil.[34] The medication was originally developed to treat hypertension (high blood pressure) and angina; the latter refers to painful episodes under the sternum (breastbone) brought on by exertion and due to insufficient oxygen reaching the heart muscle.[d]

---

[d] In medical circles, the term "angina" is pronounced with the accent on the first syllable (an'-gi-na), but the stress is often placed on the second syllable (an-gi'-na) by persons with the chest discomfort.[35] No matter how it is pronounced, the discomfort of typical angina is usually described as a heavy or pressure-like sensation — not a "sharp" or "knife-like" discomfort — located in the center of the chest under the sternum (breastbone), not directly over the heart. It is typically brought on by relatively mild exertion such as climbing stairs.

Fig. 10.5. I-70 as it meanders through Glenwood Canyon in an "amazing series of breathtaking curves."[31] The highway has been called "an engineering marvel," because of the "extremely difficult terrain and narrow space."[32] The 15-mile stretch was the last and most expensive segment of I-70 to be completed in 1992.

(*Source:* Reproduced with permission from the publisher.) Reference 31.

Angina was first brought to medical attention by a British physician, William Heberden, in a classic paper published in 1772. Because of its historic importance, an excerpt from that paper is recorded below.[e]

Early phase 1 trials (Chapter 28) of Viagra found that it had "little effect on angina"[37] but that it did "induce marked penile erections."[37] That led Pfizer to wisely redirect its marketing of the drug. It was released in the spring of 1998 and sales quickly soared to more than US$1 billion annually.[37] The Viagra

---

[e]"There is a disorder of the breast, marked with strong and peculiar symptoms, considerable for the kind of danger belonging to it, and extremely rare of which I do not recollect any mention among medical authors. The seat of it, and sense of strangling and anxiety, with which it is attended, may make it not improperly be called angina pectoris. Those who are afflicted with it are seized, while they are walking, and more particularly when they walk soon after eating, with a painful and most disagreeable sensation in the breast, which seems as if it would take their life away, if it were to increase or to continue: the moment they stand still all this uneasiness vanishes."[36]

boom led the normally staid *British Medical Journal* to editorialize that it had become the "fastest selling drug ever".[37] Because of the burgeoning sales, the American College of Cardiology (ACC) and the American Heart Association (AHA) charged an 18-person, expert committee with the task of developing a "consensus opinion" regarding the use of the then-new drug for the treatment of "male erectile dysfunction."[38] The panel's report was published in January 1999 in both the *Journal of the American College of Cardiology* and *Circulation*, the official journal of the A.H.A.[38,39] To ensure a gold-plated review of the expert panel's consensus opinion, the ACC and AHA submitted its report to "10 external referees," the manufacturer (Pfizer Pharmaceuticals), and the FDA.

The key opinions of the blue-ribbon panel were that:

1. "Viagra should not be prescribed to patients receiving any form of nitrate therapy."
2. If someone develops angina and "has taken Viagra within the past 24 hours, administration of nitrates should be avoided."
3. "[Viagra] is absolutely contraindicated in patients undergoing any long-acting nitrate therapy or using short-acting nitrates because of the risk of developing potentially life-threatening hypotension," that is "circulatory shock." (This is, in effect, a paraphrase of opinion No.1.)[38,39]

Those were the panel's views. Users — or potential users — of Viagra may also appreciate knowing that the blood level of Viagra peaks within just 30–120 min (mean 60 min) after taking the medication, that the sequence of taking Viagra and a nitrate preparation doesn't matter, and that there are many, many long-acting nitrate preparations commercially available in the U.S — too many to list here, but a full list is readily available at http://content.onlinejacc.org/cgi/content/full/33/1/273, if you're not sure whether you are using a nitrate preparation.

A Clever Nonclinical Application of Viagra

Within six months of the drug's release, in a rare application of human research to plants, Israeli and Australian researchers "... discovered that small concentrations of the drug dissolved in a vase of water can also double the shelf life of cut flowers, making them stand up straight for as long as a week beyond their natural life span".[40] They didn't wilt nearly so fast. There's even a plausible scientific explanation for the curious connection: Viagra slows "the breakdown of cyclic guanosine monophosphate (cGMP) [the production of which is mediated by nitric oxide]" in plants just as it does in humans. Explained Israeli professor Yaacov Leshem, "Viagra costs much more [than nitric oxide] but . . . it's easier to use in cut flowers."

The molecule of the year, 1992

For all the molecule's diversity — from atmospheric pollutant to neurotransmitter to making flowers stand up straighter — nitric oxide was crowned the Molecule of the Year for 1992 (Table 10.1). That moved the editor of the prestigious journal *Science,* Daniel Koshland, to say, "This year's Molecule of the Year once again shows that scientific rewards can come from pursuing unconventional thinking".[41] And for their work establishing "nitric oxide as a signaling molecule in the cardiovascular system," three pharmacologists shared the Nobel Prize in Physiology or Medicine for 1998. Robert Furchgott (State University of New York, Brooklyn), Louis Ignarro (University of California, Los Angeles), and Ferid Murad (University of Texas, Houston) (Fig. 10.4). Congratulations, gentlemen!

Table 10.1   Nitric Oxide's Accolades.

- Molecule of the Year, 1992;
- Publication of a new scientific journal bearing its name;
- 1998 Nobel Prize in Physiology or Medicine awarded to Robert Furchgott, Louis Ignarro, and Ferid Murad for discovering the signaling properties of nitric oxide.

Fig. 10.4.   The 1998 winners of the Nobel Prize in Physiology or Medicine. (*Source*: Reproduced with permission from the publisher.) Reference 42.

References

1. Rappoport SM, Schewe T, Wiesner R, Italangk W, Ludwig P, Janicke-Hohne M, Tannert C, Hiebsch L, Klatt D. (1979) The lipoxygenase of reticulocytes. Purification, characterization and biological, dynamics of the lipoxygenases. Its identity with the respiratory inhibitors of the reticulocyte. *Eur J Biochem* **96**: 545–561.

2. Kuhn H, Borngraber S. (1998) Mammalian 15-lipoxygenases. Enzymatic properties and biological implications. In: Nigam E, Pace-Asciak CR (eds.), *Lipoxygenases and Their Metabolites. Biological Functions.* Kluwer/Plenum, New York, Boston, Dordrecht, London, Moscow, pp. 5–28.

3. Yla-Herttuala S, Rosenfeld, MR, Parthasarathy S, Glass CK, Sigal E, Witzum JL, Steinberg D. (1990) Colocalization of 15-lipoxygenase mRNA and protein with epitopes of oxidized low density lipoprotein in macrophage-rich areas of atherosclerotic lesions. *Proc Natl Acad Sci USA* **87**: 6959–6963.

4. Tsimikas S, Glass CK, Steinberg D, Witzum JL. (2004) Lipoprotein oxidation, macrophages, immunity, and atherogenesis. In: Chien KR (eds.), *Molecular Basis of Cardiovascular Disease,* 2nd. ed. Saunders, Philadelphia, pp. 385–413.

5. Yla-Herttuala S, Rosenfeld ME, Parthasarathy S, Sigal E, Sarkeja T, Witzum JL, Steinberg D. (1991) Gene expression in macrophage-rich human atherosclerotic lesions. 15-lipoxygenase and acetyl low density lipoprotein receptor messenger RNA colocalize with oxidation-specific lipid-protein adducts. *J Clin Invest* **87**: 1146–1152.

6. Yla-Herttuala S, Luoma J, Viita H, Hiltunen T, Sisto T, Nikkari T. (1998) Transfer of 15-lipoxygenase into rabbit iliac arteries results in the appearance of oxidation-specific lipid-protein adducts characteristic of oxidized low density lipoprotein. *J CIin Invest* **95**: 2692–2698.

7. Feinmark SJ, Cornicell JA. (1997) Is there a role for 15-lipoxygenase in atherogenesis? *Biochem Pharma* **54**: 953–959.

8. Shen J, Henderick E, Cornhill JF, Zsigmond E, Kim HS, Kuhn H, Valentinova W, Chan L. (1996) Macrophage-mediated 15-lipoxygenase protects against atherosclerosis. *J Clin Invest* **98**: 2201–2208.

9. Sendobry SM, Cornicelli JA, Bocan T, Tait B, Trivedi BK, Colbry N, Dyer RD, Feinmark SJ, Daugherty A. (1998) Attenuation of diet-induced atherosclerosis in rabbits with a highly selective 15-lipoxygenase inhibitor lacking specific antioxidant properties. *Br J Phar Macol* **120**: 1199–1206.

10. Personal Communication from Steven Feinmark with Pfizer Pharmaceuticals, May 11, 2010.

11. Personal Communication from Sarai Spohn with BIOMOL on May 11, 2010.

12. Cyrus T, Witztum JL, Rader DJ, Tangirala R, Fazio S, Linton MF, Funk C. (1999) Disruption of the 12/15-lipoxygenase gene diminishes atherosclerosis in apoE-deficient mice *J Clin Invest* **103**: 1597–1604.

13. Steinberg D. (1999) At last, direct evidence that lipoxygenases play a role in atherogenesis. *J Clin Invest* **103**: 1487–1488.

14. Glass CK, Witzum JL. (2001) Atherosclerosis: the road ahead. *Cell* **104**: 503–516.

15. Sparrow CP, Parthasarthy S, Steinberg D. (1998) Enzymatic modification of low density lipoprotein by purified lipoxygenase plus phospholipase A_2 mimics cell-mediated oxidative modification. *J Lipid Res* **29**: 745–753.

16. Parthasarathy S, Wieland E, Steinberg D. (1989) A role for endothelial cell lipoxygenase in the oxidative modification of low density lipoprotein. *Proc Natl Acad Sci USA* **86**: 1046–1050.

17. http://en.wikipedia.org/wiki/Nitric_oxide

18. Su Y, Han W, Giraldo C, De Li Y, Block ER. (1998) Effect of cigarette smoke extract on nitric oxide synthase in pulmonary artery endothelial cells. *Am J Resp Cell Molec Biol* **19**: 819–825.

19. Culotta E, Koshland DE, Jr. (1992) NO news is good news. *Science* **258**: 1862–1864.

20. http://en.wikiedia.org/wiki/Biological_function-of-nitric-oxide

21. http://nobelprize.org/nobel_prizes/medicine/laureates/1998/illpres

22. Roman LJ, Masters BSS. (2006) The cytochromes P450 and nitric oxide synthases. In: Devlin TM (eds.), *Textbook of Biochemistry with Clinical Correlations*, 6th ed. Wiley-Liss, Hoboken, NJ, Chap. 11, pp. 413–442.

23. Aust AE. (2004) Reactive oxygen/nitrogen species. Generation and reactions in the lung. In: Vallyathan V, Castranova V, Shi X (eds.), *Oxygen/Nitrogen Radicals and Lung Injury and Disease.* Marcel Dekker, New York, Basel, Chap. 1, pp. 1–34.

24. http://en.wikipedia.org/wiki/Ascanio_Sobrero

25. http://en.wikipedia.org/wiki/Diatomaceous_earth

26. http://en.wikipedia.org.wiki/Dynamite

27. http://www.constructioncompany.com/historic-construction-projects/interstate-highway

28. http://enwikipedia.org/wiki/Eisenhower_tunnel

29. http://en.wikipedia./wiki/Nitroglycerin

30. http://inventors.about.com/od/dstartinventions/a/Alfred_Nobel.htm

31. http://realtravel.com/e-178945-glenwood_springs_entry_i70_west_across_colorado

32. http://en.wikipedia.org/wiki/Interstate_70

33. Beckman JS, Koppenol WH. (1996) Nitric oxide, superoxide, and peroxynitrite: the good, the bad, and the ugly. *Am J Physiol* (Cell Physiol 40): C1424–C1437.

34. http://www.wikipedia.org/wiki/Sildenafil

35. *Stedmen's Medical Dictionary*, 28th ed. Lipincott Williams & Wilkins, Philadelphia, Baltimore, New York, London, Buenos Aires, Hong Kong. Sydney, Tokyo.

36. Heberden W. (1772) Some account of a disorder of the breast. *Med Trans R Coll (London)* **2**: 59–67.

37. Gregoire A. (1998) Viagra: on release. Evidence on the effectiveness of sildenafil is good. *Br Med J* **317**: 759–760.

38. Cheitlin HD *et al.* (1999) ACC/AHA Expert Consensus Document. Use of sildenafil (Viagra) in patients with cardiovascular disease. *Circulation* **99**: 168–177.

39. Cheitlin HD *et al.* (1999) ACC/AHA Expert Consensus Document. Use of sildenafil (Viagra) in patients with cardiovascular disease. *J Am Coll Cardiol* **33**: 273–282.

40. Siegel-Itzkovich J. (1999) Viagra makes flowers stand up straight. *BMJ* **319**: 274.

41. Koshland DE, Jr. (1992) The Molecule of the Year. *Science* **258**: 1861–1.

42. http://nobelprize.org/nobel_prizes/medicine/laureates/1998/illpres

11

WHAT IS THE RAISON d'être FOR OXIDIZED LDL?

A question that inevitably comes up is: What is the reason — the physiological function — for *oxidized* LDL and its receptor? The reason for the receptor is easier to give than the reason for oxidized LDL itself.

A Possible Rationale for Scavenger Receptors

Scavenger receptors are normally involved with the removal of oxidized LDL(s) that was (were) generated as waste products during normal cellular metabolism. One (physiological) reason for the scavenger receptor may be that by taking up and storing oxidized LDL in foam cells, the vital overlying endothelial cells are protected from the potentially cytotoxic (cell-damaging or cell-destroying) effects of oxidized LDL. (Macrophages also have receptors, by the way, for other "debris" they ingest, such as bacteria.) So, protection of the delicate endothelium from oxidized LDL may account for its "normal" function in man.[1] But that doesn't explain why the receptor evolved and persisted long before man arrived on Earth, or long before atherosclerosis existed. Nonetheless, since evolution is geared to strengthen traits that promote survival and to weaken those that do not, why have oxidized LDL receptors, which promote atherosclerosis (hardly a survival tactic), persisted? Or, put another way, since atherosclerosis occurs only in humans (a late arriver on Earth), how and why did oxidized LDL receptors pass the test of time when atherosclerosis wasn't around? The ubiquity of scavenger receptors (even in herbivores, such as rabbits and cattle) implies that they have some function other than the removal of oxidized LDL. Teleologically, why else would the substance (and its receptor) have evolved and persisted?

113

One such function may be related to the fact that red blood cells, which are also ubiquitous in mammals, have lipids in their cell membranes that undergo oxidation as red blood cells age. The fatty acid–amino acid bonds in red blood cell membranes are closely related — if not identical — to the chemical bonds found in oxidized LDL after linoleic acid and lysine have interacted. This similarity suggests that the "evolutionary rationale for the preservation of the scavenger receptor" could be "recognition of damaged cell membranes"[2] regardless of genus, species, or tissue type. If so, the recognition of oxidized LDL by the scavenger receptor may be a human deployment of a receptor initially designed for another purpose in other animals. "Its removal by macrophages may prevent damage to arterial endothelium. Interference with its removal might do more harm than good."[2]

Because "survival of the species" is a fundamental principle of evolution, the scavenger receptor may have evolved as a defensive measure. Support for such a protective function in man comes from the recent documentation of scavenger receptors in fruit flies, of all things. These insects have scavenger receptors with binding properties almost identical to the binding properties of human scavenger receptors! Presumably, "host defense" — be the host a fruit fly or a human being — may be the *raison d'être* for scavenger receptors both in insects and in humans.[3] Thus, the scavenger receptor may have a universal function of host defense in many forms of life, as well as the unique function of ingesting in oxidized LDL in man.

What is the Rationale for Oxidized LDL?

Coming up with a reason for oxidizing LDL in the first place is harder to envision. There are, however, a few examples of oxidized LDL producing an effect that could be viewed as slowing the process of atherogenesis. Blocking the production of platelet-derived growth factor is one such example.[4] This might be considered antiatherogenic (preventing or retarding the process of atherosclerosis), because platelets are required to form the blood clot in a coronary artery that ultimately leads to a heart attack. However, a list of the proatherogenic effects of oxidized LDL is far longer than a list of its antiatherogenic effects.

Finding it hard to pinpoint a reason for why oxidized LDL exists, one might settle on Steinberg's logic: "Teleological thinking is out of place when discussing atherogenesis. To be sure, there are 'good' genes and there are 'bad' genes regulating lipoprotein metabolism. However, those genes are 'good' or 'bad' by chance, not by selection. This point is particularly important when thinking about the function of. . .oxidatively modified LDL."[5]

References

1. Steinberg D. (1990) Arterial metabolism of lipoproteins in relation to atherogenesis. *Ann N Y Acad Sci* **598**: 125–135.
2. Sambrano CR, Parthasarthy S, Steinberg D. (1994) Recognition of oxidatively damaged erythrocytes by a macrophage receptor with specificity for oxidized low density lipoprotein. *Proc Natl Acad Sci USA* **91**: 3265–3269.
3. Pearson A, Lux A, Krieger M. (1995) Expression pf dSR-CI, a class C macrophage-specific scavenger receptor from *Drosophila melanogaster. Proc Natl Acad Sci* **92**: 4056–4060.
4. Fox PL, Chisolm GM, Di Corleto PE. (1987) Lipoprotein-mediated inhibition of endothelial cell production of platelet-derived growth factor–like protein depends on free radical lipid peroxidation. *J Biol Chem* **262**: 6046–4054.
5. Tsimakis S, Glass CK, Steinberg D, Witzum JC. (2004) Lipoprotein oxidation, macrophages, immunity, and atherogenesis. In: Chien KR (eds.),*Molecular Basis of Cardiovascular Disease*, 2nd ed. Saunders, Philadelphia, pp. 385–413.

12

FATTY STREAKS AND FOAM CELLS

The earliest evidence of atherosclerosis visible to the naked eye endures the unexciting name "fatty streak." A fatty streak looks, perhaps not entirely by coincidence, as if someone had swiped butter, using a fingertip, on the inner lining of an artery, leaving a yellow, slightly raised imprint that is more easily seen than felt (shown soon).[1] Although fatty streaks had been recognized for nearly a century, proof that they are actually caused by oxidized LDL carried by foam cells was established only in 1989, by extracting the lipoprotein from the aortas of young (40+/−12 years) organ donors for transplantation.[2] These relatively clear cells within the intima resemble the honeycomb appearance of a sheet of foam or the frothy bubbles on a glass of freshly poured beer (Figs. 12.1 and 12.2). Under a microscope, the cytoplasm of cells stuffed with oxidized LDL appears pale and swollen (in contrast with the dark, dense nuclei of the same lipid-laden cells) (Fig. 12.1).

More important than their microscopic appearance is the fact that foam cells — the hallmark of atherosclerosis[a] — are like tiny time bombs primed to explode. The oxidized LDL within such cells is cytotoxic, meaning that it will severely damage or kill other cells upon release. In essence, foam cells

---

[a]The term "atherosclerosis" refers to the disease process that causes the deposition of LDL in arteries. It describes the mushy consistency of cholesterol in the wall of arteries during the early and intermediate stages of the disease. The term comes from the Greek words "*athero*" (meaning "porridge") and "*sclerosis*" (meaning "hardening"). The term "arteriosclerosis" refers to the more advanced stage of the disease, when the arteries become brittle. That stage of the disease process is often not reversible. "Arteriosclerosis" comes from the Greek words "*arterio*" and "*sclerosis*" (meaning "artery" and "hardening," respectively). The pharse "hardening of the arteries" is often applied to such vessels. In this book, word "atherosclerosis" is used almost exclusively to discuss (for the most part) the early and middle phases of the disease.

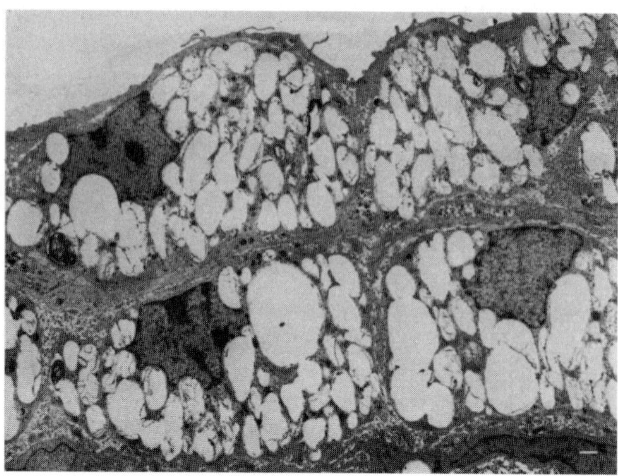

Fig. 12.1.   A fatty streak with two layers of foam cells in a monkey after eating a high-cholesterol diet for just two months. The serum cholesterol level rose from 160 mg/dL to 380 mg/dL during that time. The lipid-laden macrophages are 4–6 times larger than lipid-lean macrophages seen in control animals whose serum cholesterol level did not change.

(*Source:* Reproduced with permission from the publisher.) Reference 3.

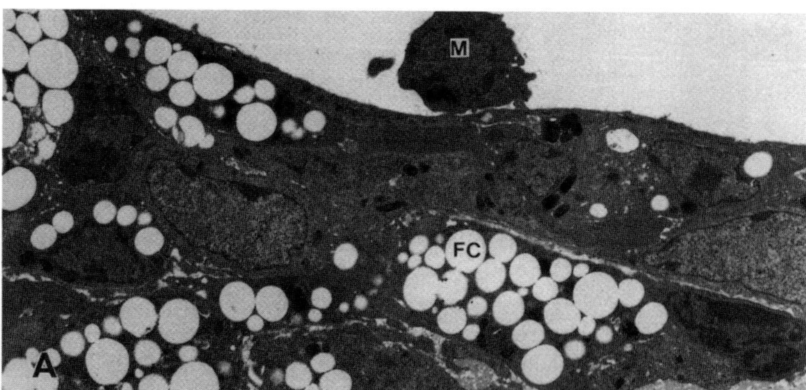

Fig. 12.2.   Another fatty streak, from a monkey fed a high-fat diet for two weeks. A monocyte (M) is perched on top of the thin endothelium. After squeezing through a gap junction (not shown) between two endothelial cells, monocytes are transformed into scavenger cells that become cholesterol-laden foam cells (FC).

(*Source:* Reproduced with permission from the publisher.) Reference 4.

The Dangerous Metamorphosis Of a Monocyte into a Foam Cell

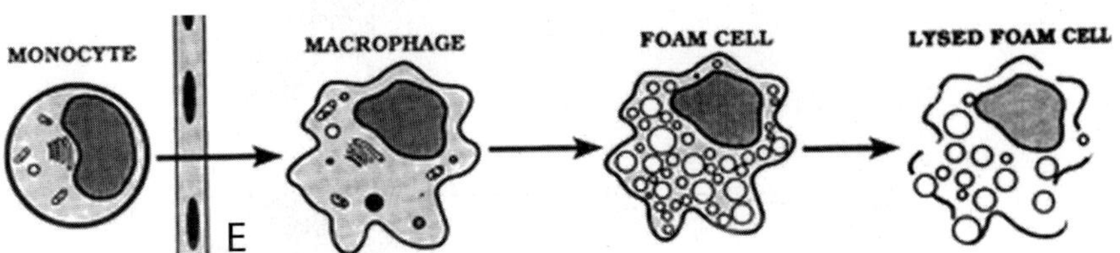

Fig. 12.3. This schematic shows the transition of a monocyte, through the macrophage–foam-cell stage, followed by lysis (dissolution) of the foam cell with release of its toxic, oxidized LDL into the subendothelial space.

(*Source*: Reproduced with permission from the publisher.) Reference 4.

commit suicide. Once lethally injured, they empty their harmful contents into the subendothelial space around other (living) lipid-storage cells still carrying their dangerous cargo. These, too, are mortally wounded and release their own toxic contents in what becomes a rare suicide–murder — ending a series of transformations (Fig. 12.3) that began when a well-intentioned monocyte squeezed through a gap junction (Fig. 8.1) in the thin endothelium of an artery.

Now under the thin protective layer, the freed, oxidized LDL turns its vengeance on to the single layer of (protective) overlying endothelial cells. The bombarded cells begin to lose their ability to ward off cholesterol. Damaged endothelial cells not only allow more LDL to enter the subendothelial space, but also allow monocytes in the blood flowing past them — on the other side — to stick to the surface facing the bloodstream (Fig. 12.2).

Finding a gap junction between two endothelial cells, monocytes squeeze through the space into the subendothelial region. There they are transformed into macrophages that begin to accumulate LDL, which enters through separate, large pores on the surface of endothelial cells (see Chap. 9). Figure 12.4 depicts the transformation of a circulating monocyte into a foam cell and the conversion of native LDL into oxidized LDL in the wall of a blood vessel. Figure 12.5 is a graph illustrating the cholesterol levels of 10 monkeys maintained on a high-saturated-fat diet comparable to the "average American diet" for 1–4 months. Their cholesterol levels rose from 90–150 mg/dL to 300–500 mg/dL. The endothelium (inner lining) of one such monkey's aorta developed a carpet of cholesterol-laden foam cells, as shown in Fig. 12.6, at a

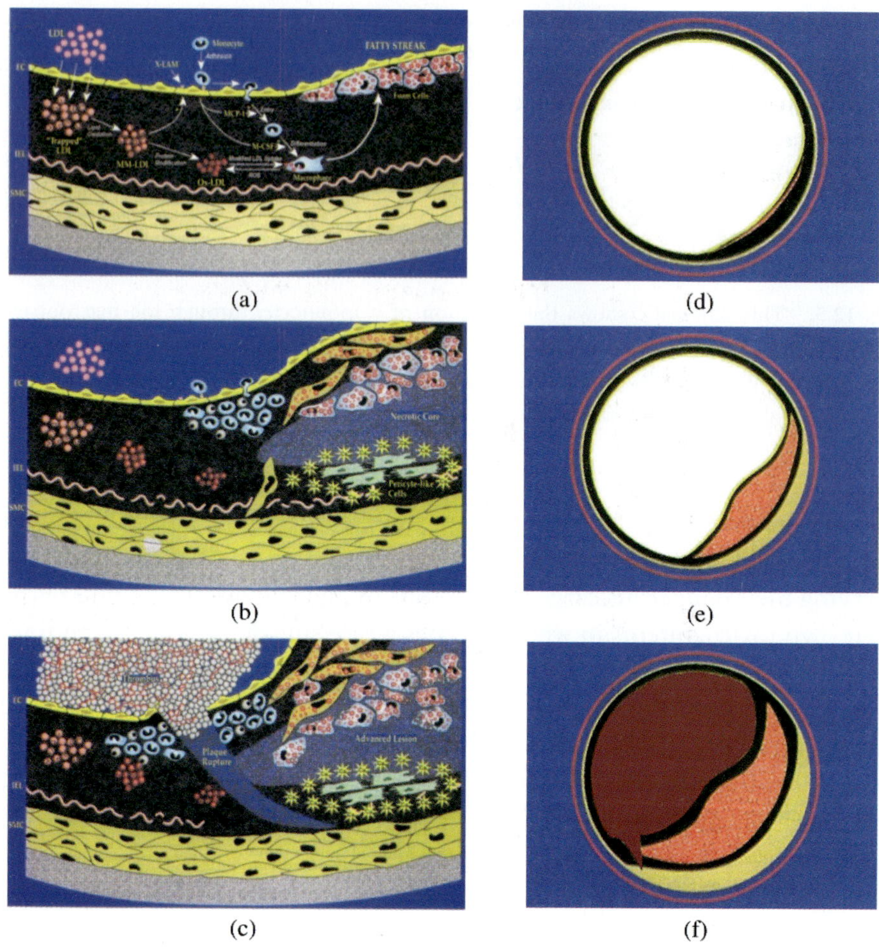

(a)   (d)

(b)   (e)

(c)   (f)

Fig. 12.4.   This schematic depicts the progression of a fatty streak to an acute myocardial infarction — a heart attack — in a longitudinal (left) and a cross-sectional (right) display. Panels A and D represent an early fatty streak, panels B and E an intermediate lesion, and panels C and F an advanced plaque complicated by plaque rupture and an acute myocardial infarction.

In panel A, progressing in a counterclockwise direction, LDL enters the subendothelial space from the bloodstream by diffusing through the single layer of endothelial cells (EC). After being "trapped" in the subendothelial space by collagen fibrils, LDL initially is oxidized to a minimal degree by free radicals resulting in MM-LDL — minimally modified (oxidized) LDL. MM-LDL induces endothelial cells to secrete adhesion molecules (labeled X-LAM) such as VCAM-1 and ICAM-1, attracting more monocytes. When the protein component of LDL is oxidatively modified, LDL becomes sufficiently altered to be recognized by scavenger receptors on macrophages. As more and more LDL is taken up by macrophages, they are transformed into fat-filled foam cells.

cholesterol level of 416 mg/dL. Figure 12.7 shows the appearances of the endothelium from a monkey with a cholesterol level of 115 mg/dL while eating a regular diet. The endothelium is smooth and contains no foam cells.

Ever so slowly — over many years — more and more LDL particles pass through the endothelium into the subendothelial space, where macrophages

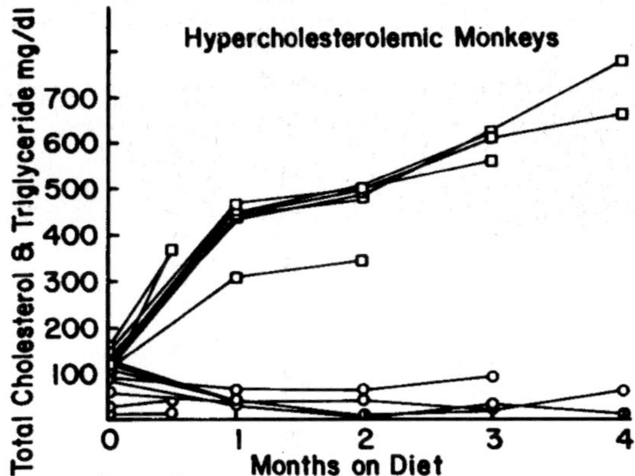

Fig. 12.5. This graph illustrates the changes in plasma cholesterol and triglyceride levels in each of 10 monkeys fed a high-cholesterol diet comparable to the average American diet. After two months, the levels rose from 90–150 mg/dL to 300–700 mg/dL The cholesterol levels of monkeys eating a standard diet are shown in the lower half of the graph.

(*Source:* Reproduced with permission from the publisher.) Reference 3.

In panel B, the Trojan horses release their necrotic core — a semisolid substance composed of cholesterol and dead/dying cells oxidized LDL — into the subendothelial space. The necrotic core is often referred to as "gruel," because its consistency resembles that of porridge. Bulging with foam cells, the plaque begins to protrude into the lumen of the coronary artery.

In panel C, a more advanced lesion undergoes plaque rupture. An occlusive thrombus (blood clot) now forms at the site of rupture, resulting in an acute myocardial infarction.

In panel D, which is a cross-sectional view of a coronary artery corresponding to panel A, a fatty streak (yellow) produces no significant obstruction of the arterial lumen.

In panel E, a more advanced plaque (red) protrudes into the arterial lumen, causing mild obstruction.

In panel F, the plaque has ruptured. A thrombus or blood clot (maroon), has formed at the site of rupture, causing complete occlusion of the arterial lumen and an acute myocardial infarction.

(*Source:* Reproduced with permission from the publisher.) Reference 6.

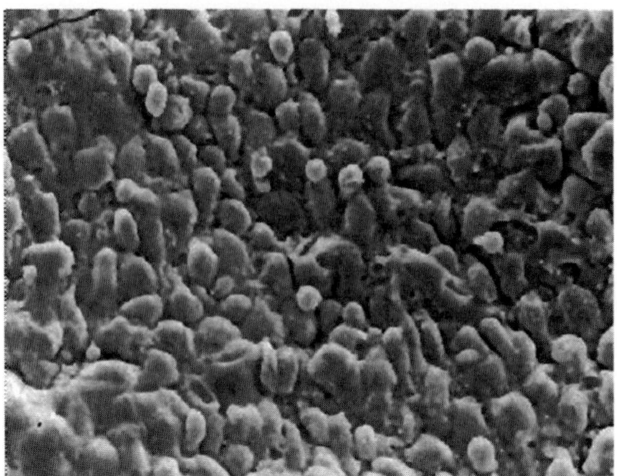

Fig. 12.6.   A carpet of foam cells as seen in a scanning electron micrograph of the endothelium of the aorta from a monkey fed a high-cholesterol diet for one month. The surface of a fatty streak has a striking lobular (lumpy) appearance, due to a large number of underlying, lipid-filled foam cells.

(*Source:* Reproduced with permission from the publisher). Reference 7.

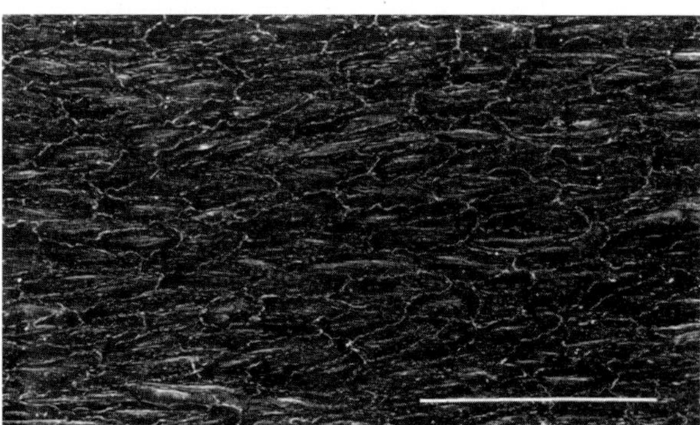

Fig. 12.7.   A scanning electron micrograph of the endothelium from a monkey while it was eating a standard diet. The endothelial surface is smooth and contains no foam cells.

(*Source:* Reproduced with permission from the publisher.) Reference 3.

oxidize more and more LDL. As time goes by, the fatty streak — which is completely reversible — is converted into an "intermediate lesion" composed of smooth muscle cells and macrophages and a central core of lipids (Fig. 12.7). With yet more time, the intermediate lesion develops into a more advanced lesion known as a "fibrous plaque" — which is much less reversible — consisting

Table 12.1

Stages of atherosclerosis	Age
Fatty streaks	Birth to 20
Intermediate lesions	20 to 40
Fibrous plaques	40 to ?
Acute myocardial infarction	40 to ?

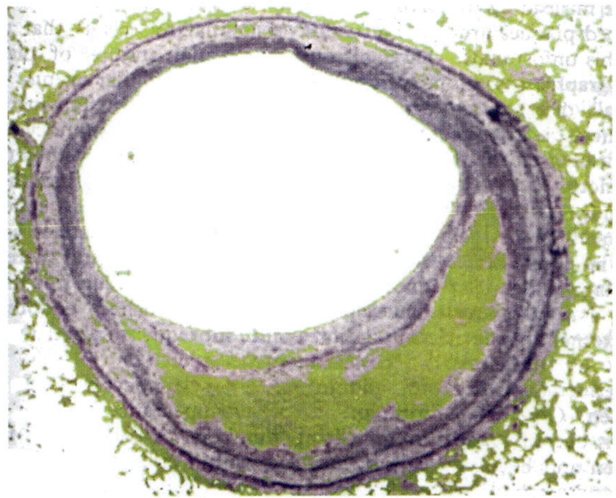

Fig. 12.8. A photomicrograph of a cross-sectioned human coronary artery. A large pool of cholesterol (yellow) derived from dead or dying lipid-laden foam cells is covered by a thick, fibrous cap (stained purple). The lumen of the artery (white) is narrowed by the crescent-shaped plaque, but is still open enough to carry adequate blood.

(*Source:* Reproduced with permission from the publisher.) Reference 8.

of a core of dead or dying foam cells encircled by a rather tough, protective fibrous cap (Table 12.1; Fig. 12.8).[7]

An "intermediate lesion" is far less reversible than a fatty streak. Such lesions generally develop between the ages of 20 and 40 (Table 12.1). As another 20 or so years go by, intermediate lesions progress to "fibrous plaques (advanced lesions)." These fibrous plaques are prone to calcify around this time.[b] Foam cells in the fibrous plaques then begin to manufacture enzymes

---

[b] The phrase "hardening of the arteries" is appropriate at this stage of atherosclerosis. Calcification of coronary artery plaques renders them brittle. A heart scan can detect calcium

that weaken the cap. Finally, one day 40 or 50 years after a fatty streak first formed, the protective cap literally ruptures, emptying its contents into the bloodstream. Platelets in the blood may then stick to the surface of the ruptured plaque. In a matter of minutes to hours, the platelets form a blood clot within the vessel. (The rationale behind taking a daily aspirin tablet is based on preventing platelets from sticking to a freshly ruptured plaque.) If the "clot that kills" is not quickly broken up by a clot-dissolving drug such as t-PA (tissue plasminogen activator) or by an emergency coronary angioplasty, an acute myocardial infarction follows.

Attempts to Prevent Foam Cells by Genetic Engineering

While mice are naturally resistant to atherosclerosis, their resistance can be further heightened by the deletion of specific genes regulating cholesterol metabolism.[11] One way to enhance their relative immunity is by deleting the gene governing lipoxygenase synthesis, as mentioned in Chap. 10. Knocking out the lipoxygenase gene markedly reduces atherosclerosis development in mice whose apolipoprotein E gene is also knocked out — a deletion making them prone to accelerated atherosclerosis. Apolipoprotein E is the principal lipoprotein carrying excess cholesterol in the blood back to the liver for degradation. Its absence leads to cholesterol levels in the 500–600 mg/dL range.

A second way to enhance atherosclerosis resistance should be by limiting the uptake of cholesterol by macrophages. Macrophages, like all other cells, need some cholesterol for their membrane synthesis. Any surplus cholesterol must be removed from the membrane and stored in a cell's cytoplasm. Because cholesterol molecules are naturally rigid (in order to stabilize cell membranes), they must be converted into a more malleable and compact form while being stored within the cell for future use. Toward that end, cholesterol is "esterified," i.e. "free" (unesterified) cholesterol is converted into cholesterol esters. The term "ester" comes from the German words "*essig*," (meaning "vinegar") and "*ather*" (meaning "ether"). An ester, then, is

---

deposits in the coronary arteries. But because coronary calcification occurs only at advanced stages of atherosclerosis, reversal of the disease process is unlikely. A positive scan means that coronary heart disease exists, but a negative scan does not imply its absence. Soft, cholesterol-rich plaques — undetectable by a heart scan — could certainly be present.

Heart scans are not recommended routinely by the American Heart Association or the American College of Cardiology.[9]

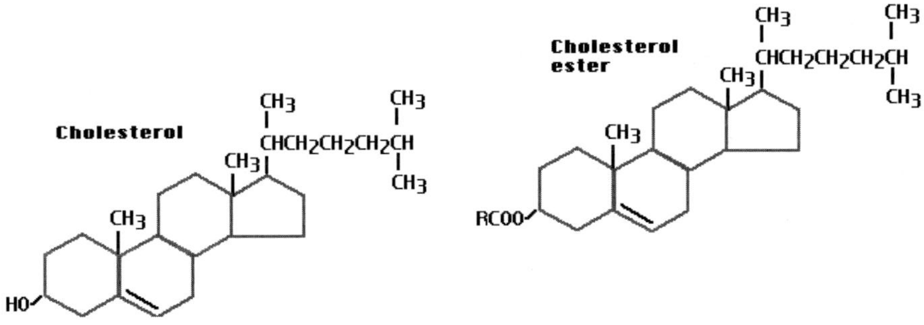

Fig. 12.9. The chemical structure of free (unesterified) (left) and esterified cholesterol (right).

a compound derived from an acid (acetic acid — the acid in vinegar) by replacing an acidic hydroxyl group (OH) with an alkaline alkoxyl group. In the case of cholesterol, its sole hydroxyl group is replaced by a long-chain fatty acid (RCOO; right panel, Fig. 12.9), leading to esterified cholesterol, as shown in Fig. 12.9. Esterified cholesterol is hydrophobic (i.e. it repels water molecules), and shifts into the core of the LDL particle to avoid the water in blood, leaving unesterified cholesterol molecules on the surface (see Fig. 9.1).

Esterification is accomplished by the enzyme acyl-CoA cholesterol acyltransferase[c] (ACAT) by attaching a long-chain fatty acid to the single hydroxyl group (HO) in cholesterol. Storage of small amounts of intracellular cholesterol is innocuous. When cholesterol is overabundant, however, "ACAT1 may promote enough lipid storage to induce conversion of macrophages into lipid-laden foam cells."[12] That line of thinking has led to several studies since 1993 aimed at assessing the impact on atherogenesis (in mice) of selectively inhibiting ACAT1.

Fazio and his colleagues[13] at Vanderbilt created double-knockout mice in which both the gene that regulates LDL receptor synthesis and the gene that controls ACAT1 synthesis were deleted. The first deletion should lead to a very high plasma cholesterol level favoring atherosclerosis, while the second deletion should lead to inhibition of ACAT1 activity discouraging its development. Yet, knockout of the ACAT1 gene resulted in animals with more extensive atherosclerosis than in control mice. The unexpected finding was

---

[c]ACAT was first isolated in 1993. Subsequently, two forms of ACAT were identified, designated simply as ACAT1 and ACAT2. Only ACAT1 is involved with cell membrane synthesis. ACAT2 serves the same function — esterification of cholesterol — in the liver.

speculated to be due to "macrophage toxicity from unesterified cholesterol," forcing them to release their LDL into the subendothelial space. Macrophages aren't immortal. Their death is due to excess intracellular cholesterol related to defects in ACAT1-mediated esterification leaving them with too much free (unesterified) cholesterol.[14] The investigators concluded that "selective. . . ACAT1 inhibition may not be a beneficial strategy for treating or preventing atherosclerosis, and may in fact be detrimental."

Fatty Streaks and Foam Cells in Infants and Children

In 1953, a landmark paper described an unexpected high frequency of coronary atherosclerosis in young — average age 22.1 years — men killed in action during the Korean War.[15] Among 300 *apparently* healthy American soldiers, 77.3% had evidence of atherosclerotic heart disease at autopsy, varying from fatty streaks in 35% to complete occlusion of a coronary artery in 3%. Another 39% had plaques narrowing their coronary arteries, ranging from 10% to 90%, yielding a total of 77.3% with some degree of coronary heart disease (Table 12.2). Only 23% were free of the disease at age 18. In November 1986, the editors of *JAMA* republished the 1953 paper,[16] because they considered it to be "clearly a landmark in our understanding of the development of coronary atherosclerosis. . . .This widely cited publication dramatically showed that atherosclerotic changes appear in the coronary arteries years and decades before the age at which coronary heart disease (CHD) becomes a clinically recognized problem."[17]

About the same time, Stary, a pathologist at Louisiana State University, published a series of papers[18–20] showing that the coronary arteries of children "as young as one month frequently contain. . .macrophage foam cells."[21] What's more, isolated fatty streaks were found in the coronary arteries of 50% of children aged 10–14. Between 15 and 20 years, the number of fatty streaks (Fig. 12.10) tripled, and by age 20 necrotic foci and fibrous plaques began to appear. Then, after 30, fatty streaks were replaced by advanced lesions (Fig. 12.11). The increase in advanced lesions was reflected 20 years later — and still is — by a surging rate of coronary heart disease morbidity and mortality.[21]

In 1983, nine university hospitals collaborated to conduct the Pathological Determinants of Atherosclerosis in Youth Study. Together, they collected 1532 cases of young people who died between the ages of 15 and 34, mostly from motor vehicle accidents, suicides, and homicides. When the initial report was

Table 12.2 Prevalence of atherosclerosis in young persons 15–34 years old.

Sex	Race	Age(yrs.)	Frequency (%)
Male	White	15–19	100
		20–24	100
		25–29	100
		30–34	99.2
	Black	15–19	100
		20–24	100
		25–29	100
		30–34	100
Female	White	15–19	100
		20–24	100
		25–29	100
		30–34	100
	Black	15–19	100
		20–24	100
		5–29	100
		30–34	100

Source: Modified and reproduced with permission from the publisher. Reference 23.

published in 1993, all but one of these young individuals had fatty streaks in their aortas and about half had such streaks in their coronary arteries Table 12.2.[23] And it made no difference whether they were male or female, black or white. Many of the adolescents and young adults had advanced lesions, such as fibrous plaques involving roughly 20% of the aorta in 15–19-year-olds and 30% of the aorta in 30–34-year-olds. The authors' concern about the extent of atherosclerosis so young in life was reflected in their summary emphasizing preventive tactics during childhood to avoid atherosclerosis as an adult. They wrote, "Prevention of adult coronary heart disease is likely to be most effective if…risk factor [modification] begins in childhood or adolescence. Many of the risk factors originate in youthful behavior with patterns of diet, physical activity, and tobacco use persisting from adolescence into adulthood; the earlier they begin the more difficult it becomes to modify them. The results of this study clearly confirm the origin of

Fig. 12.10.   The aorta of a child opened longitudinally. The aorta is the large artery in the thorax and abdomen that carries blood from the heart to other organs before dividing into the two iliac and femoral arteries in the lower extremities. The white arrow points to one of a number of fatty streaks. They're hard to see. The one at the arrow tip runs horizontally and is about 3/8 of an inch in length. The darker, irregular area adjacent to the white asterisk is relatively free of fatty streaks, for comparison. Atherosclerosis begins subtly early in life. Had the child lived to age 50 or more, the aorta would have resembled the one shown below.

(*Source:* Reproduced with permission from the University of Utah.) Reference 22.

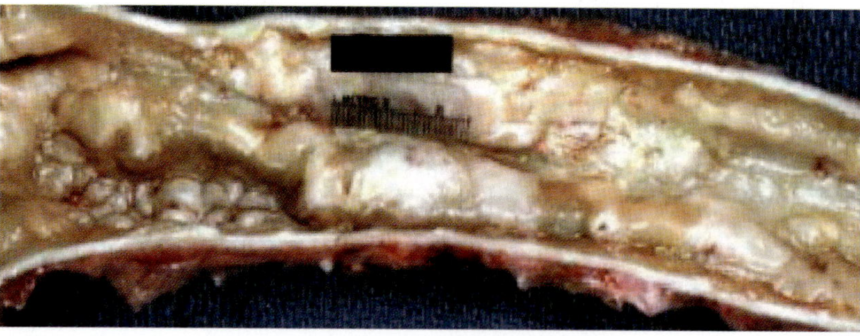

Fig. 12.11.   The aorta of an adult with advanced atherosclerosis. Over time, fatty streaks are converted into large, fibrous plaques, often calcified, as in this photograph. The phrase "hardening of the arteries" refers to vessels such as this.

(*Source:* Reproduced with permission from the publisher.) Reference 10.

atherosclerosis in childhood and its progression toward clinically significant lesions in young adulthood."[23]

In a commentary accompanying the article, the writer emphasized that while risk factor modification is worthwhile in midlife, "for maximum benefit, preventive efforts should be directed towards the young."[24] We all know that it's easier to form healthy habits early in life than it is to break unhealthy ones in later years. We, as parents, need to teach our children the virtues of eating right, regular aerobic exercise, staying lean, and avoiding cigarettes. And since children learn more through our examples than from what we say, we need to practice what we preach.

The "practice what we preach" principle may even extend to children before they are born! An important paper published in 1997 in the *Journal of Clinical Investigation*, from the University of Naples and the University of California (La Jolla), pointed out that fatty streaks were present in the aortas of 70% of 82 spontaneously aborted fetuses (mean fetal age 6.3 months) and premature infants who died within 12 h of birth (Fig. 12.12).[25] The percentage was a bit lower in fetuses and newborns of mothers with a normal plasma

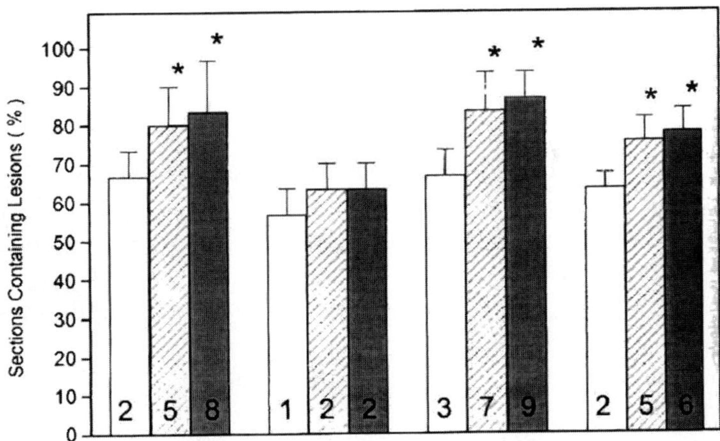

Fig. 12.12. This bar graph shows that 63% of aortas from fetuses of mothers with normal cholesterol levels had fatty streaks. The frequency rose to 78% of aortas from fetuses of mothers with transient hypercholesterolemia (i.e. mothers with transiently elevated blood cholesterol that returned to normal following delivery) and of mothers with persistent hypercholesterolemia (continued hypercholesterolemia after pregnancy).

(*Source:* Reproduced with permission from the publisher.) Reference 25.

cholesterol level (63.3%) than in those of mothers with an elevated cholesterol level (77.8%). Nonetheless, the startling finding that "early lesions are present in a large percentage of cross-sections of fetal aortas suggests that many lesions observed in children and young adults may begin much earlier than previously assumed."[25]

References

1. Halliwell B, Gutteridge JMC. (1990) Role of free radicals and catalytic metal ions in human disease: a review. In: Packer L, Glazer AN (eds.), *Methods in Enzymology.* Academic, New York, Boston, London, Sydney, Tokyo, Toronto, Vol. 186, pp 1–85.

2. Yla Herttuala S, Steinberg D. (1989) Evidence for the presence of oxidatively modified low density lipoprotein in the atherosclerotic lesions of rabbits and man. *J Clin Invest* **84**: 1086–1095.

3. Faggiotto A, Ross R, Harker L. (1984) Studies of hypercholesterolemia in the nonhuman primate. 1. Changes that lead to fatty streak formation. *Arteriosclerosis* **4**: 323–340.

4. Lusis AI, Rotter JI, Sparkes RS (eds.), *Molecular Genetics of Coronary Artery Disease: Candidate Genes and Processes in Atherosclerosis.* Karger, Basel, Freiburg, Paris, New York, New Delhi, Bangkok, Singapore, Sydney, Chap. 1.

5. Hoff HF, O'Neil J, Pepin M, Cole TB. (1990) Macrophage uptake of cholesterol-containing particles derived from LDL and isolated from atherosclerotic lesions. *Eur Heart J 11 (Suppl E)*: 105–115.

6. Berliner JA, Navab M, Fogelnman AM, Frank JS, Demer LL, Edwards PA, Watson AD, Lusis AJ. (1995) Atherosclerosis: basic mechanisms. Oxidation, inflammation, and genetics. *Circulation* **91**: 2488–2496.

7. Ross R. (1993) The pathogenesis of atherosclerosis: a perspective for the 1990s. *Nature* **362**: 801–809.

8. Davies MJ, Woolf N. (1993) Atherosclerosis: What is it and why does it occur? *Br Heart J 69 (Suppl 1)*: S3–S11.

9. Mayo Clinic staff. MayoClinic.com. Coronary calcium scans: heart scans mired in controversy.

10. http://peir2.path.uab.edu/pdl/dbr.cgi?db=images&uid=default&Collection=PEIR&Collection2=%7BPEIR%7D&Type=Gross&Type2=%5BGross%5D&Description=atherosclerosis&syn=off&mh=20&hits=%6020%21&File=&view_records=View+Records&view_records=Search

11. Glass CK, Witzum JL. (2001) Atherosclerosis: the road ahead. *Cell* **104**: 503–516.

12. Rudel LL, Shelness GS. (2000) Cholesterol esters and atherosclerosis — a game of ACAT and mouse. *Nat Med* **6**: 1313–1314.

13. Fazio S, Major AS, Swift LL, Gleaves LA, Accad M, Linton MF, Farese RV, Jr. (2001) Increased atherosclerosis in LDL receptor-null mice lacking ACAT1 in macrophages. *J Clin Invest* **107**: 163–171.

14. Tabas I. (2005) Consequences and therapeutic implications of macrophage apoptosis in atherogenesis. The importance of lesion stage and phagocytic efficiency. *Arterioscler Thromb Vasc Biol* **25**: 2255–2264.

15. Enos WF, Holmes RH, Beyer J. (1953) Coronary disease among United States soldiers killed in action in Korea: preliminary report. *J Am Med Assoc* **152**: 1090–1093.

16. Enos WF, Holmes RH, Beyer J. (1986) Coronary disease among United States soldiers killed in action in Korea: preliminary report. *JAMA* **256**: 2859–2862.

17. Strong JP, Connary LA. (1986) Coronary atherosclerosis in soldiers: a clue to the natural history of atherosclerosis in the young. *JAMA* **1256**: 2863–2866.

18. Stary HC, Letson GD. (1983) Morphometry of coronary artery components in children and young adults. *Arteriosclerosis* **3**: 485a.

19. Stary HC. (1983) Macrophages in coronary artery and aortic intima and in atherosclerotic lesions of children and young adults. In: Schetter FG, Gotto AM, Middlehoff B, Hebenicht AJR, Junutka KR (eds.), *6th International Symposium on Atherosclerosis.* Springer-Verlag, New York, pp. 462–466.

20. Stary HC. (1983) Evolution of atherosclerotic plaques in the coronary arteries of young adults. *Arteriosclerosis* **3**: 471a.

21. McGill HC, Jr. (1984) Persistent problems in the pathogenesis of atherosclerosis. *Arteriosclerosis* **4**: 443–451.

22. http://library.med.utah.edu/WebPath/CVHTML/CV114.html

23. Pathobiological Determinants of Atherosclerosis in Youth (PDAY) Research Group. (1993) Natural history of aortic and coronary atherosclerotic lesions in youth. Findings from the PDAY study. *Arterioscler Thromb Vasc Biol* **13**: 1291–1298.

24. Reference needed.

25. Napoli C. D'Armiento FP, Mancini FP, Postligione A, Witzum JL, Palumbo G, Palinski W. (1997) Fatty streak formation occurs in human fetal aortas and is greatly enhanced by maternal hypercholesterolemia. *J Clin Invest* **100**: 2680–2690.

13

EMBRYONIC STEM CELLS

Stem cells are those cells from which all other cells originate. They are the biological origin of life in mice and men. . .or, at least, that is the implication of the term. In reality, the one-celled, fertilized egg is *the* cell from which all other cells arise, whether in mice or men.[1] Stem cells do not "appear" until the one-celled egg has divided five or six times during the first 3–4 days after an egg is fertilized. And even then, stem cells cannot be recognized by looking at a freshly fertilized egg under a microscope. There are none. These powerful cells must be cultivated from such an egg in a laboratory using meticulous cell culture techniques. In short, as a recent NIH report pointed out, it is not "certain that embryonic stem cells exist as such in the embryo. Instead. . .they are derived from. . .the early embryo"[2] in a tissue culture laboratory. The key word is "derived."

Although the single-celled fertilized egg is the only cell from which all others evolve — by dividing repeatedly into more and more cells — the one-celled egg has a shortcoming: it cannot generate more cells to replace damaged or diseased cells that develop after birth.[3] The single-celled fertilized egg and its earliest descendents — the 2-, 4-, 8-, and 16-celled eggs — can only form duplicate, healthy cells with unlimited developmental potential during the first two or three days of embryonic life. After the one-celled egg has divided five times, the cells of the then 32-celled egg begin to acquire characteristics different from one another. This allows fifth-generation cells — from which stem cells can be grown in a laboratory — to start forming the various specialized tissues of the body when they are introduced into the human body after being cultured in a laboratory on a specific culture medium for many months. Thus, although the single-celled fertilized egg has the potential to form a complete human being, it cannot form new tissues after childbirth. Stem cells, on the other hand, though they must be cultured from early human embryos several days after the one-celled fertilized egg has disappeared, have the unique ability to divide into two cells, one of which is a duplicate cell — ensuring perpetugtion of the stem cell — while the other is a distinctly different cell that can form any tissue of the body, just as a single-celled egg can. They are fascinating cells!

Development of the Fertilized Egg During the First Week of Life

Mouse Versus Man

The speed at which the single-celled egg develops into a multicelled egg differs between mice and men, reflecting their different gestational periods. In mice, the single-celled, fertilized egg — which contains no stem cells in either mice or men — develops into a 64–128-celled egg (after 5 or 6 divisions, over 3–4 days), as a result of the animals' 21-day gestational period. In (wo)men, with a nine-month pregnancy, the same process takes 4–5 days (Fig. 13.1).

The 64–128-celled egg consists of two parts an outer shell, called a trophoblast, and an inner group of cells referred to, unpretentiously, as the "inner cell mass[a]." The outer shell will become the placenta. The inner cell mass of the human, 4–5-day-old embryo will become the fetus[b] several weeks later.

The Crucial Blastocyst

Because after five or six divisions the egg has developed into a cystlike structure, it is called a "blastocyst," (Fig. 13.2), the prefix "blasto-" referring to the resemblance of a five-day-old egg to, for example, a (fruit tree's) bud in springtime. The cavity of the blastocyst is filled with fluid, essentially water, while the inner cell mass is located at one pole of the cavity. The inner cell mass, itself, consists of a clump of 15–20 cells, none of which are yet "stem cells."[2] Those 15–20 cells must first separate into two layers a thin lower layer — the hypoblast — that forms the "yolk sac"[c] and a thicker upper layer — the epiblast — from which

---

[a] The inner cell mass does, in fact, have a corresponding scientific name: An embryoblast. But scientists, more often than not, refer to this early stage of development as simply the "inner cell mass." Scientists can use plain English. . .once in a while, anyway.

[b] In man, an embryo — by convention — is usually considered to represent the stage of development from the time an egg is fertilized until about the eighth week of intrauterine life, when organs appear and the features of a human being become recognizable. An embryo becomes a fetus at that point in time, again by long usage, rather than by man-made law.

[c] The yolk sac is a small, spherical structure — about the size of this font-size-10 letter "o" — that does contain yolk in chickens, but does not in humans. In hens, the yolk furnishes the embryo

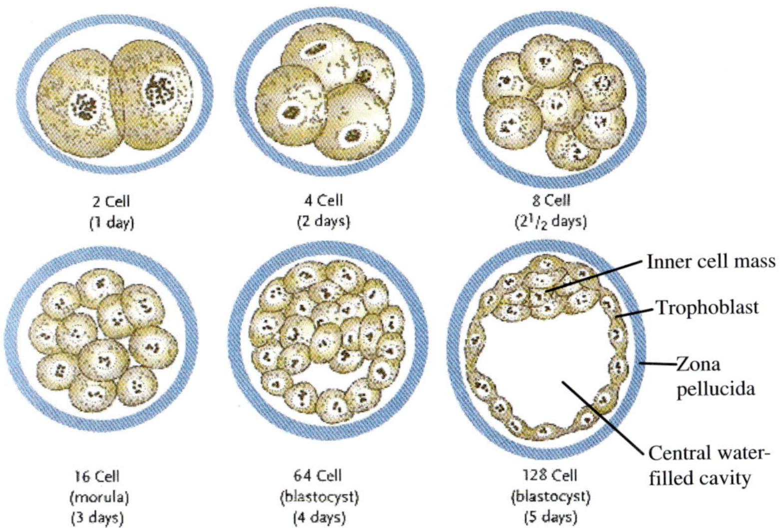

Diagrams of the Early Stages of Human Embryonic
Development at the Cellular Level

Fig. 13.1.   The drawings of the 2–8-celled-stage fertilized eggs (1–2½ days old) show the eggs as a conglomerate, composed of the stated number of cells.

The drawings of the 16- and 128-celled-stage eggs (3–5 days old) represent sections through the egg and, therefore, they show fewer cells than are present in the whole embryo. The inner cell mass consists of a group of 15–20 cells, depicted as 7 cells, that form the embryo. The outer shell — the trophoblast, from the Greek word "*trophe.*" meaning "nutrition" — becomes the placenta. A clear membrane — the zona pellucida — encloses the fertilized egg while it passes through the Fallopian tube, functioning as a shell. The zona pellucida will be shed when the egg reaches the uterus.

(*Source;* Reproduced with permission from the publisher.) Reference.

10–15 coveted stem cells can later be derived in a laboratory.[5] The entire blastocyst is covered by a clear, relatively thick protective layer — the zona pellucida — that originally formed around the unfertilized egg when it was in the

---

with nutrients. In humans, devoid of a yolk, the yolk sac is a vestigial structure, no longer vital to the nutrition of the embryo.[4]

The yolk sac's main importance today resides in its being visible by (transvaginal) ultrasound within 4–6 weeks after an egg is fertilized. As a result, it can be used to confirm pregnancy. However, since a standard urine pregnancy test becomes positive before the yolk sac can be visualized by ultrasound — at a fraction of the cost — the diagnostic value of seeing a yolk sac by ultrasound is limited.

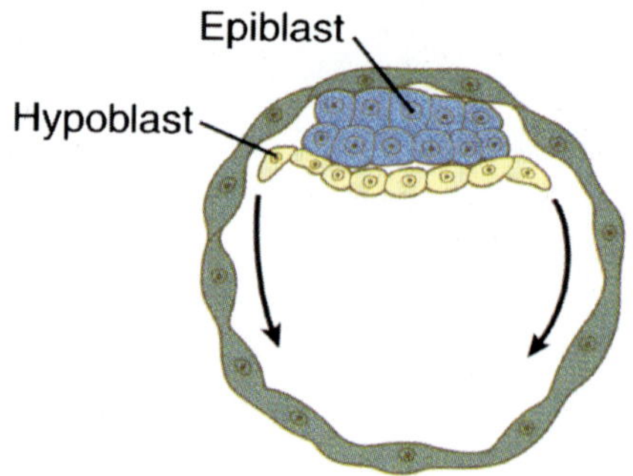

Fig. 13.2.   A drawing of a human blastocyst on day 6 following fertilization.

The inner cell mass has divided into a lower layer, the hypoblast, and an upper layer, the epiblast. The epiblast is the source of stem cells when the inner cell mass is cultured in a laboratory, but none are present at this stage of actual development.

The arrows indicate that the hypoblast will soon spread in a circular fashion to form a separate ring of cells within the trophoblast. The zona pellucida is no longer present in this diagram of the six-day-old human blastocyst because it was shed hours earlier, when the egg reached and attached to the inner lining of the uterus.

(*Source:* Reproduced with permission from the publisher.) Reference 3.

ovary to (usually) ensure the entry of only one sperm into the egg as it enters the Fallopian tube, thereby markedly reducing the chances of identical twins being born to 4 per 1000 births.[6] The zona pellucida will be shed as the fertilized egg exits the Fallopian tube, to allow its firm attachment to the uterine lining (Fig. 13.3).

The Progression of an Egg from the Ovary to the Uterus

As the blastocyst approaches the uterus, it escapes from the zona pellucida through a process referred to as "hatching." The zona pellucida gradually thins, while the blastocyst expands, and finally squeezes through a hole in the zona pellucida. The egg is now "free" and could implant into the animal's uterus — be it the womb of a mouse or a human. If the womb is human, the crucial blastocysts from which a few — typically 10–15[9] — precious stem cells can be derived are far more difficult to obtain than if the animal is a mouse. The reason, quite simply, is that mice can be sacrificed to acquire their five-day-old

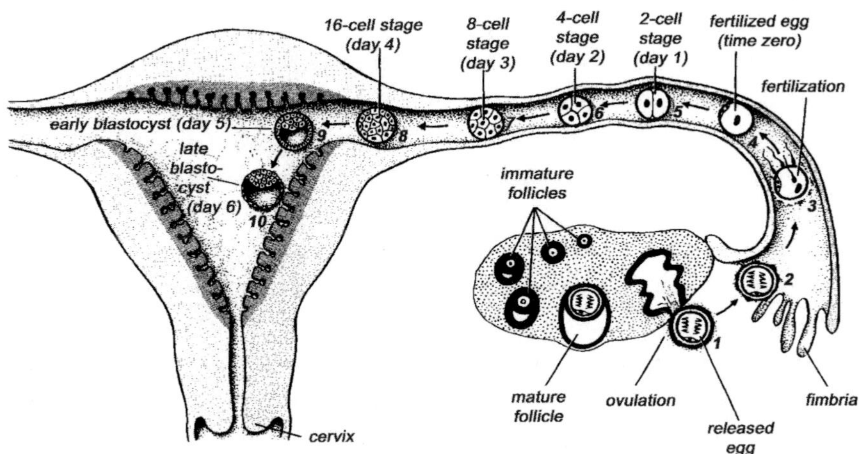

Fig. 13.3. The stepwise progression of an egg from the ovary to the uterus. (1) A mature egg is released from the ovary into the mouth of the Fallopian tube. (2) The unfertilized egg is swept by the finger-like fimbria into the mouth of the Fallopian tube. (3) The egg is charged at by many sperm, but only one is able to penetrate the protective zona pellucida (not shown in this figure), to produce fertilization. Fertilization is a stepwise, not instantaneous, process that occurs over 6–12 h after the egg is released from the ovary.[d] (4) The now-fully fertilized egg — the earliest stage of the embryo — is designated as time zero, although the unfertilized egg was released from the ovary several hours earlier. (5) The 2-celled egg at approximately 30 h after fertilization. (6) The 4-celled egg. (7) The 8-celled stage. Three days are required to reach the 8-celled stage, but only three more days are needed for the 256-celled blastocyst to implant into the uterus. (8) The 16-celled egg approaching the uterus. (9) The 64–128-celled blastocyst entering the uterus. The zona pellucida (not shown) dissolves to prepare the egg for implantation. (10) Implantation into the endometrium of the hormonally prepared uterus.

(Source: Modified and reproduced with permission from the publisher.) Reference 8.

eggs,[e] and humans cannot. Therefore, the procurement of a human blastocyst hinges on a rare event. A uterus with its Fallopian tubes must become fortui-

---

[d]Since the process of fertilization takes a number of hours, it is probably "unjustified to speak of 'the moment of fertilization.'"[7]

[e]Mice have another reason for their popularity in stem cell research: they are the only animals other than humans and monkeys from which embryonic stem cells can be cultivated in the laboratory.[10] And mice trump monkeys for dollars-and-cents reasons. A mouse costs US$2; a monkey, US$1000. Add to the acquisition cost the expense of feeding and housing a mouse versus a monkey, and mice win the popularity contest.

tously available either as a result of a surgically indicated hysterectomy done 4–5 days after coitus or as a result of an autopsy (!) being done 4–5 days after coitus associated with conception. And then — even if either of those scenarios should occur — recovery of the blastocyst and its potential stem cells would require that the pathologist examining the reproductive organs intentionally search for a fertilized egg in the Fallopian tubes. Although the Fallopian tubes are routinely inspected with the naked eye, they are not routinely studied under a microscope for a five-day-old egg. Consequently, it would take the very unlikely coincidence of coming upon a uterus and its tubes on the fifth day after intercourse associated with conception and a pathologist specifically searching for a five-day-old egg to discover a recently fertilized human egg. In 1953, an embryologist noted that the "earliest human embryo described at [that] time...[was] about seven days old".[11] Sixty years later, another embryologist wrote, "Very few. . .human embryos have been taken. . .from the [time] period just preceding implantation. . .." of a fertilized egg into the uterus.[12] As a result, a naturally fertilized, five-day-old human egg and its treasure of potential stem cells is valued by embryologists and stem cell biologists alike.

An Alternative Source of Stem Cells

In view of the extremely few embryonic stem cells that can be obtained from the inner cell mass of a five-day-old human egg fertilized *in vivo*, human embryonic stem cells have been derived almost exclusively from eggs acquired by *in vitro* fertilization. The initial isolation of human embryonic stem cells in 1998 by Dr. James Thomson at the University of Wisconsin was a landmark discovery made through the culturing of eggs produced by *in vitro* fertilization.[f,13] Subsequent studies worldwide employing human embryonic stem

---

[f]The first isolation of human stem cells was, in fact, in 1994, by Dr. Ariff Bongso in Singapore.[14] However, his stem cells survived as such for only two weeks before differentiating into more specialized cells. The stem cells isolated by Thomson in 1998, on the other hand, continued to grow laterally — akin to a small Midwestern town expanding on to adjacent farmland — without "growing up" for many months. The dissimilar survival times apparently were related to the use or nonuse of a "feeder layer" — to be described more fully — by the two investigators. When a feeder layer was used it allowed firm attachment of the inner cell mass to the feeder layer. This anchoring effect of the feeder layer on the inner cell mass led to a similar bond of the subsequent stem cells to the feeder layer. The stem cells formed in groups around the inner cell mass, and yet individual cells composing the group retained their identity. Later, the feeder layer inhibited the groups of stem cells that arose from the inner cell mass from developing into

cells to — one day — create replacement tissues for diseased human tissue have also relied on eggs acquired through *in vitro* fertilization. In Singapore, human blastocysts procured through *in vitro* fertilization have been used to establish stem cells lines.[15] In China, eggs fertilized *in vitro* that were judged three days after fertilization to be of such "poor quality" that they had a less than 10% likelihood of resulting in a live birth if implanted into a uterus for fetal development, have been sources of stem cells for research purposes.[16] And in the United States, a report in 2004 from Harvard and the Howard Hughes Medical Institute described the doubling of human stem cell lines from blastocysts derived by *in vitro* fertilization.[17] Women who have donated unfertilized eggs for *in vitro* fertilization no longer need to await the birth of a child after artificial insemination to deserve recognition for their altruism. Without their donation there would be no human stem cells for either research or clinical purposes.

Quality Versus Quantity

But are five-day-old human eggs (blastocysts) derived though *in vitro* fertilization the same as five-day-old eggs developed *in vivo* — i.e. naturally? And are all human eggs obtained by *in vitro* fertilization of the same grade (not in comparison with eggs derived *in vivo,* but in comparison with other eggs derived *in vitro*)?

An answer to the first question is not available, for the reason implied by the virtual nonexistence of naturally fertilized (five-day-old) human eggs for stem cell extraction. A comparison of human *in vitro* with human *in vivo* blastocysts can only be inferred from comparative observations made on mice. However, two findings indicate that they probably are not of equal quality.

---

the specialized cells of a particular tissue, such as muscle or nervous tissue. When a feeder layer was not used, the inner cell mass gave rise to single stem cells that differentiated quickly in Bongso's laboratory, as opposed to groups of stem cells that retained their identity much longer in Thomson's laboratory.[13] Single stem cells naturally interact with one another — as if they were being "social" — during early embryonic development to form specific tissues, each with a specific function. Groups of stem cells — produced by their sticking to a feeder layer — prevent individual stem cells from interacting with each other. They remain "unspecialized" while being grown in the laboratory. Yet they retain the ability to grow up if the feeder layer is removed and the stem cells are placed into a living tissue.

First, mouse blastocysts developed *in vitro* contain fewer cells within their inner cell masses — from which stem cells can be derived — than do those developed *in vivo*. Some have "no inner cell mass cells, at all."[18] This could lead to *fewer* stem cells derived from blastocysts obtained by *in vitro* fertilization. Second, mouse blastocysts developed in culture rarely, "if ever," have a hypoblast (see Fig. 11.2) on the inner aspect of the inner cell mass.[18] Since stem cells arise only from the epiblast,[3] this difference could favor the derivation of *more* embryonic stem cells from *in-vitro*-fertilized eggs. Considered together, the two findings may have a zero net effect on the issue of *in vitro* versus *in vivo* blastocyst quality — in mice. But, again, there are no comparable data in man, in whom the matter really counts.

An answer to the second question does exist. The development of a scoring system to assist in the "optimal selection of embryos" derived by *in vitro* fertilization for implantation into the uterus with the expectation of a full-term pregnancy[19] implies that all human eggs obtained through artificial insemination are not of equal quality[g]. The system grades five-day-old eggs by taking into consideration the symmetry of their cells and their "fragmentation" — i.e. are the cells broken to some degree, or are they intact? (see Fig. 13.4.) The blastocyst is assigned a score of zero to fifty, the higher the score the greater the likelihood of pregnancy. This system became the standard for assessing embryo quality. A later report, describing a modified scoring system, confirmed the increased probability of pregnancy from a "high-scoring" blastocyst.[20]

Isolation of Inner Cell Masses and Cultivation of Human Embryonic Stem Cells

Future stem cells nestled within the tiny inner cell mass of a blastocyst are surrounded by an outer ring of trophoblastic cells, themselves wrapped in the

---

[g]The implication is further supported by the disappointingly low 10–20% implantation rate of human embryos conceived through *in vitro* fertilization.[20] The low implantation rate is probably due to the fact that *in vivo* the day 2 or day 3 embryos reside in the Fallopian tube, not the uterus. When such embryos are transferred to the uterus prematurely, they often do not survive.[20]

The dilemma might appear to be solvable by simply delaying implantation until the egg is 4–5 days old. However, it is not solved because "deficiencies. . .in the culture media" used to grow embryos in the laboratory result in "an implantation and live birth rate of 7%."[20].

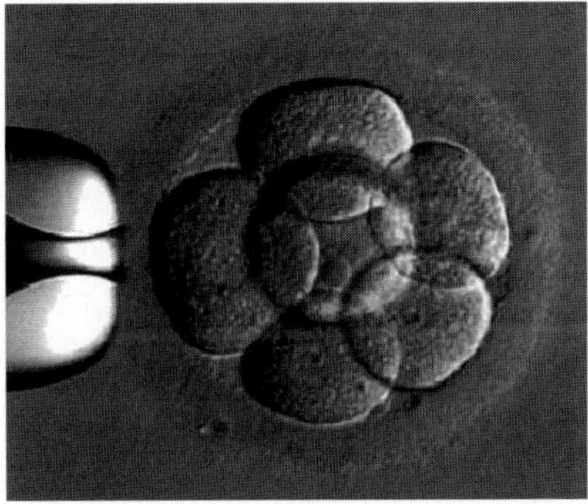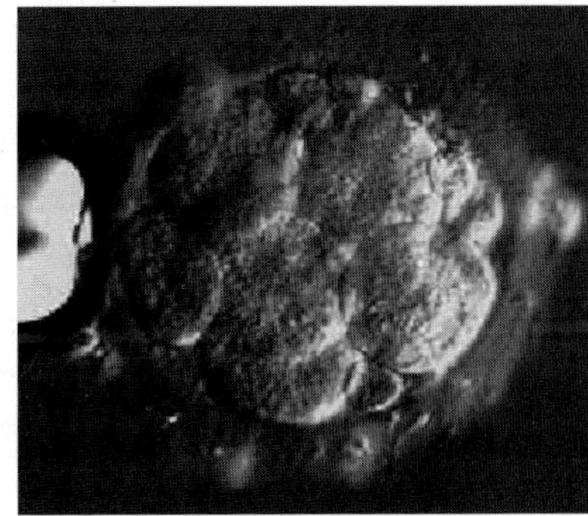

Fig. 13.4.   *Left*: An eight-celled human egg, with six of the cells clearly visible in this microscopic plane. None of the cells is fragmented. *Middle.* A moderately fragmented human egg. The outlines of its cells are much less distinct. *Right.* A severely fragmented human egg, with only one intact cell at 7 o'clock. The edges of all other cells are not clear.

(*Source:* Reproduced with permission from the publisher.) Reference 18.

protective zona pellucida, as shown in Fig. 13.1. Therefore, the zona pellucida must be removed and the trophoectoderm then eliminated to get to its inner cell mass — much like harvesting the milk of a coconut by first breaking its outer shell, then removing its "meat."

The zona pellucida can be dealt with rather easily by, literally, dissolving it with a digestive enzyme. Once the zona has been removed, the ring of trophoblastic cells is exposed. They are then eliminated by either of two methods, leaving only the inner cell mass, which — now isolated — can be extracted.[h]

---

[h]The outer ring of trophoblasts can be removed and the inner cell mass can be extracted by either of two methods microsurgery or immunosurgery.

Microsurgery — also known as microdissection — entails the use of a very-fine-tipped needle to open a blastocyst (by cutting its trophoectoderm), followed by the precise removal of its inner cell mass, while the blastocyst is observed under a microscope. Microsurgery yields inner cell masses that may be less "contaminated" by adjacent trophoblastic cells. However, the microdissection method is more time-consuming and costly than the alternative method.[21]

Immunosurgery makes use of antibodies to the ring of trophoblastic cells surrounding the inner cell mass, located at one end of the blastocyst. The protective zona pellucida is dissolved using a weak acid, exposing the trophoblastic cells. A drop of antibodies to the cells is added. This leads to irreversible damage to the outer circle of cells, while leaving the inner cell mass unharmed. It is then gently lifted, through the use of a micropipette, on to a culture dish, where it is allowed to propagate.[22]

The young inner-cell-mass cells are next cultured — i.e. grown — in Petri dishes in the laboratory. Stem cells "grow" by replicating themselves about every 30 h, not by getting bigger, in the manner of a growing child. Or, put another way, they grow by multiplying, rather that getting much larger in size. On a culture medium, they quickly spread over its surface, similar to a small town stretching on to nearby open space.

The bottom of a Petri dish is first covered with a thin "feeder layer" of cells known as fibroblasts.[i] Fibroblasts are large, undifferentiated cells (particularly) present in skin, muscle, and cartilage. Although embryonic stem cells can give rise to all other cell types when they are in a fertilized egg, they are critically dependent on a feeder layer for their growth in the laboratory. Fibroblasts are the key to their growth. Curiously, fibroblasts facilitate the growth of stem cells while preventing their "growing up" too soon. They do so by synthesizing and releasing a marvelous substance — fibroblast growth factor[j] — that initially promotes the spread of stem cells on the surface of a Petri dish and later prevents their differentiation into more mature, specialized cells. Fibroblasts also contribute to the growth of stem cells by depositing another substance, known as fibronectin[26] — the "sticky" com-

---

[i]The fibroblasts are typically generated from mouse embryos, introducing the possibility of contamination of human embryonic stem cells by feeder cells from a nonhuman source. Such contamination would preclude the use of human stem cells grown on a mouse-derived feeder layer for therapeutic purposes. In 2001, Ren-He Xu, a developmental biologist working in the same laboratory at the University of Wisconsin that first isolated human embryonic stem cells in 1998, developed a method to grow human embryonic stem cells in the absence of a mouse-derived feeder layer.[23] Instead of a murine feeder layer, a new synthetic medium known as Matrigel was used to grow human embryonic stem cells (Fig. 11.5).

[j]A growth factor is simply a chemical that stimulates cells to grow. In addition to fibroblast growth factor, which leads to the proliferation of fibroblasts (the principal cell type in connective tissue), there are other growth factors for cells in the liver, skin, and blood vessels, among others. While the first growth factor was not discovered until 1974,[24] more than 100 are now known to exist. Others will likely be found. Although the number of tissues is set at four, the number of growth factors may exceed four because a tissue may have more than a single growth factor.

Fibroblast growth factor has received considerable attention because it is closely involved with the regeneration of limbs in salamanders and newts and with tail regeneration in lizards.[24] In man, the regeneration of tissue is limited to the skin, blood cells, and cornea, and, to a lesser extent, the liver. If the secret of salamanders and lizards could be learned, its application to humans with lost or damaged limbs or with blindness due to lens cataracts would be an unimaginable blessing.

pound mentioned in footnote 57 . Since stem cells grow by spreading on a flat surface, it makes sense that they will grow more easily if they adhere to that surface, rather than being separated from it, even by a short distance. By adhering to the surface they gain the traction needed for mobility. But the number of contact points must be kept to a minimum, since increasing the number will immobilize the cell.[27]

The feeder layer itself serves two purposes. It provides a sticky surface[k] for the inner cell mass to attach itself to, and it releases nutrients into a separate, thin layer of culture medium just above the fibroblast feeder cell layer.[l,28] The overlying layer contains glucose, amino acids, and vitamins — at just the right pH — needed by the inner cell mass to grow. Together, the two layers provide a milieu similar to the environment a five-day-old egg would encounter after it becomes attached to the inner lining of the uterus and begins to derive the nutrients it needs to grow. The inner cell mass is then examined daily under a microscope. After 5–10 days pass, small "clumps" of cells become visible.[29] These clumps sprouting from an inner cell mass often form relatively thick, overlapping layers, making identification of individual cells difficult. To separate such clumps into their component cells, they are gently broken up using a very-fine-tipped needle under a microscope. The smaller groups of cells are then placed on fresh feeder layers and returned to the incubator. Over 7–10

---

[k]Most human cells require adhesion to an underlying tissue — a basement membrane — for their growth and survival. The undersurface of the skin, for example, is anchored to a layer of connective tissue by "sticky" compounds and tough filaments. The "stickiest" compound is a protein — fibronectin. Fibronectin is made by fibroblasts themselves. As such, it is a self-serving protein that helps fibroblasts adhere to one another, which, as the principal cells in connective tissue, is their function. Fibronectin molecules in the fibroblast feeder layer provide a scaffold for embryonic stem cells to grow on.

The synthesis of fibronectin is impaired when the skin is cut or burned, leading to slow healing of the damaged skin.

[l]In addition to these two functions, fibroblasts in the feeder layer synthesize and release a substance — fibroblast growth factor — that (paradoxically) inhibits the differentiation of embryonic stem cells into any of the 210 specific cell types in humans.[28] While the ultimate differentiation, of embryonic stem cells into specific cell types — such as cardiac, neural, or blood — is hoped for when stem cells are injected into damaged organs, their *premature* differentiation must be prevented in the laboratory to preserve them in their embryonic, unspecialized form. In the absence of inhibition, a pure population of stem cells will differentiate too soon and lose its potential to develop into healthy tissues.

How a "growth factor" prevents stem cells from differentiating is not known.

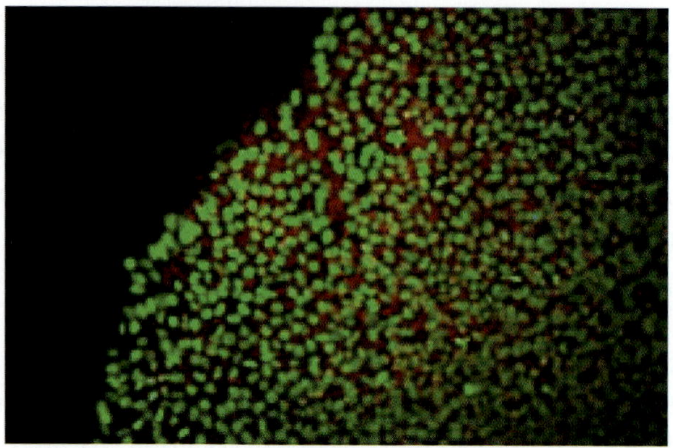

Fig. 13.5. A colony of human embryonic stem cells grown over a period of 10 months. The cell nuclei are stained green, while the cell surfaces are red. On the periphery, at 7–9 o'clock, where there is less overlap, several individual stem cells can be seen.

(*Source:* Reproduced with permission from the University of Wisconsin Stem Cell Research Laboratory.) Reference 31.

days, these cultured cells replicate enough times — they divide every 30 h — to form a new collection of cells, now referred to as a colony (Fig. 13.5). On the periphery, where the colony is thinner, individual, small cells with a large nucleus can be seen. These cells — finally — are embryonic stem cells.

Addendum

On November 27, 2007, the scientific world was rocked by twin reports tantamount to science fiction come true. Two independent "teams of scientists [from Kyoto University in Japan and the University of Wisconsin in Madison] reported that they had turned human skin cells into what appear to be embryonic stem cells without having to make or destroy an embryo — a feat that could quell the ethical debate troubling the field."[30] They did it "simply" by introducing four genes that "reprogrammed" the human cells' chromosomes "into blank slates that should be able to turn into any of the 220 cell types of the human body, be it heart, brain, blood or bone."

The achievement drew instant applause from leaders of the stem cell research field. Said Jamie Thomson, the University of Wisconsin scientist who first cultured human embryonic stem cells in 1998, "By any means we test them they are the same as embryonic stem cells."[30] Doug Melton, the coordinator of the Stem Cell Institute at Harvard, commented, "It is ethically uncomplicated." And Leonard Zon, director of the stem cell program at Boston's Children's Hospital, also affiliated with Harvard Medical School, chimed in, "It really is amazing."[30] Even President George W. Bush was "very pleased"![30]

There is a glitch connected with turning human skin cells into functional embryonic stem cells by inserting outside genes into a healthy cell's chromosomes. It is that the foreign genes are inserted into the host's chromosomes using a "retrovirus." A retrovirus is a virus wrapped in a protective shell that enables the virus to penetrate into a host cell's DNA, in this case human skin cells' DNA. The prefix "retro-" refers to the virus's ability to synthesize DNA from RNA, which is the reverse of the usual, DNA-to-RNA transcription. The human immunodeficiency virus, identified in 1984 as the cause of AIDS, is a good example of a retrovirus. The conversion enables the virus to enter the host cell's DNA. Once in the host cell's chromosomes, the retrovirus converts the host chromosomes into what is, in effect, a blank check that can be cashed by any of the body's 200-plus cell types.

One potential shortcoming of the introduction of foreign genes into human cells is the introduction of a cancer gene. However, *The Times* concluded, ". . .stem cell researchers say they are confident that it will not take long to perfect the method and that today's drawbacks will prove to be temporary."[30]

References

1. Sell S. (2004) Stem cells. What are they? Where do they come from? Why are they here? When do they go wrong? Where are they going? In: *Stem Cells Handbook.* Sell S. (ed.), Humana, Totowa, New Jersey, pp. 1–18.
2. http://stemcells.nih.gov/info/scireport/execSum.asp
3. Bongso A, Lee EH. (2005) Stem cells: their definition, classification and sources. In: *Stem Cells: From Bench to Bedside.* Bongso A, Lee EH (eds.), World Scientific, Singapore, Hackensack (NJ), London, India, Shanghai, Beijing, Taiwan, Hong Kong, Chap. 1.
4. Carlson BM. (2004) Placenta and extraembryonic membranes. In: *Human Embryology and Developmental Biology.* Carlson BM (ed.), 3rd ed. Mosby, Philadelphia, PA, pp. 43–63.
5. Brook FA, Gardner RL. (1997) The origin and efficient derivation of embryonic stem cells in the mouse. *Proc Natl Acad Sci USA* **94**: 5709–5712.

6. National Vital Statistics Report, 1999.

7. O'Rahilly R, Muller F. (2001) *Human Embryology and Teratology,* 3rd ed. John Wiley & Sons, New York, Chichester, Weinheim, Brisbane, Singapore, Chap. 3, pp. 19–35.

8. Carlson BM. (2004) Transportation of Gametes and Fertilization. In: *Human Embryology and Developmental Biology.* Carlson BM (ed.), 3rd ed. Mosby, Philadelphia, PA. pp. 27–41.

9. The human embryonic stem cell and the human embryonic germ cell. http://stem cells.nih.gov/info/scireport/chapter3.asp

10. *Ibid.,* Chapter 2.asp

11. Barth LG. (1953) *Embryology.* Holt, Rinehart and Winston, New York, Chicago, San Francisco, Toronto, London, pp. 369–401.

12. Carlson BM. (2004) Cleavage and implantation. In: *Human Embryology and Developmental, Biology.* Carlson BM (ed.), 3rd ed. Mosby, Philadelphia, PA. pp. 43–63.

13. Thomson JA, Iskovitz-Eldor J, Shapiro SS, Waknitz MA, Swiergiel JJ, Marshall VS, Jones JM. (1998) Embryonic stem cell lines derived from human blastocysts. *Science* **282(5391)**: 1145–1147.

14. Bongso A, Fong CY, Ng SC, Ratnam S. (1994) Isolation and culture of inner cell mass cells from human blastocysts. *Hum Reprod* **9**: 2110–2117.

15. Bongso A, Richards M, Fong C-Y. (2005) From human embryos to clinically compliant embryonic stem cells: blastocyst culture, xeno-free derivation and cryopreservation, properties and applications of embryonic stem cells. In: *Stem Cells: From Bench to Bedside.* Bongso A, Lee EH. (eds.), World Scientific, Singapore, Hackensack, (NJ), London, India, Shanghai, Beijing, Taiwan, Hong Kong, p. 146.

16. Chen H, Qian K, Hu D, Lu W, Yang Y, Wang D, Yan H, Zhang S, Zhu G. (2005) The derivation of two additional human embryonic stem cell lines from day 3 embryos with low morphological scores. *Hum Reprod* **20**: 2201–2206.

17. Cowan CA, Klimanskaya I, McMahon J, Atienza J, Witmyer J, Zuckur JP, Wang S, Morton CC, McMahon AP, Powers D, Melton DA. (2004) Derivation of embryonic stem-cell lines from human blastocysts. *N Engl J Med* **350**: 1353–1356.

18. Thomson J. (2003) The blastocyst and inner cell mass cells. In: *Human Embryonic Stem Cells: An Introduction to the Science and Therapeutic Potential.* Kiessling AA, Anderson SC (eds.), Jones and Bartlett, Sudbury (MA), Boston, Toronto, London pp. 95–107.

19. Steer CV, Mills CL, Tan SL, Campbell S, Edwards RG. (1992) The cumulative embryo score: a predictive embryo scoring technique to select the optimal number of embryos to transfer in an *in-vitro* fertilization transfer programme. *Hum Reprod* **7**: 117–119.

20. Gardner DK, Lane M, Stevens J, Schlenker T, Schoolcraft WB. (2000) Blastocyst score affects implantation and pregnancy outcome: towards a single blastocyst transfer. *Fertil Steril* **73**: 1155–1158.
21. Lin TP. (1969) Microsurgery of inner cell mass of mouse blastocysts. *Nature* **222**: 480–481.
22. Solter D, Knowles BB. (1975) Immunosurgery of mouse blastocysts. *Proc Natl Acad Sci USA* **2**: 5099–5102.
23. Xu C, Inokuma MS, Denham J, Golds K, Kundu P, Gold J, Carpenter MK. (2001) Feeder-free growth of undifferentiated human embryonic stem cells. *Nat Biotechnol* **19**: 971–974.
24. Baird A. (1994) Fibroblast growth factors: activities and significance of non-neurotrophin neurotrophic growth factors. *Curr Opin Neurobiol* **4**: 78–86.
25. Martin GR. (1998) The roles of FGFs in the early development of vertebrate limbs. *Genes Develop* **12**: 1571–1586.
26. Pollard TD, Earnshaw WC. (1996) Membrane structure and dynamics. In *Cell Biology*. Saunders, Philadelphia, London, New York, St Louis, Sydney, Toronto, 268.
27. Aota S, Nagai T, Olden K, Akiyama SK, Yamada KM. (1991) Fibronectin and integrins in cell adhesion and migration. Biochem Soc. Trans **19**: 830–835.
28. Schuldiner M, Benvenisty N. (2004) *In vitro* culture and differentiation of human embryonic stem cells. In: *Gene Targeting and Embryonic Stem Cells*. Thomson A, McWhir J (eds.), Garland Science/B105 Scientific, London New York, pp: 141–169.
29. Odorico JS, Kaufman DS, Thomson JA. (2001) Concise Review. Multilineage differentiation from human embryonic stem cell lines. *Stem Cells* **19**: 193–204.
30. Kolata G. (2007) Scientists bypass need for embryo to get stem cells. *The New York Times*, No. 27.
31. *http://www.news.wisc.edu/newsphotos/stemcell2005.html*

14

ADULT STEM CELLS

"Adult" Stem Cells Arise from Fetal Tissue

Although a baby is many years from becoming an adult, by all developmental and societal standards, a baby nonetheless has adult stem cells. This scientific–societal disjunction regarding the use of the word "adult" appears to be the result of the somewhat arbitrary use of "eight weeks" to distinguish between an embryo and a fetus, referred to in footnote b of Chap. 10. Thus, stem cells obtained at birth are considered to be " 'adult' stem cells since they are derived from nonembryonic sources such as bone marrow. . .umbilical cord blood. . .liver. . .skin. . .and other fetal. . .tissues."[1] But they're certainly not "adult" in the conventional meaning of the word.

Adult stem cells can be found in umbilical cord blood — routinely obtained at birth to establish an infant's ABO blood type — which indicates that they were also present in the fetus for months before birth. However, exactly when and how adult stem cells develop in the fetus is not known,[2] as surprising as that may sound. In fact, "no one knows the origin of adult stem cells in any mature tissue" ("mature" meaning "after birth"), the National Institutes of Health asserts on its website.[3] Their roots remain obscure in spite of their ubiquity — yet paucity — in the human body. Or, restated, even though adult stem cells exist in all human organs other than (possibly) the heart and the islet cell region of the pancreas, there are very few of them in any given organ. They are ubiquitous, but not prevalent. Even in the bone marrow where stem cells are most abundant, it is estimated by the NIH that perhaps one in 100,000 marrow cells is an adult stem cell.[3] That may be a best-case estimate by "Dr. Pangloss," since some authorities judge that "perhaps 1 in 10 billion marrow cells have such versatility."[4] But at least in a bone one knows where to begin looking for adult stem cells — in its marrow. And at least in a five-day old fertilized egg one also knows where to begin looking for its embryonic stem cells — in its inner cell mass. But when hunting for adult stem cells in most tissues, even a stem cell sleuth doesn't know where to begin the search.

Why "Adult" Stem Cells in Children. . .and Younger?

Apparently, the designation of "adult" stem cells arose from stem cell studies in the 1960s showing that stem cells from fetuses, newborns, and adults are limited (equally) in their capacity to form tissues to the single tissue in which they reside. This tissue specificity contrasts with embryonic stem cells, which can form any tissue in the body. Hence, to make a sharp distinction between embryonic stem cells and all other stem cells — including those found in newborn babies — one was called "embryonic" and the other "adult." The terms "embryonic" and "nonembryonic" might be less vivid, but they would be more accurate from both a scientific and a semantic point of view. Alternatively, the designations of "embryonic" and "tissue" stem cells might be used. However, since tissues begin to form during embryonic life, this would not be a wholly accurate distinction. In addition, "embryonic" refers to *when,* while "tissue" refers to *where* the stem cells are found. The terms "embryonic" and "nonembryonic" stem cells make the most sense, but inasmuch as "embryonic" and "adult" stem cells are used widely by the print and television media, they will also be used here for consistency — with the acknowledgment that "adult" stem cells form during fetal life and without another peep of protest.

Cells, Tissues, and Organs

Stem cells whether of embryonic or adult origin give rise to a variety of other cell types, which form tissues, constituting an organ. The human body harbors 220 cell types, 4 tissues, and 14 organs — not counting organs such as the organ of Corti and the organ of Zuckerkandl, which few people have ever heard of and still fewer care about. Yet, only 4 tissues exist, in spite of there being 220 individual cell types, because tissues are derived from only three *layers* of cells in the early embryo, while *individual,* specialized cells types are formed later, during embryonic development. The three layers — labeled simply as the inner layer (endoderm), the middle layer (mesoderm), and the outer layer (ectoderm) — then differentiate into the 20,000–25,000 genes

Table 14.1   The 3 layers of cells — the outer, middle, and inner layers — in a 1–2-week-old cmbryo give rise to 4 tissues forming 14 organs composed of 220 cell types, each containing 20,000–25,000 genes. Among the 220 cell types, every one has a specific function. For example, nerve cells conduct electrical impulses, heart cells contract, and digestive cells absorb nutrients.

3 layers
4 tissues
14 organs
220 cell types
20,000–25,000 genes

(Table 14.1). Among the 220 cell types, everyone has a specific function. For example, nerve cells conduct electrical impulses, heart cells contract, and digestive cells absorb nutrients.

Most adult organs are formed from more than one of the three embryonic cell layers and usually all three are present in a single organ. Of greater practical importance is the fact that all four adult tissues[a] — epithelial, connective, muscle, and neural ("nervous") — contribute to each organ. Thus, while the heart is composed mostly of (cardiac) muscle, it relies on nervous tissue to provide electrical impulses to make its muscle contract in an orderly, effective manner. In the absence of a biological pacemaker — the sino-atrial node — the heart would either beat at a rate of less than 30 beats per minute or beat very erratically — ventricular fibrillation. (At 30 or less beats per minute, most of us would be lightheaded when upright, and some not conscious, though still alive. Ventricular fibrillation means death, if not terminated quickly

---

[a]Epithelial tissue, connective tissue, muscle tissue and neural ("nervous") tissue are the four tissues that compose the human body. Epithelial tissue forms the epithelium, the single layer of cells that is in contact with the outside environment (i.e. air), including the outer layer of the skin, and the inner layer of the respiratory and digestive tracts. Connective tissue holds other tissues together, making them stronger. The inner layer of the skin, ligaments, and bone exemplify connective tissue. Muscle tissue, of course, forms the various contracting muscles of the body. There are three distinct types of muscle tissue: smooth muscle forms the middle layer of all arteries, skeletal muscle forms all the muscles attached to bone used every day for many ordinary activities, and cardiac muscle. Neural tissue makes up the central nervous system, including the brain, the spinal cord, and the many nerves enabling muscles to contact. Some biologists consider the blood to be such a highly specialized form of connective tissue that they designate it as the fifth tissue.[5]

through a jolt of electricity.) It is striking to see individual heart muscle cells contracting in a Petri dish (see http://whyfiles.org/189stem_cell/3.html for a short segment of a video showing spontaneously beating cardiac muscle grown from human embryonic stem cells in Thomson's Lab at the University of Wisconsin), but even groups of such contracting cells can't support life without a biological pacemaker prodding them into beating faster and more synchronously than they do in a culture dish. In addition to nervous tissue making heart muscle tissue contract efficiently, the muscle requires oxygen carried by the blood through blood vessels, both derived from a third tissue type — connective tissue. As a result of the complex interactions between different tissues forming the heart as an organ, the generation of heart muscle tissue to replace damaged muscle is not the same as creating a new heart. A few beating cardiac cells while an impressive start are a far cry from a beating heart.

Transdifferentiation: An Exciting Concept

Almost as if in response to the headlines grabbed by embryonic human stem cells at the time of their isolation in 1998, their adult counterparts snatched the limelight back a year later. After 30 years of use to revitalize the ailing bone marrow of patients with leukemia — through marrow transplantation — a new use for an old cell was discovered in 1999. In fact, the timing was a coincidence. However, the new development would not only rekindle interest in adult stem cells, it would also launch intensive investigations into how such mature cells could change their behavior in a way that contradicted a long-held belief.

A dogma of stem cell biology had been that once a stem cell had developed — had matured — into its adult, specialized form, it could not change its mind, so to speak, and adopt a new specialty. It could not "decide" — or even be coaxed — to change its destiny. Yet, in 1999, Dr. Christopher Bjornson and his associates at the NeuroSpheres Institute in Calgary reported just that.[6] They demonstrated that adult stem cells from brain tissue could, in fact, form red and white blood cells if the brain cells were introduced into the bone marrow.[b] A commentary in the same issue of *Science* pointed out that the new

---

[b]It is certainly reasonable to wonder why anything so ostensibly preposterous as turning brain into blood would even be considered as a possibility worthy of investigation. The incentive was a paper published in an obscure journal in 1995.[9] The investigators from Case Western Reserve University in Cleveland found that rat bone marrow cells could be transformed into muscle

study showed that "the brain can perform tasks of a completely different tissue."[7] Twenty-three months later, researchers at the National Institutes of Health (U.S.A.) reported, even more amazingly, that blood cells could be converted into brain tissue.[8] A true stem cell biologist might not find the conversion of blood into brain as astonishing as the reverse conversion, since bone marrow stem cells can give birth to a variety of mature blood cells — red blood cells, white blood cells, and platelets. This has been known for years. Yet, the formation of brain from blood seemed to contradict conventional wisdom among biologists in general that adults have an allotment of brain cells that can decrease over time, but not increase. To keep these phenomena within the range of plausibility, it is appropriate to point out that the blood–brain interconversions described in 1999–2000 were in mice. . .not humans. . . not yet, anyway.

The phenomenon of a specialized (differentiated) adult stem cell from one tissue adopting the characteristics of another tissue upon being relocated in a foreign tissue was termed "transdifferentiation." Soon a synonym was coined referring to the newly discovered ability of adult stem cells to change their fate — their "plasticity." While the metaphorical term is a reasonable one, its necessity seems questionable since "transdifferentiation" is really not difficult to pronounce and the word is in line with other common terms denoting a change of character, such as "transformed" or "transgendered." Besides that, use of the term "plasticity" with regard to the malleability of adult stem cells leads to use of the more unsettling phrase "plastic stem cells,"[12] which does take a stretch of the imagination. The idea of an adult stem cell adapting to its new environment — or being reprogrammed by its new surroundings if relocated in a different organ — is similar, in principle, to that of an adult immigrant — or even a visitor — to a foreign country adopting some of the cultural characteristics of his or her new residence. Perhaps he or she could be described as a "plastic person," but the word "transcultured" — if there were such a word — would somehow sound more dignified.

---

cells *in vitro* upon being exposed to a chemical compound known as 5-azacytidine. That observation led other investigators in Milan to wonder if the same transformation might occur *in vivo*. The Italian molecular biologists found that bone marrow cells from mice migrated into areas of damaged muscle where they "regenerated normal muscle".[10] That led to an accompanying commentary in the journal *Science* saying that the findings opened up "a whole area of potential therapies that didn't exist before."[11]

Semantics aside, the ability of adult stem cells to change their fate offers the hope of generating healthy tissues to repair tissues severely damaged by certain inherited or acquired diseases using cells from nonembryonic sources, thereby bypassing the ethical issue. And if stem cells from an individual with an inherited or acquired disease are "donated" to him- or herself, the issue of (probably) rejecting a truly donated tissue is also bypassed. From a functional perspective, if organ-specific (adult) stem cells can, indeed, cross over to form the tissue of another organ, then there would be "essentially no (functional) difference between organ-specific (i.e. adult) stem cells and embryonic stem cells."[13] If there were no functional difference between the two types of stem cells, that "would remove the need to collect stem cells from human embryos for clinical purposes, thus overcoming many of the political and ethical barriers to stem-cell therapy."[14]

Limitations of Adult Stem Cells

At the same time that the unexpected versatility of adult stem cells was being recognized, their limitations were being appreciated. Unfortunately, adult stem cells are hard to find in the human body and hard to grow in the laboratory. Although most organs have a few of these parent cells, they do not occur in any predictable location in any given organ.[15] Couple their uncertain whereabouts with their rarity in an organ the size of a kidney or the liver, and finding adult stem cells becomes a huge undertaking. Add to their scarcity the fact that adult stem cells look no different under a microscope than any of the other cells in a particular organ, and the task of finding even one in an organ becomes essentially impossible. The idiom about searching for a needle in a haystack comes to mind. In spite of the critical role adult stem cells play in tissue healing and regeneration, they blend in with the crowd of cells around them. They are, in effect, members of a royal family dressed in workingmen's clothes.

In order to tell whether or not a cell is a stem cell, its function has to be tested. Only a stem cell's *function* reveals its identity.[2,16] To establish a potential stem cell's function requires that it be scrutinized in the laboratory. And, unfortunately, that action "inevitably demands that the cell must be manipulated experimentally which may alter its [identifying] properties."[17]

The challenges posed in trying to identify adult stem cells in life are mimicked by those met in trying to grow them in the lab. As the NIH has pointed out, adult stem cells are not only difficult to "identify" [find], but they are also hard to "isolate and purify" [culture] in the laboratory with available technology.[18] Given the technical difficulty of growing adult stem cells, the government concluded that "an important limiting factor for the use of adult stem cells in future cell-replacement strategies is that there are insufficient numbers of [such] cells available for transplantation."[18]

Transdifferentiation: Suddenly a Controversial Concept

Just as it appeared that adult stem cells had attained near-equal respect with embryonic stem cells — even if not equal availability — a shoe dropped. It fell in the U.S. and the U.K., but was heard in Canada as well as at the N.I.H., where the initial reports claiming to show the conversion of blood into brain, and vice versa, had originated. In April 2002, back-to-back papers from the University of Florida[19] and two esteemed British universities Oxford and Edinburgh[20] showed that adult stem cells spontaneously fused with — literally merged with — embryonic stem cells (again in mice) to form a hybrid stem cell with more features of the latter stem cell than the former.[21] The principle of cell fusion with an exchange of nuclear DNA was not new. The original fusion of a male and a female nucleus to form a fertilized egg with a single nucleus on the first day of embryonic life is a poignant example of cell (and nuclear) fusion. And, as with adult and embryonic stem cell fusion, the end result of a sperm fusing with an egg is a hybrid — in this case a baby — with features resembling those of both parents. The unexpected application of cell fusion to embryonic and adult stem cells quickly prompted the need for a "reappraisal" of the transdifferentiation concept.[22]

The summer of '02 was not a happy time for adult stem cells. Month after month they took a beating in the pages of scientific journals from experts in the cell biology world who doubted the importance — even the existence — of transdifferentiation (Table 14.2; note the sharp change in sense after 2001). Other authorities came to the defense of the beleaguered cells. Adding to the spirited exchanges in the pages of highly respected journals over the issue of whether or not adult stem cells could change their colors, a story within a story unfolded, involving two of the principal investigators in the field. Both worked on the same university campus Stanford though in different departments.

Table 14.2　Article titles, 1998–2003.

Year	Title	Journal
1998	Embryonic Stem Cell Lines Derived from Human. . ..	*Science* **282**: 1145 (1998)
1999	Turning Brain into Blood	*Science* **283**: 534 (1999)
2000	Turning Blood into Brain	*Science* **290**: 1779 (2000)
2001	Can Stem Cells Cross Lineage Boundaries?	*Nat. Med.* **7**: 393 (2001)
2001	Stem Cell Potential: can Anything Make Anything?	*Curr. Biol.* **11**: R7 (2001)
2002	Studies Cast Doubt on Plasticity of Adult Stem Cells	*Science* **295**: 1289 (2002)
2002	Cell Fusion Causes Confusion	*Nature* **416**: 485 (2002)
2002	Plasticity: A Time for Reappraisal?	*Science* **290**: 2126 (2002)
2002	Is Transdifferentiation in Trouble?	*J. Cell. Biol.* **157**: 15 (2002)
2003	Transdifferentiation — Fact or Fiction?	*J. Cell. Biochem.* **8**: 829 (2003)

In one building, cell biologist Helen Blau promoted the idea of transdifferentiation. She did so having seen in late 2000 bone marrow cells injected intravenously, in mice, give rise several months later to cells with some functional "characteristics of central nervous system neurons," i.e. brain cells.[23] They had transdifferentiated. She proved the change by injecting bone marrow cells which had been labeled before injection with a fluorescent protein into the bloodstream. The purpose of the fluorescent protein was to make the bone marrow cells turn green — no matter where in the body they ended up — when they were looked for later through a microscope while under light from a laser source.

Sure enough, "several" months later some cells in the brain glowed green, indicating that somehow they had inherited the color of their parent cells from the bone marrow (Fig. 14.1). Not very many cells made the change, to be sure. In fact, a paper in the same issue of the journal confirming Dr. Blau's results found that only 5% of all brain cells were from the green bone marrow cells.[8] Nonetheless, although the number of cells that had changed their tissue type was small, the transformation vividly made the point that transdifferentiation can occur.

Meanwhile, across campus, Dr. Irving Weissman in the Department of Developmental Biology (cellular biology and developmental biology are closely related) placed just one green-labeled stem cell into the bone marrow of 22 mice, and then did nothing more until he examined the animals' bone

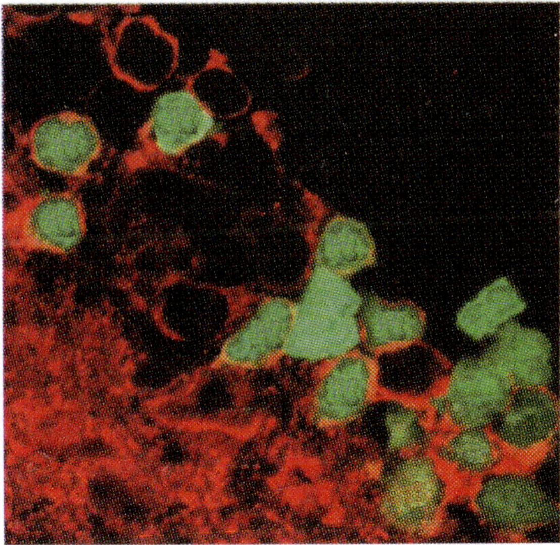

Fig. 14.1. A cluster of green cells with functional characteristics of brain cells. These central nervous system cells were derived from bone marrow cells injected several months earlier. (*Source:* Modified and reproduced with permission from the publisher.) Reference 23.

marrow and brain tissues under a laser light source three and one-half months later. Although the bone marrow of the recipient mice was heavily populated with fluorescent cells, only one brain cell of the many analyzed fluoresced green (Fig. 14.2).[24] The conclusion of the cross-campus group was that "transdifferentiation. . .is an extremely rare event, if it occurs at all" (Table 14.3).

That conclusion was soon supported by a paper from the University of Toronto in 2002.[25] The authors wrote that in spite of infusing more than "12×10^6 neural stem cells" intravenously in 128 mice, "we never observed" transdifferentiation. A guest commentary in the same issue of *Nature Medicine* pointed out that despite "extensive attempts to replicate the [1999 reports of transdifferentiation]. . .their experiments failed to show a single event of [transdifferentiation]."[26] Even so, as one of the Toronto researchers graciously put it, although "our own data fail to replicate transdifferentiation. . .there are so many reports out there, [that] I am unwilling to believe all of them are false."[21]

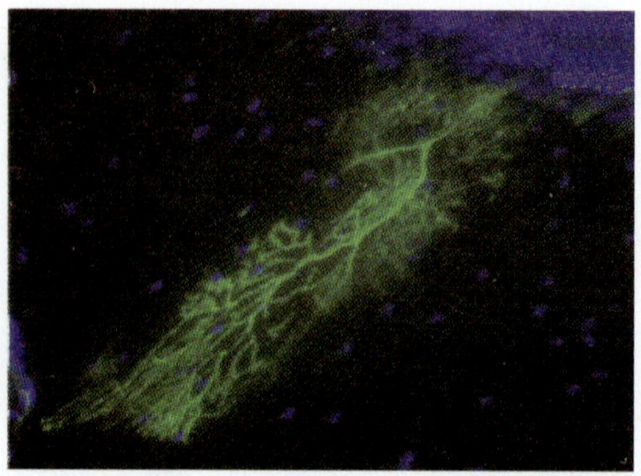

Fig. 14.2.   The single branching nerve cell among 60 sections of the brain analyzed several months following bone marrow transplantation. This was the only instance of transdifferentiation.

(*Source*: Modified and reproduced with permission from the publisher.) Reference 24.

Table 14.3   Concerns about the significance of transdifferentiation.

Low frequency
In consistent reproduction
Cell fusion

How Does an Adult Stem Cell Transdifferentiate?

To address the second question regarding *how* transdifferentiation of an adult stem cell occurs, the critical related issue raised by Dr. Weissman of *whether* transdifferentiation occurs should logically be taken into account first. Concerning the whether-it-occurs subject, Dr. Weissman later suggested that "five very stringent conditions be met" to verify transdifferentiation.[27] First, that "one cell of a specific tissue lineage" be able to convert "into a different lineage" without assistance from any other cell in the tissue. Second, that "'reprogramming' of the transdifferentiated nucleus. . .be silenced. . .[while] genes specific to the new cell fate are activated." Third, that the cell be "minimally manipulated" during culture since manipulation could affect the cell's "gene expression profile and/or chromosome configuration." Fourth, that "cell fusion" be excluded. And fifth (a criterion no one would likely disagree with) that the "experiment. . .be independently replicated by more than one laboratory using more than one experimental model."

Let us to paraphrase these proposals for proving transdifferentiation. First, that a *single* cell — not one of 50 cells that look alike in a given tissue — be shown to give rise to cells of a totally different tissue. This is important, because a group of lookalike cells may behave differently than its individual members, just as a crowd of people may act differently than its members act when alone. Second, that cells in the new tissue function differently than they did in the original tissue, not only look differently. (Although "individuality" was at first an asset that made a single stem cell stand out from the crowd, now the new cell should blend in with its new neighbors.) Third, that the single cell being evaluated for its transdifferentiation potential be treated with TLC in the laboratory to minimize any harm to its genes and chromosomes. Fourth, that cell fusion be ruled out to avoid the possibility that two cells join to form a new one, rather than that one cell adopts a different function to become a "new" one. And fifth, that the study be confirmed by another person or persons for common-sense reasons. The transdifferentiation bar is high enough that "only some of these criteria have been met in the avalanche of reports that have been generated by the 'stem cell plasticity field'. . .and none of the (*in vivo*) reports meets the rigorous criteria for plasticity. . .."[2]

The complex question of *how* transdifferentiation happens at the molecular level is easy to answer — no one knows.[c] Interesting possibilities exist, however. For example, an adult stem cell might regress in its chameleon-like behavior to that of a less mature cell. After all, an adult stem cell itself came from an even more primitive cell — presumably one of the cells within the clump of cells forming the inner cell mass of a five-day-old embryo. And that group of cells came from a one-cell, fertilized egg hours after a sperm joined an egg. Thus, the less developed a cell is the more developmental choices it has. Newts and salamanders as mentioned in footnote b of Chap. 13, figured out how to go back in developmental time to a more primitive cell with the ability to generate a specialized cell, which could then regenerate a limb, a tail, or an eye. So maybe an adult stem cell really doesn't transdifferentiate. Maybe it "dedifferentiates" from a more exclusive set of cells into a less "prejudiced" cell — akin to societal desegregation.

Finding the Few Adult Stem Cells That May Transdifferentiate

A number of the "how can adult stem cells transdifferentiate?" issues reraise the question brought up much earlier regarding where adult stem cells are in

---

[c]The answer to the equally interesting question of what happens at the molecular level to keep an adult stem cell from transdifferentiating until the "right" time is not known either.

the body and how they got to where they are. There are, essentially, two places where they may be, each reflecting the mobility or immobility of an adult stem cell. The answer could be "either place" or "both places." They may lurk in tiny nests of embryonic stem cells waiting in each organ for the "right time" to grow up. (And if that's the case, a related question is: How in the world are they kept in such an untainted state for so long in the presence of specialized, adult stem cells all around them in an organ?) Or, they may circulate in the blood for a long time before taking up permanent residence in a particular tissue, where they finally have children that grow up acting like all the other kids in the neighborhood.

There is yet a third possibility apropos of where adult stem cells are in the body — and how they got there to begin with. Could they be neither static in a tissue nor constantly in motion in the blood? Could they spend their early days tucked safely in the tissue of their origin, then leave the nest during their adolescence to migrate through the bloodstream and years later take up final residence in another tissue? Given that scenario, they might start out with their indigenous genes, then modify their genes' function during their middle years and later fine-tune their midlife changes for existence in one last tissue.

Such a peripatetic, changing life is consistent with the principle that the essence of a stem cell — its nucleus — can't change or be permanently inactivated. But it can be modified, i.e. reprogrammed. Any nucleus "exposed to the appropriate constellation of proteins" — meaning the type and number of proteins — in its cytoplasm should be able to redirect its behavior.[28] Mechanistically, its nuclear genes can be silenced or aroused by different tissues *or* by the same tissue under new circumstances (such as injury) *or* even at different times during the life of the cell, no matter where it may be. Modifying a gene's function without changing its chromosomes is the biological basis of transdifferentiation. How the genes in an adult stem cell are reprogrammed is unknown.

But several facts are known about adult stem cells: they are rare, their transdifferentiation is rarer, and the mechanism or mechanisms underlying their change from one cell type to another are mysterious. Although they are rare in a given tissue — 1 in 100,000 to 1 in 10 billion bone marrow cells, where they are most common — the good news is, the brain, the cornea, and the retina made the list of tissues blessed with adult stem cells. The bad news is, the heart didn't make it. (See Chap. 24 for a discussion on cardiac stem cells.)

A Stem Cell Highway

To exemplify an adult stem cell's organ-to-organ excursion through the bloodstream, Dr. Blau has suggested a "stem cell highway" paved with blood (Fig. 14.3). One or more tissue stem cells could enter the highway at certain times over a lifetime using on-ramps, travel in the bloodsteam until they are needed in another tissue, and then leave the bloodstream using off-ramps to

Fig. 14.3. This diagram depicts adult stem cells (blue disks) rolling down a stem cell highway (the bloodstream). The cells enter the highway using on-ramps serving various tissues and exit the highway using off ramps to enter a different tissue, where they adopt the function of that tissue without changing their appearance.

(*Source;* Reproduced with permission from the publisher.) Reference 16.

make a final stop in another tissue. Once at their destination, they come in contact with cells residing in that organ, which induce them to take on the characteristics of that organ and to participate in its function. Such adaptive behavior is akin to that of immigrants — or even visitors — to a foreign country taking on certain characteristics of and participating in some functions of their new residence, as mentioned earlier.

The ability of adult stem cells to migrate and to repair tissue damage in a distant organ was confirmed in 1998. The first report showed that cells in the bone marrow — of mice — could be converted into muscle in mice with too little muscle,[10] and it was based on a report three years earlier showing *in vitro* that bone marrow cells could form muscle cells under certain laboratory conditions, even when the cells were removed from the animal.[9] From that basic laboratory work sprang elegant studies by cell biologists around the world providing evidence that blood (from humans) can form brain, lung can form brain, liver can form muscle, and vice versa.[15,16,29]

The Issue of "So What If Adult Stem Cells Differentiate in a Laboratory?"

More to the point than what adult stem cells do in a laboratory is what they do in the body. From a purely academic standpoint, the impact of transdifferentiation and nuclear fusion has been profound. From a therapeutic standpoint, the question of whether cells change their behavior by two cells fusing to form one new one (with 46 chromosomes) or one cell "dividing" to form another with a different function (with the same 23 chromosomes) is a less weighty issue. More to the point is whether useful tissues can be generated from adult stem cells, even ones from another tissue.

Judging from studies in mice, it looks like they can. And observations in humans seem to support the conversion of one cell type to another, just as in mice. In transgenic mice with an enzyme deficiency causing liver failure — akin to alpha$_1$-antitrypsin deficiency in man leading to liver failure — the intravenous injection of bone marrow stem cells led to the generation of enough healthy liver tissue after seven months to correct their liver function tests and produce islands of normal, fluorescent tissue under a microscope.[30] While the laboratory success was later found to be due to cell fusion, rather than transdifferentiation,[31,32] the exciting news was, it worked. It worked in humans as well. Bone-marrow-derived stem cells differentiated into functional liver cells in a total of 26 men and women from the U.S. and the U.K.

undergoing bone marrow transplantation.[33,34] As the Yale and New York University pathologists and surgeons put it, although "it is impossible to know of the human relevance of [laboratory] findings without studying human beings directly. . .we conclude that [transdifferentiation] of bone marrow cells into liver cells does in fact occur in humans."[34]

References

1. Shamblott MJ, Sterneckert JL. (2005) Characteristics of human embryonic stem cells, embyonal carcinoma cells and embryonic germ cells. In: *Human Embryonic Stem Cells.* Odorico J, Zhang S, Pedersen R (eds.), BIOS Scientific, Abingdon, Oxon, New York, pp. 29–44.

2. Doyonnas R, Blau HM. (2004) What is the future for stem cell research? Whether entity or function? In: *Stem Cells Handbook.* Sell S (ed.), Humana Totowa, New Jersey, pp. 491–499.

3. http://stemcells.nih.gov/info/scireport/chapter4.asp

4. Vogel G. (2001) Can old cells learn new tricks? *Science* **287**: 1418–1419.

5. Cooper GM. (2000) *The Cell: A Molecular Approach,* 2nd ed. ASM Washington, DC, pp. 3–39.

6. Bjornson CRR, Rietze RL, Reynolds BA, Magli MC, Verscovi AL. (1999) Turning brain into blood: A hematopoetic fate adopted by adult neural stem cells *in vivo. Science* **283**: 534–537.

7. Strauss E. (1999) Brain stem cells show their potential. *Science* **283**: 471.

8. Mezey E, Chandross KJ, Harrta G, Maki RA, Mc Kercher SR. (2000) Turning blood into brain: cells bearing neuronal antigens generated *in vivo* from bone marrow. *Science* **290**: 1779–1782.

9. Wakitani S, Saito T, Caplan AI. (1995) Myogenic cells derived from rat bone marrow mesenchymal cells exposed to 5-azacytidine. *Muscle Nerve* **18**: 1417–1426.

10. Ferrari G, Cusella C, Coletta M, Paolucci E, Stornaivelo A, Cossu G, Mavillo F. (1998) Muscle regeneration by bone marrow–derived myogenic progenitors. *Science* **279**: 1528–1530.

11. Pennisi E. (1998) Muscle disease: bone marrow cells may provide muscle power. *Science* **279**: 1456–1457.

12. Pera J. (2005) In: *Stem Cells: From Bench to Bedside.* Bongso A, Lee EH. (eds.), World Scientific, Singapore Hackensack, (NJ), London, p. 63.

13. Anderson DJ Gage FH, Weissman IL. (2001) Can stem cells cross lineage boundaries? *Nat Med* **7**: 393–395.

14. Wurmser AE, Gage FH. (2002) Cell fusion causes confusion. *Nature* **416**: 485–487.

15. Rosenthal N. (2003) Prometheus's vulture and the stem-cell promise. *N Engl J Med* **39**: 267–274.

16. Blau H, Brazelton TR, Weimann JM. (2001) The evolving concept of a stem cell: entity or function? *Cell* **105**: 829–841.

17. Loeffler M, Roeder I. (2002) Tissue stem cells: definition, plasticity, heterogenicity, self-organization and models — a conceptual approach. *Cells Tissues Organs* **171**: 8–26.

18. http://stemcells.nih.gov/info/scireport/execSum.asp

19. Terada N, Hamazaki T, Oka M, Hoki M, Mastalerz DM, Nakano Y, Meyer EM, Morel L, Petersen BE, Scott EN. (2002) Bone marrow cells adopt the phenotype of other cells by spontaneous fusion. *Nature* **416**: 542–545.

20. Ying Q-L, Nichols J, Evans EP, Smith AG. (2002) Changing potency by spontaneous fusion. *Nature* **416**: 545–548.

21. Vogel G. (2002) Studies cast doubt on plasticity of adult stem cells. *Science* **295**: 1989–2001.

22. Vogel G. (2002) Plasticity: a time for reappraisal? *Science* **290**: 2126.

23. Brazelton TR, Rossi FMV, Keshet GI, Blau HM. (2000) From marrow to brain: expression of neuronal phenotypes in adult mice. *Science* **290**: 1775–1779.

24. Wagers AJ, Sherwood RI, Christensen JL, Weissman IL. (2002) Little evidence for developmental plasticity of adult hematopoietic stem cells. *Science* **297**: 2256–2559.

25. Morshead CM, Benveniste P, Iscove NN, van Der Kooy D. (2002) Hematopoietic competence is a rare property of neural stem cells that may depend on genetic and epigenetic alterations. *Nat Med* **8**: 268–273.

26. D'Amour KA, Gage FH. (2002) Are somatic stem cells pluripotent or lineage-restricted? *Nat Med* **8**: 213–214.

27. Wagers AJ, Weissman IL. (2004) Plasticity of adult stem cells. *Cell* **116**: 639–648.

28. Blau HM, Baltimore D. (1991) Differentiation requires continuous regulation. *J Cell Biol* **112**: 781–783.

29. Morrison SJ. (2001) Stem cell potential: Can anything make anything? *Curr Biol* **11**: R7–R9.

30. Lagasse E, Connors H, Al-Dhalimy M, Reitsma M, Dohse M, Osborne L, Wang X, Finegold M, Weissman IL, Grompe M. (2000) Purified hematopoietic stem cells can differentiate into hepatocytes *in vivo*. *Nat Med* **11**: 1229–1234.

31. Wang X, Willenbring H, Akkari Y, Torimaru Y, Foster M, Al-Dhalimy M, Lagasse E, Finegold M, Olson S, Groupe M. (2003) Cell fusion is the principal source of bone-marrow-derived hepatocytes. *Nature* **422**: 897–901.

32. Vassilopoulos G, Wang P-R, Russsell DW. (2003) Transplanted bone marrow regenerates liver by cell fusion. *Nature* **422**: 901–904.

33. Alison MR, Poulsom R, Jeffrey R, Dhillon, AP, Quaglia A, Jacob J, Novelli M, Prentice G.Williamson J, Wright NA. (2000) Hepatocytes from non-hepatic adult stem cells. *Nature* **406**: 257–260.
34. Theise ND, Nimmakayalu M, Gardner R, Illei PB, Morgan G, Teperman L, Henegariu O, Krause DS. (2000) Liver from bone marrow in humans. *Hepatology* **32**: 11–16.

15
TRANSGENIC MICE

The term "transgenic" implies that a foreign DNA is transferred into the DNA of a host organism. Or, restated, the term implies that DNA is *transferred* from *genes* residing in the chromosomes of a donor organism to genes of a recipient organism. The host may be as small an organism as a tiny, soil-dwelling worm known as *Caenorhabitis elegans* (not the garden earthworm) or the common fruit fly, known as *Drosophila melanogaster*.[1] But it may also be a larger animal, such as a dog, cat, rabbit, pig, sheep, or even mouse. In fact, mice have emerged as the principal laboratory animal currently used for the introduction of outside DNA into a host animal. Dogs and cats are infrequently used, because they raise ethical issues and emotions, being common household pets. Ethical considerations also limit the potential for application of transgenic technology to human subjects, for self-evident reasons.

Methods Used to Create Transgenic Mice

Although the technology used to create transgenic mice is somewhat esoteric, it is also quite comprehensible. There are two ways to make a mouse whose genes differ from those of another mouse: a gene can be *added* — to a fertilized egg — or a gene can be *replaced.* When a gene is replaced, another gene is "added," in the sense that it is inserted. But the total number of genes remains the same, since the deleted gene is replaced by another gene. The two methods will be explained in the order they were introduced to science. But, first, the focus of both techniques — DNA — will be considered.

DNA: The Universal Determinant of Human Genetic Material

DNA is, in effect, the computer of the human body. It would be presumptuous of a cardiologist trained primarily in the diagnosis and treatment of clinical disorders to write about the burgeoning field of genetics. It has become a vast specialty impacting virtually all other disciplines of medicine. In deference to its scope and importance, only a smidgen of the massive amount of published information related to genetics will be mentioned here. (Appendix A records several books devoted to the subject. These books vary in depth, allowing their classification in a range from a primer to a reference piece.) DNA stores information that can be later downloaded to either generate new cells or replace old cells as they reach their life expectancy. Many cells are replaced days (red blood cells) to months (liver) after they are first synthesized — or sustain damage (prematurely) by disease or injury. Our skin heals after being cut by a razor blade; the lining of our stomach heals after being irritated by the ingestion of certain foods (or alcohol). Both the repair and the re-creation of cells are the result of using data stored in the DNA archive.

Each DNA molecule is a huge structure consisting of many individual units linked sequentially like beads on a string. Each *bead* — referred to as a nucleotide in the jargon of molecular biology — is made up of only five elements, namely oxygen, hydrogen, nitrogen, carbon, and phosphorous, all plentiful in nature. These elements, in turn, are joined together to form four relatively uncomplicated compounds, namely adenine, cytosine, guanine, and thymine, constituting the beads (Fig. 15.1). Each *string* — the string itself is also composed of the same four compounds, arranged differently — is made up of many beads connected at specific sites, much like boxcars coupled on a freight train. And each *coupling site* is exactly the same, regardless of how long the string — the DNA chain — turns out to be. It's that simple.

What's not so simple is how the DNA becomes packaged into smaller units known as genes. The DNA molecule is simply too large to exist in one long, continuous strand. Therefore, it is divided into smaller segments, i.e. genes, themselves carried on chromosomes. Normally, in humans, there are 46 H- or X-shaped chromosomes in a cell's nucleus that carry the cell's genes (Fig. 15.2). All 46 chromosomes are the same except for the crucial sex chromosome called XY in males and XX in females, to underscore their difference. Each chromosome contains 20,000–25,000 genes and each chromosome has a match — a copy — yielding 23 pairs of chromosomes. Each of the copies is said to be *homologous*, i.e. of the same (*homo*) legend (*logos*) or kind. And each of the two copies exists as a single string — a strand — with the two strands interconnected in a spiral manner (Fig. 15.3) known as a double helix, in which each successive loop is at a constant distance from the last, in the

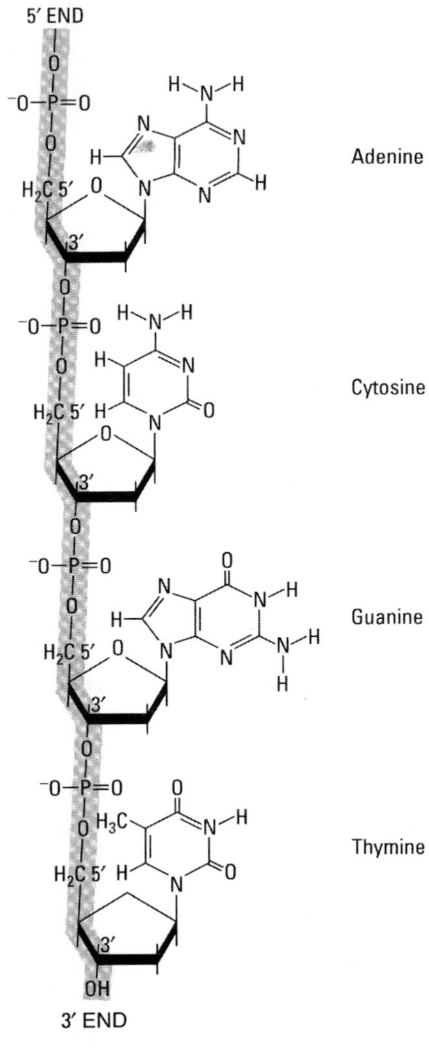

Fig. 15.1. Structure of a DNA bead. Each bead consists of four bases — adenine, cytosine, guanine, and thymine — linked together by the same coupling unit (shaded green). Adenine always combines with thymine to form a base pair, while cytosine always combines with guanine to form another base pair.

(*Source*: Reproduced with permission from the publisher.) Reference 2.

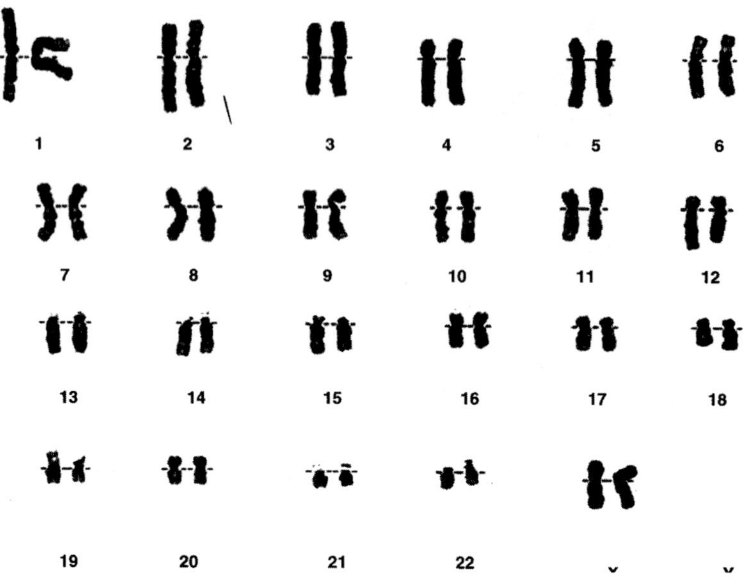

Fig. 15.2.	A set of human chromosomes.

(*Source:* Reproduced with permission from the publisher.) Reference 3.

manner of a spring.[4] Finally, the two strands are twisted together precisely in the style of a Swiss braid, not haphazardly, like two tangled telephone cords.

In the exact same manner, each gene "codes"[a] for a specific polypeptide[b] in a DNA strand. Each gene also occupies a specific place on a given chromosome. And, as the functional unit of heredity, each gene reproduces itself every time the cell divides into "daughter" cells. A second distinctive characteristic of genes is the "one gene–one enzyme" rule. Or, restated, each of the in excess of 50,000 enzymes in the human body has its own gene regulating its synthesis. And, finally, cells have the machinery to turn genes on and off and to splice them into a remarkable storage-design system — like a 21st century computer.

---

[a]The term "code" in genetic terminology refers to the system whereby a particular combination of three successive nucleotides in a DNA molecule controls the insertion of specific amino acids in the "right" place of a protein molecule, such as an enzyme. As such, a genetic code is analogous to a communication system code wherein printed or spoken words are transmitted in a secret manner or printed words are transferred onto a punchcard for storage purposes.
[b]A polypeptide is a (usually) large molecule formed by the union of many amino acids by so-called peptide links (–HN–CO–) composed of the same four elements.

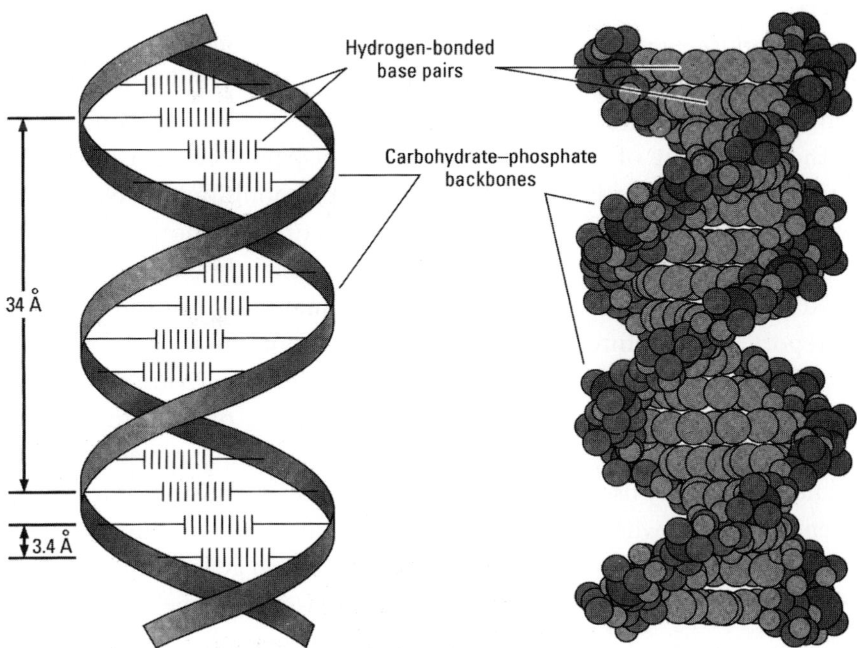

Fig. 15.3. A three-dimensional model of DNA. *Left:* The two green strands are twisted in a spiral manner to form a double helix, akin to a spiral staircase. *Right:* The parallel strands are linked by *base pairs* (gray) running through the space between the two chains. A base pair, in the language of molecular biology and genetics, refers to a pair of nucleotides that are connected by a simple hydrogen bond. In spite of the enormous size of a DNA molecule, there are just four bases: adenine, thymine, guanine, and cytosine. It is a law — dictated by spatial considerations — of genetics that adenine always pairs with thymine while guanine always pairs with cytosine. These four molecules encode the genetic information encrypted in the genetic code.

(*Source:* Reproduced with permission from the publisher.) Reference 2.

Background

The first report of modifying later-born mice by the injection of foreign DNA into a single-celled, fertilized egg[c] appeared in 1980.[5] In that paper, DNA from

---

[c] Such single-celled eggs are not synonymous with stem cells. Nor are they the immediate source of stem cells. Stem cells appear five days later — in the mouse — after the one-celled egg has divided 5 or 6 times, yielding 32–64 cells. The resulting stem cells are grouped at one pole of the fertilized egg in the inner cell mass, as shown in Fig. 11.1.

bacteria — yes, bacteria, too, have DNA — was injected into one-day-old, fertilized mouse eggs. Only 2 of the 87 mice whose eggs had been injected went on to produce newborns with the foreign genes, reflecting both the (often) lethal effect of outside DNA on recipient DNA and the small chance of a short segment of injected DNA being incorporated at a specific site within the much larger DNA molecule. Although nonhuman DNA — DNA from bacteria — was used to modify the structure of another nonhuman DNA — from mice — it nonetheless became instantly clear that it might (someday) be possible to replicate the feat in man. The bottom line of the 1980 report was that foreign DNA *could* be incorporated into the genes of a mammalian organism at a very early stage of its development.

While the initial report — from Yale University — of inserting a foreign gene into the pronucleus[d] of a single-celled egg staggered the scientific world,[e] it did not specifically use the term "transgenic mice." Coining that term didn't happen until 1983, when the authors of the original paper defined transgenic mice as "mice into which have been transferred cloned genetic material."[6] The three-year-later report was, in effect, a step-by-step recipe for producing transgenic mice. It included particulars ranging from the time of day the lights in the mice cages were turned on and off (8 a.m. and 10 p.m.) to a description of how to handmake the crucial holding pipette used to fix the position of the single cell under the microscope for microinjection of DNA into its pronucleus. Since gene addition to create a transgenic mouse is made easier to understand through figures, an illustration and a detailed account of the popular microinjection technique will follow.

Technique

Following the mating of a male and a female mouse mate (Fig. 15.5), the one-celled, fertilized eggs are flushed — within 12 h — from the Fallopian tubes,

---

[d]The "pronucleus" refers to either of the two nuclei — one from the male, one from the female — in a single-celled, fertilized egg. The prefix "pre-" — rather than "pro-" — could be used, since both denote "before." Within hours, the two pronuclei fuse, forming a single nucleus containing all the genetic information from both parents.

[e]The impact of the first paper describing the production of mice with modified genes was magnified by a startling report two years later.[7] In it, the authors injected growth-hormone-molecule genes from a rat into the pronucleus of fertilized eggs from mice. Mice with the inserted gene grew to twice the size of those without it (Fig. 15.4).

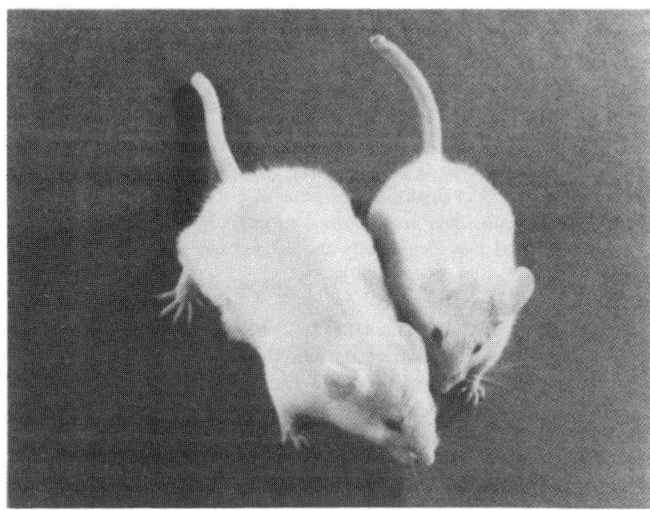

Fig. 15.4. The mouse on the left was developed by injecting a rat gene-regulating growth hormone into the pronucleus of a fertilized mouse egg. The mouse on the right did not receive the rat growth hormone gene.

(*Source*: Reproduced with permission from the publisher.) Reference 7.

after the tubes have been surgically removed[f] from the "egg donor" and placed under a microscope.[8] Characteristically, 10–20 eggs are obtained from each female.

While an egg is held in place by a pipette under slight, negative pressure (i.e. suction), it is injected with several hundred copies of the short DNA segments being studied suspended as they are in a tiny quantity (1 picoliter) of a solution adjusted to match body pH and temperature, held in a separate micropipette.[g] Both the holding and the injection pipette are precisely and mechanically

---

[f]Some laboratories prefer to procure the oviducts — Fallopian tubes — following sacrifice of the egg donor,[9,10] possibly to the objection of animal rights groups.

[g]The DNA to be injected into the pronucleus of a fertilized egg is prepared by incubating complete DNA molecules with a "restriction enzyme" that "cuts" the long DNA chain into much shorter segments — fragments. Such fragments are then further purified by a relatively straightforward (chemical) extraction–centrifugation process to yield about 2000 copies/pl of the specific genes — transgenes — to be microinjected. A large number of copies are injected into the egg, so that if a "near-miss" of the pronucleus occurs, there are a sufficient number of copies nearby to ensure insertion of the injected DNA into the egg DNA.[11] The preparation of recombinant DNA has become a routine procedure when done by specialized companies.[12]

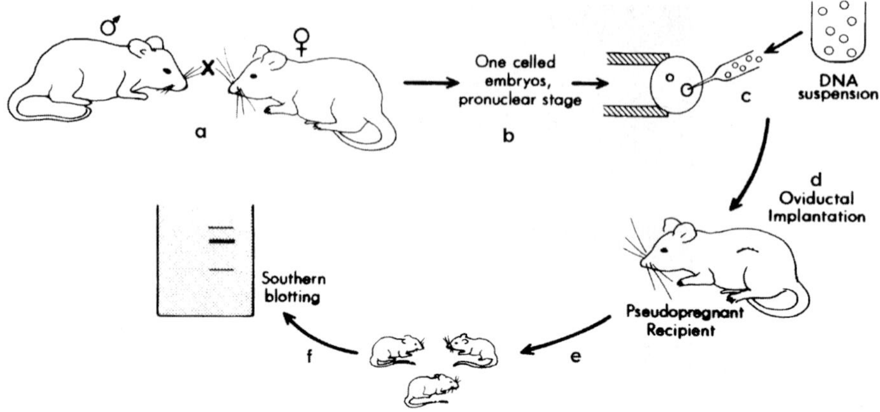

Fig. 15.5. (a) A female mouse is hormonally induced to make 30–40 eggs. She is then mated with a male mouse. (b) One-celled embryos are flushed from the Fallopian tubes and placed in a drop of medium under a microscope. (c) While an embryo is held in place by suction, the larger, male pronucleus is microinjected with several hundred molecules of DNA in suspension. (d) The embryo is then transferred into the oviduct (Fallopian tube) of a foster mother (pseudopregnant recipient), where it develops (e) till the birth of a pup. (f) After birth, the presence of the injected DNA is sought through a procedure known as Southern blotting. The presence of the foreign (injected) DNA establishes the existence of a transgenic mouse.

(*Source*: Modified and reproduced with permission from the publisher.) Reference 8.

controlled using a "micromanipulator," the essential piece of equipment for microinjection of eggs. The device permits simultaneous fixation of the egg by one pipette as the egg is injected with DNA held in another pipette (Fig. 15.6). At this very early point in time following coitus — less than 12 h — the fertilized egg has two nuclei: one from the male and one from the female mouse. The nucleus from the male is usually injected with DNA, simply because it is bigger and hence easier to inject. The two pronuclei will soon fuse to form one nucleus carrying genes from both parents (see footnote k of Chap. 13).

Mice, unlike humans, normally produce 10–12 eggs[13] at a time and can be induced — by fertility hormones — to make up to 20 at once.[14] Of these, an average of 15 eggs will be fertilized at the same time in any given female mouse. Although an unthinkable number in human terms, 15 is still not enough fertilized eggs to extract sufficient DNA for the microinjection of foreign DNA. Therefore, fertilized eggs from 10 mice are used to obtain

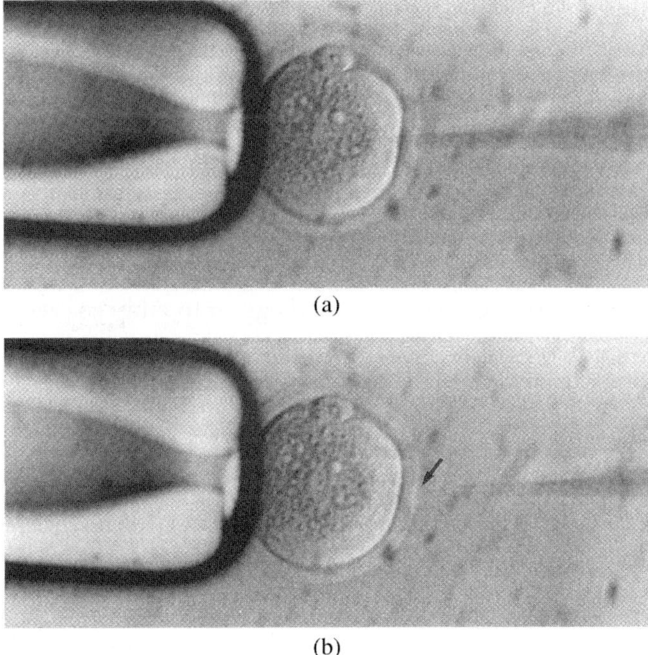

(a)

(b)

Fig. 15.6. Pronuclear injection of a one-celled, fertilized egg. (a) While the egg is held by slight suction applied through a holding pipette on the left, the male pronucleus (enclosed in the solid circle) of a single-celled, one-day-old, fertilized egg is injected with DNA through a microneedle. The male pronucleus is used because it is larger than the female pronucleus, (enclosed in the dashed circle). (b) When slight swelling of the nucleus is observed, the needle is withdrawn. The point of entry into the zona pellucida produces a dimple (arrow).

(*Source:* Reproduced with permission from the publisher.) Reference 11.

approximately 100 one-celled eggs for microinjection. Such a large number is required because only 10%–20% of eggs will survive the injection[h] and of those that do only 1%–5% will incorporate the foreign DNA into the host DNA. (The "inadvertent removal of pronuclear DNA" — due to adherence of DNA to the microneedle upon withdrawal of the needle — is the most

---

[h]Survival of microinjected, single-celled eggs is determined within 30 min of injection simply by their color under a microscope. Healthy cells retain their pink color. Nonsurviving cells become pale.[11] The color of a cell is used widely in pathology to infer its viability, obviating the need for more expensive methodologies.

frequent cause of embryo death.[11] Consequently, only 1–5 out of 100 mice will end up forming an egg that results in a genetically modified mouse.[15] The surviving eggs are then transferred into the Fallopian tubes of a surrogate mother. Customarily, 15 eggs are implanted into each Fallopian tube of the anesthetized foster mother through a small incision in the animal's back. Only one incision is necessary — in the midline — because it can be easily shifted, using forceps, to overlie either the left or the right ovary. The incision provides a movable window to visualize both ovaries (Fig. 15.7).

Because the development of implanted eggs into embryos and finally term mice — called pups — will still require the hormonal support of a pregnancy following coitus, the surrogate mothers are prepared by mating a mature female with a vasectomized male. As the male is sterile, the egg produced by the foster mother will not be fertilized and cannot compete with the microinjected eggs that are implanted surgically. However, as the foster mother has been hormonally stimulated by coitus, she can carry the pregnancy through the birth of a number of transgenic mice 20 days later.

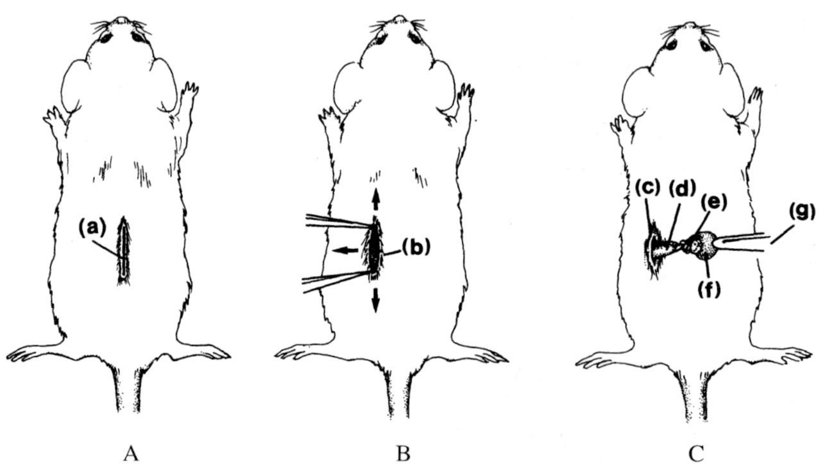

Fig. 15.7. Exposing the Fallopian tube to transfer fertilized eggs after microinjection of DNA. (A) A small midline incision (a) over the back is made. (B) The incision can be easily shifted over either Fallopian tube, serving as a movable window (b). (C) Through the incision (c), a small portion of the uterus (d), the Fallopian tube (e), and the ovary can be pulled out of the abdomen using forceps (g). After the transfer of 15 eggs into the left Fallopian tube, the window can be shifted over the right ovary for the transfer of 15 additional eggs.

(*Source*: Reproduced with permission of the publisher.) Reference 16.

The technique is basically the same for other animals, such as rabbits, pigs, sheep, and cows. However, the success rate in other animals is much lower, because they are "less prolific and because integration of foreign DNA occurs with a lower frequency."

Major equipment and supplies

Although the technique of injecting a single-celled egg was daunting, initial investigators met the challenge. By the early 1990s, the pioneers in microinjection technology were quite comfortable with the technique and were teaching a new wave of researchers, who were persuaded by their mentors that it was "relatively easy to master."[17] Consequently, the expense of the equipment needed to inject DNA into the eggs became the limiting factor in more widespread use of the technique. Yet the cost of the indispensable micropipette holder and microscope (approximately US$50,000) "pales when compared to [the costs] incurred for [larger] animal purchase and *per diem* charges," a leader in the field of microinjecting eggs correctly pointed out.[17]

Add to those dollars saved through the purchase of mice for research purposes, the additional dollars saved by their requiring less laboratory space for cages and food — compared with other species — plus a short gestational period, and mice become even more attractive potential models for studying human diseases. ("Potential" because in 1990, although transgenic mice had been created, they were free of disease and it was unknown if such healthy

Fig. 15.8. A laboratory mouse.
(*Source:* Reproduced with permission from the publisher.) Reference 22.

mice could be further modified genetically to develop disease processes common in man.)

Similarity of the mouse genome and the human genome

But their appeal doesn't stop with fewer dollars and less laboratory space. There's one other huge factor: mice — believe it or not — have a genetic makeup very similar to that of human beings. Ninety-nine percent of mouse genes are the same as human genes. Both the entire human genome and the entire mouse genome (the total genetic information contained in a cell) were established between 2001 and 2004. In two historic papers that appeared in February 2001, the "first drafts" of the human genome were published.[18,19] The projects moved forward faster than expected, due to the competition produced between a public firm, the Human Genome Sequencing Consortium,[18] and a private firm, the Celera Corporation.[19] The competing groups essentially completed the extraordinary feat of identifying and sequencing all the genes in the human genome by October 2004.[i] There are only 20,000–25,000 genes in the human genome, as it turns out,[20] and establishing "these sequences. . .revolutionized molecular biology."[21]

Meanwhile, the mouse genome was also being established. That collaborative effort found that "ninety-nine percent of mouse genes have a sequence match in the human genome.. . ."[22] Or, restated, "The proportion of mouse genes without any homologue. . .in the human genome (and vice versa) seems to be less than 1%."[23] This is important because, as an accompanying commentary put it, "there can scarcely be a major area of mammalian biology or medicine to which mouse studies have not contributed in some way, often as surrogates for human studies."[22]

The genetic similarities between mice and men may be looked at in another way. In both mice and men the genome comprises the same four nucleotides (a molecule formed from the combination of one base containing nitrogen, a sugar, and a phosphate group) arranged in the same sequence in every cell of the body.[j] For that reason, DNA from skin, hair, nails, and saliva — as well as

---

[i]If a 1000-piece jigsaw puzzle is challenging, how daunting might a 20,000–25,000-piece genetic puzzle be to assemble?

[j]Although all cells in an individual have the same DNA, different cell types use different sections of their DNA to carry out their particular functions. For example, the stomach has genes that control the production of its digestive enzymes, while the ovaries have genes that regulate the synthesis of estrogens.[12]

any other tissue of the body — can be used to physically identify people. And for *that* reason, pathologists and police can determine who's who from an individual's genetic fingerprint, provided a previous specimen of that person's DNA is available. Such specimens are increasingly available as "DNA libraries" grow, the result of obtaining a DNA "fingerprint" from all convicts released from prison since 1994[24] and all newborns in the U.S. since 2007.[25]

Gene Replacement

The creation of a transgenic mouse was a landmark event in the early 1980s, which proved to be a stepping stone to an even more consequential event in the late 1980s.[k] Four limitations of the first transgenic mice created by the pronuclear injection technique led to further studies to make the genetically engineered animal even more useful. The first restriction was imposed by the natural — almost universal — incorporation of foreign (donor) DNA into host (recipient) DNA *randomly*. Although several hundred copies of the foreign gene were injected into one of the two nuclei of a one-celled egg, the injected genes were still a small fraction of the (roughly) 20,000 genes in the host nucleus. The nonparental gene and its copies were, in effect, "lost" in the total gene pool and found their way into the host DNA only by chance.

The second limitation was a practical reflection of the random insertion of microinjected DNA into the host DNA. Because the DNA was injected into one single-celled egg at a time, and the injected DNA entered the cell's DNA in a random fashion, it was necessary to later look at many, many cells to find the rare ones with the injected DNA.[6]

The third limitation was related to the pronuclear injection technology used to generate prototype transgenic mice. It could only add — not subtract — genes. Consequently, any one transgenic mouse was genetically different from any other (all other) mouse (mice).

The final limitation of first-generation transgenic mice was simply the fact that they were healthy mice. They represented a biotechnological feat. Yet, these animals did not help to solve the mysteries of the causation of various

---

[k]Despite the now 25 years of creating transgenic mice, exactly how injected genes are integrated into the genes carried on the recipient's chromosomes has not been established. It is known, however, that the injected genes — also known as transferred genes or "transgenes" — are usually integrated in a head-to-tail fashion at a single site. It is also recognized that — for undetermined reasons — 50 or more copies of the injected genes line up at the single integration site.[5,13]

diseases and/or their treatments, because the mice were disease-free. A more sophisticated mouse model was needed — one with a better chance of incorporating outside DNA into its nucleus and one with a genetic defect that mimicked the genetic defect in humans with inherited diseases — one model for each disease to be studied.

Gene Targeting

Such was the setting in 1980 when Dr. Mario Capecchi made a critical observation. As the saying goes, "Discovery favors the prepared mind." Dr. Capecchi was working in his laboratory at the Howard Hughes Medical Institute in Chevy Chase, Maryland, when "a surprising phenomenon captured [his] attention."[1] He saw that not one, but many molecules of the foreign DNA he had injected into the nucleus of a recipient mouse's egg had, somehow, inserted themselves end to end, in the same direction at a single site. This defied the odds. It might be analogous to the odds of a handful of chocolate chips tossed into a bowl of batter aligning themselves end to end (in tandem), rather than becoming scattered throughout the flour-and-water mixture. Similarly, the inserted DNA molecules had been "stitched" together in the same direction, rather than being left to distribute by chance!

This led to a series of studies to replicate in the laboratory what the cells had done "on their own." A gene was chosen for injection. Knowing that finding the few cells — among the one million or more cells to choose from — which integrated the gene into their DNA would be tedious, if not impossible, a method was devised to improve the chances of finding them. It was reasoned that those chances should be increased if the injected DNA could somehow be highlighted, making it stand out from the crowd. Accordingly, an ingenious method of "targeting" — in the DNA vernacular — or selecting the DNA to be transferred was conceived.

Homologous Recombination

The method involved the insertion of yet another gene. The third gene inserted was one that protects the cell from a certain antibiotic. (To be sure, it sounds unnatural to us to seek protection from an antibiotic. However, the extra gene was made by a bacterium whose job security depends on its ingenuity.) The novel gene was placed so that DNA in the host cell flanked it.

When the inserted DNA was later mixed (in a test tube) with the antibiotic, the antibiotic *killed* those cells that lacked the protective gene. Only those cells carrying both the protective gene *and* the modified gene survived, making them easily identifiable.

This technique was termed "homologous recombination," implying that when two strands of homologous — i.e. the same — DNA face each other, they may be broken by enzymes within the cell, crossed over, and rejoined. The basis for its development was the fact that when two DNA strands recombine with each other following the repair of an injury to one or both strands, whether they do so randomly or specifically depends on how the DNA in one strand lines up with the DNA in the other. If they align with the injected DNA facing DNA molecules in the host chromosome with the same sequence, then the injected gene(s) is (are) cut by restriction enzymes and replace(s) the host gene(s). This same-to-same attraction is, of course, 180° different than the attraction of opposite poles by a magnet; and yet the bond it creates results in the absolutely precise replacement of the host gene by the foreign gene.

References

1. Capecchi MR. (1994) Targeted gene replacement. *Sci Am* **270(3)**: 52–59.
2. Sedivy JM, Joyner AL. (1992) *Gene Targeting*. W.H. Freeman, New York, pp. 1–12.
3. Schuldiner M, Benvenisty N. (2004) *In vitro* culture and differentiation of human embryonic stem cells. In: Thomson AJ, McWhir J (eds.), *Gene Targeting & Embryonic Stem Cells.* BIOS Scientific, London and New York, pp. 141–169.
4. Singer M, Berg P. (1991) *Genes and Genome: A Changing Perspective.* University Mill Valley, CA. Science Books.
5. Gordon JW, Scangos GA, Plotkin DJ, Barbosa JA, Ruddle FH. (1980) Genetic transformation of mouse embryos by microinjection of purified DNA. *Proc Nat Acad Sci USA* **77**: 7380–7384.
6. Gordon JW, Ruddle FH. (1983) Gene transfer into mouse embryos: production of transgenic mice by pronuclear injection. *Methods Enzymol* **101**: 411–433.
7. Palmiter RD, Brinster RL, Hammer RE, Trumbauer ME, Rosenfeld MG, Birnberg NC, Evans RW. (1982) Dramatic growth of mice that develop from eggs micro-injected with metallothionein-growth factor hormone genes. *Nature* **300**: 611–615.
8. Isola LM, Gordon, JW. (1988) Transgenic animals: a new era in developmental biology and medicine. In: First N, Haseltine FP. (eds.), *Transgenic Animals.* Butterworth-Heinemann, Boston, London, Singapore, Sydney, Toronto, Wellington, pp. 3–20.

9. Abbondanzo SJ, Gadi I, Stewart CL. (1993) Derivation of embryonic stem cells lines. *Methods Enzymol* **225**: 803–823.

10. Ramirez-Solis R, Davis AC, Bradley A. (1993) Gene targeting in embryonic stem cells. *Methods Enzymol* **225**: 855–878.

11. Gordon JW. (1993) Production of transgenic mice. In: Wassarman PM, De Pamphilis ML. (eds.), *Methods in Enzymology.* Academic, San Diego, New York, Boston, London, Sydney, Tokyo, Toronto **225**: 747–771.

12. Preparation of DNA routine procedure in footnote.

13. Holt M, Vangen O, Farstad W. (2004) Components of litter size in mice after 110 generations of selection. *Reproduction* **217**: 587–592. Normal # of mice in litter.

14. Superovulated # of mice.

15. Houdebine LM. (2002) Transgenic animals. In *Encyclo pedia of. Life Sciences* **18**: 449–454. Nature, London, New York, Tokyo.

16. Mann JR, Jr. (1993) Surgical techniques in production of transgenic mice. In: Wassarman PM, De Pamphilis ML (eds.), *Methods in Enzymology.* Academic San Diego, New York, Boston, London, Sydney, Tokyo, Toronto; **225**: 782–793.

17. Field LJ. (1993) Transgenic mice in cardiovascular research. *Ann Rev Physiol* **55**: 97–114.

18. International Human Genome Sequencing Consortium. (2001) Initial sequencing and analysis of the human genome. *Nature* **409**: 860–921.

19. Venter JC, Adams MD, Myers EW *et al.* (2001) The sequence of the human genome. *Science* **291**: 1304–1351.

20. International Human Genome Sequencing Consortium. (2004) Finishing the euchromatic sequence of the human genome. *Nature* **431**: 931–945.

21. Stein LD. (2004) End of the beginning. *Nature* **431**: 915–916.

22. Boguski MS. (2002) The mouse that roared. *Nature* **420**: 515–516.

23. Mouse Genome Sequencing Consortium. (2002) Initial sequencing and comparative analysis of the mouse genome. *Nature* **420**: 520–562.

24. http://en.wikipedia.org/wiki/Combined_DNA_Index_System. Accessed June 10, 2010.

25. http://www.govtrack.us/congress/bill.xpd?bill=s110–1858&tab=summary. Accessed June 10, 2010.

Appendix A

1. Emery AEH, Malcolm S. (1995) *An Introduction to Recombinant DNA in Medicine,* 2nd ed. John Wiley & Sons, Chichester, New York, Brisbane, Toronto, Singapore.

2. Waldman AS. (2004) *Genetic Recombination: Reviews and Protocols.* Methods in molecular biology. Humana, Totowa, NJ.

3. *The Recombination of Genetic Material.* Low KB (eds.), (1988) Academic, Harcourt Brace Jovanovich, San Diego, New York, Berkeley, Boston, London, Sydney, Tokyo, Toronto.
4. Hofker MH, van Deursen J. (2003) *Transgenic Mouse. Methods and Protocols.* Humana, Totowa, NJ.
5. *Transgenesis Techniques.* Cartwright EJ (ed.). (2009) Humana, New York.

Part II

16

TREATMENT OF CORONARY HEART DISEASE WITH ANTIOXIDANTS: AN INTRODUCTION

Before beginning this discussion regarding the use of antioxidants to prevent atherosclerosis specifically, an overview of *non-disease-causing* oxidative processes and the use of supplemental (i.e. nondietary) antioxidants seems in order.

Cellular oxidative processes, in general, are an integral part of normal (physiological) cellular metabolism. Cells use oxidative processes as a defense mechanism. Similarly, lipoxygenases are called upon in response to outside allergic antigens and foreign materials. Consumption of large amounts of antioxidants could interfere with these biological defense functions. Accordingly, an LDL-specific antioxidant would, in theory, be a good idea (assuming oxidized LDL has few or no beneficial effects). But a *nonspecific* antioxidant (such as vitamin E) could lead to problems by interfering with protective, non-LDL oxidative processes, such as fighting bacteria. Or, in lieu of a specific natural or synthetic LDL antioxidant, wouldn't it be wonderful to have available a nonspecific antioxidant whose effect is limited in its site of action to the wall of an artery? In other words, an antioxidant that functions *only* in the subendothelial space, where defending the region against bacteria, for example, isn't nearly so important as defending against micro-organisms in the *alveolar* space (in the lungs) or the *subcutaneous* space, (beneath the skin). Pneumonia and abscesses can develop in those two regions, respectively. But an infection within the wall of an artery is rare. We want to ward off bacteria in the lungs or under the skin before the micro-organisms reach the bloodstream, where, even if they do become bloodborne, they must face

187

the endothelial barrier, which regulates entry of noxious molecules and potential pathogens into the body.

After a decade of research in the 1980s, the time had come to begin the large, clinical trials testing the LDL modification hypothesis. Vitamin C, beta-carotene, and vitamin E were selected, because of their known antioxidant properties and safety. On the eve of the birth of the antioxidant trials (in 1992), Steinberg wrote, "What is generally regarded as the best evidence for a protective effect of an antioxidant is that the rate of progression of lesions [in laboratory animals] is slowed 50–80%."[1] Across America, people were randomized in a double-blinded fashion to take one of the three compounds or a placebo. The end points would be the frequency of nonfatal and fatal heart attacks over several years. The studies would be performed on subjects cared for at a number of U.S. university-affiliated hospitals. Hope was high that a simple, harmless way to reduce the number of heart attacks and deaths from coronary heart disease was just around the corner.

On the other hand, hope was balanced by the knowledge that it may not do any good to take antioxidants to treat a disease process that began in childhood. And since antioxidants affect the formation and early growth of fatty streaks — not the more mature plaque which ruptures, leading to a heart attack — would it do any good to take an antioxidant? Yet another variable in a large clinical trial of supplemental antioxidants would be the effect of diet. How would the varied diets consumed by study participants influence the study outcomes? An investigation completed in 1991 showed that what we eat affects the composition of the fatty acids in our LDL. Foods high in polyunsaturated fatty acids lead to LDLs high in polyunsaturated fatty acids, which are more easily oxidized. Foods with more monounsaturated fatty acids promote LDLs containing higher amounts of *those* fatty acids, which are more resistant to oxidation. And so, with much anticipation, the antioxidant trials began amidst lofty expectations that one of the three compounds would take us down the Yellow Brick Road to fewer heart attacks and longer lives.

References

1. Reaven P, Parthasarathy S, Grasse BJ, Miller E, Almazan F, Mattson JC, Khoo JC, Steinberg D, Witzum JL. (1991) Feasibility of using an oleate-rich diet to reduce the susceptibility of low-density lipoprotein to oxidative modification in humans. *Am J Nutr* **54**: 701–706.

17

THE MEDITERRANEAN DIET

The story of the Mediterranean diet began humbly, under the spectator seats at the University of Minnesota's football field — the Memorial Stadium in Minneapolis. There in his office Ancel Keys, a physiologist with a Ph.D. in Biology (Berkeley) as well as Physiology (Cambridge), was impressed by the number of heart attacks he saw personally among local business executives who had the finances to eat well and (presumably) avoid diseases associated with a poor diet. At the same time Keys was struck by the wave of heart attacks across the U.S. after the end of WWII, and began to wonder if there was some connection between food and a heart attack. (It is hard for us to imagine a time when someone did not know about the relationship of cholesterol to heart disease, but in the 1950s no one had heard of the association!) His initial curiosity was piqued further during a sabbatical year at Oxford in 1951–1952 and by travel in continental Europe "that opened his eyes to cultural differences in diet, behavior, and disease risk, and put him in touch with nutritional and clinical scientists [also] beginning to consider such differences."[1]

Also in the 1950s, the N.I.H. began awarding more frequent and larger research grants to look into what was behind the epidemic of heart attacks in the U.S. Keys received funding and designed the so-called Seven Countries Study. The investigation was aimed at exploring the association between diet and the risk of coronary heart disease in seven contrasting countries. Greece (in particular, the Mediterranean island of Crete off the southeast coast of Greece) Finland, Italy, Japan, The Netherlands, the United States, and Yugoslavia (now Serbia and Montenegro) formed the nucleus of the Seven Countries Study.

Crete became the epicenter of the investigation. While tiny in area (3219 square miles, Sicily 9926 square miles) and population (625,000; Sicily 5,200,000), its inhabitants enjoyed an exceedingly low rate of coronary heart disease. They ate a diet with an abundance of grains, greens, fresh fruit, fish, olive oil, and red wine, and that, it was postulated, might be the reason for their well-being. In contrast, the frequency of coronary heart disease was alarming in the more well-to-do Western European countries and in the United States. Based on those observations, the Seven Countries Study was born in 1958 to answer the question "What causes the striking differences in the frequency of coronary heart disease in different populations?".[2] It would become "one of the finest scientific adventures of our time".[1]

The study enrolled more than 12,000 healthy men, aged 40–59 at entry (between 1958 and 1964), personally recruited by Keys, his wife (herself a biochemist), and a handful of other committed investigators. At the study's inception, Finland and the United States had the highest coronary heart disease mortality rates, while Greece, Japan, and Yugoslavia had the lowest. The participants' cholesterol levels at entry varied from an average of 164 mg/dL in Serbia and 165 mg/dL in Japan to 240 mg/dL in the United States and 253 mg/dL in Holland and Finland. The country-to-country entry cholesterol levels are listed in Table 17.1.

When death rates were determined 25 years later (in 1995), the differences were striking.[3] [Remarkably, only 56 of the 12,467 men (0.4%) recruited in the late 1950s and early '60s were lost to followup after 25 years.] Japan boasted the lowest coronary heart disease mortality rate, while Finland had the highest, the variations in death rates largely mirroring the variations in cholesterol concentrations measured 25 years earlier (Table 17.1). The mortality rate was 5–6 times higher in Finland and the United States than in Japan (Fig. 17.1).[3]

Table 17.1   Entry cholesterol levels and 25-year coronary heart disease mortality of 12,763 men in the seven countries study.

Country	Cholesterol level (mg/dL)	25-year coronary mortality (%)
Finland, The Netherlands	253	20.3
United States	240	16.0
Inland southern Italy	204	9.1
Mediterranean Italy, Greece	200	4.7
Serbia	164	7.7
Japan	165	3.2

Source: Reference 3.

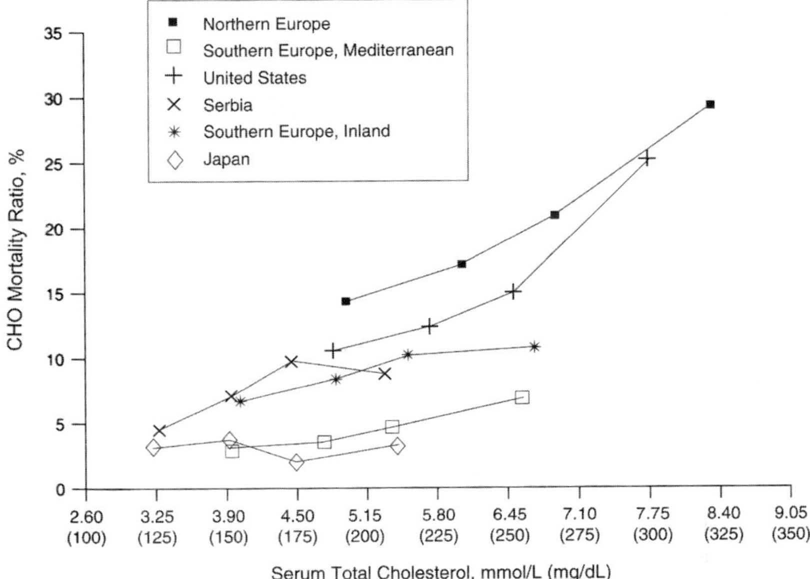

Fig. 17.1. The relationship of the 25-year coronary heart disease (CHD) mortality rates to the entry blood cholesterol level measured in 1958–1964 in the Seven Countries Study. Japan and the mediterranean countries fared well, while northern Europe (Finland and The Netherlands) and the United States had the highest CHD mortality rates. Cholesterol levels are expressed both in mmol/L the units commonly used in Europe and Japan and in mg/dL, the units customarily used in the U.S.

(*Source*: Reproduced with permission from the publisher.) Reference 3.

Japanese longevity is probably not due to genetics. When Japanese migrate to Hawaii and take on a Western diet (more saturated fats and *less* alcohol), their coronary heart disease mortality rises.[4] At any given entry cholesterol level there was a range of coronary death rates 25 years later. For example, at a cholesterol concentration of 200 mg/dL, mortality rates varied from 4% to 5% in Japan and the Mediterranean countries to 10% in the United States and 14% in Finland (Fig. 20.1). This implied that an additional factor — or factors — must be operative.

It probably was. Dietary flavonoid (explained in Chap.18) intake was twice as high in the Mediterranean countries as in Finland and the United States, and was highest in Japan, where the coronary heart disease mortality rate was lowest. This observation led to the speculation that "it may be impossible to achieve a reduction of coronary heart disease mortality rates in Northern Europe and the United States [comparable] to the Mediterranean level by a cholesterol reduction alone; the reduction in cholesterol must be accompanied by changes in other factors to achieve a profile more typical of the 'low risk' countries."[3] The "low-risk countries" included Crete, Sicily, and Greece, where poorer people used olive oil generously and were little affected by coronary heart disease. Ancel Keys, in 1961, decided to explore the seeming paradox he saw in Minnesota — well-to-do businessmen ate heartily but often suffered an acute myocardial infarction in their 40s, 50s, or 60s. For so doing, his image graced the cover of the January 13, 1961, edition of *Time Magazine* (Fig. 17.2). In the related article, titled "Fat of the Land," Keys is depicted as a "doggedly inquisitive scientist" with tremendous energy.[5] So much so that it would take him, with his wife, all over the world conducting the Seven Countries Study. All that energy after an eggless breakfast and a spartan lunch consisting of "a sardine sandwich, an olive, a cookie and a glass of skim milk," which would keep him fit. He dismissed "appetite suppressants such as amphetamines (Dexedrine, Benzedrine) as dangerous crutches for a weak mind."[5]

The *Time* article came on the heels of his popular book *Eat Well and Stay Well* (first published in 1959), which championed healthy eating and was among the first to warn Americans of the perils of obesity and too much cholesterol. His campaign to limit cholesterol earned him the sobriquet "Mr. Cholesterol."[a] During a 1959 interview soon after his book was published, Keys noted with "characteristic bluntness" that the wave of coronary heart disease sweeping across the U.S. and Canada was due to "the North American habit of making the stomach a garbage disposal ... for a long list of harmful foods."[6] All this just one year after Cassius Clay (later Muhammad Ali) won a gold medal in boxing at the 1960 Olympics (held in Rome) and one year after John F. Kennedy became president of the United States.

Why Does the Mediterranean Diet Work? Is It the Marine-Derived Omega-3 Fatty Acids (EPA and DHA) or the Plant-Derived Omega-3 Fatty Acid (Linolenic Acid)?

Because there are 21 culturally distinct countries (Fig. 17.3) embracing at least three religions — Christianity, Islam, and Judaism — with coastlines on the

---

[a]Keys also is credited with developing a compact, nonperishable, yet nutritionally adequate food package — at the request of the U.S. government — in 1942 for distribution to troops in Europe and Japan during World War II. The original rations consisted of "biscuits, dry sausage and chocolate" in a packet stamped with the letter "K" in recognition of Keys and became known overseas as "K rations."[6]

Fig. 17.2. Ancel Keys physiologist and epidemiologist.

(Reproduced with permission from *Time Magazine.*[5])

Fig. 17.3. The Mediterranean Basin.

(*Source*: Reproduced with permission from the publisher.) Reference 7.

Table 17.2   Countries with a coastline on the Mediterranean sea, from west to east.

Spain	Serbia	Lebanon
France	Montenegro	Israel
Italy	Albania	Egypt
Malta	Greece	Libya
Slovenia	Turkey	Tunisia
Croatia	Cyprus	Algeria
Bosnia	Syria	Morocco

Table 17.3   Common features of the several Mediterranean diets.

A high consumption of seafood
A low consumption of red meat
A high consumption of olives and olive oil
A high consumption of vegetables, fruits, and legumes
A high consumption of grains, cereals, and bread
A moderate consumption of dairy products, mostly as cheese
A moderate consumption of alcohol, mainly as red wine

Mediterranean Sea (Table 17.2), there is no single "Mediterranean diet." However, the variants do share some features: a high consumption of seafood; a high consumption of vegetables, fruits, and legumes; a high consumption of grains and cereals, including bread; a moderate consumption of dairy products, mostly in the form of cheeses; a low consumption of red meat; and a moderate consumption of alcohol, mainly as red wine[8] (Table 17.3).

Seafood and olive oil are central to the several variants of the Mediterranean diet. Because they have somewhat different fatty acids as their backbone, there has been an upswing in both scientific and public interest in the importance of omega-3 fatty acids (found in seafood and seafood oils) and omega-9 fatty acids (found in olive oil). A third group of fatty acids, found in the grains, cereals, and bread component of the Mediterranean diet, is made up of omega-6 fatty acids.

What Are Omega-3 Fatty Acids?

Omega-3 fatty acids are long-chain (18–22 carbon atoms in chain length) polyunsaturated fatty acids, with the first of several double bonds located

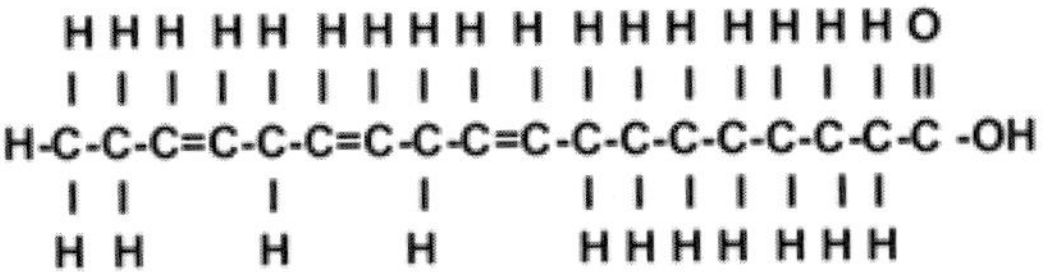

Fig. 17.4.   Linolenic acid, an omega-3 fatty acid prevalent in vegetable oils. (*Source*: Reproduced with permission from the publisher.) Reference 9.

between the third and fourth carbon atoms (counting from the CH_3 end of the fatty acid molecule) (Fig.17.4). (Note that the numbering scheme for double bonds starts at the omega or methyl — CH_3 — end of the fatty acid, whereas the numbering of carbon atoms starts from the opposite carboxyl — COOH — end of the carbon backbone.) Linolenic acid (discussed in Chap. 9) is the most plentiful omega-3 fatty acid in our diet. There are two other dietary omega-3 fatty acids (both with challenging-to-pronounce names — eicosapentaenoic acid and docosahexaenoic acid) mercifully referred to as "EPA" and "DHA." (These two abbreviations will be used from this point on.) The typical North American diet provides about 20 times more linolenic acid than EPA and DHA.[10] Fortunately, EPA and DHA also can be synthesized from linolenic acid.

The common dietary sources of EPA and DHA — all in seafood — are listed in Table 17.4.[12] Mackerel, herring, and salmon top the list, while tuna, shrimp, and cod are further down the line.[b] Vegetables and vegetable oils do not have any EPA or DHA, but some do contain lesser amounts of linolenic acid.[c] Flaxseed oil, canola oil, and English walnuts are plentiful sources of linolenic acid. Vegetable oils in general and linolenic acid in particular promote healing and health.[11]

What Are Omega-6 Fatty Acids?

Omega-6 fatty acids are a group of polyunsaturated fatty acids having in common a carbon double bond in the omega-6 position, i.e. the sixth carbon

---

[b] Fish are often categorized as fatty fish or lean fish. In the fatty category are salmon, rainbow trout, Baltic herring, whitefish, mahi-mahi, and tuna. Lean fish include pike, perch, pike–perch, cod, and pollock, known in Scandinavian countries as saithe.[12]

[c] There are two types of linolenic acid: alpha- and gamma-linolenic acid. Since alpha-linolenic acid is the predominant form in vegetable oils, the unmodified term "linolenic acid" will be used for simplicity.

Table 17.4 Omega-3 fatty acid (EPA and DHA) content of selected fish or seafood.

Fish of seafood	Combined concentrations of EPA and DHA
Mackerel	2500 mg/100 g
Herring	1700 mg/100 g
Salmon	1200 mg/100 g
Trout	500 mg/100 g
Halibut	400 mg/100 g
Tuna	400 mg/100 g
Shrimp	300 mg/100 g
Cod	300 mg/100 g

Source: Modified and reproduced with permission from the publisher. Reference 11.

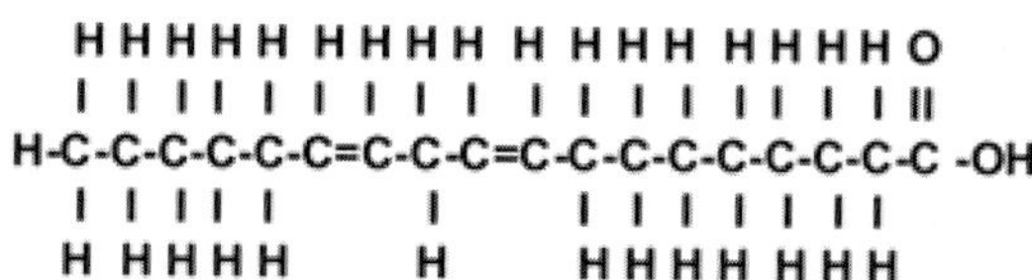

Fig. 17.5. Linoleic acid, an omega-6 fatty acid abundant in corn, sunflower, safflower, and soybean oils.

(*Source*: Reproduced with permission from the publisher.) Reference 9.

atom from the CH_3 end of the fatty acid. (The Greek word "*omega*" means "last." The omega-6 carbon atom is the sixth carbon atom when we count inward from the last carbon atom, a part of the CH_3 group.) These fatty acids tend to promote inflammation, opposing the healing effects of omega-3 fatty acids. *Linoleic,* as opposed to *linolenic,* acid is the principal omega-6 fatty acid (Fig. 17.5).

In humans, the effects of omega-6 fatty acids "are largely mediated by their interactions with omega-3 fatty acids."[13] For example, because omega-6 and omega-3 fatty acids use largely the same enzyme pathway for their incorporation, competition between the two develops that leads to unequal amounts of

Fig. 17.6. Animal meats, eggs, and dairy products are good sources of arachidonic acid for the human body.

(*Source*: Reproduced with permission from the publisher.) Reference 14.

EPA and DHA (linolenic acid's eicosanoids) and arachidonic acid (linoleic acid's eicosanoid). [Red meats, eggs, and dairy products have abundant arachidonic acid, an essential fatty acid for humans (Fig. 17.6).] As a result, the *ratio* — as opposed to the absolute amount — of the two fatty acids becomes "critical to disease prevention."[15] An imbalance of omega-6 and omega-3 fatty acids — favoring omega-6 — has been linked to coronary heart disease as well as several other serious disorders (cancer, diabetes, Alzheimer's disease, and attention deficit disorder).

In effect, "good cholesterol" and "bad cholesterol" have fatty acid counterparts — "good fatty acids" (omega-3 fatty acids) and "bad fatty acids" (omega-6 fatty acids) — although in the case of fatty acids, it's the ratio of omega-6 to omega-3 fatty acids that counts, not the avoidance of all omega-6 fatty acids. At this time, the "ratio of n-6 to n-3 fatty acids (consumed by most Americans) is still much higher than that recommended (i.e., 2.3:1)."[16]

In the U.S. and Canada, vegetable oils (such as corn, safflower, and sunflower oils) are often "enriched" with omega-6 fatty acids, mainly as linoleic acid. Enrichment is done because a higher intake of linoleic acid provides "some modest cholesterol lowering."[9] But because linoleic acid and linolenic acid use mostly the same enzyme pathways, such supplementation simultaneously augments the rivalry between the two fatty acids, resulting in the formation of less healthful omega-3 fatty acids — EPA and DHA — skewing the balance, unfavorably.[9]

A healthy diet should include up to four omega-6 fatty acids for every one omega-3 fatty acid. Yet, a typical American diet has 10–30 times as many omega-6 fatty acids as omega-3 fatty acids.[15] (To attain the recommended ratio of 2.3 to 1 would require "an approximate 4-fold increase in fish consumption in the United States."[16]). In contrast, the Mediterranean diet has a "healthy balance between omega-3 and omega-6 fatty acids. It emphasizes whole grains, root and green vegetables, fruit, fish and poultry, olive and canola oils, and alpha-linolenic acid, along with discouragement of ingestion of red meat and total avoidance of butter and cream."[11] The classic Spanish dish — one of the author's favourites — paella is a sumptuous microcosm of the Mediterranean diet, although some recipes include chorizo sausage or pepperoni.

What Are Omega-9 Fatty Acids?

Omega-9 fatty acids are long-chain (18–22 carbon atoms in chain length) monounsaturated fatty acids, with the only double bond between the ninth and tenth carbon atoms (counting from the CH_3 end of the fatty acid molecule) (Fig. 17.7). The only two biological omega-9 fatty acids are oleic acid and erucic acid. The former is a cousin of the linolenic–linoleic fatty acid brothers, while the latter is a member of the rapeseed group of plants, themselves kindred of the mustard family.[17]

Olive oil, the vegetable oil widely used in the Mediterranean region, has a singular fatty acid composition. Monounsaturated oleic acid is its most abundant fatty acid, while polyunsaturated linoleic acid accounts for less than 20%. In addition, olive oil (Fig. 17.8) contains several minor ingredients, including polyphenols, which impart the typical taste and aroma of extra virgin olive oil and confer its resistance to oxidation.[19] The world's top eight olive-oil-producing and-consuming countries are in the Mediterranean Basin (Table 17.5). Spain and Italy put out the most, while Greece is by far the leading consumer.

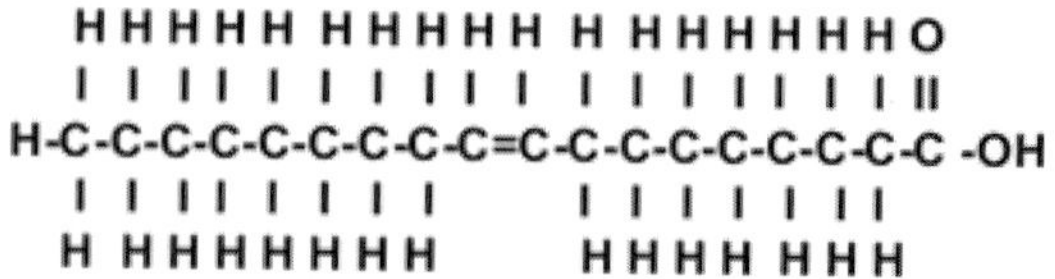

Fig. 17.7.   Oleic acid, an omega-9 fatty acid abundant in olive oil.
(*Source:* Reproduced with permission from the publisher.) Reference 9.

Fig. 17.8.

Source: Reference 18.

Table 17.5   The world's leading olive-oil-producing and -consuming countries.

Global producers of olive oil	2005 production	Annual per capita consumption (kg)
Spain	36%	13.6
Italy	25%	12.3
Greece	18%	23.7
Turkey	5%	1.2
Syria	4%	6
Tunisia	8%	9.1
Morocco	3%	1.8
Portugal	1%	7.1
U.S.	0	0.56
France	0	1.34

Source: Reference 20.

Effect of the Mediterranean Diet on LDL Oxidation

A study conducted at the University of California, San Diego aimed to see if there was a difference in the susceptibility to oxidation of LDL from Greeks eating a typical Mediterranean diet and Americans living in Southern California eating a typical American diet. Indeed there was. The extent of LDL oxidation was 42% less in the Greek group compared with the American subjects.[21]

In a second study done by the same research group, healthy Americans (with mild hypercholesterolemia, i.e. a mildly elevated cholesterol level) consumed a liquid diet enriched with either oleic acid (the major fatty acid in olive oil) or linoleic acid (the major fatty acid in sunflower oil) for eight weeks. The LDL from the oleic-acid-enriched group was much more resistant to oxidation than that from the sunflower oil group. These findings were consistent with those of another investigation in which healthy people drank 50 g (about $1\frac{2}{3}$ ounces) per day of olive oil for two weeks. In the laboratory, scavenger cells later turned up their noses at LDL from olive-oil-supplemented subjects.[22] The researchers decided that "dietary enrichment of LDL with oleic acid is realistic" and that such enrichment can be achieved through the Mediterranean diet (for those who choose to chew what they eat, rather than drink it).[22]

While these studies establish the ability of olive oil to lessen LDL oxidation, they don't determine if the beneficial effect is due to the monounsaturated oleic acid in the oil or to its antioxidant polyphenols. To figure that out, (different) subjects consumed either virgin olive oil or oleic-acid-enriched sunflower oil for three weeks. LDL oxidation decreased only in the olive oil group, indicating that an ingredient(s) in olive oil other than its oleic acid was responsible for its antioxidant properties.[23] The beneficial ingredient was most likely its antioxidant polyphenols. So now you can explain to a dinner partner that the healthful effects of the Greek salad he or she is relishing are due to the polyphenoic antioxidants in the olive oil and feta cheese (Fig. 17.9), if the oil is not too processed.

Grades of olive oil

There are four grades of olive oil, as determined by the European Community. They are extra-virgin, virgin, pure, and extra-light.[18] The differences among them are listed below.

Extra-virgin olive oil is the top grade. It comes from the olive juices extracted during the first pressing of olives, without using heat or chemicals.

Virgin olive oil comes from the juices extracted during the second pressing. Virgin and extra-virgin olive oil have high amounts of polyphenolic compounds and vitamin E — both powerful antioxidants — that are lost when the oil is further refined into the pure and extra-light grades.[25,26]

Pure olive oil has undergone chemical extraction of what's left — called "mash" — after the second pressing. Some virgin or extra-virgin olive oil is then added to the chemically extracted oil.

Extra-light olive oil has undergone considerably more processing.

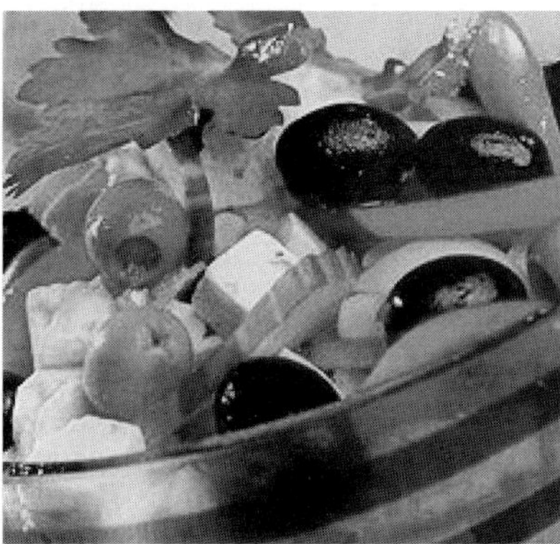

Fig. 17.9.   The makings of a healthful Greek salad laced with polyphenolic antioxidants. (*Source*: Reproduced with permission from the publisher.) Reference 24.

It's a good idea to replace margarine and butter with virgin or extra-virgin olive oil, lightly salted and sprinkled with Parmesan cheese. Freshly baked, warm Italian bread dipped in the oil — and a red wine, such as Chianti, to maintain the ethnic theme — are a healthy accompaniment to a spaghetti or ravioli meal.

Canola oil

Canola oil is sometimes used as an alternative to olive oil. The term "canola oil" comes from "Canadian oil," because it was first derived from the rapeseed plant in Canada, in the 1970s. It contains a low concentration of erucic acid, the other omega-9 fatty acid in the human diet. The rapeseed plant has a relatively large amount of erucic acid, which can be mildly toxic to humans, necessitating the development of a hybrid with a lower erucic acid concentration.

In addition to monounsaturated erucic acid content, canola oil has a healthful ratio of beneficial omega-3 to omega-6 fatty acids. In a study carried out at the Human Nutrition Center on Aging at Tufts University in 1993, canola oil (and corn oil) lowered the plasma LDL cholesterol level a bit more (3%) than did olive oil. On the other hand, canola oil (and corn oil) also lowered the plasma HDL cholesterol level (a common effect of all vegetable oils) more than did olive oil. The bottom line was that "none of [the three] oils had a significant advantage in terms of altering the overall lipoprotein profile."[27]

Table 17.6   Sources of omega-3, omega-6, and omega-9 fatty acids.

Omega-3 fatty acids	Omega-6 fatty acids	Omega-9 fatty acids
Flaxseed oil	Safflower oil	Olive oil
Canola oil	Sunflower oil	Canola oil
Soybeans	Corn oil	Avocados
English walnuts	Flaxseed oil	Almonds
Pumpkin seeds	Pistachio nuts	Pistachio nuts
Sesame seeds	Pumpkin seeds	Cashew nuts
Avocados	Sunflower seeds	Pecans
Salmon	Eggs	Peanuts
Halibut	Milk	Hazelnuts
Tuna	Butter	Macadamia nuts
Tofu	Red meats	

Source: References 26 and 29.

Like olive oil, canola oil extracted without heat or chemicals is superior to processed and refined canola oil.[28] Wholesome, natural olive oils and canola oils are found in health food stores and specialty outlets, more so than in supermarkets. Some food sources of omega-3, omega-6, and omega-9 fatty acids are listed in Table 17.6.

Coronary Heart Disease in Arctic Circle Countries and Japan

The French paradox (Chap. 1) had not yet been described in 1972 when two physicians from Copenhagen set out to explain an apparent contradiction. A recently concluded autopsy study of 339 Alaskan Eskimos who died between January 1, 1959 and December 31, 1969 (Alaska gained statehood on January 3, 1959) had shown just four people with evidence of a heart attack.[30] (There were actually more deaths from congenital heart disease than from coronary heart disease in Alaska.) Yet, in the lower 48 states coronary heart disease was, by far, the leading cause of death. Cholesterol had been indicted as the culprit in the 1960s and the American Heart Association had begun its campaign to educate the populace about the diet–heart-disease link.

Dyerberg, Bang, and Greenlandic Eskimos

Enter Jorn Dyerberg and Hans Bang — two Danes studying Greenlandic Eskimos for practical (proximity) reasons — who would throw a wrench into the American Heart Association's anticholesterol campaign. In Greenland, as in Alaska, coronary atherosclerosis is also "almost unknown" among Eskimos.[31] Yet, Dyerberg and Bang — both M.D. and Ph. D. — noted that in spite of an "extraordinarily high intake of fat," derived from whale, seal, and walrus in the winter and caribou in the summer, Greenlandic Eskimos had low plasma cholesterol concentrations.[32] [Eskimos, by the way, not only eat their meat raw or minimally cooked in sea water (Chap. 1), but they eat *all of it* — the skin, bone marrow, brain, blood, abdominal organs, and stomach contents.[33]] And they don't have heart attacks.

But, alas, their "cardioprotection" — as we call it in medicine — is lost if Eskimos from Greenland migrate to the mainland in Denmark and adopt a northern European diet, signifying that it's not genetic. It's also not more cholesterol in the Scandinavian diet escalating the rate of coronary heart disease. The average cholesterol intake among Eskimos living in Greenland was actually 1½ times *greater* than among Eskimos living in Denmark (because seafood is cholesterol-rich), yet the average plasma cholesterol level was 25% *lower* in Greenlandic Eskimos than in Denmark Eskimos.[31] But … their average plasma omega-3 fatty acid (EPA and DOC) levels were increased strikingly. Seems like a lot of beneficial dietary omega-3 fatty acids can overwhelm a lot of dietary cholesterol.

Effect of diet on blood EPA and DOC levels in coastal and Inland Alaskan Eskimos

In the summer of 1985, a team of epidemiologists and nutritionists from the Oregon Health Sciences Center in Portland traveled up the coast to interview two groups of Eskimos living in southwestern Alaska. One group resided along the banks of a river 20 miles from Bethel, Alaska; the other lived in a coastal village on the Bering Sea. Both were tiny — 620 and 370 people, respectively — and villagers in both led a simple lifestyle, gathering food by fishing, hunting, trapping, and picking wild berries. Children and adults from a University of Oregon medical clinic served as a control population.

The dietary patterns of the two Alaskan sets reflected, not surprisingly, the foods linked to their respective aquatic environments.[34] Coastal dwellers caught and ate saltwater fish and marine mammals (whales, walruses, and seals), while river dwellers consumed freshwater fish (trout, halibut, and salmon[d]) and land mammals (beavers, moose, caribou, and reindeer). Both groups used seal oil for cooking, coastal residents to a greater extent than inland residents.

Their diets determined their blood lipid profile. Blood concentrations of good-for-you omega-3 fatty acids (EPA and DOC) were 5–10 times higher in the coastal dwellers than in the river dwellers and paralleled their greater intake of seafood and use of seal oil for cooking, if cooked. And concentrations of EPA and DOC were 6–13 times the concentrations of the Oregon residents. Perhaps it is no wonder that the death rate from coronary heart disease in the 1980s was 3½ times greater among U.S. (Caucasian) residents in the lower 48 states than among Eskimos living in southwestern Alaska.[35] In all fairness, however, while the lower *coronary heart disease* mortality rate among Alaskan Eskimos is largely related to their higher seafood intake, the death rate from *all* cardiac diseases combined is only slightly less among Alaskan Eskimos than the national average because of their greater alcohol intake, leading to alcoholic cardiomyopathy.[36]

Coronary Heart Disease in Japanese

In the Seven Countries Study, Japan alone had a 25-year coronary heart disease mortality rate below 4%, more than five times lower than the rate in the three countries with the highest rates — Finland, The Netherlands, and the United States (Table 2.1).[3]

The question is: Why?

Those startling figures raised the question of causality — is it heredity or diet, or both? At any given cholesterol level, there was a range of coronary heart disease prevalence among the seven countries. At a cholesterol level of 210 mg/dL, for instance, the 25-year coronary mortality rate was 3.2% in

---

[d] Salmon are born in fresh water, then swim downstream to live in salt water, before returning to the river, where they breed…and die.

Japan, 11% in the U.S., and 15% in Finland (Fig. 17.1), implying that heredity was an important factor ... well, maybe. Genes aside, the Japanese diet also happened to be the lowest in saturated fats and the richest in antioxidants, implying that diet might be a key ingredient. Yet, genetics could still be chipping in to the resistance of the Japanese to coronary heart disease. Is it genes, is it diet, or both?

An experiment of nature

Light was thrown on the question through what was tantamount to an experiment of nature. After WWII, some Japanese began to migrate to Hawaii (not yet a state) and to the west coast of the continental U.S., particularly the San Francisco Bay region. With that geographical shift came an opportunity to assess the relative importance of genetics and the Japanese diet to coronary heart disease mortality by determining the incidences of coronary heart disease in these now-Japanese Americans — whose genes hadn't changed one bit — who adopted different degrees of a Western diet after migrating to the U.S. When broken down into those who retained a traditional Japanese diet (low in fat, high in fish cooked in rapeseed oil, i.e. canola oil, with liberal amounts of sake and beer), compared with those who ate almost equal amounts of Japanese and American foods, and those who totally adopted a Western diet, there was a gradient of coronary heart disease rates. It was lowest in those who maintained a traditional Japanese diet, highest in those who fully adopted American food, and intermediate in those consuming roughly equal proportions of each cuisine.[e,37]

Another way to gauge the relative influence of genetics and diet on coronary heart disease mortality in Japanese is to measure mortality among natives consuming different amounts of traditional Japanese foods. For example, inhabitants of Kohama, an isolated island in the Okinawa prefecture, have the lowest incidence of coronary heart disease in the

---

[e]In 1968, a team of three Japanese and American physicians assigned by the Atomic Bomb Casualty Commission to monitor the long-term health of the survivors of Hiroshima and Nagasaki reported that coronary heart disease had increased since 1953 as the intake of animal fats "doubled" and cholesterol levels soared.[36] At the time, the cholesterol–coronary-heart-disease connection was still unsettled, leading the authors to quip, "In view of this impressive national dietary alteration, Japan will probably become the proving-grounds for theories on the role of dietary fat in human atherogenesis" (the formation of atherosclerotic plaques).

country. The island's residents are almost all fishermen, consuming more fresh fish a day than for any other region in Japan. Among these fishermen, there are three times as many nonagenarians in Kohama than on the mainland.[38] When they were studied in 1985, a full 22.5% were over age 65 but nary a soul had evidence of a heart attack by the EKG. Their longevity and near-absence of coronary heart disease appear due to a blood EPA concentration, on average, 40% higher than that of residents from a Japanese farming village and 20% higher than that of residents from another fishing village on the mainland. To boot, Kohama fishermen had an average total cholesterol level of 188 mg/dL, of which nearly two-thirds was of the good cholesterol variety![38]

In a similar comparative study of fish intake and plasma EPA levels — this time in a Japanese fishing village and a Japanese farming village — the EPA levels were higher in the fishermen than in the farmers and were several times higher than in European countries, yet far lower than in Eskimos.[39] Similarly, in the University of Oregon study, the plasma EPA levels of Alaskan coastal dwellers were twice the levels of river dwellers and 14 times higher than those of Oregon residents.[34] This good effect of seafood is likely due both to slowing (or preventing) cholesterol-rich plaque formation (over many years) and to the blood-thinning effect of EPA. Recall that it is a small blood clot forming atop a plaque that ultimately triggers an acute myocardial infarction.

Social and cultural factors

But diet isn't the sole reason for the dramatic differences in coronary heart disease between Japanese, Japanese-Americans, and Americans. "Social and cultural factors" contribute a lot. Those Japanese who continue to observe traditional Japanese culture with its emphasis on "group cohesion and achievement and social stability" — reducing stress along the way — have less coronary heart disease — 3–5 times less — than those who become "most acculturated" (most culturally assimilated).[37] In contrast, those Japanese who adopt a typical American lifestyle with its "emphasis on social and geographic mobility" — often characterized as that of a "type A or coronary-prone behavior pattern" — develop coronary heart disease more often than do those who continue to practice traditional Japanese customs.[37]

Benefitting from fish Inland

You don't have to live in a Japanese fishing village or along the coast of Alaska to benefit from more seafood. In a Dutch town — which is likely far closer to the sea than the place many Americans call home — the "mortality from coronary heart disease was more than 50 percent lower among those who consumed at least 30 grams (1 ounce) of fish per day than among those who did not eat fish" at all.[38] (Remember the Japanese eat about 100 g of fish a day and the Eskimos put away about 400 g a day.[f,38]) While the daily consumption of fish may seem a bit too often to many inland Americans, the habitual "consumption of as little as one or two fish dishes per week may be of [some] preventive value [with regard] to coronary heart disease."[38] Seeing a fish on the plate once or twice a week may not be out of the question even for those of us living in cattle country U.S.A.

Salmon fillets and fish oil capsules

Salmon is an easy-to-take fish and at the Oregon Health Sciences Center, which salmon swim near every day, a salmon fillet and a fish oil capsule (MaxEPA) three times a day made the blood cholesterol and triglyceride levels nosedive (Fig. 17.10). In contrast, a diet with lots of vegetable oils led to "a rapid and significant rise in plasma triglycerides".[41] But there's a catch. These 20 people had outrageously elevated cholesterol and triglyceride levels before they started eating salmon fillets and taking fish oil capsules several times a day. (Note that cholesterol concentrations of 200–700 mg/dL mean 373 mg/dL on the ordinate, and triglyceride concentrations of 475–2500 mg/dL mean 1353 mg/dL on the abscissa.) With cholesterol and triglyceride levels this high, EPA works brilliantly. But when there are smaller fish to fry, fish oil capsules don't work so well.

---

[f] The "440 grams per day" quantity of fish consumed by Eskimos cited in the Dutch report is a bit of a "good news, bad news" story. The authors of the report cited the two Danish researchers — Dyerberg and Bang — as the source of the figure. The good news is that in a letter to the editor of the *New England Journal of Medicine*, Dyerberg and Bang pointed out that it is "not quite correct" to say that Eskimos eat that much fish a day because the figure of 400 g refers to the total daily amount of "meat from Arctic sea mammals, such as seal and whale," not fish.[40] So you don't need to eat *that* much fish to get the heart-healthy omega-3 fatty acids you need. The bad news is that only about "20%" of the total daily Eskimo seafood intake comes from fish. So you *do* need to eat about 300 g a day of seal and whale blubber to keep your heart ticking like an Eskimo's.

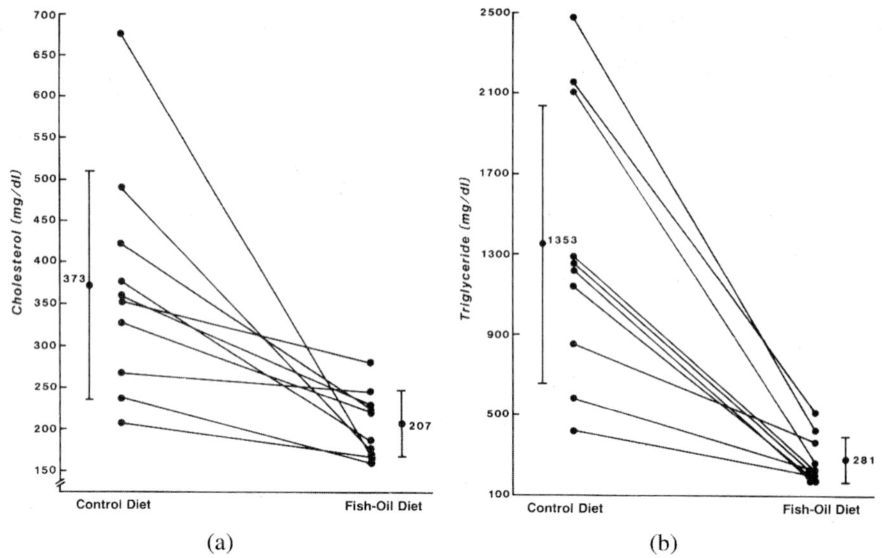

Fig. 17.10.   Graphs showing the blood cholesterol (a) and triglyceride (b) levels while on a control diet and a fish oil diet in 20 subjects. Note that the higher the cholesterol and triglyceride levels, the greater the drop, after a month of a fish oil diet, while the closer to "normal" the baseline levels are, the smaller the drop is.

(*Source*: Reproduced with permission from the publisher.) Reference 41.

Or, put another way, if blood cholesterol levels are in the mildly elevated, 200–240 mg/dL range, so common in many Americans who eat often at McDonald's and the like, then "when fish oil is fed and saturated fat intake is [held] constant, LDL [cholesterol] levels either do not change or may increase."[42] Or, put yet another way, you can't wash down a Big Mac with a swig of fish oil and expect that your cholesterol and saturated fats will be "neutralized." Unfortunately, that's not the way it works. If it did there'd be a liquid EPA dispenser in carbonated beverage machines at McDonald's everywhere. McDonald's and the many EPA manufacturers would likely have pounced on "an EPA dispenser" as a big profit-maker long ago. And if EPA could cancel out cholesterol some folks might put away two Big Macs for lunch with the expectation of impunity.

Eat Meat and Take 18 EPA Capsules a Day

But if you just don't like eating fish and can't imagine a day without red meat, there is another way to get your lipids in line. Or…if you really don't want to eat salmon fillets three times a day, there is another option. It may not be the most mouth-watering option, but it works. You can eat whatever you want and take 18 Max-EPA capsules a day, rich in omega-3 fatty acids. That'll knock your cholesterol level down in a hurry — if it's *really* elevated.[43] The American Heart Association recommends the consumption of about 1 g of fish oil daily for patients with coronary heart disease. So 18 capsules amounts to 5400 g of EPA and DOC, which is an intake in the Eskimo league. The American Heart Association also points out that "a dietary (i.e. food-based) approach to increasing omega-3 fatty acid intake is preferable" to taking omega-3 supplements.[44] But, for people who cannot readily consume enough marine fish to get the recommended amount of about 1 g (total) of DHA and EPA per day (range 0.5–1.8 g per day), supplements may be used.[g,44]

Coronary Heart Disease in Japan and the United States in the Late 20th and Early 21st Centuries

While total cholesterol levels decreased in the U.S. population during the later years of the 20th century, they increased strikingly in the *Japanese* population — as Western ways ranging from diet to dancing in a mosh pit infiltrated the country — such that the mean levels in Japanese men and women rivaled those in U.S. adults.[46] These increases seem to be associated with the marked rise in coronary heart disease mortality observed in Japanese natives in the early 1990s, soon after McDonald's metastasized to Japan. Figure 17.11 discloses the mean blood total cholesterol levels in Americans and Japanese by decade of life according to 1972–1976 data from the U.S. and 1980 data from Japan. Japanese men and women who were around 20 years of age, at the time of the study, had higher cholesterol levels than did their U.S. counterparts. As this segment of the population ages, there is concern that the elevated cholesterol levels may translate into more heart attacks and strokes.[46] Also shown in the figure is the distribution of

---

[g] On September 8, 2004, the F.D.A. announced the "availability of a qualified health claim for reduced risk of coronary heart disease (CHD) on conventional foods that contain eicosapentaenoic acid (EPA) and docosahexaneoic acid (DHA) omega-3 fatty acids."[45] A qualified-health-claim status means that there is "credible scientific evidence" on hand to support the claim. Lester Crawford, the Acting F.D.A. Commissioner, said, "This new qualified health claim for omega-3 fatty acids should help consumers as they work to improve their health by identifying foods that contain these important compounds."[45]

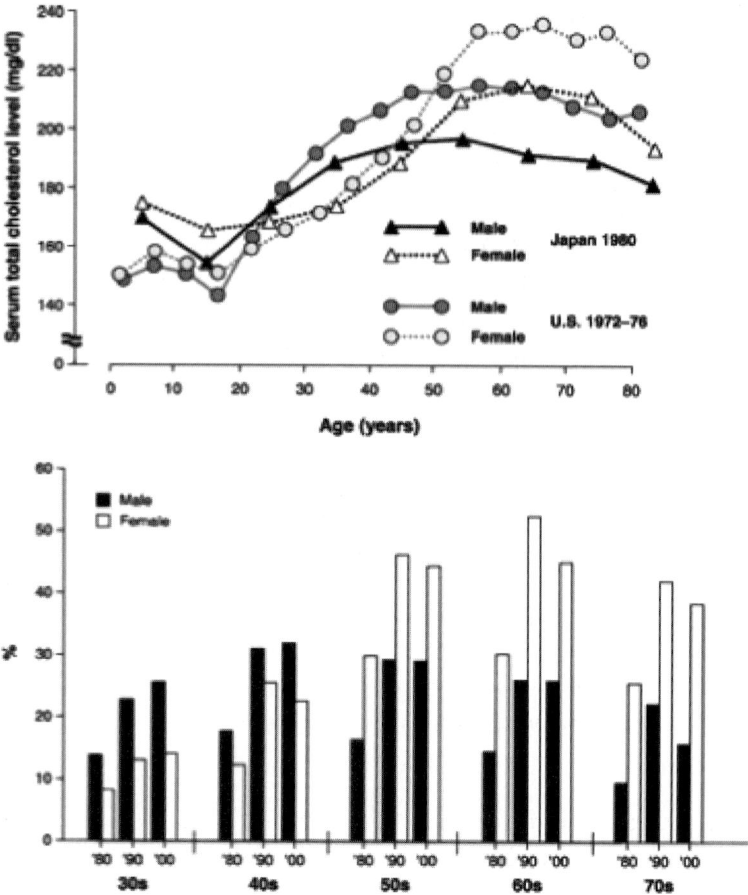

Fig. 17.11.   *Top*: The mean blood total cholesterol levels according to age in the Japanese and U.S. populations. The U.S. men and women win the cholesterol contest in a battle they'd rather lose. *Bottom*: The proportion of Japanese men and women with a total cholesterol ≥220 mg/dL (5.69 mmol/L) by decade of age (from the 30s to the 70s) in surveys in 1980 ('80), 1990 ('90), and 2000 ('00).

(*Source*: Adapted with permission from, the publisher. Data from Ministry of Health, Labor and Welfare.) Reference 46.

Japanese adults with total cholesterol levels >220 mg/dL (5.69 mmol/L) by decade of life in surveys performed in 1980, 1990, and 2000. (In addition to an increasing prevalence of elevated total cholesterol levels between 1980 and 1990 for each decade of life, the data also suggest a higher prevalence of elevated cholesterol levels in older, postmenopausal women.[46])

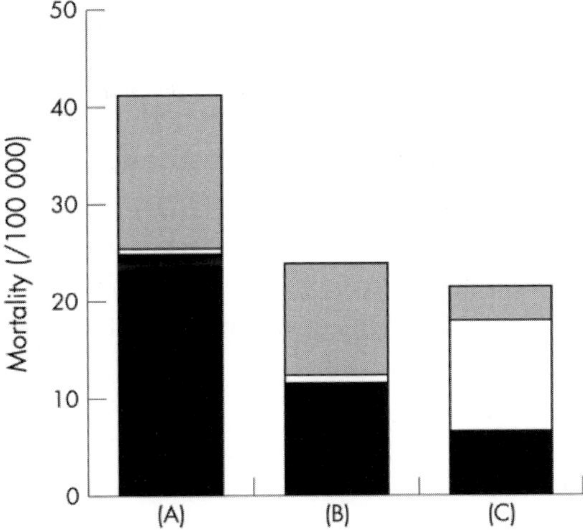

Fig. 17.12. The mortality from coronary heart disease (black), heart failure (white), and other causes of heart disease (gray) per 100,000 men aged 35–44. (a) Caucasian men in the U.S. (b) Japanese-American men in Hawaii. (c) Japanese men in Japan. Caucasian American men often die from heart attacks, Japanese men living in Japan die more often from heart failure related to high blood pressure or alcoholic cardiomyopathy (discussed in Chap. 18), and Japanese men in Hawaii straddle the aisle.

(*Source*: Reproduced with permission from the publisher.) Reference 47.

Nonetheless, the coronary heart disease mortality rate in "Japanese men in Japan and Japanese-American men in Hawaii remain(s) substantially lower than in U.S. white men" (Fig. 17.12).[47] But the rates of lung cancer and chronic obstructive pulmonary disease, i.e. emphysema are far greater in Japanese men than in American men. Both of these conditions are agonizing ways to die. I can still recall inveterate smokers standing on the roof top of a Manhattan hospital smoking cigarettes through their tracheotomy!

After WWII, lifestyle changes (less smoking, more exercise, and lower cholesterol intake) cut the prevalence of coronary heart disease in the U.S., while lifestyle changes in Japan (more smoking and higher cholesterol intake) had the opposite effect on coronary heart disease prevalence. Now, Japanese men smoke more, while Americans weigh more. And with the awareness that omega-3 fatty acids are the biochemical basis for the health benefits of cold-water fish the American Heart Association — after years of urging Americans to eat lean meats — was (ironically) obliged to recommend that we eat "fat fish."

Epilog

Ancel Keys, the father of the Seven Countries Study, practiced what he preached. In the 1970s, he shifted his office from under the bleachers at the Memorial Stadium in Minneapolis to a villa on the coast of the Tyrrhenian Sea in southern Italy. (the Tyrrhenian Sea is an arm of the Mediterranean Sea situated between Italy and the islands of Corsica, Sardinia, and Sicily.) There, he and his wife, Margaret, enjoyed the beauty of the sea and the benefits of the Mediterranean diet — until his death in 2004, at age 100.

References

1. http://www.epi.umn.edu/research/7countries/overview.shtm
2. Keys A, Kimura N, Kusukawa A, Bronte-Stewart B, Larsen N, Keys MH. (1958) Lessons from serum cholesterol in Japan, Hawaii, and Los Angeles. *Ann Intern Med* **48**: 83–94.
3. Verschuren WMM, Jacobs DR, Bloemberg BPM, Krornhoput D, Menotti A, Aravanis C, Blackburn H, Buzina R, Dontas AS, Fidanza F, Karvonen MJ, Nedeljikovic S, Nissinen A, Toshima H. (1995) Serum total cholesterol and long-term coronary heart disease mortality in different countries. *JAMA* **274**: 131–136.
4. Yano K, Rhodes GC, Kagan A. (1977) Coffee, alcohol, and risk of coronary heart disease among Japanese men living in Hawaii. *N Engl J Med* **297**: 405–409.
5. The fat of the land. *Time Magazine,* January 13, 1961.
6. http://www.the-aps.org/membership/obituaries/ancel_keys.htm
7. http://en.wikipedia.org/wiki/Mediterranean_Basin
8. Trichopoulou A, Lagiou P. (1997) Healthy traditional Mediterranean diet: an expression of a culture, history, and lifestyle. *Nutr Rev* **55**: 383–389.
9. http://www.thepaleodiet.com/nutritional_tools/fats.shtml
10. Holub BJ. (2002) Clinical nutrition: 4. Omega-3 fatty acids in cardiovascular disease. *CMAJ* **166**: 608–615.
11. http://www.umm.edu/altmed/articles/flaxseed-oil-0003
12. de Mello DF, Erkkila AT, Schwab US, Pulkkinen L, Kohlehmanen M, Atalf M, Mussalo H, Lankinon M, Oresik M. Lehto S, Uusitupa M. (2009) The effect of fatty fish or lean fish intake on inflammatory gene expression in cells of patients with coronary heart disease. *Eur J Clin Nutr* **48**: 447–455.
13. Lee KW, Lip GYH. (2003) The role of omega-3 fatty acids in the secondary prevention of cardiovascular disease. *QJM* **96**: 465–480.
14. http://en.wikipedia.orgwiki.Arachidonic_acid
15. Simpoulos AP. (2002) The importance of the ratio of omega-6/omega-3 essential fatty acids. *Biomed Pharmacother* **56**: 365–379.

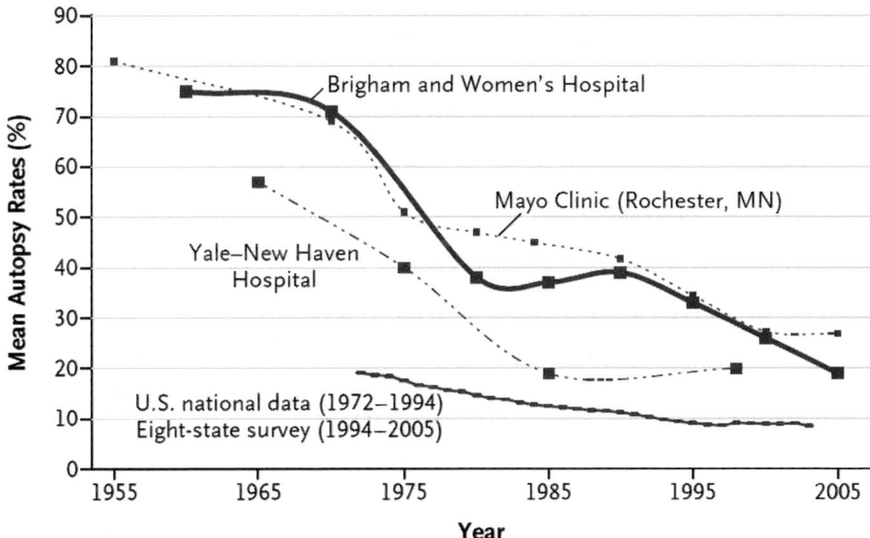

Fig. 18.2. Beginning in the mid-1950s, autopsy rates at smaller "community hospitals" round the country began falling from an average of about 50% in 1955 to 10% in the late 1990s. In 1970, the Joint Commission for Accreditation of Hospitals dropped its requirement that a given hospital need an autopsy rate of 20% to be accredited. Even the mighty Mayo Clinic was not immune to the decline, as its rate dropped from 80% in 1955 to 40% in 1995, with a further dip to less than 30% by 2005.

For coronary heart disease, the death rate among both men and women was halved between 1980 and 2000. (While the death rate for coronary heart disease fell during those years, the death rate for obesity and diabetes rose. Given the current alarming prevalence of obesity in children, there may well be a corresponding increase in the death rate 50-plus years from now.) About half the decline in coronary heart disease deaths is attributable to reductions in the major risk factors — cigarette smoking, lower cholesterol levels, and blood pressure control in hypertensive adults — while the other half is attributable to therapy, including the widespread use of clot-dissolving medications at the moment of a heart attack, coronary angioplasty, coronary artery bypass surgery, stents, cholesterol-lowering drugs including the "statins," angiotensin-converting enzyme (ACE) inhibitors, and β blockers. Nonetheless, coronary heart disease remains the No. 1 cause of death in the U.S. because of its sheer prevalence.

(*Sources*: Reproduced with permission from the publisher.) References 2–11.

Fig. 18.3. The Mayo Clinic logo. The three shields refer to the Clinic's three principal activities of patient care, research and education. The larger center shield refers to patient care, its main focus. The two flanking shields reflect its research and education, activities scholarly endeavors that "keep (its) patient care at the forefront of excellence," according to the Clinic's Public Relations Department.

(*Source*: Reproduced with permission of the Clinic.) Reference 15.

centers not only perform more autopsies[c] and do special studies on the derived material, but they often save the specimens for years and years for future research in unanticipated ways.

I had the opportunity, in the early and mid-1990s, to benefit from such an anatomical archive at the Mayo Clinic (Fig. 18.3). The question raised was: Why do hearts, often hearts "too good to die," rupture — literally break — developing a through-and-through rent with all the contained blood escaping into the delicate sac surrounding the heart, rendering it unable to pump? This catastrophe is largely limited to the first seven days after an acute myocardial infarction, with all-but-certain death quickly following. I suspected that there were subtle electrocardiographic clues portending cardiac rupture. I could

---

[c]A new radiological technique may obviate the need for an autopsy in the near future. Known as multidetector computed tomography (MCDT), the procedure is "emerging as the foremost cross-sectional imaging modality in forensic medicine. Its speed [and] ease of use…make it an excellent complement to autopsy" (Fig. 18.4).[13] To do the procedure following death, the cadaver is simply placed in the scanner, which will then obtain "4 to 128 slices with each rotation of the X-ray tube."[13] The technique yields a "high-quality almost-instantaneous noninvasive coronary [angiogram] with high diagnostic accuracy for the assessment of coronary stenoses [narrowings] and the identification and characterization of coronary plaques."[14] And it does so with a sensitivity and specificity of 91% and 93%, respectively.[14] Naturally, it is terribly expensive. Pathologists may need to collaborate with radiologists to make cadaveric scanning a reality. A so-called "virtual autopsy" could replace a traditional autopsy in specific instances.

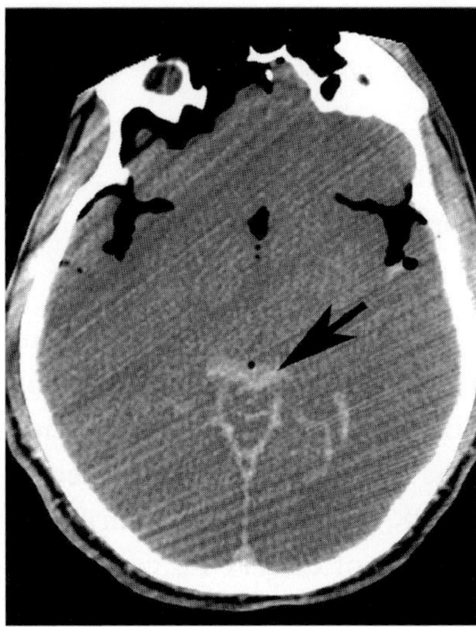

Fig. 18.4. A multidetector computed tomographic scan (MDCT) of the brain from a fatal motor vehicle accident. It shows a subarachnoid hemorrhage at the arrow. Given the high quality of the image and the ease of the procedure, it is possible that "MDCT could replace autopsies altogether," said Angela Levy, a radiologist at Georgetown University.

(*Source*: Reproduced with permission from the publisher.) Reference 13.

hear the hoof beats, but couldn't tell if they were those of a horse, llama, donkey, or even a galloping camel. Or, reput, I couldn't pinpoint the clues predicting cardiac rupture — or prove the point — without having hearts from unfortunate individuals who had died from this disaster to examine. The Clinic had the hearts — hundreds of them preserved in a tissue registry dating back 50 or more years — all readily retrievable, and with typical Mayo class the Clinic offered them for current restudy.[d] Through close examination of the serial electrocardiograms made during the week or so before death and re-examination of the preserved hearts, the answers came forth. The truth rested in a short segment of the EKG known as the "T wave." Images of those

---

[d]The only "catch" was that I had to come to Rochester to re-examine them and Minnesota winters can be trying. And, yes, winters in Colorado's mountains can also be trying. But I live in Boulder at the very foot of the Rocky Mountains, where winters are entirely livable (even appealing), spending summers in Aspen, where summertime activities far surpass the wintertime activity the town is renowned for.

EKGs and their correlation with the preserved hearts may be viewed in the September 1993 issue of the *Journal of the American College of Cardiology*, if wanted.[16]

Cardiac Mortality in Norway During WWII

Around the same time, two Norwegian hygienists determined the cardiac mortality rate in Norway from 1927 to 1948 from data derived from government records. During WWII, food was scarce in many European counties — other than Germany. In Nazi-occupied Norway, the death rate plummeted between 1940 and 1945 as calorie, fat, and cholesterol consumption dropped, and then rose in 1946–1948, when the diet returned to pre-WWII standards. In that country, the majority of dietary fat comes from butter, milk, cheese, and eggs. Figure 18.5 shows that until 1940 there was a "steady rise in mortality from circulatory diseases."[6] But, after the start of the war, there was "a sharp fall in mortality [from circulatory diseases], which was lowest in 1943–1945. When the war ended, cardiac mortality rose sharply."[13] Figure 18.6 shows a close correlation between the curve for consumption of fat and the curve for mortality from circulatory diseases between 1938 and 1948. (The term "circulatory diseases" is the equivalent of the current term "coronary heart disease," which came into play in the 1970s.) During the war, there was a severe shortage of red meats, whole milk, cream, cheese, and eggs, which led to the consumption of more fish, skim milk, cereals, potatoes, and vegetables.[15]

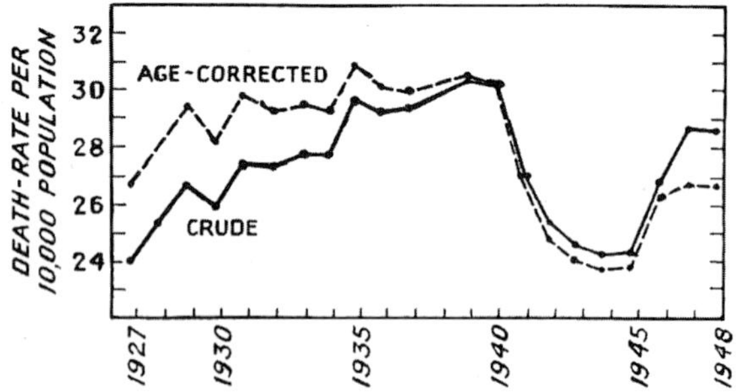

Fig. 18.5.   The death rate for circulatory diseases per 10,000 persons in Norway from 1927 to 1948. A slow rise between 1927 and 1940 was followed by a sharp fall during the war years and a sharp rise immediately after the war.

(*Source*: Reproduced with permission from the publisher.) Reference 17.

No question the war-imposed Norwegian diet was healthier than the native diet, and government records show fewer cardiovascular deaths at all ages in both sexes during the war.

These striking graphs beg the question: *Why should such an abrupt drop in cardiac mortality occur when dietary cholesterol is unexpectedly slashed if atherosclerosis is a disease that advances very slowly over many, many years?* It would be anticipated that a lessening of cholesterol intake in the order of 10–30 years — not weeks or a few months — would be needed to have an impact on the numbers of heart attacks noted clinically.

The answer to the question likely lies in a separate effect of cholesterol on the coagulation (clotting) system of the body. While plaque growth slowly narrows coronary arteries and sets the stage for an acute myocardial infarction, it is a tiny blood clot forming *within minutes* atop a long-standing coronary plaque that actually triggers a heart attack (see Chap. 12). The blood clot is the *coup de grace.* The well-known procoagulant effects of cholesterol and other fats promote "the clot that kills." It's an "it takes-two-to-tango" scenario. Cholesterol and other lipids are the initial culprits, while sticky platelets — also fostered by cholesterol and other fats — lead to a blood clot in one of the

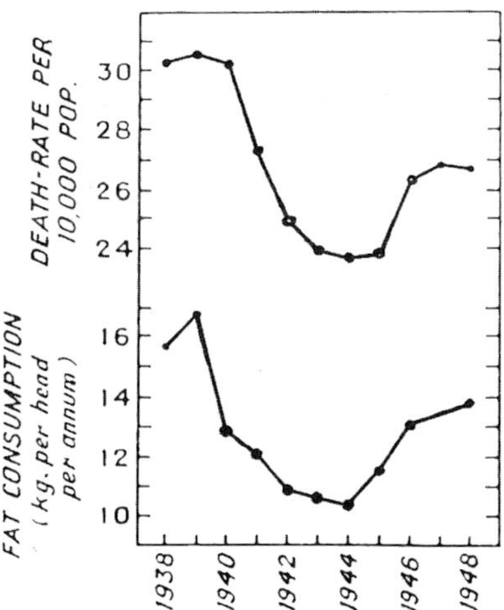

Fig. 18.6.   The mortality rate for cardiovascular diseases in Norway from 1938 to 1948 as a function of fat intake in the form of butter, milk, cheese, and eggs. Cardiovascular mortality mirrors fat consumption.

(*Source*: Reproduced with permission from the publisher.) Reference 17.

three coronary arteries. When the Norwegian/Danish diet changed from one with lots of "fat — especially from dairy products, meat, and eggs" to one with generous amounts of fish with lots of omega-3 fatty acids — blood platelets weren't nearly so "sticky," so clots couldn't form as easily and cardiac mortality plummeted.[18] The war-years Danish diet was (qualitatively) more like the Greenland Eskimo diet. While the Danes didn't eat seal and whale blubber, they did consume far more fatty fish with their high DOC and EPA content during the war. And blood clots didn't only tend to form less frequently in the coronary arteries between 1940 and 1945. They also tended to form less frequently in leg veins (thrombophlebitis) — a problem that follows many types of surgery — as fish consumption trebled during the war.[16]

References

1. Kimura N. (1956) Analysis of 10,000 postmortem examinations in Japan. In: Keys A, White PD. (eds.), *World Trends in Cardiology,* selected papers from the 2nd World Congress of Cardiology and the 27th Annual Scientific Meeting of the American Heart Association, Washington, DC. Hoeber-Harper, Medical Book Department of Harper & Brothers, New York, Vol. 1, pp. 22–33.

2. http://www.medicinenet.com/autopsy/page3.htm

3. Shojania KG, Burton KC. (2008) The vanishing nonforensic autopsy. *N Engl J Med* **358**: 872–875.

4. Guariglia P, Abrahams C. (1985) The impact of autopsy data on DRG reimbursement. *Hum Pathol* **16**: 1184–1186.

5. McPhee SJ. (1996) Maximizing the benefits of autopsy for clinicians and families. *Arch Pathol Lab Med* **120**: 743–748.

6. Regal KM, Graubard BI, Williamson DF, Gail MH. (2005) Excess deaths associated with underweight, overweight and obesity. *JAMA* **293**: 1861–1867.

7. Gregg FW, Cheng VJ, Cadwell BL. (2005) Risk factors according to body mass index in US adults. *JAMA* **293**: 1868–1875.

8. Ogden CL, Carroll MD, Curtin LR, Mc Dowell MA, Tabak CJ, Flegal KM. (2006) Prevalence of overweight and obesity in the United States. *JAMA* **295**: 1549–1555.

9. Reichert CM, Kelly VL. (1985) *Health Aff* **4**: 82–92. Prognosis for the autopsy.

10. http://www.mayoclinicproceedings.com/content/75/6/562.

11. Ford ES, Ajani UA, Croft JB, Critchley JA, Labarthe DR, Kottke TE, Giles WH, Capwell S. (2007) Explaining the decrease in U.S. deaths from coronary disease, 1980–2000. *N Engl J Med* **367**: 2388–2398.

12. Nemetz PN, Roger VL, Ransom JE, Bailey KR, Edwards WD, Leibson CL. (2008) Recent trends in the prevalence of coronary disease: a population-based autopsy study of nonnatural deaths. *Arch Intern Med* **168**: 264–270.
13. Levy AD, Collins KA. Postmortem radiology and imaging.
14. Lewis BS, Halon DA. (2007) Integrating multidetector computer tomography into clinical practice. *JACC* **47**: 960–962.
15. http://sharing.mayoclinic.org/2009/03/09/logo-logic-what-do-those-shields-mean
16. Oliva P, Hammill SC, Edwards WD. (1993) Cardiac rupture: a predictable complication of acute myocardial infarction. *J Am Coll Cardiol* **212**: 720–726.
17. Strom A, Jensen RA. (1951) Mortality from circulatory diseases in Norway, 1940–1945. *Lancet* **1**: 126–129.
18. Bang HO, Dyerberg J. (1972) Plasma lipids in Greenlandic West Coast Eskimos. *Acta Med Scand* **192**: 85–94.

19

WINE, BEER, AND SPIRITS AS ANTIOXIDANTS

Although wine was thought for centuries to reduce social inhibitions and to enhance a person's mood — when consumed in a light-to-moderate amount — no scientific evidence to corroborate a healthful benefit of wine existed until the late 20th century. The harmful effects of heavy alcohol drinking on the liver were well known. Now, "a consistent body of evidence supports the hypothesis that moderate alcohol consumption reduces the risk of coronary heart disease."[1]

A Brief History of Wine

About 4000 B.C., natives of Mesopotamia — the region of current-day Iraq — made wine by stomping on grapes, collecting the juice in a stone trough or tank with a spout, and then storing the liquid in a vat, where the process of fermentation began.[2] In the culture of ancient Egypt, where wine production spread between 3000 and 2000 B.C., wine was a popular beverage limited to the privileged upper class.[3] When the drink reached Greece, Dionysus — the Greek god of wine — lauded the virtues of some wine and the dangers of too much. Plato wrote, long before the science of medicine reached other European shores, that "wine. . .is a medicine given with the aim of securing. . .strength of the body."[4] And by the time the Roman Empire flourished — between 260 B.C. and 476 A.D. — common people, as well as the aristocracy, were enjoying its pleasant effects, when consumed moderately. The natives of all three Mediterranean countries saw wine as a "healthful beverage" providing "calories and vitamins."[5]

Plato's view of wine as a "medicine" found limited therapeutic application until after the Renaissance. In spite of the surge in laboratory research following the Renaissance during the 17th and 18th centuries, few effective drugs became available soon. Quinine (to treat malaria) and opium (to lessen pain) were notable exceptions, both having been developed at an earlier time. Alcohol like opium was recognized as a painkiller among the culture-centered Greeks and Romans. The drink's anesthetic properties were put to practical use in the mid-to-late 19th century, when large doses of whiskey and/or wine were sometimes given before physicians or laymen removed bullets from gun-slingers in the Wild West. (Ether became a more sophisticated means of reaching the same end point, following its introduction as an inhaled general anesthetic on October 16, 1846.)[a] Opium was clearly a more potent anesthetic, but whiskey and wine prevailed, being far more available.[b]

Fast forward to 1988, when whiskey, wine, and beer — consumed moderately — were found to reduce the number of deaths from heart disease among nurses in the U.S.[6] In that report from the Harvard School of Public Health — called the Nurses' Health Study — the 59,431 RNs who drank "moderate amounts of alcohol"[c] during the four-year study had fewer heart attacks and strokes than did the 28,095 nondrinkers. The most plausible explanation for the survey's findings seemed to be the known beneficial effect of alcohol on the HDL cholesterol level. But that interpretation was reached before flavonoids were shown, in 1990, to significantly slow LDL oxidation,[7] and that red wine was a rich source of flavonoids, in 1993.[8] Suddenly, a new explanation for the French paradox (see chap. 1) — independent of its alcohol content — was at hand.

---

[a]A general anesthetic can also be given intravenously, subcutaneously, or intramuscularly. The inhaled form is used most often, although a short-acting, sleep-inducing drug, such as Valium, may be administered intravenously just before the introduction of an inhaled anesthetic.

[b]Opium and its principal derivative morphine are unfortunately very addictive; a solution known as "tincture of opium" contains 10%–20% opium and is used occasionally in modern medicine. Morphine remains a mainstay in American coronary care units. It is regularly given to patients during the first six hours of a heart attack, to reduce the excruciating pain and the inevitable apprehension.

[c]A "moderate" amount of alcohol was defined as roughly one drink a day. A "drink" was defined as a 12-ounce container of beer, a 4-ounce glass of wine, or a "standard" mixed drink, presumably made by stirring a "jigger" (1.5 ounces) of an 80-proof liquor with a carbonated beverage or a juice.

Flavonoids: What Are They? Where Are They? Why Did They "Appear" in the Late 20th Century?

Origin of the term

The term "flavonoid" refers to a group of compounds made by nearly all plants.[9] Fungi and algae are the sole exceptions. The absence of flavonoids in fungi makes flavonoids effective antifungal agents.[10] Flavonoids are concentrated in the skin and seeds of fruits, and in petals. None are made by the human body.

The term "flavonoid" comes from the Latin word "*flavus*" meaning "yellow". More to the point of this chapter, flavonoids come from a colorless plant pigment flavone whose *derivatives* appear yellow and thus are "flavonoids." Although general awareness of these compounds has occurred only since their documentation in red wine (in 1993), chemists have known about them since 1826, when a flavonoid was discovered in citrus fruits.[11] A century later (in 1936), a (relatively) modern chemist, Dr. Albert Szent-Gyorgyi, found a substance in lemon peels that decreased capillary permeability — which made them less "leaky."[d] He called it "vitamin P."[e,12] Oranges particularly the peels were soon shown to be a rich source of the new vitamin. A new nutrient coupled with the then-new process of manufacturing frozen, concentrated orange juice led to a boom in the Florida citrus industry. Not only could the juice be sold in a frozen form, — the seemingly worthless peels could also be used to extract vitamin P![13]

Unfortunately, research on vitamin P during the 1940s did not confirm that the leaky blood vessels were due to a flavonoid deficiency.[14] During the same time period, vitamin P was also shown not to be "essential" for human health since no symptoms of deficiency developed when it was removed from the diet.[15] The dual setback led the Joint Committee on Nomenclature to recommend, in 1950, that the term "vitamin P" be replaced by "bioflavonoids."[16] Years later in 1986 a British scientist concluded that the "aftermath of the controversial claims of 1936. . .undoubtedly discouraged [further] medical

---

[d]One of the most visible consequences of "leaky" capillaries is edema — swelling of the ankles and feet. However, pedal edema is not likely to resolve by eating citrus fruits or taking vitamin P, because the swelling is not due to an inherent reduction in capillary permeability. The edema is related to a high pressure in the capillaries of the feet, usually from congestive heart failure.

[e]Vitamin P was the second discovery by Szent-Gyorgi. In 1932, he found vitamin C. The twin discoveries led to his being awarded the Nobel Prize in Physiology or Medicine in 1937.

experiments [on flavonoids] at the time."[17] Research on flavonoids was rejuvenated by the French paradox in 1979 and by the discovery that red wine had "potent antioxidant properties toward [the] oxidation of human LDL," due to its "natural flavonoids," in 1993.[8]

5000 Types, 5 classes, and 4 chemical structures

Because of the ubiquity of flavonoids in the vast plant kingdom, a huge number of individual flavonoids exist. In 1986, approximately 3000 flavonoids had been identified. By 1996, about 4000 types of flavonoids were known, and in 2006, roughly 5000 had been discovered. Despite the large number of individual flavonoids, "a limited number of basic structures exist".[18](Fig. 19.1).

Where Flavonoids are Found

Although flavonoids were first discovered in citrus fruits, they were later found in the skins of dark grapes, in black and green tea, and in

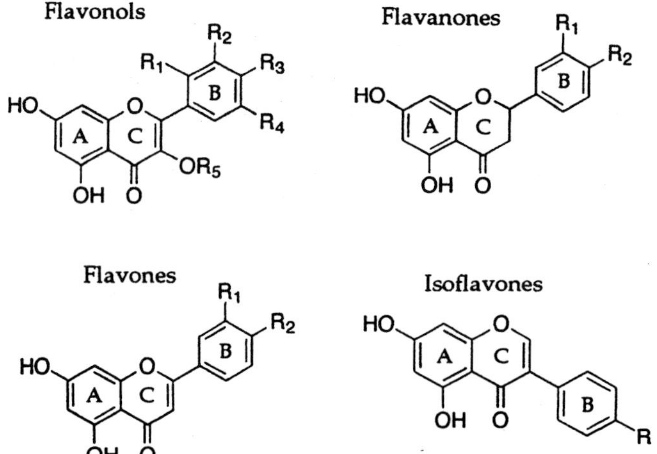

Fig. 19.1. The chemical structures of the four basic groups of flavonoids that give rise to more than 5000 individual flavonoids. The antioxidant effect of various flavonoids increases with the number of hydroxyl groups (OH) that each of the four basic structures contains.

(*Source*: Modified and reproduced with permission from the publisher.) Reference 19.

Fig. 19.2.   A stack of cocoa (or cacao) pods.

chocolate — a substance made by man from the cocoa tree. The tree, which grows well in the tropical climate of Central America and similar latitudes in Africa and Southeast Asia, bears pods filled with seeds — called beans — and their flavonoids. The pods are usually yellow, but may be red or purple, depending on which of the three main tree varieties they grow on (Fig. 19.2).

The pods grow on trees between 10° north and 10° south of the Equator. Each tree bears 25–50 pods per year, and each pod yields 20–40 seeds. The pods take 5–6 months to mature. As they ripen, their color changes from green to yellow to red to purple. After the pods are harvested, the "beans" inside are placed into unrefrigerated outdoor wooden boxes covered with banana leaves, and allowed to ferment — literally rot, as tropical bacteria and yeast inevitably smother them. After fermenting for 3–6 days, the beans are sun-dried. During the drying process they take on the familiar brown color of chocolate. The bitter taste characteristic of flavonoids disappears after milk and sugar are added later by chocolate companies to convert the cocoa beans into candy.

Tea

In Japan, where tea consumption is high, the incidence of coronary heart disease is relatively low — in spite of heavy cigarette smoking. Tea, as a member of the plant kingdom, also contains flavonoids. And tea flavonoids slow LDL oxidation *in vitro.* Adding these three observations up, one might make the reasonable conclusion that tea is the wholesome agent.

However, that conclusion could be a false syllogism, for several reasons. In Japan, at least, seafood consumption is also high; lots of omega-3 fatty acids might be responsible.

What is the Best Beverage to Reduce the Chances of Coronary Heart Disease: Wine, Beer, or Spirits?

It is now accepted that "moderate (alcohol) intake appears to exert a protective effect against coronary heart disease, as compared with drinking no alcohol."[1] Nearly 100 studies document the benefit of moderate alcohol consumption with regard to the risk of coronary heart disease.[22] The data are convincing enough to reach the textbook level. One such cardiology textbook simply acknowledges as a matter of fact that "moderate alcohol intake is associated with a protective effect against coronary artery disease."[23] The principal question today is: Which type of alcoholic beverage statistically confers the most cardiac protection? Or are they equal?

Wine Is More Cardioprotective than Beer or Spirits

Most of the many articles published in the medical literature — now totaling BLANK — looking at the effect of alcoholic beverages on coronary heart disease have either found wine to be more cardioprotective than beer or distilled spirits or found there to be no difference among the three types of beverages for the incidence of heart attacks. But a few reports give beer or distilled spirits the edge. The explanation for the advantage of wine in some reports — mostly from Europe — not showing equality appears to lie in its flavonoid content, as a grape derivative. All three types of beverages contain alcohol — the compound that increases the concentration of HDL in blood — but only wine has a significant amount of flavonoids.

A paper from Copenhagen published in the *British Medical Journal* concluded, "Low to moderate intake of wine is associated with lower mortality from cardiovascular disease and cerebrovascular disease…." However, a "similar intake of spirits implied an increased risk, while beer drinking did not affect mortality" (Fig. 19.3).

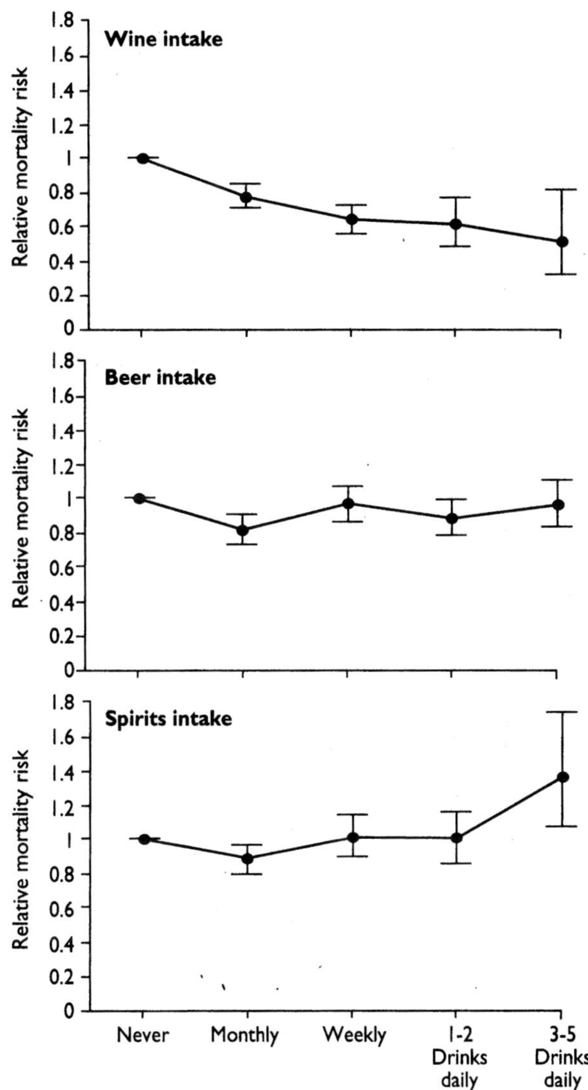

Fig. 19.3. Among more than 13,000 men and women aged 30–70, those subjects who drank as many as 3–5 glasses of wine a day had half the risk of dying over the next 10 years compared with those who did not drink wine. Beer drinkers had not a different mortality rate than abstainers, while those people who drank 3–5 glasses of spirits a day experienced an increased mortality rate.

(*Source*: Reproduced with permission from the publisher.) Reference 24.

In the famed Framingham study, wine also had a "strong" inverse association with coronary heart disease mortality.[25] Mortality was cut by about 40% in men who did not smoke and nearly 50% in men who smoked heavily. In women who smoked, a similar, though less striking, reduction in mortality was seen.

Those findings were reinforced by an analysis of mortality data gleaned from the 1990 World Health Statistics Annual Report. Upon reviewing a number of reports on the relationship of diet and alcoholic beverage consumption to coronary heart disease mortality rates in the U.S. and Europe, Criqui[26] learned that France "had the highest wine intake and the highest total alcohol intake [among 21 countries], and the second lowest coronary heart disease mortality rate." Only Japan fared better (Fig. 19.4).

The "strongest and most consistent correlation" was between wine consumption and mortality from coronary heart disease. Beer and spirits were only "weakly" inversely correlative. He summed up, "...wine was the strongest

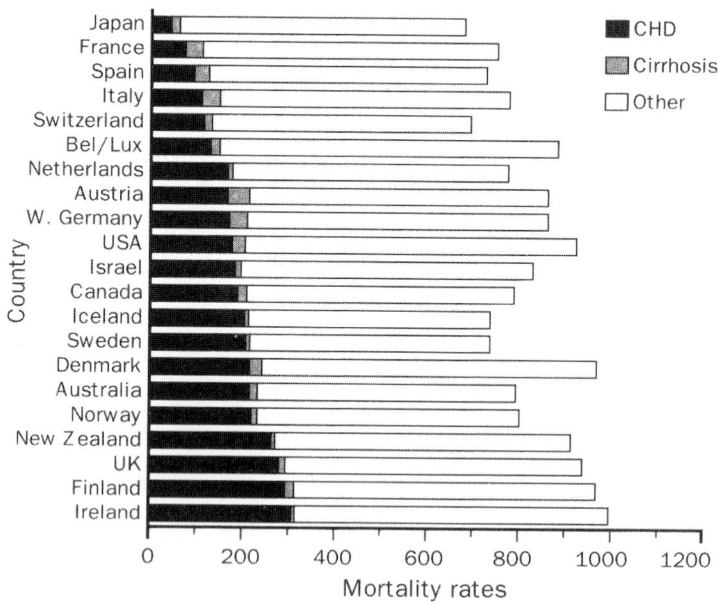

Fig. 19.4. The mortality rates for coronary heart disease per 100,000 population in 21 countries. The age range of each population is 35–74 years. Eight of the nine lowest mortality rates are in European countries, where wine consumption is high.

(*Source*: Reproduced with permission from the publisher.) Reference 26.

dietary correlate of coronary heart disease." That conclusion was shared by a report from the British Regional Heart Study. It found that "wine drinkers showed substantially lower risks of CHD and all-cause mortality than those consuming other types of alcohol."[27]

Wine, Beer, and Spirits are Equally Cardioprotective

On the other hand, E.B. Rimm, an epidemiologist at the Harvard School of Public Health, and A. Klatsky, a cardiologist at the Kaiser Permanente Medical Center in the Bay Area, pointed out that of 10 published prospective cohort studies (to be defined), four showed a "significant inverse association between wine consumption and coronary heart disease, four reported such an association for beer, and four reported an association for spirits."[28] Hardly a consensus. (The sum is greater than its parts, because two studies found a significant association for two of the three types of alcohol.) Mustering more evidence, they pointed out that if all 10 studies are added up the total number of participants is "more than 305,000 men and women followed up for more than 1.8 million person years." Considering those figures, they concluded that "if any type of drink [provides] extra cardiovascular benefit apart from its alcohol content, the benefit is likely to be modest at best...."[28] To back that opinion up, they pointed out that among the nearly 100 published studies (using any of the three methodologies to be described) concerning the relationship of alcohol consumption to coronary heart disease, alcohol of "any type" reduces the risk of coronary heart disease.[22,28] The Nutrition Committee of the American Heart Association shares the belief that wine, beer, and spirits are equally beneficial. Its Science Advisory says that the "consumption of one or two drinks [of any type] per day is associated with a reduction in risk [of coronary heart disease] of approximately 30–50%."[29] The committee attributes "approximately 50%" of the cardioprotective effect from alcohol to an elevation of the HDL cholesterol level and the other half to its ability to thin out the blood making formation of a blood clot in a coronary artery less likely, and to its antioxidant properties. The Advisory ends up recommending "one or two drinks per day" provided "no contraindications to alcohol are present."

How can These Differences Between Studies be Reconciled?

The differing results may be due to the use of different methods to collect data in different epidemiological studies. For example, in an *ecological study* a population of people living within a geographical region is followed to determine the frequency of an outcome over time. The 1993 study which first described the French paradox was an ecological study.[30] The population of people studied consisted of the inhabitants of 18 countries, while the outcome measured was the death rate. The per-capita consumption of wine, beer, and spirits in the 18 countries was the variable correlated with the death rate. Wine had a strong inverse correlation with mortality from coronary heart disease (Fig. 1.1); beer and spirits did not. There was also a strong positive association of saturated and monounsaturated fat consumption with death from coronary heart disease.

In a *case control* study, individuals who have had an outcome of interest (such as a heart attack) are compared with individuals who have not had the outcome of interest. The two groups are then juxtaposed to determine what characteristics were more prevalent in the group with the outcome of interest. Case control studies are relatively easy to do and are inexpensive to carry out, but they may have inaccuracies because such studies rely on an individual's recall of the past.

A *cohort study* takes a very different approach. In such a study, as an observational study,[f] a (usually) large group of people — a cohort — who have not yet developed signs or symptoms of a disease are followed over an extended time (usually years) to determine the frequency of a particular outcome, such as death from a heart attack. The prevalence of certain lifestyle characteristics (such as diet and exercise) or external factors (such as divorce or loss of a job) in those individuals who later develop an outcome of interest can then be compared with the prevalence of those characteristics and factors in those who do not develop the outcome of interest. Importantly, the cohort is identified *before* the appearance of the disease under investigation.[31] Prospective cohort studies bypass the problem of accurately recalling the past, because they collect the data from study participants before the outcome of interest occurs. The Nurses' Health Study[32] and the Framingham Study[g,33] are two

---

[f]A cohort study may also be referred to as an "observational study" because the cohort can be "observed" over a long period of time while no attempt is made to influence its members' medical outcomes; for example, no treatment for the condition being studied is given.

[g]The Framingham (Massachusetts) Study is an ongoing study of the natural history of coronary heart disease. It was begun in 1948 under the direction of the N.I.H. and in collaboration with Boston University. A group of 5209 men and women then aged 30–62 were extensively interviewed, with particular emphasis on their lifestyle characteristics. Much of what is now

Table 19.1 Three types of commonly used epidemiological studies.

Type of study	Methodology
Ecological study	A population of people is followed for years to determine the frequency of an outcome.
Case control study	A population that has an outcome of interest is compared with a population that does not have the outcome of interest.
Prospective cohort study	A population of people with a characteristic of interest, such as smoking or alcohol consumption, is compared with a population of people without the characteristic of interest to determine an outcome of interest.

well-known cohort studies. The three study designs are shown in Table 19.1. Each has its strengths and weaknesses.

The three different study designs used in the many studies that have correlated the type of alcoholic beverage with coronary morbidity and mortality largely account for the different conclusions regarding the question of whether wine is more cardioprotective than beer and spirits or whether the three drinks are equally effective. For example, most ecological studies favor wine. And the majority of the ecological studies come from central Europe — France, Italy, Spain, and Switzerland, the chief wine-producing and wine-consuming countries of the world. It makes sense that wine would statistically offer more cardioprotection than beer or distilled spirits in those nations.[34] Countries such as France, where wine is the preferred alcoholic drink, also "have the largest overall alcohol consumption."[26] That brings up the possibility that the protective effect of wine may simply reflect the (primary) protective effect of its alcohol. Or, the protective effect could be a sign of the healthier lifestyle characteristics of wine drinkers, in general (to be discussed soon). And the flavonoids in wine might add cardioprotection to that of the alcohol itself.[26]

In contrast to the ecological studies that found wine to be more cardioprotective than beer or spirits, most case control and prospective cohort studies

---

common knowledge about the risk factors for coronary heart disease — high blood pressure, high cholesterol level, smoking, obesity, insufficient exercise, and diabetes — was learned through the Study.

do not show that wine is more effective than beer or spirits.[28] In 10 prospective cohort studies, four found an inverse association between the risk of coronary heart disease and wine consumption, four others found that the association also held true for beer, and another found that it applied to spirits as well.[28] (The sum of 12 associations exceeds the number of studies because two showed an association for more than one type of beverage.) To boot, "most cohort studies have been reported from countries such as the U.S., U.K. and Australia where wine consumption does not exceed other (alcoholic) beverages."[26] Therefore, studies from these countries are not likely to find a difference among the three beverages in their cardioprotective effect.

Beer and/or Spirits are Better than Wine

But wait. Beer lovers and even those who prefer "mixed drinks" will find encouragement from what follows. There are a few reports favoring beer or liquor over wine for cardioprotection. Among Japanese men now living in Hawaii, beer drinkers had fewer nonfatal heart attacks and deaths from coronary heart disease than did wine or liquor drinkers.[35] In a 1999 study of American men and women in the Boston area, the risk of a heart attack was also lower in beer drinkers than in wine or spirits drinkers.[36] Two other reports found a slightly lower risk of a nonfatal heart attack[37] or coronary death[38] in beer drinkers than in wine or spirits drinkers.

At least one report favors liquor. In the Health Professionals Study[39] — covering dentists, pharmacists, veterinarians, optometrists, osteopaths, and podiatrists aged 40–75 — the risk of a fatal or nonfatal heart attack during a 12-year followup was lowest among those who drank one-half to two drinks containing liquor per day.[h] The finding could simply reflect the higher consumption of liquor than beer or wine in this group of healthcare providers. It is harder to show a statistical advantage in any underrepresented group or groups in a given study. Thus, the alcoholic "beverage consumed most widely by a given population is the one most likely to be inversely associated with the risk of [a] myocardial infarction in that population."[40] For example, in France, where 79% of the alcohol consumed is in the form of wine, wine is more beneficial than beer or spirits.[41] On the other hand, in Northern Ireland, where 95% of

---

[h]The need for either angioplasty or coronary bypass surgery was also reduced in those men who drank liquor moderately.[39]

the alcohol consumed is *not* in the form of wine, i.e. beer and spirits predominate, a cardioprotective effect of both drinks comparable to the effect of wine in France is the rule.[42]

So What's the Verdict? Wine, Beer, or Spirits?

Taste aside, given the evidence that all three beverages may be equally beneficial or that one may offer more cardioprotection than another, the choice may come down to which one provides more (or less) potentially helpful or harmful substances, other than alcohol. The potentially helpful substances are flavonoids. Here red wine gets the nod. The potentially harmful substances are nitrosamines. These compounds are produced through the combination of *nitrates* and *amines* (amino acids) under the acidic conditions of the stomach.[43] They are linked to cancer of the mouth, pharynx, and esophagus.[44] Nitrosamines are found in all three types of alcoholic drinks.[i] But their concentrations differ. Wine has an "insignificant" amount of nitrosamines, while beer contains the highest concentration and distilled spirits have an intermediate quantity.[45] For that reason, here again wine gets the nod. And while it sounds like drinking a moderate amount of alcohol regularly may be trading a heart attack for cancer, the chances of the latter are far less than the chances of the former. In almost all studies, the *all-cause* mortality — as well as the cardiac mortality — is lower in individuals who drink a moderate amount of alcohol regularly.

How do Moderate Drinkers Fare Compared with Nondrinkers?

According to a 2001 editorial in *JAMA*, "numerous. . .studies have shown that light to moderate drinkers of alcohol are at lower risk than abstainers for fatal or non-fatal coronary heart disease."[46] In a report from Great Britain, of 7735 middle-aged men, the highest mortality rate was seen in nondrinkers and occasional drinkers, while the lowest mortality rate was seen in light-to-moderate drinkers (Fig. 19.5).[47] Ditto in a study of 22,071 male U.S. physicians, 40–84 years of age, followed for an average of 10.7 years; those who

---

[i]Nitrosamines are also found in fish and in most meat and cheese products preserved with a nitrite pickling salt. The U.S. government is aware of this and strictly limits the amount of nitrite used in meat and cheese products.[43]

ALCOHOL INTAKE AND MORTALITY

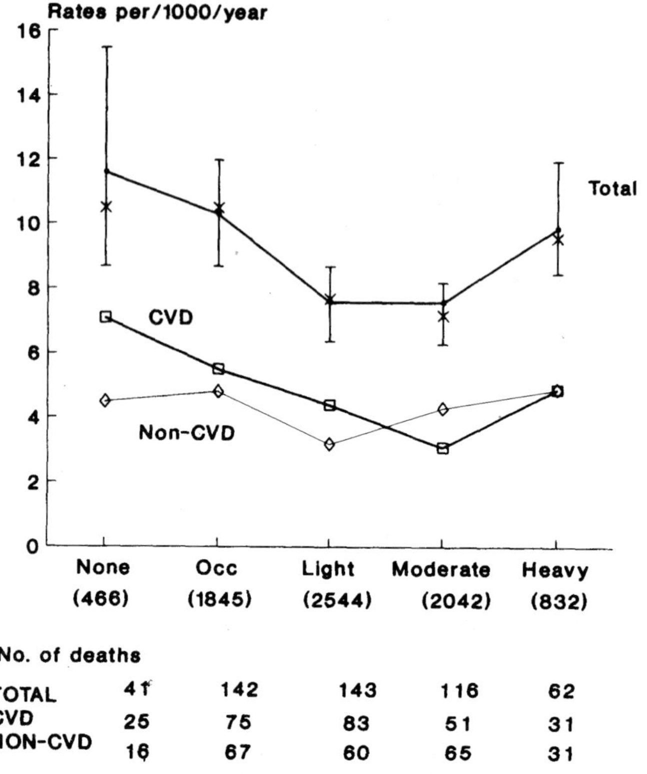

No. of deaths	None	Occ	Light	Moderate	Heavy
TOTAL	41	142	143	116	62
CVD	25	75	83	51	31
NON-CVD	16	67	60	65	31

Fig. 19.5. The total, cardiovascular (CVD), and noncardiovascular (non-CVD) mortality over 7.5 years in five alcohol intake groups ranging from none to heavy. The highest mortality rates were seen in nondrinkers and occasional drinkers. Light and moderate drinkers had the lowest mortality rate.

(*Source*: Reproduced with permission from the publisher.) Reference 47.

consumed 5–6 drinks per week had fewer heart attacks than those who consumed less than one drink per week or did not drink at all.[48]

At this time, there is a substantial amount of data supporting the regular consumption of some alcohol to lessen the likelihood of a heart attack. Yet few physicians would recommend that a person drink alcohol to reduce the risk of coronary heart disease if that person objects to alcohol for religious reasons or has had a child or other close family member killed by a drunk driver or if the person is pregnant. Nor should a person who drinks heavily use "it's good for my heart" as an excuse to continue drinking. Heavy drinking runs a big risk of cirrhosis, as well as cancer of the mouth, pharynx, esophagus, liver, and breast[49] — in addition to a serious motor vehicle accident.

Does the Cardioprotective Effect of Alcohol Apply to Diabetic Patients?

Apparently yes. A 1999 study from the University of Wisconsin found that among 983 older diabetics (30–84 years old; mean age 68.6) followed for a mean of 12.3 years, those who drank 1–2 drinks a day had a "considerably reduced risk of death due to CHD compared with never drinkers."[50] The authors said that their results were "not inconsistent" with the guidelines of the American Diabetes Association, which recommends that "the same precautions regarding the use of alcohol that apply to the general public also apply to people with diabetes," i.e. "no more than 1 drink/day for women and no more than 2 drinks/day for men."[51] An editorial in the same issue of *JAMA* cautioned that "while evidence is mounting that light to moderate alcohol consumption may be associated with reduced insulin resistance and lower CHD risk," it is still risky to recommend that diabetics drink alcohol because "even at relatively low doses, alcohol intake may induce hypoglycemia (leading to insulin shock). . . ."[52]

Can a Couple of Drinks a Day be Taken Instead of an Aspirin a Day to Ward off a Heart Attack?

Probably not. If the estimate is correct that "approximately 50% of the protective effect of alcohol is mediated through increased levels of HDL cholesterol,"[29] while "the other half" is mediated through its antioxidant and "blood-thinning" properties, then only the effect of alcohol on the latter would be similar to the established effect of aspirin on platelets — thinning the blood. And even if the effect is qualitatively similar, is it quantitatively equivalent?

Aspirin and alcohol act at different points in time during the long period between plaque initiation and a heart attack (see Chap 12 regarding foam cells and plaque development). Alcohol acts early on by increasing the HDL level, thereby slowing — or perhaps reversing — the growth of a plaque. Aspirin acts much later, when, if despite leading a healthy lifestyle and drinking a moderate amount of alcohol, a plaque continues to grow, setting the stage for plaque rupture and clot formation. Through its effect on platelets, aspirin cuts down the chances that a dreaded blood clot will form on the surface of a recently ruptured, cholesterol-laden plaque.

In the Physicians' Health Study, "no interaction" between aspirin and alcohol was found (Fig. 19.6).[53] The protective effect of aspirin — at a low dose of one adult aspirin (325 mg) every other day — was not affected by alcohol. (Such a relatively low dose of aspirin generally avoids stomach irritation. Even one baby aspirin a day (82 mg) thins the blood effectively.) The men who took randomly assigned aspirin had 44% fewer myocardial infarctions and an "exceptionally low cardiovascular death rate" compared with those who took placebo. The benefit was so striking that the trial was terminated early because the Data Monitoring Committee felt it was unethical to any longer withhold aspirin from the placebo group.[54]

Alcohol consumption did not statistically affect the remarkable reduction in myocardial infarctions and cardiovascular deaths in the physicians who took aspirin. Nonetheless, according to the program's Project Director, that did not exclude "synergy or overlap of biological mechanisms between aspirin and alcohol" because synergy might simply not have reached a statistically detectable level.[55] Therefore, the study group found it "difficult to determine

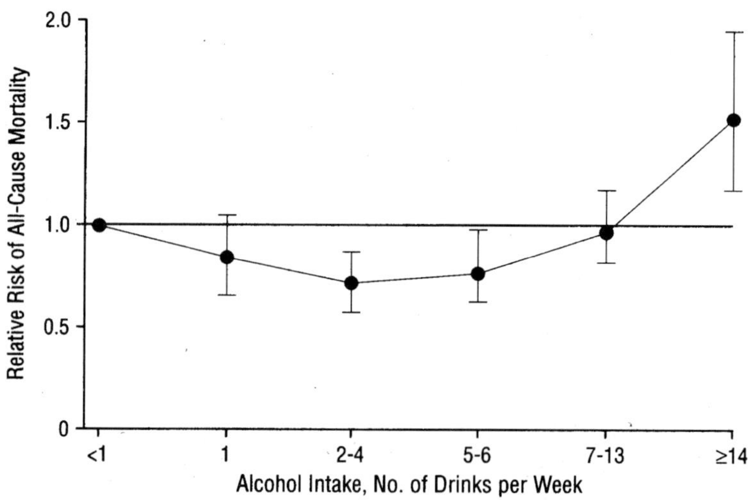

Fig. 19.6.   The relative risk of all-cause mortality in the Physicians' Health Study, according to alcohol intake. Vertical bars represent 95% confidence limits. Data are adjusted for aspirin intake, age, smoking, exercise, and diabetes.

(*Source*: Reproduced with permission from the publisher.) Reference 48.

whether drinking 2–6 drinks a week is…'equivalent' to taking 325 milligrams of aspirin every other day in terms of reducing the risk of a myocardial infarction." Because the study was terminated early and because synergy between aspirin and alcohol — not detected statistically — could have been present, "aspirin and alcohol (when used in moderation, of course) may each potentially reduce the risk of coronary events!"

Can Alcohol Make a Difference in Older Adults?

Few studies have examined the effect of alcohol on longevity in older people. One that did found that among 490,000 "middle-aged and elderly adults" (age range 30–104 years, mean 56 years) who consumed 1–2 drinks a day, mortality from heart disease and strokes was 30%–40% lower over nine years than among those who did not drink (Fig. 19.7).[49] In that investigation, sponsored by the American Cancer Society and the World Health Organization, there were substantially more deaths from cirrhosis, and cancer of the liver, mouth, pharynx, larynx, and breast (in women) — all having a well-established association with excessive alcohol consumption — among both men and women consuming four or more drinks a day. And, importantly, the health benefit of modest alcohol consumption "was far smaller than the large increase in risk produced by tobacco,"[49] negating the notion that a few drinks a day can offset the harm done by smoking cigarettes.

Similarly, in a separate study of 4410 men and women aged 65 or more, there were fewer heart attacks and cardiac deaths over an average of 9.2 years among those who consumed two drinks a day — regardless of the type of alcoholic beverage — than among abstainers.[56] Finally, among 12,321 middle-aged or older British men (50–90 years of age) the lowest mortality rate over 13 years was in those who imbibed 1–2 alcoholic beverages per day.[57] And, once again, the type of beverage consumed — wine, beer, or spirits — made not a whit of difference.

This information may be applicable and reassuring to the families of elderly adults residing in an assisted living center where the amount of alcohol is monitored by the nursing staff. But its applicability to elderly adults living alone or with an elderly mate is questionable. It might be easier to "have another nip" if unsupervised or if living with an elderly mate, who also "nips," and a fractured hip following a fall for any reason in a person over 70 is too often the "kiss of death."

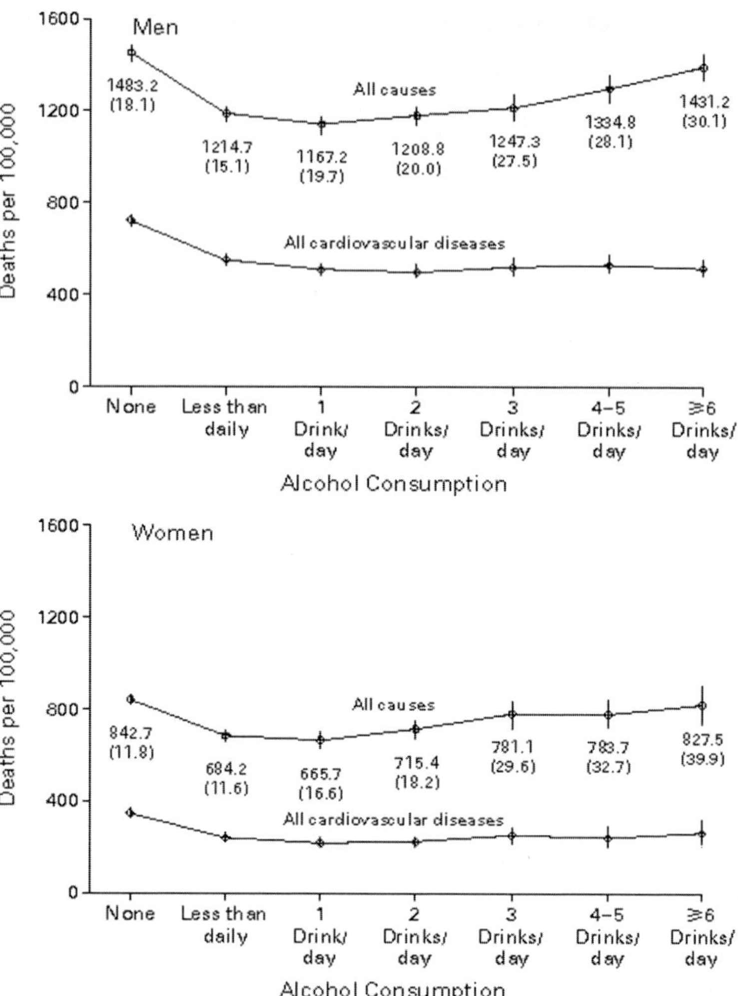

Fig. 19.7. This graph plots the death rate among 490,000 men and women (mean age 56 years) for cardiovascular diseases and all other causes on the ordinate against the amount of alcohol consumed on the abscissa. The rate of death from cardiovascular diseases is 30–40% lower among those individuals who drank alcohol than the rate in nondrinkers. The reduction prevailed irrespective of the type of alcoholic beverage, i.e. wine, beer, and liquor were equally beneficial. However, the death rate for "all causes" began to climb when more than two drinks daily were consumed because of a rise in mortality due to cirrhosis as well as cancer of the mouth, esophagus, pharynx, larynx, liver, and breast (in women) — all well-known consequences of excess alcohol. The optimal alcohol consumption, therefore, is 1–2 drinks per day.

(*Source*: Adapted and reproduced with permission from the *New England Journal of Medicine*.) Reference 49.

Do Wine Drinkers Lead a Healthier Life Style than Beer or Liquor Drinkers?

The answer is "Yes, usually they do." Wine drinkers, in general, are leaner, exercise more often, smoke less, eat more fruits, fish, vegetables, salads, and olive oil[58] (Table 19.2), and are better educated than beer and liquor drinkers[58,59] (Table 19.3). An editorial published in the American Heart Association's official journal, *Circulation*, agreed: "Studies typically have found that wine drinkers have a healthier lifestyle profile than do beer drinkers."[22] Although these healthy lifestyle characteristics are admirable, they tend to "confound" the beneficial cardiac effects of wine, scientifically. Is it the wine alone or is it

Table 19.2   Drinking wine was associated with eating a healthy diet in the danish diet, cancer, and health study. Wine drinkers consumed more fruits, fish, vegetables, and salads and used olive oil for cooking more often than beer or spirits drinkers.

	Dietary habits by alchohol beverage drinking pattern			
	Abstainers	**Wine preference**	**Beer preference**	**Spirits preference**
		(%)		
Women (n = 25,479)				
Fruits	30.6	33.5	23.6	26.1
Fish	30.5	41.4	35.5	42.2
Vegetables	55.1	69.6	57.3	57.3
Salads	44.9	66.2	49.7	49.6
No fats on bread	25.5	25.9	16.4	22.4
Olive oil for cooking	5.6	12.0	6.4	5.3
Men (n = 23,284)				
Fruits	21.6	19.0	12.1	17.8
Fish	31.5	42.2	31.9	37.3
Vegetables	47.5	66.0	48.2	50.7
Salads	33.4	61.2	35.5	38.8
No fats on bread	18.8	18.4	10.0	14.4
Olive oil for cooking	7.1	13.4	5.1	6.4

(*Source*: Modified and reproduced with permission from the publisher.) Reference 58.

Table 19.3   Common lifestyle characteristics of wine, liquor, and beer drinkers.

Wine. — Young and middle-aged women, more nonsmokers, better-educated, exercise more often, eat a healthier diet.

Liquor. — Middle-aged and older men, more smokers, less formal education, exercise less often, eat less healthful foods.

Beer. — Young men with characteristics intermediate between wine and liquor preferrers; better-educated than beer or liquor drinkers.[58,59]

the healthier lifestyle associated with drinking wine that lowers the risk of a heart attack or stroke?

In the Health Care Professionals Follow-up Study,[60] covering 38,077 physically active U.S. male healthcare professionals — dentists, pharmacists, veterinarians, optometrists, osteopaths, and podiatrists, but no physicians — who were "free of major illnesses," while they were being followed for 12 years, those who consumed "one half to two drinks per day" had fewer heart attacks than those who did not drink alcohol at all or who consumed larger amounts of alcohol (Fig. 19.8). Notably, beer and liquor were consumed more consistently than wine among this group of American healthcare providers with four healthy lifestyle characteristics.[39] The cohort was mostly — approximately 85% — nonsmokers, whose weight was proportional to their height, who ate a diet high in fruits, vegetables, cereal, fish, and fowl, and who were "physically active." Exercise included "brisk walking [>3.0], running, bicycling, swimming, squash and racquetball." The question addressed by the investigation was whether moderate alcohol consumption in men leading a healthy lifestyle could further lower the risk of an acute myocardial infarction. The study concluded that "even in men already at low risk [for a heart attack] on the basis of body mass index, physical activity, smoking, and diet, moderate alcohol intake [was] associated with a lower risk for MI" — i.e. a heart attack — "with the lowest risk in men who drank. . .approximately one half to 2 drinks per day." The risk of a heart attack was "weakest" for red wine, "intermediate" for white wine, and "strongest" for beer and liquor.

Although the results of the male Health Care Professionals Study are impressive, the conclusions reached by a similar analysis of lifestyle-related risk factors for coronary heart disease in the Nurses' Health Study were even more striking.[32] In that report of 84,129 registered U.S. nurses ranging from 30 to 55 years of age who were followed for 14 years for major coronary events (a coronary event was defined as a heart attack or death due to coronary heart disease), the

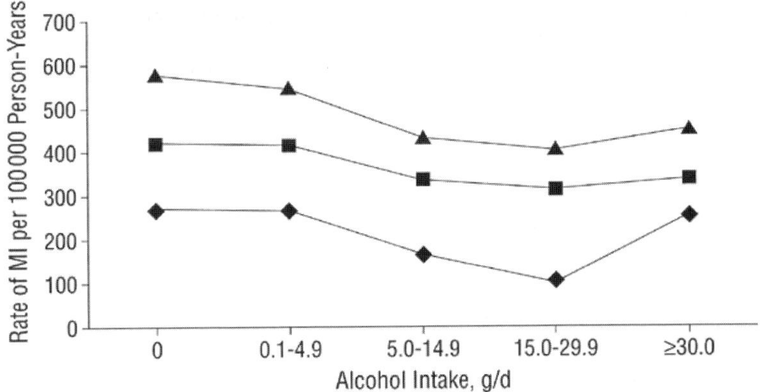

Fig. 19.8.   A graph relating the number of heart attacks to the amount of alcohol consumed on a daily basis among men aged 40–75 who had a variable number of healthy lifestyle characteristics. (See text for description of these characteristics.) Diamonds, squares, and triangles, respectively, indicate the number of men with all 4, with 3 or 2, and with 1 or none of the favorable lifestyle features. Even among men leading the healthiest lifestyle (represented by diamonds) those who drank one-half to two drinks per day (alcohol intake of 5.0–29.9 g/d) had fewer heart attacks than those who drank less or not at all.

(*Source*: Adapted and reproduced with permission from the publisher.) Reference 60.

investigators found that "middle-aged women who did not smoke cigarettes, were not overweight, maintained a healthful diet, exercised moderately or vigorously for half an hour a day, and consumed alcohol moderately (the equivalent of one glass of wine every other day) had an incidence of coronary events that was more than 80 percent lower than in the rest of the population" (Table 19.4).[32] Unfortunately, only 3% of these healthcare providers fell into this lowest-risk group (out of four groups). Eighty-two percent of the coronary events occurred in women who fell into one of the three higher-risk groups.

How Does Cigarette Smoking Factor In?

Cigarette smoking was the "most important single factor" for the development of a nonfatal or fatal heart attack. And while light smokers had a lower risk of a coronary event than heavy smokers, "even smoking 1 to 14 cigarettes per day tripled the risk" of coronary heart disease. Impressively, "41 percent of the coronary events could be attributed to smoking."

In addition, a gradient of increasing risk was associated with increasing unfavorability of each modifiable risk factor. Women who ate the healthiest

Table 19.4   The five modifiable risk factors for coronary heart disease in the Nurses' Health Study. They are arranged as a gradient, with the least favorable factor first and the most favorable last. For example, smokers of more than 15 cigarettes per day had the highest relative risk of a coronary event. The relative isk of a coronary event decreases as each risk factor increases in favorability. See Table 19.5 for conversion of alcohol expressed as grams per day to drinks per day.

The data were collected by investigators at the Brigham and Women's Hospital amd the Harvard Medical School.

Distribution of individual modifiable risk factors and relative risk of coronary events in the nurses' health study, 1980–1994

Factor	Relative risk (95% CI)	Percentage in each category
Dietary score (quintile)		
1	1.90 (1.55 – 2.34)	20
2	1.50 (1.21 – 1.88)	17
3	1.57 (1.29 – 1.91)	28
4	1.23 (0.98 – 1.55)	16
5	1.0 (reference)	20
Exercise (h /wk)		
<1.0	1.41 (1.15 – 1.75)	20
1.0 – 2.2	1.23 (0.99 – 1.53)	15
2.3 – 3.5	1.18 (0.94 – 1.47)	18
3.6 – 5.5	1.05 (0.82 – 1.34)	18
>5.5	1.0 (reference)	17
Body mass index		
>30.0	1.57 (1.30 – 1.91)	12
25.0 – 29.9	1.33 (1.12 – 1.57)	24
23.0 – 24.9	1.16 (0.95 – 1.41)	18
<23.0	1.0 (reference)	33
Smoking (cigarettes/day		
>15	5.48 (4.67 – 6.42)	15
1 – 14	3.12 (2.50 – 3.90)	7
Former smoker	1.55 (1.31 – 1.82)	34
Never smoked	1.0 (reference)	44

(Continued)

Table 19.4   (*Continued*)

**Distribution of individual modifiable risk factors and relative risk
of coronary events in the nurses' health study, 1980–1994**

Factor	Relative risk (95% CI)	Percentage in each category
Alcohol consumption (g/day)		
0	1.65 (1.39 – 1.95)	34
0.1 – 5.0	1.41 (1.18 – 1.68)	33
5.1 – 10.0	1.26 (1.00 – 1.60)	11
>10.0	1.0 (reference)	22

(*Source*: Reproduced with permission from the publisher.) Reference 32.

diet and exercised for at least 5½ h per week had the leanest body mass, and those who never smoked or had stopped smoking and drank one glass of wine (or its equivalent) every other day had the lowest risk of a coronary event during the 14-year study (Table 19.4). Smokers who make the tough decision to stop smoking can look forward to a risk of coronary heart disease that "approximates the [risk] level of those who have never smoked after 10 to 14 years."

A breakdown of all the "coronary events" described in the initial report of the Nurses' Health Study on the relative risks of a nonfatal heart attack or of fatal heart disease (from any cause) showed that the risk of a nonfatal myocardial infarction steadily decreased with an alcohol intake of up to 25 g of alcohol per day (Table 19.5).[62]

The risk of any form of fatal heart disease decreased similarly with increasing alcohol intake up to the highest level of intake, at which point the risk of death from heart disease rose, probably because of deaths from so-called alcoholic cardiomyopathy — a disorder in which excessive alcohol use over time leads to weakening of the cardiac muscle to the point where it cannot pump blood effectively. Congestive heart failure consistently follows.

The Nurses' Health Study was published in July 2000, just six months after a similar study appeared correlating mortality with alcohol consumption in a group of 89,299 male physicians. In the so-called Physicians' Health Study, a 26% reduction in mortality over 5.5 years was seen among those men who

Table 19.5 Alchohol equivalents.

Grams/day	Drinks/day
0	0
<1.5	<¼ of a 12-ounce bottle of 4.5% beer
	<a 1-ounce glass of 12% wine
	<a 0.2-ounce glass of 80-proof spirits
1.6–4.9	¼–½ of a 12-ounce bottle of 4.5% beer
	A 1–2-ounce glass of 12% wine
	0.2–0.5 ounces of 80-proof spirits
5.0–14.9	½ to all of a 12-ounce bottle of 4.5% beer
	A 2–4-ounce glass of 12% wine
	0.5–1.5 ounces of 80-proof spirits
15.0–24.9	1–2 12-ounce bottles of 4.5% beer
	A 4–8-ounce glass of 12% wine
	1.5–2.5 ounces of 80-proof spirits
<25	>two 12-ounce bottles of 4.5% beer
	>an 8-ounce glass of 12% wine
	>2.5 ounces of 80-proof spirits

consumed 1–14 drinks per week (Fig. 19.9). The reduction in cardiac mortality was "primarily driven by a significant 32% to 47% reduction in MI" (myocardial infarction).[63] A comparison of the three studies of healthcare professionals is shown in Table 19.6.

Can Sedentary People Raise Their HDL Concentration to a Level Comparable to the HDL Level in Athletic People by Drinking Alcoholic Beverages?

Possibly. At least one small study from Baylor University found just that. Among 16 marathon runners, 15 joggers (averaging 19 miles per week), and 13 inactive people, the relatively sedentary people who drank three 12-ounce cans of beer a day for three weeks had a "significant increase" in their HDL levels.[64] It doesn't seem fair, but the marathoners and joggers drinking three cans of beer a day during the three-week study experienced no "significant impact" on their HDL levels. Fair or not, the authors concluded that "the

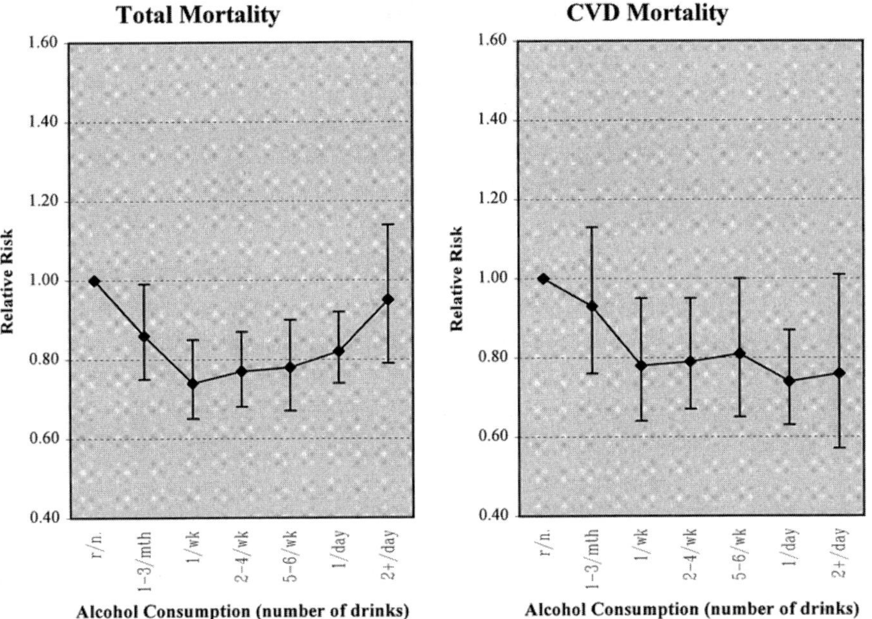

Fig. 19.9. A plot of total and cardiovascular (CVD) mortality against alcohol consumption. Compared with nondrinkers, men who drank one drink per day had a significant reduction in death over 5.5 years. Those who consumed two drinks a day had a 34%–53% reduction in CVD mortality, due primarily to there being significantly fewer myocardial infarctions. The lower CVD death rate persisted after adjustment for aspirin intake.

(*Source*: Modified and reproduced with permission from the publisher.) Reference 63.

consumption of alcohol in moderation seems to be associated with increased HDL levels in inactive men but not in men who engage in regular running or jogging."

Do the Benefits of Alcohol for Heart Disease Offset the Dangers of Nicotine?

This is an interesting question often asked by smokers and the families of smokers urging them to give up cigarettes. Studies related to this question don't support the notion that a beer with a cigarette lessens the risk of lung cancer or chronic obstructive pulmonary disease. A 1988 report from the Royal Free Hospital in London found that smokers aged 40–59 who drank

Table 19.6 Studies correlating the incidence of heart attacks with alcohol consumption in healthcare professionals.

Study	Number of participants	Study composition	Age range	Results
Health Care Professionals Study[39,60]	38,077	Male dentists, pharmacists, veterinarians, optometrists, osteopaths, podiatrists.	40–70 years	Fewer heart attacks over 12 years among men who drank a moderate amount of alcohol. Beer and liquor more effective than wine.
Nurses' Health Study[32,62]	84,129	Female nurses	30–55 years	Fewer heart attacks and deaths due to coronary heart disease over 14 years among women who drank 1–2 glasses of wine a day.
Physicians' Health Study[48,53,54,63,66]	89,299	Male physicians	40–84 years	Fewer heart attacks among those who drank 1–2 alcoholic beverages per day than among those who drank less than one per week.

lightly to moderately had no different total mortality than smokers who didn't drink. And it stands to reason that smokers who drank heavily had the highest death rate during the 7.5-year study[47] (Fig. 19.10).

The highest total and cardiovascular mortality rates were seen among *ex-smokers* who didn't drink, while the lowest mortality rates were in men who had *never smoked or drunk*. The high death rate among men who quit smoking and didn't drink suggests that they gave up both smoking and drinking because of poor health. On the other hand, smokers who gave up the cigarette habit but drank moderately fared better than those ex-smokers who didn't drink at all — consistent with the cardioprotective effect of alcohol in people who have never smoked — inferring that moderate drinking might serve as a "reward" for giving up nicotine. The low(est) death rate among men who never smoked or drank seems at odds with all the evidence that moderate alcohol intake leads to fewer coronary deaths than does no alcohol intake. This appears to be a statistical aberration simply due to the fact that in this

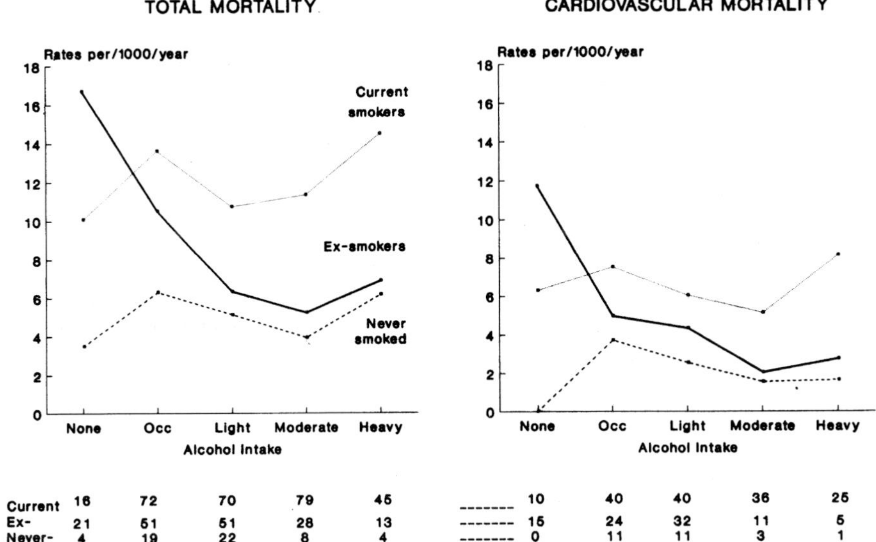

	None	Occ	Light	Moderate	Heavy		None	Occ	Light	Moderate	Heavy
Current	16	72	70	79	45	--------	10	40	40	36	25
Ex-	21	51	51	28	13	-------	15	24	32	11	5
Never-	4	19	22	8	4	-------	0	11	11	3	1

Fig. 19.10. The total and cardiovascular mortality rates among five groups of men whose alcohol consumption varied from none to heavy. "None" means drinking no alcohol during the mean followup time of 7.5 years of the study; "occasional," 1 drink per week, "light," 2–15 drinks per week; "moderate," 16–42 drinks per week (hardly moderate by American standards); and "heavy," more than 42 drinks per week. A drink was defined as half a (British) pint (9.5 U.S. ounces) of beer, a single "measure" (a "jigger") of spirits, or an 8-ounce glass of wine.

Note that that the absence of cardiovascular deaths among men who did not drink or smoke cigarettes does not imply immortality. It simply indicates that during the 7.5 years of the study there were no men who did not drink or smoke, yet died from a cardiovascular disease.

(*Source*: Reproduced with permission from the publisher.) Reference 47.

particular study there were no men who never smoked and never drank. There can't be any deaths in a nonexistent group.

A report from the American Cancer Society and the World Health Organization[49] corroborated those findings. In adults between the ages of 35 and 69, those men and women who smoked and drank moderately had a slightly lower probability of death than those who smoked and did not drink, *but* the smokers who drank had a death rate nearly double that of those men and women who did not smoke and drank moderately (Fig. 19.11). Drinking alcohol did "not compensate for the large increase in risk [of death] produced by smoking."

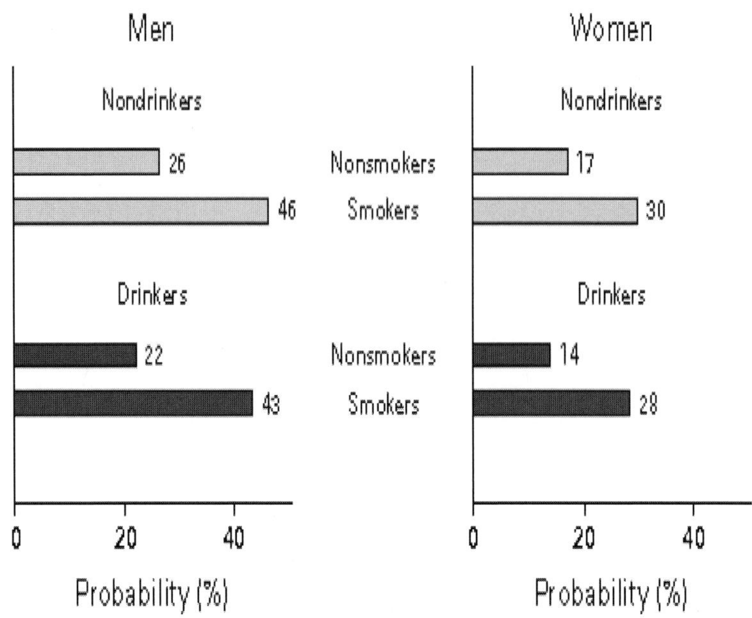

Fig. 19.11. This graph depicts the probability of death from any cause in the U.S. population between the ages of 35 and 69 for four combinations of alcohol consumption and smoking. The probabilities reflect the smoking of about one pack per day and the consumption of 1–2 drinks per day by those who drank alcohol. While the probability fell from 26% to 22% in non-smoking men and from 17% to 14% in nonsmoking women who drank alcohol, the probability rose from 22% in nonsmoking men who drank alcohol to 43% in men who smoked and drank alcohol. Similarly, the probability rose from 14% in nonsmoking women who drank alcohol to 28% in women who smoked and drank alcohol.

(*Source*: Reproduced with permission from the publisher.) Reference 49.

Does Alcohol Lessen the Chances of a Second Heart Attack After the First?

Too few studies have been done to answer this question convincingly. Confirmation or refutation of initial reports — whether they are "positive" or "negative" — is generally needed in medicine before conclusions are reached. But existing reports suggest that the answer may be "Yes."

In the Lyon (France) Diet Heart Study, men who drank wine moderately (two glasses a day) after their first heart attack fared better than those who did not. Those who drank wine had 59% fewer deaths and recurrent acute myocardial infarctions, and needed fewer angioplasties and bypass operations over the next four years than those who did not drink wine after their first coronary event.[65]

That was backed up by the Physicians' Health Study in the U.S. In that report, 5358 of the total 90,150 men had had a prior myocardial infarction. Among them, the cardiovascular mortality over the next five years was lower in those who drank 2–6 drinks per week than in those who drank less or not at all (Fig.19.12).[66]

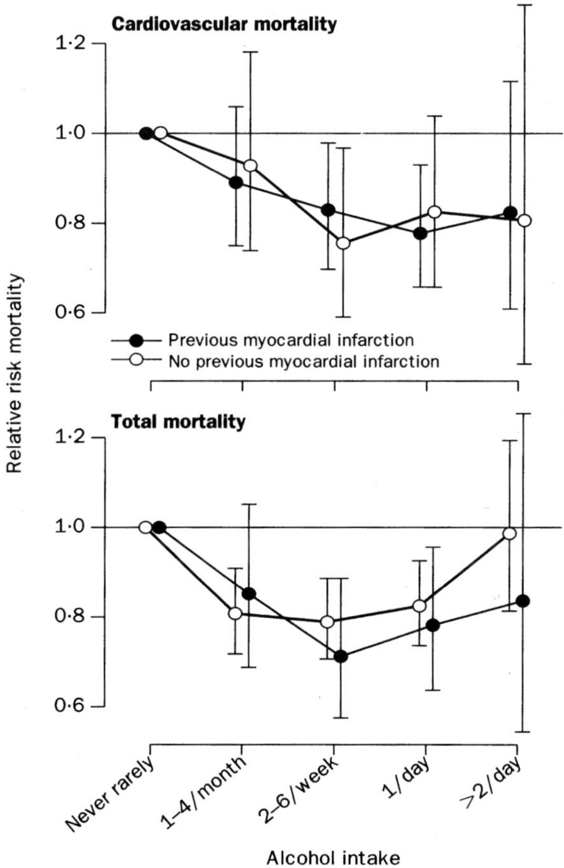

Fig. 19.12.   The cardiovascular and total mortality rates in the Physicians' Health Study over five years among the 5358 men with a previous infarction and the 84,792 without. For both groups, the lowest mortality occurred in those who consumed 2–6 alcoholic drinks per week. (*Source*: Reproduced with permission from the publisher.) Reference 66.

An editorial in *JAMA* in 2001[46] posed the following question: ". . .should persons who have had [an] MI. . .be advised to drink small amounts of alcohol" following their first heart attack to reduce the chances of a second one? The answer proffered by the editorialist was "Yes," based on a paper from the Harvard School of Public Health in the April 18, 2001 issue showing a better six-year survival rate after the first heart attack among men and women who had at least one drink a day before the heart attack than among those who did not drink any alcohol during the preceding year (Fig. 19.13).[67] In fact, the Harvard authors concluded that "adults who abstained from alcohol prior to AMI appeared to be at particularly high risk of long-term mortality," and that "light or moderate alcohol use following AMI is safe."[67] However, the editorialist — Arthur Klatsky, a cardiologist at the Oakland branch of Kaiser Permanente — took a more conservative stance, saying that "the data do not

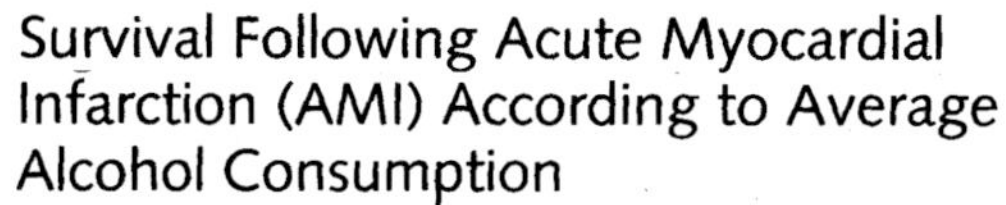

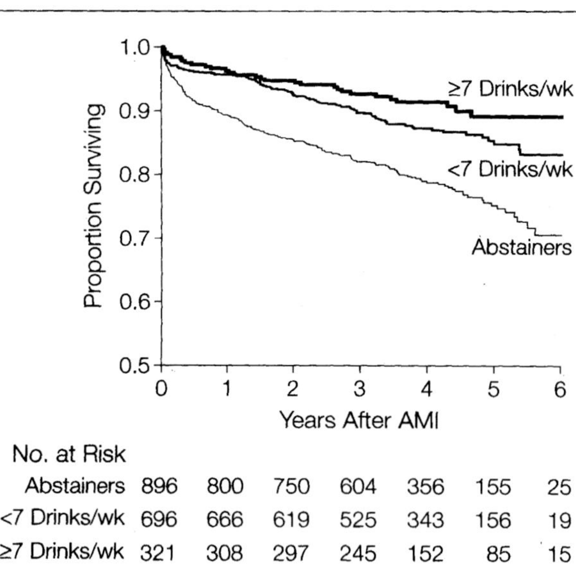

No. at Risk							
Abstainers	896	800	750	604	356	155	25
<7 Drinks/wk	696	666	619	525	343	156	19
≥7 Drinks/wk	321	308	297	245	152	85	15

Fig. 19.13. The six-year survival after a heart attack among 1913 men and women. Survival was highest among those who drank at least one drink a day before their first heart attack and lowest among abstainers.

(*Source*: Reproduced with permission from the publisher.) Reference 67.

justify advising lifelong nondrinkers with [an AMI] to start drinking for health."[46] Bottom line is, if you drank alcohol before a heart attack, don't stop afterward; but if you didn't, you have the option of hoisting a celebratory glass or two a day afterward.

Does Drinking Alcohol on Weekends Only Protect Against Heart Attacks?

This question hasn't been addressed scientifically. Yet it's a practical one to ask. Many Americans limit their drinking to TGIF get-togethers after work on Friday evenings; or, simply, to Saturday night dinners with family or friends.

Virtually all studies on the effect of alcoholic beverages on cardiac mortality and morbidity have included people who drank moderate amounts regularly. Surely there must have been days when they "missed" having a drink for various personal or professional reasons. But, for the most part, they drank something daily, not only on weekends.

So-called "binge drinking" — defined as five or more alcoholic beverages on one occasion — is not a healthy idea, for noncardiac reasons. Binge drinkers — comprising largely college-aged males.[68] — are "14 times more likely to drive while impaired compared with nonbinge drinkers."[69] It is doubtful that a study on the cardiac effects of binge drinking will be done for apparent reasons.

References

1. Gaziano JM, Buring JE, Breslow JL, Goldhaber SZ, Rosner B, Van Denburgh M, Willett W, Hennekens CH. (1993) Moderate alcohol intake, increased levels of high-density lipoprotein and its subfractions, and decreased risk of myocardial infarction. *N Engl J Med* **329**: 1829–1834.
2. Younger W. (1966) The ancient Middle East and Egypt. In *Gods, Men and Wine*. The Wine and Food Society in association with the World Publishing Company, Chap. 2 pp. 27–78.
3. Phillips R. (2002) *A Short History of Wine*. HarperCollins, New York, On the trail of the earliest wines: from the Fertile Crescent to Egypt. Chap. 1, pp. 1–29.
4. Plato. In *The Symposium*, Penguin, Baltimore, Maryland. — translated by Hamilton W.
5. Amerine MA, Cruess WV. (1960) *The Technology of Wine Making*. Avi, Westport, Connecticut.

6. Stampfer MJ, Colditz G, Willett WC, Speizer FE, Hennekens CH. (1988) A prospective study of moderate alcohol consumption and the risk of coronary disease and stroke in women. *N Engl J Med* **319**: 267–273.

7. De Whalley CV, Rankin SM, Hoult JRS, Jessup, W, Leake DS. (1990) Flavonoids inhibit the oxidative modification of low density lipoprotein. *Biochem Pharmacol* **39**: 1743–1750.

8. Frankel EN, Kanner J, German JB, Parks E. (1993) Inhibition of human low density lipoprotein by phenolic substances in red wine. *Lancet* **341**: 454–457.

9. Kuhnau J. (1976) The flavonoids. A class of semi-essential food components: their role in human nutrition. *World Rev Nutr Diet* **24**: 117–191.

10. Harborne JB. (1988) Flavonoids in the environment. Structure–activity relationships. In *Plant Flavonoids in Biology and Medicine. II. Biochemical, Cellular, and Mechanical Properties.* Alan R Liss, New York, pp. 17–27.

11. Lebreton PJ. (1826) Note sur la matiiere crystalline des organettes et amalyse de ces fruits des Hesperidees. *J Pharm Chim Paris* **14**: 377–392.

12. Ruszynak St, Szent-Gyorgyi A. (1936) Vitamin P: flavonols as vitamins. *Nature* **138**: 27.

13. Attaway JA, Buslig BS. (1998) Antithrombogenic and antiatherogenic effects of citrus flavonoids. In *Flavonoids in the Living System.* Pages 165–173.

14. Montanari A, Chen J, Widmer W. (1998) Citrus Flavonoids: a review of past biological activity against disease. In *Flavonoids in the Living System.* Pages 103–116.

15. Cody MM. (1841) Substances without vitamin status. In: *Handbook of Vitamins:* Cadenas E, Packer L (eds.), pages: 571–585.

16. Joint Committee on Nomenclature; Vickery HB, Nelson EM, Almquist HJ, Elvehjem CA. (1950) Term "vitamin P" recommended to be discarded. *Science* **112**: 628.

17. Harborne JB. (1986) Nature, distribution and function of plant flavonoids. In *Plant Flavonoids in Biology & Medicine: Biochemical, Pharmacological, and Structure–Activity Relationships.* Alan R Liss, New York, pp. 15–24.

18. Bors W, Heller W. Michel C. (1998) The chemistry of flavonoids. In: *Flavonoids in Health and Disease.* Rice-Evans CA, Packard L (eds.), Marcel Dekker, New York, Basel, Hong Kong, Chap.4, pp. 111–136.

19. Vierheilig H, Bago B, Albrecht C, Poulin M-J, Piche Y. (1998) Flavonoid and arbuscular-mycorrhizal fungi. In: *Flavonoids in the Living System.* Manthey JA, Buslig BS (eds.), Plenum, New York, pp. 9–33.

20. http:\\www.yachanagovernment.com

21. Riesmersma RA, Rice-Evans CA, Tyrrell RM, Clifford MN, Lean MET. (2001) Tea flavonoids and cardiovascular health. *Quart J Med* **94**: 277–282.

22. Rimm EB, Stampfer MJ. (2002) Editorial. Wine, beer, and spirits: Are they really horses of a different color? *Circulation* **105**: 2806.

23. Tall AR, Breslow JL. (1996) Plasma high-density lipoproteins and atherogenesis. In: *Atherosclerosis and Coronary Artery Disease*. Vol. 1. Fuster V, Ross R, Topol EJ. (eds.), Lippincott-Raven, Philadelphia, Chap. 7.

24. Gronbaek MA, Deis A, Sorenen TT, Becker U, Jensen G. (1995) Mortality associated with moderate intake of wine, beer, or spirits. *Br Med J* **310**: 1165–1169.

25. Friedman LA, Kimball AW. (1986) Coronary heart disease mortality and alcohol consumption in Framingham. *Am J Epidemiol* **124**: 481–489.

26. Criqui MH, Ringel BL. (1994) Does diet or wine explain the French paradox? *Lancet* **344**: 1719–1723.

27. Wannamethee SG, Shaper AG. (1999) Type of alcoholic drink and risk of major coronary heart disease events and all-cause mortality. *Am J Pub Health* **89**: 685–690.

28. Rimm EB, Klatsky, A, Grobee D, Stampfer MJ. (1996) Review of moderate alcohol consumption and reduced risk of coronary heart disease: Is the effect due to beer, wine, or spirits? *Br Med J* **312**: 731–735.

29. Pearson TA. (1996) Alcohol and heart disease. *Circulation* **94**: 3023–3025.

30. St Leger AS, Cochrane AL, Moore LF. (1979) Factors associated with cardiac mortality in developed countries with particular reference to the consumption of wine. *Lancet* **313**: 1017–1020.

31. http://en.wikipedia.org/wiki.Cohort_study

32. Stampfer MJ, Colditz GA, Willett WC, Speizer FE, Hennekens CH. (1988) A prospective study of moderate alcohol consumption and the risk of coronary disease and stroke in women. *N Engl J Med* **319**: 267–273.

33. http://www.nhlbi.nih.gov.about/framingham/design.htm

34. Vogel RA. (2002) Vintners and vasodilators. Are French wines more cardioprotective? *Circulation* **41**: 479–481.

35. Yano K, Rhodes GC, Kagan A. (1977) Coffee, alcohol and risk of coronary heart disease among Japanese men living in Hawaii. *N Engl J Med* **297**: 405–409.

36. Gaziano RJ, Hennekens CH, Godfried SL, Sasso HD, Glynn RJ, Breslow JL, Buring JE. (1999) Type of alcoholic beverage and risk of myocardial infarction. *Am J Cardiol* **83**: 52–57.

37. Kaufman DW, Rosenberg l, Helrich SP, Shapiro S. (1985) Alcoholic beverages and myocardial infarction in young men. *Am J Epidemiol* **121**: 548–554.

38. Klatsky AL, Armstrong MA, Freiedman GD. (1990) Risk of cardiovascular mortality in alcohol drinkers and non-drinkers. *Am J Cardiol* **66**: 1237–1242.

39. Mukamal KJ, Conigrave KM, Mittleman MA, Camargo CA, Jr, Stampfer MJ, Willett WC, Rimm EB. (2003) Roles of drinking pattern and type of alcohol consumed in coronary heart disease in men. *N Engl J Med* **348**: 109–118.

40. Rimm EB, Giovannucci EI, Willett WC, Colditz GA, Ascherio A, Rosner B. (1891) Prospective study of alcohol consumption and risk of ccoronary disease in men. *Lancet* **338**: 4664–4668.

41. Rimm EB. (1996) Alcohol consumption and coronary heart disease: good habits may be more important than just good wine. *Am J Epidemiol* **143**: 1094–1098.

42. Marques-Vidal P, Dueumetiere P, Evans A. 1996) Alcohol consumption and myocardial infarction: a case-controlled study in France and Northern Ireland. *Am J Epidemiol* **143**: 1089–1093

43. http://en.wikipedia.org/wiki/Nitrosamine

44. Gronbaek M, Becker U, Johansen D, Tonnesen H, Jensen G, Sorensen TIA. (1998) Population-based cohort study of the association between alcohol intake and cancer of the digestive tract. *Br Med J* **317**: 844–848.

45. Berger A. (1998) Science commentary: Why wine might be less harmful than beer or spirits. *Br Med J* **317**: 848.

46. Klatsky AL. (2001) Should patients with heart diseases drink alcohol? *JAMA* **285**: 2004–2006.

47. Shaper AG, Wannamethee MW, Walker M. (1988) Alcohol and mortality in British men: explaining the U-shaped curve. *Lancet* **2**: 1267–1273.

48. Camargo CA Jr, Hennekens CH, Gaziano M, Glynn RJ, Manson JE, Stampfer MJ. (1997) Prospective study of moderate alcohol consumption and mortality in U.S. male physicians. *Arch Intern Med* **157**: 79–85.

49. Thun MJ, Peto R, Lopez AD, Monaco JH, Henley J, Heath CW, Doll R. (1997) Alcohol consumption and mortality among middle-aged and elderly U.S. adults. *N Engl J Med* **337**: 1705–1714.

50. Valmadrid CT, Klein R, Moss SE, Klein BEK, Crukkshands KJ. (1999) Alcohol intake and the risk of coronary heart disease mortality in persons with older-onset diabetes mellitus. *JAMA* **282**: 239–246.

51. American Diabetes Association. (1999) Nutrition recommendations and principles for people with diabetes mellitus. *Diabet Care* **22**(Suppl 1): S42–S45.

52. Criqui MH, Golomb BA. (1999) Should patients with diabetes mellitus drink to their health? *JAMA* **282**: 279–280.

53. Camarrgo CA, Stampfer MJ, Glynn RJ, Grodstein F, Gaziano JM, Manson JE, Burina JE, Hennekens CH. (1997) Moderate alcohol consumption and risk for angina pectoris or myocardial infarction in U.S. male physicians. *Ann Intern Med* **126**: 372–375.

54. Steering Committee of the Physicians' Health Study Research Group. (1989) Final report on the aspirin component of the ongoing Physicians' Health Study. *N Engl J Med* **321**: 129–135.

55. Sesso H, Project Director, Physicians' Health Study. Personal communication hsesso@harvard.edu

56. Mukamal KJ, Chung H, Jenny HS, Kuller LH, Longstreth WT, Jr, Mittleman MA, Burke GL. (2006) Alcoholic consumption in older adults: The Cardiovascular Health Study. *J Am Geriatr* **54**: 30–37.

57. Doll R, Peto R, Hall E, Wheatley K, Gray R. (1994) Mortality in relation to consumption of alcohol in male British doctors. *Br Med J* **309**: 911–918.

58. Tjonneland A, Gronbeck M, Stripp C, Overvad K. (1999) Wine intake and diet in a random sample of 48763 Danish men and women. *Am J Clin Nutr* **69**: 49–54.

59. Klatsky AL, Armstrong MA, Kipp H. (1990) Correlates of alcoholic preference: traits of persons who choose wine, liquor or beer. *Br J Addict* **85**: 1279–1289.

60. Mukamal KJ, Chiuve S, Rimm EB. (2006) Alcohol consumption and risk for coronary heart disease in men with healthy lifestyles. *Arch Intern Med* **166**: 2145–2150.

61. Department of Agriculture. (1982) Provisional table on the nutrient content of beverages. Human Nutrition Information Service.

62. Steinberg D, Pearson TA, Kuller LH. (1991) Alcohol and atherosclerosis. *Ann Intern Med* **114**: 967–976.

63. Gaziano JM, Gaziano TA, Glynn RJ, Sesso RJ, Ajani UA, Stampfer MJ, Manson JE, Hennekens CH, Buring JE. (2000) Light-to-moderate alcohol consumption and mortality in the Physicians' Health Study. *J Am Coll Cardiol* **35**: 96–105.

64. Hartung GH, Foreyt JP, Mitchell RE, Reeves RS. Gotto AM, Jr. (1983) Effect of alcohol intake on high density lipoprotein cholesterol in runners and inactive men. *JAMA* **249**: 747–750.

65. de Lorgeril M, Salon P, Martin J-L, Boucher F, Paillard F, de Leiris J. (2002) Wine drinking and risks of cardiovascular complications after recent acute myocardial infarction. *Circulation* **106**: 1465–1469.

66. Muntwyler J, Hennekens CH, Burung JE, Gaziano JM. (1998) Mortality and light to moderate alcohol consumption after myocardial infarction. *Lancet* **352**: 1882–1885.

67. Mukamal KL. (2001) Prior alcohol consumption and mortality after acute myocardial infarction. *JAMA* **285**: 1965–1970.

68. Wechsler H, Davenport A, Dowdall G, Moejens B, Castillo CS. (1994) Health and behavioral consequences of binge drinking in college: a national survey of students at 140 campuses. *JAMA* **272**: 1672–1677.

69. Naimi TS, Brewer RD, Mokad A, Denny C, Serdula MK, Marks JS. (2003) Binge drinking among US adults. *JAMA* **289**: 70–75.

20

A PERSPECTIVE ON ANTIOXIDANT VITAMINS FOR THE TREATMENT OF CORONARY HEART DISEASE

Before launching a discussion on the use of antioxidant vitamins to treat coronary heart disease, a perspective concerning the good and bad features of *both* antioxidants and free radicals may be appropriate. We have a propensity to label free radicals as evil and antioxidants as good. That's not always the case. Antioxidants aren't always divine, nor are free radicals always the devil. Here's why.

Free radicals are produced constantly by the human body, and they are not in cahoots with the pharmaceutical industry to make us (more) susceptible to certain diseases — such as coronary heart disease, cancer, and cataracts — but simply as a byproduct of all normally functioning human cells. The cells are merely taking care of business, be it to make the heart beat or the brain perform all of its extraordinary functions. The cells' business requires glucose — their fuel — and as glucose is burned, free radicals are unavoidably generated. The oxidation of glucose takes place in the cytoplasm — the watery part of a cell surrounding its semisolid nucleus.[1] But the cytoplasm is not the major source of intracellular free radicals. The mitochondria are.[2]

Mitochondria are sometimes called cellular "power plants." It is within these organelles that powerful adenosine triphosphate (ATP) molecules are generated. During the stepwise transfer of electrons along the mitochondrial electron transport chain, 34 molecules of ATP are formed[2] — compared to just 2 ATP molecules during the burning of glucose in the cytoplasm!

261

The Beauty of the Human Body

The human body is a truly magnificent product, irrespective of whether it was created long ago in the Garden of Eden or evolved slowly from earlier animals, in a Darwinian way. Either way, free radicals and antioxidants alike have their merits and demerits. For example, free radicals destroy bacteria aiming to infect us, inactivate toxins, and assist in apoptosis (see Chap. 23), which eliminates would-be cancer cells and other life-threatening critters, such as viruses, fungi, and parasites (Table 20.1). On the other hand, while antioxidants scavenge free radicals with less noble intentions, "excessive antioxidants could dangerously interfere with [such] protective functions [of free radicals], while [even] temporary depletion of antioxidants can enhance [the] anti-cancer effects of apoptosis," as Rudolph Salganik — an M.D. and Ph.D. from my alma mater's fiercest basketball rival (the University of North Carolina at Chapel Hill) — put it in a well-written paper.[4] The two basketball powerhouses — the Blue Devils and the Tar Heels — lie just 11 miles apart in central North Carolina.

Salganik went on to say, "Intake of exogenous antioxidants [vitamin E, C, and beta-carotene, and others] could protect against cancer and other

Table 20.1   The virtues and hazards of free radicals and antioxidants.

Free radicals	Antioxidants
Good	*Good*
Destroy pathogens such as bacteria, fungi, viruses, and parasites.	Scavenge excessive free radicals.
Inactivate certain toxins.	Offer some protection against coronary heart disease, cancer, and cataracts.
Apoptosis.	
Bad	*Bad*
Coronary heart disease	β-carotene may increase the incidence of lung cancer in smokers.
Cancer	Vitamin E may increase the incidence of strokes.
Cataracts	Vitamin E may cause complex congenital heart diseases.

degenerative diseases, particularly in people with innate or acquired high levels of ROS [reactive oxygen species, meaning oxygen-derived free radicals]. However, abundant free radicals might suppress these protective functions, particularly in people with a low innate baseline level of ROS." He pointed out, "Screening human populations for [oxygen-derived free radical] levels could help identify groups with a high level of [oxygen-derived free radicals] that are at a [greater] risk of developing cancer and other degenerative diseases," including coronary heart disease. "It also could identify groups with a low level of [oxygen-derived free radicals] that are at a risk of down-regulating [oxygen-derived free radical]-dependent anti-cancer and other protective reactions. [Such] screening [of] populations could provide a scientifically [sound] application of antioxidant supplements, which could significantly contribute to the nation's health."

The North Carolinian added that humans are "heterogeneous with regard to their [oxygen-derived free radical] production," just as we are with regard to so many other biological traits. That's good, because it keeps physicians on their toes, attempting to differentiate among clinical hoof beats, which at first all sound the same, but with attentiveness allow the clear separation of horse, zebra, donkey — even camel — hoof beats. In short, we need a healthy dose of free radicals!

Later, Salganik mused, "Endogenous [natural] antioxidants do not completely remove free radicals in the body." This raises the following (rhetorical) question: Why not? "…why, despite the existence of a powerful cellular system of antioxidants, are the short-lived [free radicals] are not removed entirely… and permanently.…" Perhaps the answer lies in Daniel Steinberg's sage comment: "Teleological thinking is out of place when discussing [science]."[5] But if that explanation seems dismissive and too philosophical, an alternative explanation could be that "continuously produced [oxygen-derived free radicals] are needed to perform some important biological functions." Cells seemingly "are tuned to remove excessive [oxygen-derived free radicals] and to leave the required level of oxidants" to "protect us from deadly microorganisms, killing them [instead] by producing an avalanche of [oxygen-derived free radicals]." Salganik noted, "Importantly…[oxygen-derived free radicals] kill not only invading bacteria, but also cancer cells. Too many antioxidants scavenge these beneficial [oxygen-derived free radicals] and can thereby interfere with the protective functions of phagocytes," the white blood cells that protect the body by ingesting bacteria as well as dead or dying cells.

If "survival of the fittest" is a fundamental principle of evolution, then it "seems plausible that [oxygen-derived free radical] generation is prevented from being entirely suppressed by endogenous [natural] antioxidants because of these beneficial functions." Apoptosis[a] (pronounced "ap´-o-to´-sis") refers to programmed cell death. Free radicals are the trigger. Apoptosis can be a good thing…if the cell is an early cancer cell. Such precancerous lesions can be killed by a blast of free radicals. Looking straight into a gun barrel loaded with free radicals can be scary, even to a cancer cell. Apoptosis, sometimes referred to as "a guardian angel," selectively targets "cancerous and other cells that threaten our health and life. The sacrifice of the 'bad' cells occurs to save the integrity and life of the whole organism," Salganik explained.[3] Chemotherapy and radiation therapy kill cancer cells by inducing apoptosis… and a barrage of free radicals.

References

1. http://en.wikipedia.org/wiki/Cytoplasm
2. http://pubs.niaaa.nih.gov/publications/arh27
3. Salganik RI. (2001) The benefits and hazards of antioxidants: controlling apoptosis and other protective mechanisms in cancer patients and the human population. *J Am Coll Nutr* **20** (Suppl): 464S–472S.
4. http://en.wikipedia.org/wiki/Apoptosis
5. Tsimakis S. Glass Ck, Steinberg D, Witzum JC. (2004) Lipoprotein oxidation, macrophages, immunity, and atherogenesis. In: Chien KT (eds.), *Molecular Basis of Cardiovascular Diseases*, 2nd ed. Saunders, Philadelphia, pp. 385–413.

---

[a] Apoptosis is a series of biochemical reactions that lead to programmed cell death. About 50–70 billion cells die in the human body each day due to apoptosis.[4] These cells simply fragment and are phagocytized — devoured — by macrophages, white blood cells that clean up cellular debris. Most are regenerated by stem cells. In contrast, the term "necrosis" refers to the pathologic, irreversible death of cells, often due to ischemia (oxygen deprivation), as occurs after an acute myocardial infarction or a stroke.

21

VITAMIN A

The term "vitamin A" is a bit of a misnomer, because it implies a single substance. In nature, the vitamin includes three closely related compounds — retinol, retinal and retinoic acid — sharing a similar chemical structure (Fig. 21.1). To make matters yet more confusing, β-carotene, a precursor of vitamin A, is often referred to as "provitamin A."[1] The name "β (beta)-carotene" implies the existence of an α-carotene — and it does exist. As do γ (gamma)-carotene, δ (delta)-carotene, ε (epsilon)-carotene, and ζ (zeta)-carotene, the first five (lower-case) letters of the Greek alphabet, designated in their order of discovery, like the order of discovery of the vitamins themselves (using the English alphabet). A substance is called a "vitamin" if it "cannot be synthesized in sufficient quantities by an organism, and must be obtained from the diet.[2] In other words, it is "vital."

There are currently 13 known vitamins. Four are fat-soluble (A, D, E, and K) and nine are water-soluble (the eight B vitamins and vitamin C). At one time there was a "vitamin P." When it was determined that it did not satisfy the criteria to be labeled a vitamin, it was reclassified as a bioflavonoid in 1950 (Chap. 18). The natural occurrence of more than 600 "carotenoids"[1] adds to the terminological confusion regarding vitamin A. And it adds the need to define "carotenoid." Aside from "carotene-like," *Webster's* defines carotenoid as " . . .any of several. . .pigments most of which are yellow, orange, or red occur[ing]. . .widely in plants. . ."[3].

Structure of Vitamin A

The structure of vitamin A's three closely related cousins and that of β-carotene are shown in Fig. 21.1. Natural vitamin A is found only in animals and

265

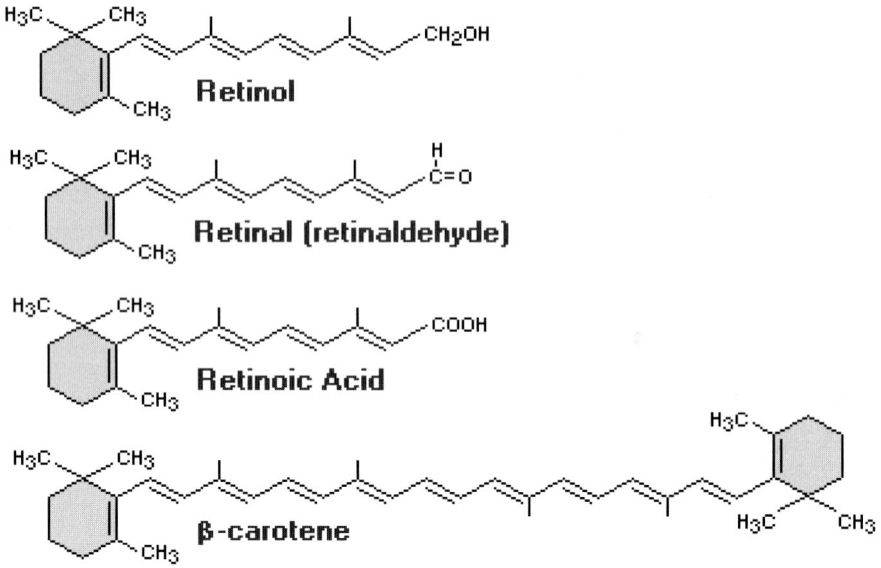

H3C   CH3   CH2OH
                    Retinol
          CH3

H3C   CH3   H
                                                   C=O
                    Retinal (retinaldehyde)
          CH3

H3C   CH3   COOH
                    Retinoic Acid
          CH3

H3C   CH3   H3C

                    β-carotene   H3C   CH3
          CH3

Fig. 21.1.   The structure of vitamin A (retinol) is depicted above. When retinol is used by cells, it gives rise to two important metabolites, retinal, which plays a critical role in vision, and retinoic acid, which has all of the functions of vitamin A, other than the ability to form visual pigment (rhodopsin), which requires retinal. Retinal is the aldehyde form of retinol, hence the suffix "-al." The terminal CHO group is the aldehyde. β-carotene, also called provitamin A, is abundant in plants and can be readily converted into vitamin A by the body. Normal amounts of vitamin A are required for sperm production and menstrual cycle regularity.

(*Source:* Reproduced with permission from the publisher.) Reference 4.

β-carotene is found only (and widely) in plants. Ingested β-carotene can be easily converted into vitamin A in the small intestine.

Sources of Vitamin A

Vitamin A is present in a few foods of animal origin.[5] It is particularly plentiful in "liver or dairy products [such as] whole milk, cheese, and butter, as well as oily marine fish, such as tuna, sardines, and herring."[1] And, OK, OK, sardines and herring are a bit too salty for some folks' taste, but tuna is pretty palatable. If it's packed in oil, rather than water, we get the added benefit of lots of high-quality omega-3 fatty acids (Chap. 17). [And it's cheap. My goodness, bought 10 cans (on sale) for US$10 just the other day!] You may be surprised to hear a cardiologist recommending "oily" fish, since we harp on

Fig. 21.2. A variety of colorful fruits.

avoiding "greasy" foods. But oily seafood is good for us, while animal-derived cooking fats, such as lard and butter, are not. If, however, the thought of eating liver or taking cod liver oil is unappetizing, a much more palatable source of carotenoids is a variety of fruits and vegetables (Fig. 21.2).

The pigment in fruits and vegetables that imparts the greatest amount of color is β-carotene. In general, brightly colored yellow, orange, and red fruits and vegetables — such as carrots, papayas, tomatoes, squash, and pumpkin — provide significant amounts of carotenoids. Green vegetables also contain some carotenoids, although "the pigment can't be seen because it is masked by chlorophyll."[1]

Absorption, Transportation, and Storage of Vitamin A and β-Carotene

Absorption

The retinol form of vitamin A is "not generally found free. . .in foods but is typically. . .bound to fatty acid esters.. . ."[1] [An ester is derived from an acid by

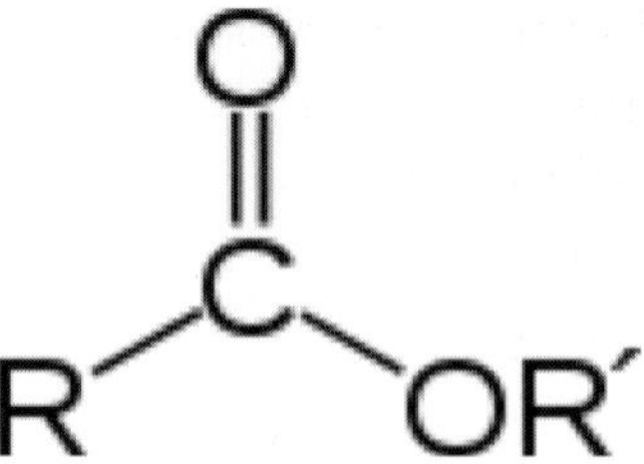

Fig. 21.3.   An ester of carboxylic acid. R and R' denote any alkyl or aryl group. (*Source*: Reproduced with permission from the publisher.) Reference 6.

replacing an acidic hydroxyl group with an alkoxyl group. If the parent compound were R–COOH, the derivative would be R–CO–O–R' (6) (Fig. 21.3)]. Cooking of "plant-food [sources of] vitamin A before consumption. . .facilitates [their] absorption by weakening the protein-carotenoid complexes," making carotenoids more available for absorption.[7] (We often think that vegetables eaten raw are better than cooked ones. But, at least when β-carotene and vitamin A are the substances being absorbed, cooking improves their absorption.) Vitamin A and β-carotene are digested enzymatically in the stomach and then absorbed downstream in the distal one-third of the 21-foot-long small intestine — the ileum. Provitamin A (β-carotene) absorption also "depends greatly on the amount of [fat] ingested with [it]; [fat] increases the uptake of [β-carotene]."[8]

Retinol is the form of vitamin A derived from animal food sources.[8] Those two truths could be important for vegetarians who rely solely on fruits and vegetables for sustenance. On the other hand, the consistent and strong association found in epidemiologic studies of a lower prevalence of coronary heart disease among persons who consume a lot of fruits and vegetables implies that those persons *do* get enough fat to absorb vitamin A. It may be that vegetarians consume enough fatty fruits and vegetables, like avocados or Brussels sprouts,[a,9,10] or cook their foods in vegetable oils such as canola,

---

[a]Recently, bioengineers in Great Britain cleated a new plant in the cabbage family by inserting two genes from algae and another from a fungus into a mother cabbage plant.[11] The new plant can produce substantial quantities of both omega-3 polyunsaturated fatty acids (PUFAs) and omega-6 PUFAs (Chap. 17). The omega-3 PUFA — eicosapentaenoic acid (EPA) — is thought to confer significant health benefits on humans, such as protecting against hypertension and a variety of inflammatory diseases. Omega-3 PUFAs are found primarily in cold-water fish, while

sunflower, safflower, and peanut oil[b]) to sustain vitamin A absorption — or take vitamin A supplements — or that the participants in the many epidemiologic studies who benefited from cardioprotection ate a lot of fruits and vegetables but did not eschew meat completely. The Nutrition and Coordinating Center at the University of Minnesota has published a sample vegan diet for one day that provides sufficient vitamin A (Table 21.1).[14] The Linus Pauling Institute has also published Recommended Dietary Allowances of vitamin A for all age groups for both sexes, including women during pregnancy and if/while breast-feeding.[15]

Another source of fat tastily available to vegetarians is salad dressings — but not fat-free salad dressings. In a small study conducted at Iowa State University and Ohio State University in 2004, healthy young men and women ate a generous salad (after an overnight fast) composed of fresh spinach, romaine lettuce, raw shredded carrots, and cherry tomatoes, all topped with Italian salad dressing. They ate the salad on three mornings, each meal separated by two weeks (during which they did not fast, but they also did not eat any carotenoid-rich foods for four days before each of the three test-salad-consumption mornings). The Italian salad dressing was concocted by mixing dry dressing with white vinegar, water, and canola oil. The dressing contained either 6 g or 28 g of canola oil. (On one of the three mornings, the canola oil was omitted and water was used in its place, thereby creating a salad dressing with zero fat.) Chowing down the yummy salad was the fun part. The not-so-fun part followed. A blood sample was drawn every hour for 12 h to measure the plasma carotenoid concentration. (Mercifully, a short intravenous catheter was inserted and left in place for the 12 h to obtain the samples with just one needle stick.)

---

omega-6 PUFAs are limited to certain grains and poultry products."[12] The achievement was hailed as "an important advance" in an accompanying editorial, because "our present reliance on marine products as the only sources of [PUFAs] is highly undesirable. Not all communities have ready access to fish supplies, either for other geographic or economic reasons; some people are unable to consume fish owing to allergic reactions; some people choose not to eat fish because of a vegetarian lifestyle; and there are increasing concerns about high levels of mercury accumulation in some fish species harvested from certain contaminated waters. Most importantly, global fish resources have declined dramatically since the commencement of industrial fishing — in some cases, to unsustainably low levels — and are unlikely to meet future demands of the human population for [long-chain] PUFAs."[13] Technology to the rescue!

[b] Vegetable sources of β-carotene are, naturally, fat- and cholesterol-free.

Table 21.1   Vitamin A intake from a vegan diet high in carotene-rich fruits and vegetables.

One-day sample diet			
Meal	**Foods consumed**	**β-Carotene equivalents intake IU (μg)**	**Vitamin A intake IU(μg)**
Breakfast	Bagel (1 medium)	0	0
	Peanut butter (2T)	10 (1)	0
	Canned pineapple, juice pack (1/2 cup)	400 (40)	30 (3)
	Orange juice (3/4 cup)	1030 (103)	90 (9)
	Total for meal	*1450 (145)*	*120 (12)*
Snack	Banana (1 medium)	280 (28)	20 (2)
	Total for snack	280 (28)	20 (2)
Lunch	Vegetable soup, prepared from ready-to-serve can (1 cup)	11,950 (1195)	1660 (166)
	Hummus (2T)	20 (2)	0
	White pita (1 large)	0	0
	Soy milk (1 cup)	0	0
	Apple, with skin (1 medium)	700 (70)	60 (6)
	Total for meal	*20,670 (2,067)*	*1720 (172)*
Dinner	Lettuce salad: romaine lettuce (1 cup) with tomato (2 wedges) and oil and vinegar dressing (2T)	8500 (850)	710 (71)
	Baked potato (1 medium)	101,950 (10,195)	8500 (850)
	Bean burrito (1 medium) with avocado (3 slices) and salsa (2T)	16,50 (165)	130 (13)
	Soy milk (1 cup)	0	0
	Total for meal	*112100 (11,210)*	*9340 (934)*

IU = international unit.

(*Source*: Modified and reproduced with permission from the publisher.) Reference 14.

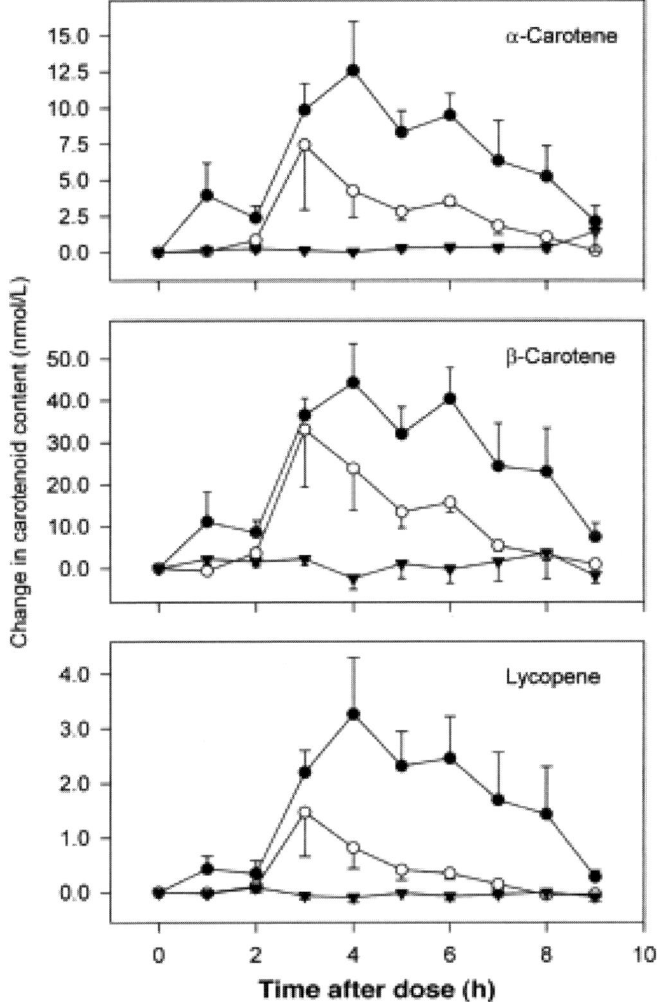

Fig. 21.4. A graph plotting the plasma carotenoid concentration at hourly intervals following a salad with a no-fat, a reduced-fat, and a full-fat Italian salad dressing. Essentially no β-carotene or α-carotene absorption occurred after a fat-free salad dressing. Substantially greater absorption occurred after a full-fat salad dressing than after a reduced-fat dressing.

(*Source*: Reproduced with permission from the publisher.) Reference 16.

The difference in plasma carotenoid concentrations following the no-fat and full-fat salad dressings was striking (Fig. 21.4). There was "essentially no absorption of carotenoids. . .when salads with fat-free salad dressings were consumed. A substantially greater absorption of carotenoids was observed when salads were consumed with full-fat than reduced-fat salad dressings."[16] The conclusion was clear-cut: "The proper intestinal absorption of carotenoids depends on the presence of fat."[16]

Transportation

In the ileum, the two compounds simply diffuse passively into mucosal epithelial cells [the intestinal wall has three layers — an endothelium, a muscular media, and an epithelium — just like the wall of an artery (Fig. 2.1)], where they are quickly re-esterified to facilitate their transportation to the liver via the lymphatic circulation[17] — as opposed to the blood circulation. In the liver, β-carotene and vitamin A are transferred to a lipoprotein for further transportation in the watery bloodstream. Curiously, they are transferred by low-density lipoprotein (LDL), the same lipoprotein that transports cholesterol. (Perhaps LDL imagines that its tarnished reputation for transporting "bad" cholesterol can be counterbalanced by chauffeuring around good molecules like β-carotene and vitamin A, eh?) The two compounds exit the liver via the lymphatic system, using the relatively large "thoracic duct," which quickly enters the far larger left subclavian vein in the upper thorax[18] (Fig. 21.5).

The principal function of the lymphatic system is "to transport fats from the digestive tract into the systemic [blood] circulatory system."[18] Unlike the circulatory system, which carries blood to and from the heart, lymph percolates through the lymphatic system in just one direction.[20] But perhaps the best-known "function" of the lymphatic system is that "it is responsible for carrying cancerous cells between the various [organs] of the body in an unwanted process called 'metastasis,'"[21] i.e. spreading. Any (malignant) cancer can metastasize throughout the body via the lymphatic system. Hodgkin's disease is the best-known form of cancer that *begins* within the lymphatic system.[22] Infectious mononucleosis, aka the "kissing disease" or — colloquially — simply "mono," is a common (usually) *benign* illness in adolescents and young adults caused by the Ebstein–Barr virus. Both Hodgkin's disease and infectious mononucleosis are characterized by enlarged lymph nodes, often first noted by the ill patient. Their treatment and outcome are distinctly different.

Storage

Vitamin A, being a fat-soluble vitamin, is fittingly stored in fat. Appreciable amounts are also stored in the liver. So, someone sporting a little more

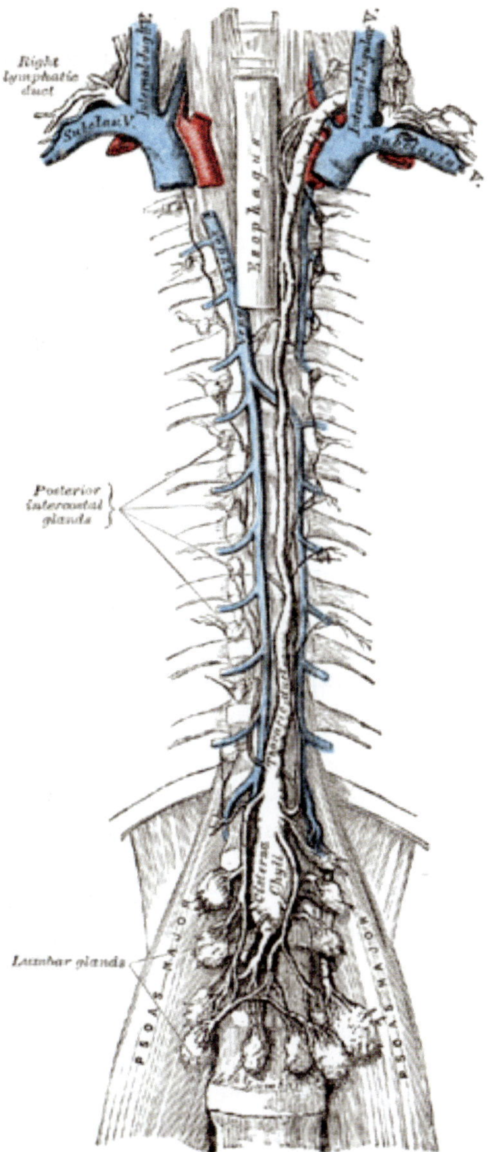

Fig. 21.5. The thoracic and right lymphatic ducts. The thoracic duct is the thin white line in the center. It drains into the left subclavian vein in the upper left corner.

(*Source*: Reproduced with permission from the publisher.) Reference (19).

adipose tissue than wanted need not be concerned about storing insufficient vitamin A. The vitamin will be mobilized and delivered to cells throughout the body, as needed.

Significance of Vitamin A

Perhaps its significance is best illustrated by what happens in its absence. Vitamin A deficiency is "quite rare in the United States," where it is usually associated with diseases that cause malabsorption, such as sprue and cystic fibrosis.[23] But it carries global significance, inasmuch as "vitamin A deficiency is. . .one of the most devastating of health problems, causing an estimated one million child deaths each year." It is estimated "to affect millions of children worldwide, some 90% of whom live in Southeast Asia and Africa" (Table 21.2).[24] In those countries, it causes an eye disease known as xerophthalmia ("dry eyes" in Greek). The name of the disease is far less important than the fact that it is ". . .the single most important cause of childhood blindness. . .affecting 250,000 to 500,000 children each year, two-thirds of whom die within months of going blind owing to their increased susceptibility to infections, also caused by the deficiency."[c,24] It is further estimated that as many as "20 million pregnant women in developing countries are also vitamin A deficient; a third have night blindness."[24] Its high prevalence in women and children highlights its significance. In addition to the vitamin's important role in vision, vitamin A also is required for reproduction, immunity against infectious diseases, and the

---

[c] While the prevention of xerophthalmia in poorer countries was the focus of medical teams in Europe and North America in the 1980s, a "drastic shift in emphasis" occurred in the 1990s.[26] A scourge of infections, often causing death, in malnourished children led to an upsurge of medical and societal interest in vitamin A's association with the immune system. It was soon learned that the vitamin has "profound effects on the immune system."[26] That led to a study of children between the ages of six months and five years living in a mountainous province of China in which half were given vitamin A (200,000 IU) and vitamin E (40 IU) for a year, while the other half were given placebo in a double-blind fashion. There was a "significant reduction" in the incidence of diarrhea and respiratory diseases in the children who took the vitamins.[27] In poverty-stricken Haiti, the mortality rate in children who did not take vitamin A was more than twice as high as in those who took the vitamin[28] — a staggering 50 deaths per 1000 children! Back in New York City, the United Nations performed a detailed meta-analysis of eight clinical trials assessing the effect of vitamin A on the mortality rate of malnourished children. With regard to the crucial question of whether vitamin A reduces the death rate of malnourished children, the answer was a "definitive yes" — a "23% reduction. . .in children between six months and five years of age."[29]

Table 21.2 Prevalence of xerophthalmia among children 0–5 years of age in developing countries.

Region	Number affected (million)	Prevalence (%)
Southeast Asia	10.0	4.2
Western Pacific (including China)	1.4	1.3
Africa	1.3	1.4
Eastern Mediterranean	1.0	2.8
Latin America	0.1	0.2
Total	13.8	2.8

(*Source*: Modified and reproduced with permission from the publisher.) Reference 25.

Table 21.3 Functions of vitamin A.

1. Vision
2. Reproduction
3. Immunity against infectious agents — bacteria, viruses, fungi, parasites
4. Differentiation of stem cells into all 210 cell types in the human body

proper differentiation of stem cells into all the other cell types in the body (Table 21.3).

Night Blindness

A second visual disturbance caused by insufficient vitamin A is night blindness, i.e. impaired dark vision, albeit it far less problematic than childhood blindness. Known technically as "nyctalopia", night blindness is "a condition making it difficult or impossible to see in relatively low light."[30] That comes about because photoreceptor cells called "rods" (referring to their long, cylindrical shape under a microscope) in the outer layer of the retina contain a pigment known as "rhodopsin" (visual purple), which is a compound in the retina that is "responsible for. . .the formation of. . .[the] photoreceptor cells."[d,31]

---

[d] Curiously, the amount of vitamin A in the retina used for vision is "only about 1% of the total amount of vitamin A in the body." Thus, it has been said that "99% of our information about the mode of action of vitamin A concerns only 1% of vitamin A in the body."[35]

Fig. 21.6.   If you're not fond of Bugs Bunny food, there are plenty of other choices from which to get enough β-carotene.

(*Source*: Reproduced with permission from the publisher.) Reference 32.

Rod cells are far more abundant than cone cells, and are also far more sensitive to light. When exposed to sunlight, the pigment immediately "bleaches," explaining why we need time to regenerate sufficient rhodopsin to see in a blackened movie theater. Conversely, when we emerge from a movie theater, the light-sensitive rods bleach, and we again need time to regain full outdoor vision.

During World War II, it was known widely that the British Air Force conducted nighttime pinpoint bombing missions. The British government spread the word that the reason for its pilots' success was that they ate a lot of carrots, *à la* Bugs Bunny (Fig. 21.6).[32] But the *real* reason for their success was "their use of advanced radar technologies."[30]

Ancient Greek and Roman physicians first recognized night blindness, as a second visual expression of insufficient vitamin A. They learned that animal liver was "effective in both the prevention and cure of the disease."[33] That led to the administration of cod liver oil to cure the disease long before vitamin A was discovered in 1913. Yet, historically, there has been widespread reluctance to accept the diet–disease association. This inspired Leslie Harris at

Cambridge in 1955 to speculate, "Perhaps the reason is that it seems easier for the human mind to believe that disease is caused by some evil agency, rather than by any mere absence of any beneficial property."[34]

How Discoveries Are Made

While scientific discoveries may appear to be made by one person or one team of persons, they are generally the product of many people and many teams. To use a reverse analogy, a pruned tree limb may appear to regrow in a relatively straight manner. Scientific discovery, however, "does not occur that way; rather it tends to follow a zig-zag course with many [investigators] contributing to many branches. In fact, the contemporaneous view of each participant may be that of a thicket of tangled hypotheses and facts. The seemingly straightforward appearance of the emergent limb of discovery is but an illusion achieved by discarding the dead branches of false starts and unsupported hypotheses, each of which can be instructive about the process of scientific discovery."[35] Or, as Albert Einstein put it (*circa* 1934), "Imagination is more important than knowledge. For knowledge is limited to all we now know and understand, while imagination embraces the entire world, and all there ever will be to know and understand."[36]

Clinical Trials Assessing β-Carotene for the Treatment of Coronary Heart Disease

While I discussed individual clinical trials in an earlier chapter (Chap. 18) dealing with the relationship between various foods/alcohol and mortality from coronary heart disease, I was daunted — "overwhelmed" may be a better word — by the literally hundreds and hundreds of published papers dealing with the relationship between antioxidants and coronary heart disease. So. . .I am choosing the easy way out for both me and you (as the reader) and will cite only review papers and meta-analyses assessing the effect of β-carotene on "hard" end points such as a nonfatal acute myocardial infarction and death. This approach will also be used in Chap. 23 when we are discussing vitamin E for the same reason. In the chapter covering vitamin C, individual reports will once again be reviewed, because there are currently (in 2009) far fewer reports on the effect of that vitamin on major coronary heart disease events.

In a meta-analysis conducted at the Cleveland Clinic, a center with a reputation comparable to that of the Mayo Clinic, researchers analyzed eight trials of β-carotene treatment to forestall death from a cardiovascular cause in a total of 138,113 individuals (all trials included 1000 people or more).[37] Four trials were "primary prevention" studies, i.e. the subjects were men and women free of any signs of coronary heart disease at the beginning of the study. Four other trials were "secondary prevention" studies, i.e. the participants had known cardiovascular disease, most often having suffered a heart attack before entering the trial. The dosage of β-carotene ranged from 15 to 50 mg daily and the followup period was 1.4–12.0 years.

The pooled data showed a *greater* chance of dying from cardiovascular disease in people who took β-carotene than in those who did not (Fig. 21.7). The greater risk occurred in both men and women who took the vitamin for primary and secondary prevention. A "harmful effect was seen in all but one of the eight studies," and spanned "diverse populations." The exception was the NSCP study (Nambour Skin Cancer Prevention Trial).[e] The authors of the meta-analysis speculated that the reason the NSCP study did not show harm from β-carotene was most likely that its "small sample size and low number of [deaths]" made "chance" a likely explanation for the findings. (Larger numbers diminish the likelihood of a "chance" finding.) The participants in the NSCP study were also younger than those in the other studies, which would favor a better outcome.

Table 21.4 records remarks made by several authorities in the β-carotene-for-coronary heart-disease field. The words "disappointing," "actively discouraged," and "increased mortality" crop up. Words like "promising" and "improved longevity" are just not used when speaking about β-carotene. Yet, you may reasonably wonder, Well, OK, β-carotene doesn't forestall a

---

[e]Nambour is a small town in Australia. The aim of the Nambour Skin Cancer Prevention Study was to see if "long-term use [of sunscreen, SPF 15-plus] can prevent skin cancer."[38] Because β-carotene "lowers skin cancer rates in [experimental animals], a second aim was to see if it also did so in humans. 1383 men and women ranging in age from 20 to 69 years (average 48 years) were the study's subjects. Over 4.5 years, the incidence of squamous cell carcinoma was reduced by about 40% by regular application of sunscreen, while the incidence of basal cell carcinoma was not affected. (Basal cell carcinomas are far more common than squamous cell carcinomas. The good thing about both is that they rarely spread.[39]). During the course of the study, there were 32 deaths, 11 among those taking β-carotene and 21 among those taking placebo.

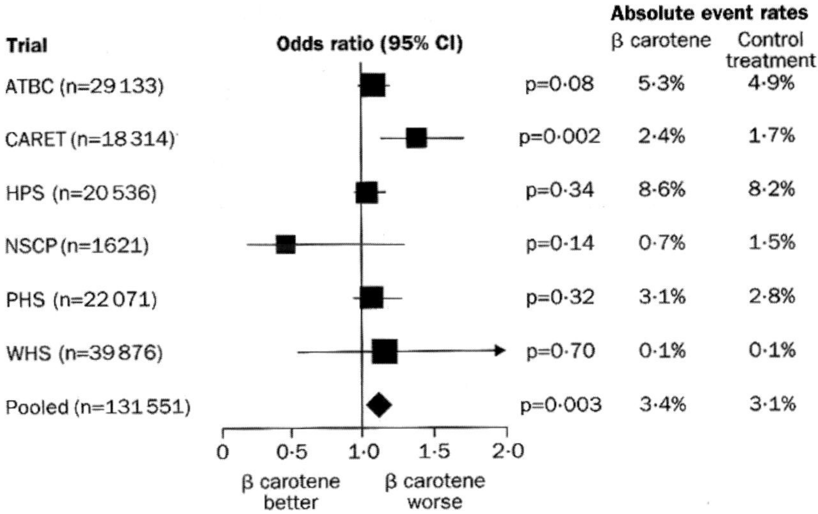

Fig. 21.7. This figure displays the "odds ratio" of a cardiovascular death in men and women treated with β-carotene or with placebo. All but one of the eight trials found more deaths with β-carotene. The NSCP trial recorded fewer cardiovascular deaths with β-carotene, probably related to the study's small size, younger patients, and fewer deaths.[38]

An odds ratio is a way of comparing whether the probability of a certain event is the same for two groups. An odds ratio of 1 implies that the event is equally likely in the two groups. An odds ratio greater than 1 implies that the event is more likely in the first group. An odds ratio less than 1 implies that the event is less likely in the first group.

ATBC = Alpha-Tocopherol, Beta-Carotene Cancer Prevention Study; CARET = Carotene and Retinol Efficacy Trial; HPS = Heart Protection Study; NSCP = Nambour Skin Cancer Prevention; PHS = Physicians' Health Study; WHS = Women's Health Study.

(*Source*: Reproduced with permission from the publisher.) Reference 37.

(nonfatal) heart attack or death from cardiovascular disease and it may even worsen the chances of lung cancer in smokers,[f] but maybe it slows how *fast*

---

[f] More bad news for smokers. In the late 1980s, the National Cancer Institute in Bethesda, MD, and the National Public Health Institute in Helsinki, Finland, commissioned two large clinical trials. The joint venture was to assess whether dietary supplements of vitamin A and β-carotene could reduce the incidence of lung cancer in smokers. The Alpha-Tocopherol, Beta-Carotene Cancer Prevention Study (ATBC) was conducted in Finland and enrolled more than 29,000 male smokers. Participants were randomly assigned to take daily supplements of β-carotene, vitamin E, both, or a placebo. The Beta-Carotene and Retinol Efficacy Trial (CARET) was conducted in the United States and enrolled more than 18,000 men and women who were current

Table 21.4   Quotes regarding the use of β-carotene for coronary heart disease.

". . .there is no further mileage in the clinical use of beta-carotene."[40]

". . .the use of vitamin supplements containing β-carotene and vitamin A. . .should be actively discouraged. . ."[41]

"The results of the [β-carotene] trials have been disappointing and failed to confirm any protective effect of [the] vitamin fo. . .cardiovascular disease."[42]

". . .in smokers. . .lung cancer developed more frequently in subjects receiving β-carotene."[43]

The results of the ATBC study "provide timely support for skepticism and for a moratorium on unsubstantiated health claims."[44]

". . .meta-analyses of. . .randomized trials have not demonstrated that beta-carotene. . .in the administered dosage lead[s] to decreased mortality.[45]

". . .β-carotene increased the risk of fatal coronary heart disease."[46]

"Beta-carotene, administered singly. . .increased all-cause mortality [mortality from any disease]."[47]

atherosclerosis progresses. That would be a good-enough reason for me to take it.

That possibility was looked into by investigators in Lille, France, who used standard ultrasound to measure the extent of atherosclerosis noninvasively in men and women taking β-carotene (in conjunction with vitamin E and vitamin C) and compared the extent with a group of men and women who did not take the antioxidant. They saw, after seven years of treatment, that "plaques [were] *more* [italics added] frequent in the group that received antioxidants,"[50] thus provid[ing] no rationale for the use of β-carotene to slow the progression of atherosclerosis.

---

smokers, former smokers, or workers exposed to asbestos. (Like smoking cigarettes, exposure to asbestos increases the risk of lung cancer.) The participants in that study were randomly assigned to take daily supplements of β-carotene and vitamin A or a placebo. In 1994, the ATBC trial was stopped early, because the incidence of lung cancer was 16% *higher* among those men who took the β-carotene than among those who took the placebo. The overall death rate from any cause was also higher — 8% — among men who took β-carotene supplements (Fig. 21.8).

In 1996, the CARET study was also stopped early for the same reason. The incidence of lung cancer was 28% higher among men and women taking β-carotene supplements than among those who took the placebo. The overall death rate from any cause was also higher — 17% — among those who took β-carotene supplements. The ATBC and the CARET studies were published in two classic papers in 1994 and 1996, respectively.[48,49]

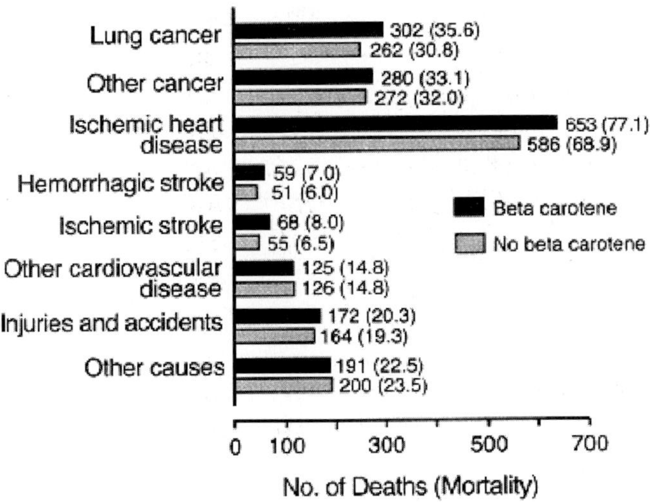

Fig. 21.8.   This figure plots the cause of death and the number of deaths in the ATBC trial. The incidences of death from lung cancer, all other cancers, and coronary heart disease were greater in men who took β-carotene. Death from all causes was 8% greater in those who took β-carotene supplements. The mortality rate is expressed as deaths per 10,000 person-years.

(*Source*: Modified and reproduce with permission from the publisher.) Reference 48.

Trials of β-Carotene in Women Only

Women may also reasonably ask, *Well, β-carotene didn't help men live longer or reduce the number of heart attacks in the ATBC study, but what about us? Maybe women who take β-carotene fare better than men?* Unfortunately, they don't. In the Women's Antioxidant Cardiovascular Study (WACS), yet another prospective, randomized, double-blind, placebo-controlled study, conducted at Boston's Brigham and Women's Hospital, women at "high risk" for an important cardiac event, who took β-carotene (50 mg every other day) for nine-plus years, had no fewer myocardial infarctions [heart attacks] or strokes than women who took a placebo.[51] They also had no less need to undergo coronary artery bypass surgery or coronary angioplasty. Nor did they live longer than those who took a placebo (Fig. 21.9). (There was actually a 14% higher mortality rate in the women who took β-carotene, which did not reach statistical significance, though it was surely significant to those who took the antioxidant.) "High risk" was defined as women who had previously suffered

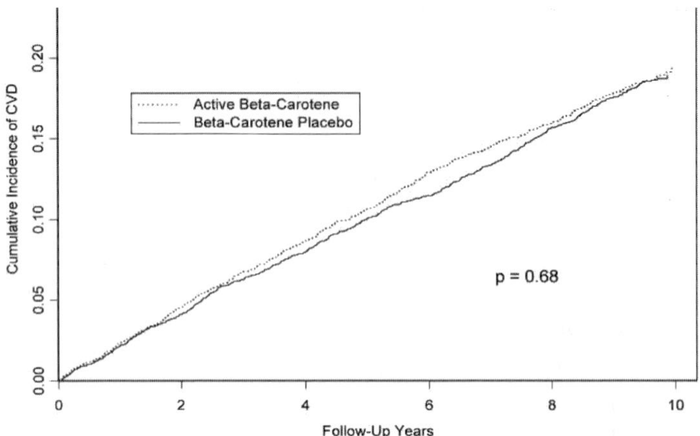

Fig. 21.9. This graph shows the cumulative incidence of major expressions of cardiovascular disease (CVD) (an acute myocardial infarction, a stroke, the need for coronary bypass surgery, a coronary angioplasty or a stent insertion or death due to CVD) in women. Between the sixth and eight years after beginning supplemental β-carotene, there was a 14% increase in the mortality rate among women taking the vitamin. The increase did not reach statistical significance and diminished over time.

(*Source:* Modified (cropped) and reproduced with permission from the publisher.) Reference 51

a heart attack or had "at least three cardiac risk factors." "Eligible" risk factors included hypertension (high blood pressure), an elevated blood cholesterol level, diabetes, a parental history of a premature myocardial infarction (before age 60), obesity, and current cigarette smoking. All women were 40 or more years of age. Eighteen percent were current smokers.

If there's any good news in the WACS report, it is that — unlike in men — β-carotene did not cause more lung cancer in women. In the ATBC study, men taking β-carotene had a 16% greater incidence of lung cancer than did those taking placebo. The bottom line of the WACS report was that "WACS and other antioxidant trials have not found consistent preventive effects on [cardiovascular disease]….[Therefore], widespread use of these…agents for cardiovascular protection does not appear warranted"…for women or men. So be it!

β-Carotene Supplements in the Elderly

How about mom and dad? Surely old folks must do better with a little extra β-carotene. With age comes "a reduced appetite and taste, poor dentition [about 25% of older people are edentulous], physical and economic barriers to food sources, [and] lack of motivation."[54] With age, it would not be surprising if older people who took vitamin supplements lived a little longer than those who did not. Turns out that what intuitively seems likely is not correct. In a study of 1214 men and women between the ages of 75 and 84 (living independently) commissioned by the UK Department of Health, in which the plasma concentration of β-carotene was measured, there was "no influence of . . . β-carotene or retinol [plasma concentration] on mortality" over the 4.4-year followup of the study.[55] In other words, later in life there was no association between the amount of β-carotene in the blood — irrespective of whether the vitamin was derived from food sources or dietary supplements — and mortality. However, a low plasma vitamin C concentration was "strongly predictive of increased mortality."[g,55]

Too Much of a Good Thing

While vitamin A can improve vision and help ensure that sperm get their job done, too much vitamin A can cause problems. Dryness of surfaces exposed to air is a common early sign of excess vitamin A — dry lips, dry eyes, dry nasal mucosa (the inner lining of the nose), dry skin, and dry nails — making them brittle.[1] Brittle nails may not be significant to some men, but brittle bones, due to vitamin A excess, becomes significant to men and women alike. In the Nurses' Heath Study, postmenopausal women taking more than 2000 IU per day of vitamin A had twice the risk of a hip fracture with only "low or moderate" trauma, compared to those taking less than 500 IU per day.[58] β-carotene was, again, exempt from this danger. In an accompanying editorial, the author pointed out that perhaps this should be taken with a grain of salt because the dietary practices of nurses "differ from the dietary practices of the U.S. population at large."[59] In general, nurses have a "higher socioeconomic status" than many other women and lead healthier lives, including the ostensibly healthier practice of taking more vitamin A. The editorialist wrapped it up saying ". . .this study serves as a reminder that vitamins are potent, essential nutrients which have effects that can precipitate harm as well

---

[g]These findings were mirrored in two other studies on the elderly. One was conducted in Massachusetts, the other in The Netherlands. Both documented no association between the plasma β-carotene concentration and mortality in older people.[56,57]

as provide benefit. The optimal source is from the foods in our diets, not the dietary supplements often taken to simplify our complex world."

Accutane

Never heard of it? That may be a good thing.

The first people ever to hear about it were physicians in London. On August 27, 1983, a report quietly appeared in *The Lancet* (Britain's version of *The New England Journal of Medicine*) concerning five newborns with absent or deformed ears or hydrocephalus (a swollen brain), all borne by mothers who had taken a then-new drug known as Accutane (Hoffman-La Roche),[60] the brand name for retinoic acid, one of the four forms of vitamin A (Fig. 21.1). (Seven other pregnancies ended with a spontaneous abortion.) So popular was the drug that 120,000 women of child-bearing age had taken it during the first 16 months following its release.[61] In 1984, the American College of Dermatology asked its members to begin reporting instances of pregnant women taking Accutane to its Adverse Drug Reaction Reporting System (ADRRS) or to the F.D.A. Among 35 such reported instances, there were "29 (83%) spontaneous abortions or infants with birth defects. The most frequently reported were severe birth defects involving the [brain and spinal cord]. . .and the cardiovascular system."[61]

That report paved the way for pediatricians and epidemiologists at the Centers for Disease Control (in Atlanta) and a host of medical schools across the United States, in 1985, to review the records of 154 pregnancies in which the mothers had taken the acne medication during their pregnancies.[62] Of those 154 pregnancies, 95 (62%) ended with an elective abortion. (The elective abortions were done because of the high incidence of spontaneous abortions or infants with serious birth defects reported through the ADRRS or to the F.D.A. in 1984.) Twelve other women (7%) had spontaneous abortions, while 21 live-born infants had severe deformities of the ears — absent or malformed ears — and 26 infants had no major abnormality. Eight newborns had an unusual congenital heart abnormality. The drug is, by the way, highly effective for the treatment of even severe acne, but the price some adolescents and young women may pay in the form of serious birth defects is high. Many dermatologists now advise women of child-bearing age not to take Accutane or to use contraceptives if they do use it.[1]

The American academy of pediatrics position on the medication

In 1992, the American Academy of Pediatrics published a position paper on the topic.[63] It made stout recommendations for all physicians, not only pediatricians. They were/are as follows:

1. Be certain that "The patient has received oral and written warning of the reproductive hazards of [Accutane] during pregnancy and has acknowledged her understanding in writing. It must be emphasized that one third of the affected infants reported. . .were born to mothers who were using contraception." Therefore, "the manufacturer recommends that two contraceptive measures be used simultaneously."
2. "[Accutane] should be prescribed only by those physicians with experience in the therapy of severe dermatologic disorders." In other words, a primary care physician [family doctor] caring for an adolescent girl with severe acne should request evaluation by a dermatologist before beginning the medication.
3. "[Accutane] should not be prescribed for women of child-bearing potential."

The public learns about accutane

In 1995, the medication came to the public's attention through an article in The *New England Journal of Medicine,* picked up quickly by the news media, reporting that 1 out of every 57 infants borne by women taking more than 10,000 IU a day of vitamin A supplements — not Accutane — had a baby with a malformation attributable to the vitamin. The principal defects again were "malformed or absent ears" and unusual congenital heart defects.[64] The paper also pointed out that the dosages of vitamin A responsible for the undesirable effect were "not far above those currently recommended."[h]

The risk of birth defects is not limited to women who take vitamin A supplements during pregnancy. It extends to consumption of supplements *before pregnancy.*[65] This is presumably a reflection of the vitamin's prolonged storage in the liver and fat,[i] requiring 2–6 weeks to be "washed out." The U.S. Public

---

[h] The U.S. recommended daily allowance for pregnant women is 8000 IU/day.
[i] The liver is the primary storage depot. But about 15%–20% of the body's vitamin A is stored in fat.[66]

Health Service recommends that "women of reproductive age should be encouraged to follow [the agency's] recommendations that they consume 0.4 mg of folic acid daily to prevent [spinal cord and brain abnormalities]." It also tells women that "they can be assured that a safe and effective way to meet this requirement is to take a single, daily multivitamin preparation."[65] ("Single" is the operative word.) That's a simple formula to steer clear of the serious consequences of too much vitamin A, don't you think?

Other Unwanted Side Effects of Vitamin A

Fatty liver

Another unwanted effect of excess vitamin A — as opposed to Accutane — is a "fatty liver." A fatty liver also occurs with excess alcohol intake and may go on to cause cirrhosis. So may a fatty liver due to excessive vitamin A. Under a microscope, it's hard to tell the two apart. Folks taking as little as 25,000 IU per day of vitamin A for 6 years or as much as 100,000 IU per day for 2.5 years may develop a fatty liver or cirrhosis.[67] A paper in the medical journal *Gastroenterology* concluded that "prolonged and continuous consumption of doses [of vitamin A] in the low therapeutic range can result in life-threatening liver damage."[67]

These problems led the Center for Disease Control to recommend that women in the U.S. "limit vitamin A in prenatal vitamin preparations to 5,000 IU [per day] and [that] the vitamin A content of all multivitamin preparations [be no more than] 10,000 IU."[65] However — and this is a big however — women also "need to understand that the daily consumption of. . .no more than the U.S. recommended daily allowance [RDA][j] of vitamin A [by] pregnant women [8000 IU/day] is beneficial."[65] A safe alternative is to take supplemental β-carotene. It seems to be an innocuous form of vitamin A. . .but not if you are a heavy smoker.[68]

The same goes for multivitamins and congenital heart disease. Current multivitamin preparations are safe if manufactured in the U.S. or Europe and taken as just *one capsule,* just *once a day.*[64] The U.S. Public Health Service recommends that women of reproductive age be encouraged to "consume 0.4 mg

---

[j]The RDA for men is 3000 IU/day, and for women who are not pregnant it is 2333 IU/day. Values for infants and children are available at http://lpi.oregonstate.edu/infocenter/vitamins/vitamin A.

of folic acid daily to prevent [congenital spinal cord and brain abnormalities]."[65] Women can be assured that "one safe and effective way to meet this requirement is to take a single, daily multivitamin preparation that contains 0.4 mg of folic acid and no more that 8,000 IU of vitamin A."[65]

Being fat-soluble, vitamin A tends to hang around the body a long time — just like fat. It's hard to get rid of surplus vitamin A. A lot of vitamin A taken over a short time or a lesser amount (yet still more than the recommended RDA) over a long time can lead to toxicity.[1]

Yellow skin

Finally, there's the real possibility of the skin turning yellow while one is taking β-carotene. β-carotene is, after all, present in considerable amounts in Bugs Bunny foods. In the ATBC Cancer Prevention Study, one third of the men developed yellow skin while taking β-carotene, just 20 mg per day.[48] This may be of little consequence to some men, but may be distressing to many women. Should your skin become yellow while you are taking β-carotene, there is no need to fret that you may have developed liver disease with "yellow jaundice." Simply look in the mirror. The sclera of the eyes — the whites of the eyes — will not be yellow. They'll still be white. The yellowing of the skin, by the way, is merely a reflection of the β-carotene stored just beneath it in the subcutaneous fat.[1] It is generally most apparent on the palms and soles. The yellow skin will become white again within 2–6 weeks after discontinuation of β-carotene.[24]

β-Carotene and Cancer

Well, what about β-carotene as an agent that may/can prevent cancer? Isn't that still a valid reason to take the vitamin? The answer, in a word is, no. In the 1990s, there was also great hope, again based on epidemiologic studies showing a lower risk of cancer in individuals who consumed a lot of fruits and vegetables, that the vitamin would prove to be a legitimate anticancer agent. But with the performance of clinical trials testing the vitamin's ability to prevent cancer, it became crystal-clear that it had no such ability. β-carotene's requiem came in December 1999, with an editorial in the *Journal of the National Cancer Institute* saying "a report. . .in this issue of the Journal makes it clearer yet that β-carotene supplements have no value as cancer chemopreventive agents. Indeed, it would now seem difficult to justify any additional investigation of β-carotene for chemoprevention….Clearly, focusing as readily as we did on β-carotene was a mistake."[k,69]

The Media's Coverage of the β-Carotene Story

As if the bad news about β-carotene in scientific journals weren't enough, the media picked up the β-carotene story. It didn't help β-carotene's quest for acceptance when *The New York Times* panned the vitamin. Wrote Gina Kolata, a superb scientific reporter, "Two large studies [the ATBC and CARET studies] have found that contrary to the beliefs or hopes of the millions of Americans who take [β-carotene]. . .a vigorously promoted vitamin supplement, is completely ineffective in preventing. . .heart disease. One of the studies found that it might even be harmful to some people."[70] She went on to speculate, as many epidemiologists have, that the reason β-carotene seemed to limit heart disease and prolong life [in the epidemiologic studies] was related more to the healthy lifestyle that vitamin supplement takers lead. They exercise more, they smoke less, they are leaner than most, and their "idea of a vegetable is a dollop of ketchup."

A Reflection on the β-Carotene Story

The β-carotene story was yet another example of a false syllogism (Chap. 18). After all, fruits and vegetables are rich sources of β-carotene (major premise), and many, many epidemiologic studies had documented fewer heart attacks and a longer life among people who consumed a lot of fruits and vegetables (minor premise). *Ergo,* taking a β-carotene supplement should reduce the number of heart attacks and lower mortality. It made total sense back in the early 1990s for physicians and many other Americans to be enthusiastic about the vitamin. Expectations were high and few — if anyone — were prescient enough to foresee the unexpected. But that's why we do large-scale clinical trials — to see if our hopes are realized in a statistically significant manner

---

[k] However, as is the case with β-carotene and its inability to prevent cardiovascular disease, the negative clinical trials with regard to the vitamin's ability to prevent cancer do not invalidate the epidemiologic findings. Fruits and vegetables contain many, many nutrients. Because of the large number of nutrients, it is simply not feasible to perform large-scale clinical trials on each nutrient. The simplest and most cost-effective (dietary) path to good health is to eat a lot of fruits and vegetables. . .and seafood.

through a prospective, double-blinded, placebo-controlled randomized trial.
That's the way the game is played — and we all lost.

One antioxidant down, two to go.

References

1. The fat-soluble vitamins. In: Groff JL, Gropper SS. Wadsworth, *Advanced Nutrition and Human Metabolism.* 3rd ed. A (eds.), Thomson Learning, Australia, Canada, Denmark, Japan, Mexico, New Zealand, Philippines, Puerto Rico, Singapore, South Africa, Spain, United Kingdom, United States, Chap. 10, pp. 316–370.
2. http://en.wikipedia.org/wiki/Vitamin
3. *Webster's Third New International Dictionary of the English Language, Unabridged.* (1993) Merriam-Webster, Springfield, Massachusetts.
4. http://www.vivo.colostate.edu/hbooks/pas/misc_topics/vitamina.html Accessed October 20, 2009.
5. http://ibdcrohns.about.com/library/fda/blvita5.htm
6. http://en.wikipedia/wiki/Ester
7. http://www.google.com/searcg?ie=UTF-8&oesourceid=navclient+8gfns=1&q=absorption+by+
8. http://en.wikipedia.org/wiki/Vitamin_A
9. Ledesma L, Munari F, Dominguez H, Montalvo C, Luna H, Juarez C, Moran Lira S. (19996) Monosaturated fatty acid (avocado) rich diet for hypercholesterolemia. *Arch Med Res* **27**: 519–523.
10. http://www.fatfreekitchen.com/nutrition/carb-counter-vegetables.html
11. Qi B, Fraser T, Mugford S, Dobson M, Sayanova O, Butler J, Napier JA, Stobart AK, Lazarus CM. (2004) Production of very long chain polyunsaturated omega-3 and omega-6 fatty acids in plants. *Nat Biotechnol* **22**: 739–745.
12. *Nature Biotechnology* online (16 May; doi: 10.1038/nbt972).
13. Green AG. (2004) From alpha to omega-producing essential fatty acids in plants. *Nat Biotechnol* **22**: 680–682.
14. http://www.ncc.umn.edu/about/scientificpublications
15. Brown MJ, Ferruzzi MG, Nguyen ML, Cooper DA, Eldridge AL, Schwartz JJ, White NS. (2004) Carotenoid bioavailability is higher from salads ingested with full-fat than with fat-reduced salad dressings as measured with electrochemical detection *Am J Clin Nutr* **80**: 396–403.
16. http://lpi.oregonstate.edu/infocenter/vitamins/vitaminA/#safety
17. Chemical and physiological properties of vitamins. In: Combs GF, Jr (ed.). *The Vitamins: Fundamental Aspects in Nutrition and Health,* 3rd ed. (2008) Elsevier, Amsterdam, Boston, Heidelberg, London, New York, Oxford, Paris, San Diego, San Francisco, Singapore, Sydney, Tokyo, pp: 35–74.

18. http://en.wikipedia.org/wiki/Lymphatic_system

19. http://en.wikipedia.org/wiki/Thoracic_duct

20. http://www.health101.org/lymph.htm

21. *Churchill's Illustrated Medical Dictionary.* (1989) Churchill-Livingstone, New York, Edinburgh, London, Melbourne.

22. National Institute of Health, National Cancer Institute. *What You Need to Know About Hodgkin's Disease.* Bethesda, MD. NIH publication No. 99-1555.

23. Owens A, Cloud HH. (2009) Special topics in toddler and preschool nutrition. Vitamins and minerals in childhood and children with disabilities. In: Edelstein S, Sharlin J (eds.), *Life Cycle Nutrition: An Evidence-Based Approach.* Jones and Bartlett, Sudbury (Mass.), Ontario, London, pp. 183–225.

24. Vitamin A. In: Combs GF, Jr (ed.). *The Vitamins. Fundamental Aspects in Nutrition and Health.* 3rd ed. Gerald F. (2008) Elsevier, Amsterdam, Boston, Heidelberg, London, New York, Oxford, Paris, San Diego, San Francisco, Singapore, Sydney, Tokyo, pp. 95–143.

25. United Nations Administrative Committee on Coordination — Subcommittee of Nutrition (1993). *SCN News* **9**: 9.

26. Bates CJ. (1995) Vitamin A. *Lancet* **345**:31–33.

27. Cheng L, Chang Y, Wang E-L, Burn T, Giessler C. (1993) Impact of large-dose vitamin A supplementation on childhood diarrhea and respiratory infections and growth. *Eur J Clin Nutr* **47**: 88–96.

28. Stansfield K, Pierre-Louis M, Leeboards GA, Augustin A. (1993) Vitamin A supplementation and increased prevalence of childhood diarrhea and acute respiratory infections. *Lancet* **342**: 578–582.

29. Benton GH, Martorelli R, Aronson KJ, Edmonton B, Mc Gabe G, Ross AC, Harvey B. (1993). Effectiveness of vitamin A supplementation in the control of young child morbidity and mortality in developing countries. United Nations: ACC/SCN State of the Art Series — Nutrition Policy Discussion Paper No. 13.

30. http://en.wikipedia.org/wiki/Nyctalopia

31. http://en.wikipedia.org/wiki/Rhodopsin

32. http://www.google.com/search?ie=UTF-8&oe=UTF-8&sourceid=navclient&gfns=1&q=bugs+bunny+photo

33. Brouzas D, Charakidas A, Vacilakis M, Nikakis P, Chataoulis D. (2001) Myctalopia in antiquity: a review of the ancient Greek, Latin, and Byzantine literature. *Ophthalmology* **108**: 1917–1921.

34. Harris LJ. (1955) *Vitamins in Theory and Practice.* Cambridge University Press, London, New York, Toronto, Bombay, Calcutta, Madras, Tokyo, pp. 1–39.

35. Discovery of the vitamins. In: Combs GF, Jr (ed.) *The Vitamins: Fundamental Aspects in Nutrition and Health,* 3rd ed: Elsevier, Amsterdam, Boston, Heidelberg, London, New York, Oxford, Paris, San Diego, San Francisco, Singapore, Sydney, Tokyo, pp. 7–33.

36. http://thinkexist.com/quotes/albert_einstein

37. Viveananthan DP, Pena DMS, Sapp SK, Hsu A, Topol E. (2003) Use of antioxidants and vitamins for the prevention of cardiovascular disease: meta-analysis of randomized trials. *Lancet,* **361**: 2617–2623.

38. Green E, Williams G, Neale R, Hart V, Leslie D, Parsons P, Marks, Gaffney P, Battistutta D, Frost C, Lang C, Russell A. (1999) Daily sunscreen application and beta-carotene supplementation in prevention of basal cell and squamous cell carcinomas of the skin: a randomized controlled trial. *Lancet* **354**: 723–729.

39. http://www.ucsfhealth.org/adult/medical_services/cancer/skin/conditions/Basal_Cell_Squamous_Cell_Carcinoma/signs.html

40. Stephens N. (1997) Anti-oxidant therapy for ischemic heart disease: Where do we stand now? *Lancet* **349**: 1710–1711.

41. SOURCE of "the use of supplements containing β-carotene should be actively discouraged."

42. Clarke R, Armitage J. (2002) Antioxidant vitamins and risk of cardiovascular disease: review of large-scale randomized trials. *Cardiovasc Drugs Ther* **16**: 441–445.

43. Need source of "in smokers. . .lung cancer developed more frequently in subjects receiving β-carotene."

44. Hennekens CH, Peto R. (1994) Antioxidant vitamins — benefits not yet proven. *N Engl J Med* **330**: 1080–1081.

45. Need source of "meta analyses of. . .randomized trials have not demonstrated that beta carotene. . ."

46. Rapola JM, Virtamo J. (1997) Randomized trial of alpha-tocopherol and beta-carotene supplements on incidence of major coronary events in men with previous myocardial infarction. *Lancet* **349**: 71–20.

47. Bjelakovic G, Neocolonial D, Gluud LL, Simonetti RG, Gluud C. (2007) Mortality in randomized trials of antioxidant supplementation for primary and secondary prevention. *JAMA* **297**:8 42–857.

48. The ATBC Cancer Prevention Study Group. (1994) The effect of vitamin E and beta carotene on the incidence of lung cancer and other cancers in male smokers. *N Engl J Med* **330**: 1029–1033.

49. Omenn GS, Goodman GE, Thornquist MD, Balmes J, Cullen MR, Glass A, Keogh JP, Mayskens FL, Valanis B, Williams JH, Barnhart S, Hammar S. (1996) Effects of a combination of beta carotene and vitamin A on lung cancer and cardiovascular disease. *N Engl J Med* **334**: 1150–1155.

50. Zuriek M, Galan P, Berterais S, Mennen L, Czernichow S, Blacher S, Ducimetiere P, Hercberg S. (2004) Effects of long-term daily low-dose supplementation with antioxidant vitamins and minerals on structure and function of large arteries. *Arterioscler Thromb HVACs Biol* **24**:1485–1489.

51. Cook NR, Albert CM, Gaziano JM, Zaharris E, MacFadyen J, Danielson E, Buring JE, Manson JE. (2007) A randomized factorial trial of vitamins C and E and beta

carotene in the secondary prevention of cardiovascular events in women. *Arch Intern Med* **167**: 1610–1618.

52. Ness AR. (2001) Commentary: Beyond beta-carotene — antioxidants and cardiovascular disease. *Int J Epidemiol* **30**: 143–145.

53. Bjelakovic G, Nikolova D, Gluud LL, Simonetti RG, Gluud C. (2007) Mortality in randomized trials of antioxidant supplements for primary and secondary prevention: systematic review and meta-analysis. *JAMA* **297**: 842–857.

54. Tsimakis S. Glass Ck, Steinberg D. Witzum JC. (2004) Lipoprotein oxidation, macrophages, immunity, and atherogenesis. In: Chien KT (eds.) *Molecular Basis of Cardiovascular Diseases,* 2nd ed. Saunders, Philadelphia pp: 385–413.

55. Fletcher AE, Breeze E, Shetty PS. (2003) Antioxidant vitamins and mortality in older persons: findings from the nutrition add-on study to the Medical Research Council Trial of Assessment and Management of Older People in the Community. *Am J Clin Nutr* **78**: 999–1010.

56. Sahyoun NR, Jacques PF, Russell RM. (1996) Carotenoids, vitamins C and E, and mortality in an elderly population. *Am J Epidemiol* **144**: 501–511.

57. Gale CR, Martyn CN, Winter PD, Cooper C. (1995) Vitamin C and risk of death from stroke and coronary heart disease in cohort of elderly people. *BMJ* **310**: 1563–1566.

58. Feskanich D, Singh V, Willett WC, Colditz GA. (2002) Vitamin A intake and hip fractures among postmenopausal women. *JAMA* **287**: 47–54.

59. Denke MA. (2007) Dietary retinol — a double-edged sword. *JAMA* **287**: 102–104.

60. Rosa FW. (1983) Teratogenicity of isotretinoin. *Lancet* **2**: 513.

61. Stern RS, Rosa F, Baum C. (1984) Isotretinoin and pregnancy. *J Am Acad Dermatol* **10**: 851–854.

62. Lammer EJ, Chen DT, Hoar RM, Agnish ND, Benke PJ, Braun JT, Curry CJ, Fernhoff PM, Grix AN, Lott IT, Richard JM, Sun S. (1985) Retinoic acid and embryopathy. *N Engl J Med* **313**: 837–841.

63. Committee on Drugs, American Academy of Pediatrics. (1992) Retinoid therapy for severe dermatological disorders. *Pediatrics* **90**: 119–20.

64. Rothman KJ, Moore LL, Singer MR, Nguyen UDT, Mannino S, Milunsky T. (1995) Teratogenicity of high vitamin A intake. *N Engl J Med* **333**: 1369–1373.

65. Oakley GP, Erickson JD. (1995) Vitamin A and birth defect — continuing caution is needed. *N Engl J Med* **333**:1 414–1415.

66. Geubel A, Galorsy C, Alues N, Rahier J, Dive C. (1991) Liver damage caused by therapeutic vitamin A administration: estimate of dose-related toxicity in 41 cases. *Gastroenterology* 1701–1709.

67. Vitamin A. In *The Vitamins. Fundamental Aspects in Nutrition and Health,* 3rd ed: Combs GF, Jr (ed.). Elsevier, Amsterdam, Boston, Heidelberg, London, New York, Oxford, Paris, San Diego, San Francisco, Singapore, Sydney, Tokyo, pp: 95–143.

68. Hathcock J. (1997) Vitamins and minerals: efficacy and safety. *Am J Clin Nutr* **66**: 427–437.
69. Marshall JR. (1999) β-carotene: a miss for epidemiology. *J Natl Cancer Inst* **91**: 2068–2069.
70. Kolata G. (1996) Studies find beta-carotene, used by millions, doesn't forestall cancer or heart disease. *The New York Times*, Jan. 19, 1996.

22

THE THALIDOMIDE SAGA

Although Accutane is a form of vitamin A, thalidomide is not. It isn't even a vitamin. And while embedding a chapter on thalidomide in the midst of the three vitamin-antioxidant chapters may qualify as a "loose association" psychiatrically, I am doing so because thalidomide — like Accutane — is (was) a drug prescribed liberally to pregnant women, although for a very different reason. In the instance of thalidomide, the medication was used as a sedative and to combat morning sickness. Plus, the thalidomide story may be the worst nightmare of a "safe" drug gone amuck.

The Roots of the Thalidomide Saga

Some readers may recall a scare/scandal in the late 1950s and early '60s related to thalidomide. A different time, a different drug, a different deformity. Infants borne then by a mother who had taken thalidomide — a drug used merely to help women (and men) fall asleep — during pregnancy had horrific birth defects. The most common malformations involved the arms and legs, which were typically shortened or absent. In a number of children, the arms and/or legs were so foreshortened that the hands and feet arose directly from the trunk, like seal flippers. (In fact the medical term for the thalidomide-induced abnormities of the limbs is the Greek word "*phocomelia*," meaning "seal-like limbs.") In deference to those individuals now alive with a thalidomide-related anomaly, their facial features will not be shown. The hands and feet were also sometimes deformed, and examples of each are presented in Fig. 22.1.

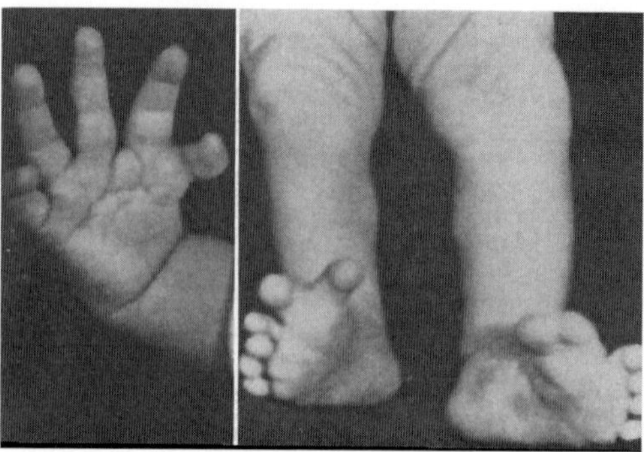

Fig. 22.1.   Thalidomide-induced a of a child's hand and feet.

(*Source:* Reproduced with permission from the publisher.) Reference 1.

Heinrich Mückter, Wilhelm Kunz and Chemie Grünenthal

The thalidomide story began in the mid-1950s, when memories of Hitlerian human experimentation atrocities were still vivid. The three key actors in the drama were Heinrich Mückter, a former army scientist with the Third Reich, Wilhelm Kunz, a sergeant under Mückter's command, and Widukind Lenz. Mückter is described as "a man of cold, hard judgment," a befitting character for the Medical Officer to the Superior Command of the German Occupation Forces.[2] Kunz, on the other hand, is pictured as a meek, obsequious sergeant quick to kowtow to Mückter.[2] Lenz was a German pediatrician and geneticist who will be linked, later in this chapter, to the ending of the thalidomide story in 1961.

Immediately after the war — in 1946 — Mückter joined the West Germany pharmaceutical firm of Chemie Grünenthal and took his trusted troll with him.[2] Although Kunz had little scientific background, he was given the title of Chief of Chemical Research for Grünenthal. His first assignment was to come up with "a simple, inexpensive method for manufacturing antibiotics from peptides — the short bonds that hold amino acids together to form. . .proteins."[2] During one of his very first experiments he simply heated an existing chemical and wound up with a different compound with lots of peptides. It didn't have a name, so he called it — guess what? — thalidomide. One pitch, one home run. Yet, thalidomide was not an "inexpensive antibiotic," which really was the aim of the research. It had no antibacterial properties at all.

In fact, it seemed to have no usefulness whatsoever. That led Stephens and Brynne — writing a book about the medication years later — to say wittily, "the only thing thalidomide had going for it was that. . .(no) dose high enough to kill a lab rat" could be found.[2] Grünenthal chemists also gave the drug to the usual array of small- and medium-sized laboratory animals — mice, guinea pigs, rabbits, cats and dogs — and found it to be free of side effects.[2] That led to coining the term "nontoxic thalidomide."

Now came the moment — the defining moment — that that would lead to the thalidomide tragedy in humans. Even though the drug had not shown any sedative effects in lab animals, chemists at Chemie Grünenthal decided to try it (as a sleep-promoting agent) in humans. The reason for even thinking it might have a sedative effect in man was that Grünenthal chemists believed thalidomide had a chemical structure similar to that of barbiturates. (The word "barbiturate" is commonly misspelled and mispronounced "barbituate.") Barbiturates, by the way, were the leading sedative prescribed worldwide at the time. But barbiturates can be lethal. An intentional or an accidental (by children) overdose can cause death. But thalidomide had not killed a single rat! (The CIBA Pharmaceutical Company in Basel, Switzerland later determined that thalidomide does not have the chemical structure the Grünenthal chemists imagined.[2]) A safe, nontoxic sedative would shock the world sleepmarketplace.

In early 1955, Grünenthal began a clinical trial in West Germany and Switzerland testing thalidomide in epileptics, although there seemed to be no reason to expect it to have an antiepileptic effect. (But people suffering from awful diseases, such as cancer and epilepsy, often are open, understandably, to nontraditional treatments.) Although it did not prevent or lessen seizures, it *did* "cause patients to go into a deep, all-night 'natural' sleep."[2] This meant that the drug was more than a sedative, which *produces relaxation* that may lead to sleep; rather, it was a hypnotic, a drug that *directly induces* sleep.

In June 1956, Aldous Huxley (of *Brave New World* fame) wrote an article in London's *Sunday Times* noting that alcohol had been used as a mind-altering drug by mankind since time immemorial. An executive at the British pharmaceutical firm of Distillers Company Ltd. read Huxley's article. He "promptly pointed it out in a memo to the company's director the next day," according to Stephens and Brynner.[2] "The ultimate target," he wrote, "would be the production of the ideal tranquilizing agent to replace alcohol among those people who would prefer to 'transform their minds' by this alternative means."

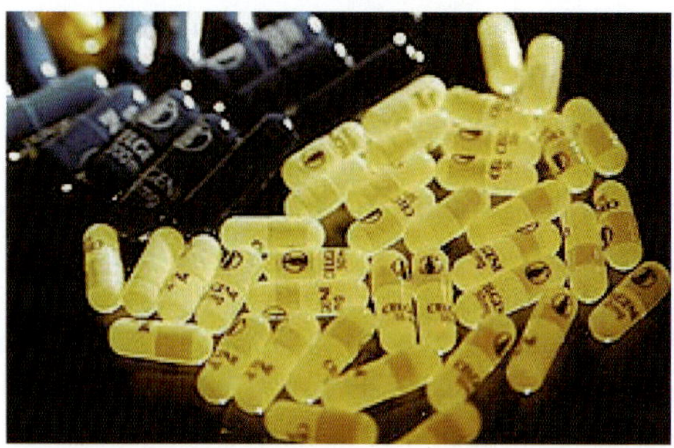

Fig. 22.2. A clutch of thalidomide capsules. Their color reflects their approximate worth between 1957 and 1961.

(*Source*: Reproduced with permission from the publisher.) Reference 3.

The company director replied ". . .it will not be long before there are as many of these things as there are brands of whiskey."[a,2]

Thalidomide released in west germany in 1957

On October 1, 1957, thalidomide was first released as a sedative in West Germany.[b] It hit the medical market through a massive promotional campaign, including advertising in 50 medical journals and mailing 250,000 personal letters to physicians highlighting its safety.[2] By 1961, thalidomide was being sold in 46 countries. So safe was the medicine considered, it was sold OTC in Germany and many other countries as a sleep aid and to combat nausea.[4] So popular was the drug, it became the best-selling sedative in Europe.[4] Quickly, thalidomide became widely used by pregnant women (both for its sedative and its antimorning-sickness effects) around the world — the U.S. being an exception. Its worldwide sales rivaled those of aspirin.[5]

Between October 1957 and December 1961, an estimated one million pregnant women were prescribed thalidomide for "morning sickness"[6] (Fig. 22.2). Then, during 1961, obstetrician William McBride, in Australia, noted an

---

[a]Within a month, Distillers contracted with Grünenthal to license and distribute thalidomide in Great Britain.[2]

[b]In April 1958, thalidomide hit the streets of Great Britain. As in Germany, it was advertised as "completely safe" and the solution to "the mounting toll of barbiturate deaths."[2]

astonishing number of infants with a similar deformity of the arms and legs, all of whom were borne by mothers he had prescribed thalidomide for either morning sickness or as a sedative months earlier.[7]

McBride quickly put two and two together and reported in the December 16, 1961, issue of *The Lancet* that "almost 20%" of the women he had given thalidomide delivered babies with severe deformities of their arms and/or legs.[c,7] The manufacturer (presumably) got wind of the impending publication and announced its withdrawal of the drug from the marketplace on December 2, 1961 in a letter to the editor of *The Lancet*. The letter read, in part, "We have just received reports from two overseas sources possibly associating thalidomide. . .with harmful effects on the foetus in early pregnancy. Although the evidence on which the reports is based is circumstantial, and there have been no reports from Great Britain. . .we feel that we have no alternative but to withdraw the drug from the market immediately.. . ."[8] Overnight, thalidomide became as popular in Germany as the *Luftwaffe* was in London some 15–20 years earlier.

Enter Widukind Lenz

The third actor in the thalidomide drama was Widukind Lenz. Lenz (Fig. 22.3), a pediatrician on the other side of the world (Hamburg, West Germany), felt a sense of impending doom "as the early outlines of a widespread epidemic began to emerge."[2] He, too, noted a number of newborns with severe deformities of the arms and/or legs that he'd not seen before. He set out to address his concern by systematically reading the birth reports of the entire city's population of 212,000 births — *sans* a computer! — between 1930 and 1955. He found just one case of phocomelia — an incidence of 0.001%. Then he looked at the birth records from 1961 and found eight such cases among 6420 newborns — an incidence of 0.8%. He quickly began interviewing mothers of infants with seal-like limb deformities.[d] Initially, he didn't

---

[c]On Christmas Day 1956, ironically, the very first baby with a thalidomide-related malformation was born before the drug became commercially available. The infant happened to be the daughter of an employee of Chemie Grünenthal who gave his pregnant wife some free tablets for her morning sickness.[4] (That case was not reported in the medical literature.)

[d]While malformations of the limbs are the major abnormalities caused by thalidomide, there are other congenital anomalies, too numerous to list. A curious reader may view these at http://www.bbc.co.uk/science/horizon/2004/thalidomide.shtml

Fig. 22.3.   William Lenz as he appeared on the cover of *Life* magazine on January 1 1962, for being "among the first to warn against use of thalidomide by pregnant women."

(*Source*: Reproduced with permission of *Life.*) Reference 9.

ask about maternal medications during pregnancy, simply because he had never heard of an association of a drug with limb malformations. But, in November 1961, during a reinterview with one mother in Hamburg, she told him she had taken thalidomide while pregnant. She hadn't mentioned it the first time because she had thought it was "so insignificant."

Continuing his sleuthing, Lenz found 14 more cases within two weeks of children with malformed limbs among women who had taken thalidomide while pregnant.[e] Images of two such infants and one other primate are shown in Figs. 22.4–22.6. At that point he called Mückter at Grünenthal and told him about his "suspicions and concerns." He told Mückter "the drug should be withdrawn immediately, and that every day the drug remained on the market was a deliberate experiment in human teratology (the study of birth defects). Grünenthal "ignored him."[2] Instead, he authorized the mailing of

---

[e]One particularly notable case was that of a 24-year-old woman with essentially no arms and legs, making her a technical challenge simply to draw blood for routine laboratory tests and to measure her blood pressure. Despite these obstacles, she went on to deliver a (normal) full-term infant.[3]

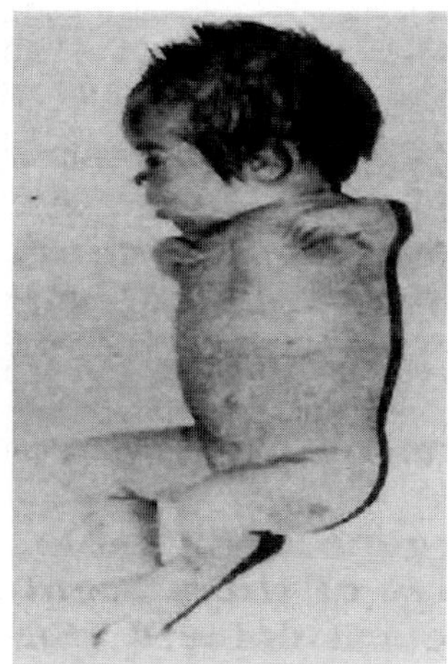

Fig. 22.4. An infant with severe phocomelia.

(*Source*: Reproduced with permission from the publisher.) Reference 10.

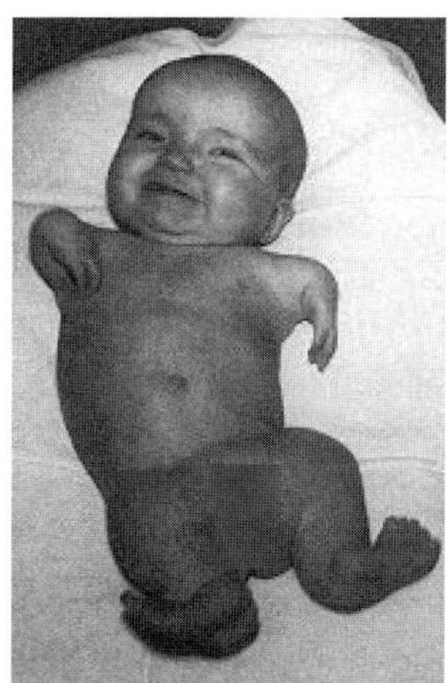

Fig. 22.5. Symmetrical deformities of all four limbs, arms more affected than legs, in an infant borne by a mother who took thalidomide during pregnancy with severe phocomelia.

(*Source*: Reproduced with permission from the publisher.) Reference 10.

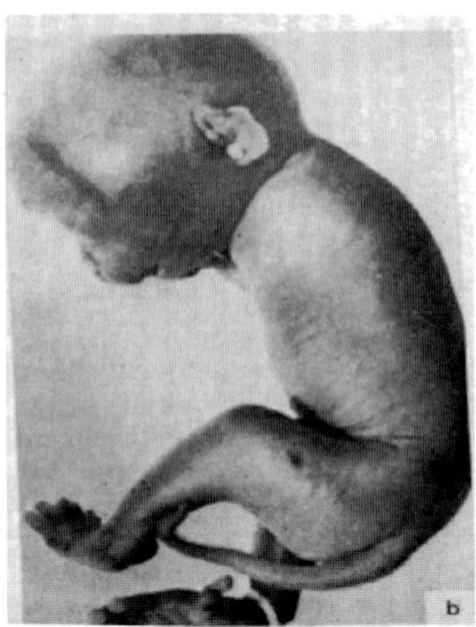

Fig. 22.6.   Thalidomide did *not* create babies with tails. But it did lead to infants with deformities of the arms and legs, as in this armless monkey with deformed legs and feet.

(*Source*: Reproduced with permission from the publisher.) Reference 11.

"almost 20,000 promotional leaflets to doctors across Germany declaring: '(Thalidomide) is a safe drug.'"[2]

Lenz persisted. Finding Mückter "nonchalant" over the phone, he wrote a letter to the pharmaceutical company the next day saying, "Every month's delay in clarification means that fifty to one hundred horribly mutilated children will be born."[2] The company again ignored him.[2] Grünenthal's intransigence made Stephens and Brynner, recounting the odyssey years later, wonder if Mückter, "like the army he served seventeen years earlier, soldiering on in face of certain defeat," was acting "with a similar contempt for the value of human life, out of loyalty to shareholders."[2]

Within days, Lenz arranged a meeting with West Germany's Ministry of the Interior. The Ministry quickly ordered Grünenthal to "withdraw thalidomide from the market, or they would ban it."[2] On November 25, 1961, Mückter again refused to withdraw the drug from the marketplace. But then, as so often happens with shoddy practices, the story hit the streets.[f] A major German newspaper, *Welt ann Sonntag*, headlined the impasse, including

---

[f]The thalidomide tragedy first reached Americans in July 1962 through an article written by Norton Mintz in Washington Post. The article fomented a public tidal wave of shock and disgust.

Lenz's assertion that "every month's delay in clarification means that fifty to one hundred horribly mutilated children will be born".[2] That did it. The next day, Mückter agreed to remove thalidomide from all British and Continental markets.

Meanwhile, Lenz published his observations, first in German,[12] then in English in a letter to the editor of *The Lancet* on February 10, 1962.[13] He also emphasized that just one thalidomide tablet was enough to cause the horrendous birth defects he'd seen. As a result of his scholarly investigative work, *Life* magazine chose Lenz for its cover photograph in January 1962 (Fig. 22.3). *The Lancet* and *Life* publications led Grünenthal to attack Lenz personally, even quoting Goethe, "Idiots and clever people are both equally harmless. Those halfwits and half-educated people who always recognize only half-truths alone are dangerous."[2] Ultimately, 5000–12000 infants (and who knows how many aborted fetuses) from 46 countries were deformed by thalidomide, of which 5,000 survived beyond childhood.[14] But the nightmare was finally over.

Thalidomide in the United States

The United States soon picked up on the European experience, largely through the alertness of then-new FDA agent Frances Kelsey. She had been on her job with the agency just one month.[15] Upon learning about the tragedies in Europe with thalidomide, she "bought time" by requiring the manufacturer to resubmit its U.S. application six times.[16] Her diligence paid off. It paid off for the U.S. by keeping the drug out of the country. It paid off for her when then-president John F. Kennedy handed her a presidential award for her vigilance (Fig. 22.7).[g]

----

[g]While the manufacturer promptly withdrew the drug from the marketplace (in late 1961) upon learning about severe thalidomide-related birth defects, it used "every legal trick" available for 12 years to avoid compensation to the parents of affected children. In 1973, the company agreed to compensate the children's families, but it did not accept responsibility for the tragedy. Finally, on November 14, 2009 — that's right, 48 years later — it published online in London's *Sunday Times* its intent to accept responsibility. Said Sir Harold Evans, "This was a disaster fathered by a lackadaisical drug regulatory system. It didn't happen in the US because they had a better regulatory system,"[17] — yet another compliment to Frances Kelsey (Fig. 22.8, 95 years old at the time!). Or, as Sir Harold put it while campaigning for more compensation for surviving thalidomide victims — based on inflation since the agreement with the manufacturer was reached in 1973 — *and* called upon the British government and Parliament to acknowledge their responsibility.[18]

Fig. 22.7.   A 1962 photograph of a pleased (and double-doctorate) Frances Kelsey, with JFK presenting her with an award for her vigilance in "protecting the world's largest market from the horrifying effects of [thalidomide]."[20] The award is the highest honor given to a civilian in the U.S. — the Presidential Award for Distinguished Federal Service.[2]

(*Source*: Reproduced with permission from the National Library of Medicine.) Reference 16.

But the awards for Frances Kelsey didn't stop in 1962. In 2005, at age 87, The National Women's Foundation honored Kelsey by naming an annual award after her.[19] In a photograph shot at an angle remarkably similar to the one in 1962, she looks younger than 91 (Fig. 22.8), doesn't she?

The Legislative Fallout from the Thalidomide Tragedy

The tragedy catalyzed "sweeping changes in drug regulation."[22] First, it was mandated that "in the future, new drugs will be tested more extensively in animals to see if they passed the placenta." Second, "the powers of the national regulatory authorities were greatly increased." Third, "systems were set up to

Fig. 22.8.   Frances Kelsey in a photograph shot at a remarkably similar angle in 2005. (*Source*: Reproduced with permission from the publisher.) Reference 21.

detect. . .side-effects at an earlier stage following release of the drug for marketing." (It is now called "postmarketing surveillance.") One such system allows physicians to easily express their concerns/suspicions about a possible adverse effect of a pharmaceutical by simply submitting a Yellow Card online (or by telephone or postal mail) in the UK. In the US, MedWatch is the equivalent of the British Yellow Card system.[23]

It's All About Balance

Physicians constantly balance possible good effects of a drug against its possible bad effects (Fig. 22.9). For example, a single aspirin tablet a day can significantly reduce the risk of a heart attack by "thinning out" the blood, making a blood clot in one of the three coronary arteries — which is the immediate cause of a heart attack — far less likely. [While a fresh blood clot precipitates a heart attack, the clot develops on top of a substantial cholesterol deposit that took years to grow (Chap. 12).] But that single aspirin tablet a day can also lead to gastric irritation and GI bleeding. Similarly, in menopausal

Fig. 22.9.   Balancing the good against the bad. This figure evokes a number of observations and interpretations by viewers. What's yours?

(*Source*: Reproduced with permission from the publisher.) Reference 24.

women estrogen replacement therapy can lessen menopausal symptoms, but also can lead to thrombophlebitis, i.e. blood clots in the legs. In the case of fertile women, there was no question about balance. Fertile women should not take thalidomide (Fig. 22.10). The FDA reminds women in bold type, "just one dose can cause severe birth defects."[25] Men too should not take thalidomide. The FDA tells men to "abstain from sexual intercourse or use a condom during intercourse while, and for one month after, taking thalidomide."[25] Thalidomide may be in the semen.

Thalidomide's Rebirth

The thalidomide story would have ended in 1961 were it not for Jacob Sheskin, an Israeli physician, working with leprosy patients in Marseilles, France in 1964. One day, "while in despair at his inability to lessen the pain and misery of his patients, he found some packs of thalidomide on the dispensary shelves.

Fig. 22.10. Using the familiar don't do this or that symbol, the FDA warns potentially reproductive females not to take thalidomide.

(*Source*: Reproduced with permission from the publisher.) Reference 25.

Deciding that this sedative might be worth a try in patients in whom other sedatives had failed, he gave two tablets to one of his patients. The man slept well, got up the following day and after two more pills, his [leprosy skin] lesions started to heal"[26] — yet another example of medical serendipity!

Worldwide Incidence of Leprosy

While leprosy — aka Hansen's disease, named after Armauer Hansen Norwegian physician who first showed [in 1873] that leprosy was caused by a mycobacterium[h,27] — is very rare in the U.S[i], there are pockets of leprosy around the world (Fig. 22.11). No doubt, there is an equatorial distribution. While climate is clearly a factor in leprosy frequency — high ambient temperatures, high humidity, and lots of rainfall promote the survival of the leprosy mycobacterium outside the body[28] — distribution is related more to substandard living conditions in those nations most affected than to climate.[28,29]

---

[h] Before then, it was believed to be a curse or related to "sinful living."[27]

[i] Fascinatingly, there is also a pocket of leprosy in the U.S. In 2008, three cases of leprosy within a 13-month period were seen in the Mississippi Delta. None had a history of "foreign travel, contact with each other, or [with] known leprosy patients."[30]

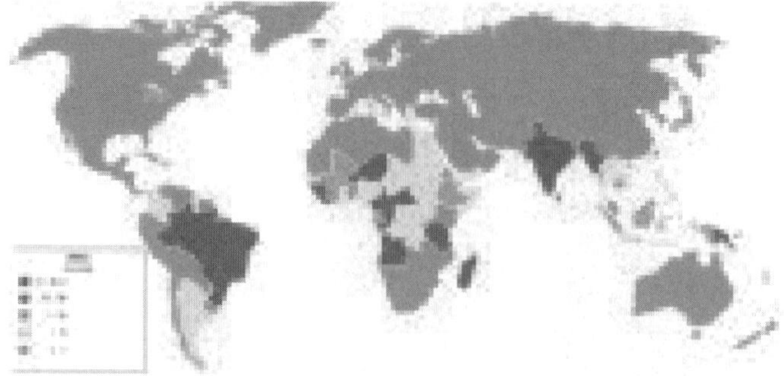

Fig. 22.11.   Worldwide distribution of leprosy in 2003. India, Brazil, and Burma are the leaders.

(*Source*: Reproduced with permission from the publisher.) Reference 28.

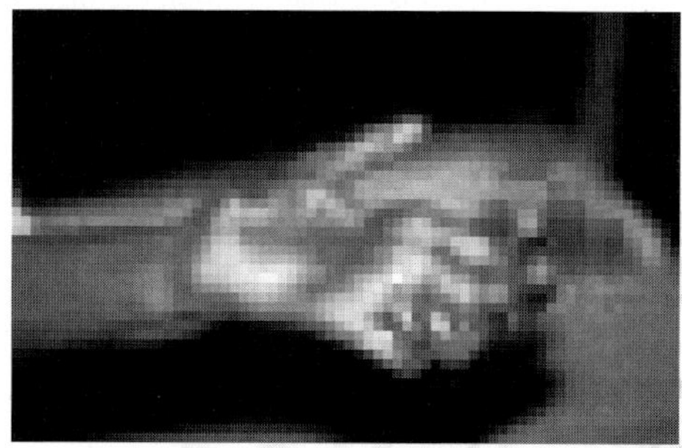

Fig. 22.12.   A typical hand deformity in a child with leprosy. The cause of the deformity is multiple injuries related to impaired or absent sensation, not thalidomide.

(*Source*: Reproduced with permission from the publisher.) Reference 31.

Physical and Social Consequences of Leprosy

The disease is limited to the skin, peripheral nerves, and upper respiratory tract, save for occasional involvement of the eyes. The scaly appearance of the skin gives rise to the term "leprosy," derived from the Greek word "*lepi*," which describes the scales of a fish.[25] Damage to the peripheral nerves often leads to poor or absent sensation in the hands, making them prone to deformative injuries (Fig. 22.12).

Fig. 22.13. Come hell or high water: Multidrug-therapy-for-leprosy — provided by the Red Cross — delivery in the Congo.

(*Source*: Reproduced with permission from the publisher.) Reference 34.

The social consequences of Hansen's disease can be as devastating as the physical consequences, though they probably need not be. Stigmatization, "community rejection, loss of employment, and sometimes forced isolation are still prevalent in both endemic and non-endemic countries."[31] Importantly, the patient becomes "non-infective within one month" of the initiation of therapy[32]. . .but the disfiguration and stigmatization may last a lifetime.

The Treatment of Leprosy

The good news is that leprosy is a "fully curable disease."[35] Since 1985, a combination of three antibiotics — dapsone, rifampin, and clofazimine[j] — has cut its prevalence to less than one case per 10,000 population.[35] The World Health Organization, WHO, reports that its official prevalence goal has been reached in 112 of the 122 countries where leprosy was endemic before the standard and widespread use of the three antibiotics.[32] Getting the drugs to where they're needed has become the new challenge (Fig. 21.13). The W.H.O. also stresses that there are "virtually. . .no recurrences of the disease after treatment is completed."[35]

---

[j]The medications have been provided free by Novartis Pharmaceuticals in Basel, Switzerland and the Sasakawa Peace Foundation in Tokyo.[33]

Some cases of leprosy, however, are complicated by a skin reaction separate from the usual cutaneous expression of leprosy. It is a serious condition. This painful "episodic reaction occurs in about half of. . .leprosy patients and mostly develops within the first two years of drug treatment."[35] On average, it lasts "about 5 years. . .[but] may persist and exacerbate even after 5–10 years.. . .Left untreated, wounds are slow to heal and often cause serious facial scarring and/or blindness. . .if they do heal."[35]

Physicians who continued to treat leprosy over the 30 years following the thalidomide scare noted that the drug remained useful for combating this leprosy reaction. The effect was usually "quick and dramatic. . .[and] remarkably consistent." So effective was thalidomide that it was recommended by the WHO in 1994 as "the preferred treatment for moderate to severe [reactions] in men and women of non-child-bearing potential."[36]

An alternative therapy is the use of methylprednisolone — a cousin of prednisone. But prednisone and methylprednisolone (both steroids) are certainly not without their dangers. Steroids can be life-saving — as in status epilepticus (repeated grand mal seizures) and severe allergic reactions — but they can be life-threatening when used for long periods of time. In general, steroids are best used in as *small* a dose as possible for as *short* a time as possible.[37]

And so it came to pass that in August 1998 thalidomide was approved by the FDA nearly 40 years after it was banned. . .but only for the treatment of this unusual, specific complication of leprosy. Round 2 of the thalidomide tragedy seems very improbable.

Thalidomide for Multiple Myeloma

Just about everyone has heard of leukemia, but not everyone knows about multiple myeloma,[k] a usually malignant tumor of the bone marrow. Myeloma is a not-so-common disorder of the blood and lymphatic systems. And while just about everyone has also heard of red blood cells and white blood cells, not everyone has heard of plasma cells. Plasma cells are one of the several types of white blood cells, and because white blood cells — including plasma cells — are made in the bone marrow and transported by the blood and lymphatic

---

[k]The term "myeloma" comes from the Greek words "myelo" (meaning "bone marrow") and "oma" (meaning tumor").

systems, multiple myeloma is fundamentally a disease of the blood and lymphatic systems.[38]

The American Cancer Society estimates that there are about 45,000 people in the U.S. with multiple myeloma and that approximately 14,600 new cases are recognized each year.[39] It also notes that the disorder is seen primarily in the "elderly. . .[and that] the disease does not occur in children."[39]

The most frequent symptoms myeloma patients experience are bone pain (particularly of the spine and ribs), infections (such as pneumonia and pyelonephritis), kidney failure, and a variety of neurologic symptoms, including headache and fatigue. (Importantly, headache and fatigue are *nonspecific* symptoms. Virtually all of us suffer an occasional headache or feel tired at times; but few of us have multiple myeloma.) Common routine laboratory abnormalities include anemia and diminished kidney function tests.[38]

Treatment with thalidomide

Thalidomide first found use for the treatment of multiple myeloma in 1999.[40] At that time, the standard therapy for myeloma was a combination of a drug called melphalan (Alkeran) and prednisone, a steroid.[41] With that therapy — which hadn't changed in 40 years — the disease was "incurable."[42] That led hematologists and oncologists at the University of Arkansas to try thalidomide as a *single* agent in a group of patients with "previously treated and progressive" myeloma.[40] While no one was cured, some (32%) went into a remission. That's not very many, but it provided a brief respite — at the price of some moderate but tolerable side effects — during the course of an incurable disease. The treating physicians considered the results to be rather extraordinary. Two editorialists also thought so saying, "Given that [all] patients in the study had relapsed after [conventional] chemotherapy, which was usually given in massive doses, [the effect] of thalidomide [was] indeed remarkable."[43]

Just *how* thalidomide works in multiple myeloma is not fully understood. The best bet may be that because even the bone marrow must have a blood supply to make fresh red blood cells, white blood cells, and plasma cells, slowing or stopping the ingrowth of new blood vessels into the tumor (a process called angiogenesis) makes it wither. No blood supply, no tumor growth. It was on the hunch that thalidomide might inhibit bone marrow angiogenesis that University of Arkansas scientists began using thalidomide in patients with drug-resistant multiple myeloma.[40]

From drug-resistant, to refractory-to-treatment (using conventional therapy), to initial therapy of new myeloma cases, the use of thalidomide expanded over the next five years. By 2004, thalidomide alone or in combination with melphalan (or a closely related drug) and prednisone became "standard therapy for relapsed and refractory myeloma."[44] And, by 2008, the FDA had approved thalidomide "in combination with [steroids]" for the treatment of "newly diagnosed patients with multiple myeloma" (Fig. 22.14).[45] Sadly, the addition of thalidomide to treat myeloma increases the chances of significant, unwanted side effects. As in the use of thalidomide for leprosy, damage to peripheral nerves is a "major concern."[46] Plus, the addition of thalidomide substantially increases the chances of thrombophlebitis (blood clots in the legs) with *its* risk of an embolus (a fragment breaking off and traveling through the bloodstream into the lungs) — a potentially fatal event. The chances of thrombophlebitis can be reduced by warfarin (Coumadin) — a blood thinner. But Coumadin has *its* drawback of possibly causing bleeding into the brain.[47] So worrisome is this that in 2006 the FDA ruled that the Coumadin label must bear a "black box," warning of this unwelcome possibility.[48] [And, yes, aspirin can be used as blood thinner, but not in patients

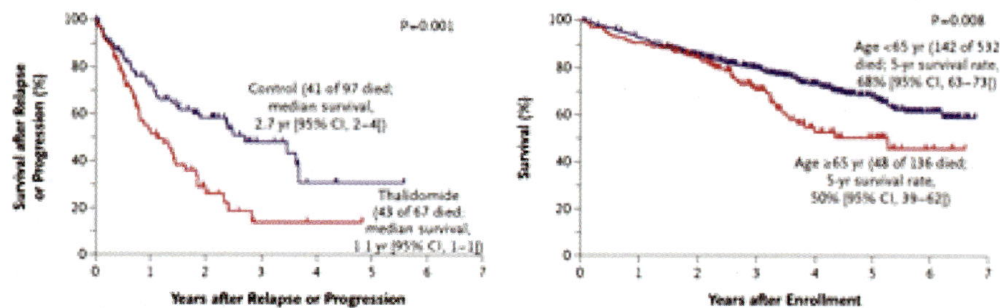

Fig. 22.14.   These graphs show the survival rates of patients with *newly diagnosed* multiple myeloma in a randomized trial comparing treatment with melphalan and prednisone plus thalidomide (red) versus melphalan, prednisone and placebo (blue) in 668 patients. Each person also underwent a stem cell transplantation. The thalidomide group had fewer relapses over five years following therapy than the placebo group (56% versus 44%), but after a relapse survival was shorter in the thalidomide group than in the control group (panel C). Panel D shows the seven-year survival curves for those patients under 65 and those 65 or more. In a commentary on healthcare carriers, the authors attributed this to the "[higher] rate of insurance denial among patients who were at least 65 years old than among younger patients." (*Source:* Modified and reproduced with permission from the publisher.) Reference 45.

at high risk of forming blood clots, such as those with an artificial (mechanical) heart valve.[49]] All in all, the treatment of a person with multiple myeloma is a challenge to the patient, the physician, and the oncology team alike.

The FDA approves thalidomide for treatment of multiple myeloma

On May 26, 2006, the FDA approved "thalidomide in combination with dexamethasone [a steroid] for the treatment of newly diagnosed [cases of] multiple myeloma."[50] With the approval of drug, the FDA mandated that the System for Thalidomide Education and Prescribing Safety (STEPS) program, mentioned earlier, be followed. This mandatory system — involving the patient, the physician, and the pharmacist — controls the marketing of thalidomide, and is designed to prevent fetal exposure to the drug during pregnancy.

Thalidomide in Comedy

A British playwright, Mat Fraser, himself a thalidomide victim, jests about a sad situation. Born with stunted arms, the result of his mother's ingestion of thalidomide during pregnancy, Fraser wrote a musical-comedy play called *Thalidomide!! A Musical.* He not only wrote the play but also composed all the songs — including clever, satirical ones like *Talk to the Flipper*, clear evidence that thalidomide babies were, mercifully, born with normal brains. The producer/director then put on his acting cape and along with another (apparent) thalidomide victim, Anna Winslet, played all the roles![51]

The story is about a romance between Glynn (played by Fraser) and Katie, a pretend thalidomide victim, played by Winslet (Fig. 22.15). Glynn wants to fall in love with Katie but can't "accept [her love] until he has come to terms with his disability."[52] The play opened in London in October 2005 and later was performed at venues across Britain and in Versailles, France. Reviews by theater critics varied widely. A play critic with the *Manchester Evening News* billed it as "one of the most life-affirming performances I've witnessed on stage."[53] while a critic for The *Scotsman* hailed it as "outrageous [and] brilliant [with] taboo busting songs.[50] But critic Emma John, with the *Manchester Guardian,* found it an "exuberantly bad taste comedy."[52] Maybe the newsletter *Disability Now* put it best, describing the play as "entertaining, witty and revealing, [a] show [that] really packs a punch, providing an ideal showcase for Fraser's many talents."[53]

Fig. 22.15. Mat Fraser and Anna Winslet kickboxing in the 2005 British musical comedy *Thalidomide!! A Musical*. Winslet appears to have no arms, an illusion created by strapping the backs of her wrists to her shoulders.[52]

(*Source*: Reproduced with permission from the publisher.) Reference 53.

But perhaps, as the king of playwriting wrote in another comedy, at another time, in another place, "All's well that ends well" — meaning problems do not matter so long as the outcome is good.

References

1. Adjust Taussig HB. (1962) A study of the German outbreak of phocomelia. The thalidomide syndrome. *JAMA* **180**: 1106–1116.
2. Stephens T, Brynner R. (2001) *Dark Remedy*. The impact of thalidomide and its revival as a vital medicine. Perseus Publishing, Cambridge, Massachusetts, Chap. 1, pp. 1–17.
3. http://www.bbc.co.uk/science/horizon/2004/thalidomide.shtml
4. Matthews JJ, McCoy C. (2003) Thalidomide: review of approved and investigational uses. *Clin Ther* **25**: 342–395.
5. Fintel B, Samars AT, Crias E. (2009) The thalidomide tragedy: lessons for drug safety and regulation. http://science in society.northwesten.edu/content/

articles/2009/thalidomide-tragedy-lessons-drug-safety-and-regulation. Accessed Nov. 16, 2009.

6. http://www.bbc.co.uk/science/horizon/2004/thalidomide.shtm
7. McBride WG. (1961) Thalidomide: congenital abnormalities. *Lancet* **2**: 1358.
8. Hayman DJ. (1961) Distaval. Letter to the Editor from the manufacturer of Distaval (brand name of thalidomide). *Lancet*, Dec. 2.
9. http://www.life.com/image/505721
10. Smithells RW, Newman CGH. (1992) Recognition of thalidomide defects. *J Med Genet* **297**: 716–723.
11. Schardien JL. (2007) Drugs affecting the central nervous system. *Chemically Induced Birth Defects.* Marcel Dekker, New York, and Basel.
12. Lenz W. (1961) Klindliche Missbildungen nach Medikament wahrend der Graviditat. *Deusch Med Wschr* **86**: 2555.
13. Lenz W, Pfeiffer RA Kosenow W, Hayman DJ. (1962) Thalidomide and congenital abnormalities. *Lancet* **279**: 45–46.
14. Tao D. (2005) *Drug Discov Today* **10**: 107–114.
15. http://www.nlm.nih.gov/changingthefaceofmedicine/physicians/biography_182.html
16. Tansey EM. (2001) Book review of "Dark remedy: the impact of thalidomide and its revival as a vital medicine." *N Engl J Med* **345**: 226–227.
17. Editorial. An end to the terrible saga of thalidomide. The Sunday Times, Nov. 14, 2009.
18. Laurang J. (2009) Thalidomide victims still fighting for justice 35 years on. *Belfast Telegraph*, Oct. 9.
19. Barber J (2005-11-10). Center ceremony honors 107 individuals, 47 groups: spring event inaugurates Frances Kelsey Drug Safety Award. Accessed Nov. 16, 2009.
20. http://www.nim.nih.gov/changingthefaceofmedicine/physicians/biography_182.htm
21. http://en.wikipedia.org/wiki/Frances_Oldham_Kelsey#cite_note-fda2005-11
22. Vandenbroucke JP. Thalidomide: an unanticipated adverse effect. The James Lund Library (http://www.jameslundlibry.org, accessed Nov. 12, 2009).
23. http://www.ashp.org/import/News/HealthSystemPharmacyNews/newsarticle.aspx?id=2058
24. Cover. *BMJ*, vol. 329; July 3, 2004.
25. Thalidomide: important patient information. U.S. Food and Drug Administration. Sep. 11, 1997.
26. Stephens T, Brynner R. (2001) Dark remedy: the impact of thalidomide and its revival as a vital medicine. Perseus, Cambridge, Massachusetts, pp. 121–135.
27. http://rarediseases.about.com/cs/infectiousdisease/a/071203.htm

28. http://en.wikipedia .org/wiki/Leprosy

29. Duncan K. (1994) Climate and the decline of leprosy in Britain. *Proc R Coll Physicians* **24**: 114–120.

30. Abide JM, Webb RM, Jones HL, Young LF. (2008) Three indigenous cases of leprosy in the Mississippi delta. *South Med J* **101**: 634–638.

31. Rinaldi A. (2005) The global campaign to eliminate leprosy. *PLoS Med* **2**: 1222–1225.

32. http://www.who.int/mediacentre/factsheets/fs101/en

33. World Health Organization. (2004) Leprosy elimination report. Status report 2003. http://www.who.int/lep/Reports/S2004.2.pdf

34. Teo SK, Reskaztak KE, Scheffler MA, Kook KA, Zeldis JB, Stirling DI, Thomas SD. (2002) Thalidomide in the treatment of leprosy. *Microb Infect* **4**: 1193–1202.

35. WHO expert committee on leprosy. WHO technical report series, #874, 7th report, World Health Organization, Geneva, Switzerland.

36. Fauci AS. The vasculitis syndromes. In Fauci AS,: *Harrison's Principles of Internal Medicine.* Braunwald E, Isselbacher KJ, Wilson JD, Martin JB, Kasper DL, Hauser SL, Longo DL (eds.), McGraw-Hill, New York, St Louis, San Francisco, Auckland, Bogotá, Caracas, Lisbon, London, Madrid, Mexico City, Milan, Sydney, Tokyo, Toronto Chap. 319, pp. 1910–1922.

37. http://en.wikipedia.org/wiki/Multiple_myeloma

38. http://en.wikipedia.org/wiki/Multiple_myeloma#cite_note-0

39. Haroussseau J-L, Moreau P. (2009) Autologous hematopoetic stem-cell transplantation for multiple myeloma. *N Engl J med* **360**: 2645–2654.

40. Singhal N, Mehta J, Desikan R, Ayers D, Roberson P, Eddlemon P, Munshib N, Anaissee E, Wilson C, Dhudapkar M, Zeldi J, Barlogie B, Siegel D, Crowley J. (1999) Antitumor activity of thalidomide in refractory multiple myeloma. *N Engl J Med* **341**: 1565–1571.

41. Palumbo A, Bringhen S, Caravita T, Merla E, Capparella E, Callea Violosi C, Rossini F, Galli M, Catalano L, Zamagni E, Petrucci MT, De Stephano V, Cecarelli M, Ambrosini MT, Avonto I, Falco P, Ciccone G, Liberati Am, Musto P, Bocadoro M; Italian Multiple Myeloma Network, GIMEMA. (2006) *Lancet* **367**: 825–831.

42. http://www.mayoclinic.com/health/multiple-myeloma/ds00415

43. Raje N, Anderson K.(1999) Thalidomide — a revival story. *N Engl J Med* **341**: 1606–1609.

44. Kyle RA, Rajkurmara SV. (2004) Multiple myeloma. *N Engl J Med* **351**: 1860–1873.

45. Palumbo A, Facon T, Sonneveld P, Blade J, Offidani M, Gay F, Moreau P, Waage A, Spencer A, Ludwig H, Boccadoco M, Harusseau J-L. (2008) Thalidomide for treatment of multiple myeloma 10 years later. *Blood* **111**: 3968–3977.

46. Handin RI. (1998) Anticoagulant, fibrinolytic, and antiplatelet therapy. In: *Harrison's Principles of Internal Medicine*, 14th ed. Fauci AS, Braunwald E, Isselbacher KJ, Wilson JD, Martin JB, Kasper DL, Hauser SL, Longo DL(eds.) McGraw-Hill, New York, St Louis, San Francisco, Auckland, Bogota, Caracas, Lisbon, London, Madrid, Mexico City, Milan, Sydney, Tokyo, Toronto.
47. http://www.nature/news.com/020682_Coumadin_blood_black_box_warning.html
48. Cannegieter SC, Rosendaal FR, Wintzen AR, van den Meer FJM, Vandenbroucke JP, Briet E. (1995) Oral anticoagulant therapy in patients with mechanical heart valves. *N Engl J Med* **333**: 1117.
49. http://www.fda.gov/AboutFDA/cenresOffices/CDER/ucm095651.htm
50. http://en.wikipedia.org/wiki/Thalidomide!!_A_Musical
51. http;//www.guardian.co.uk/stage/2005/nov/07/theatre
52. http://www.cdp.org.uk/arts/art_events_jobs.htm
53. http://www.channel4.com/life/microsite/B/bornfreak/birthday.html

23
VITAMIN C

Sandwiched between the two fat-soluble antioxidant vitamins — vitamin A and vitamin E — is the water-soluble vitamin C. Being water-soluble, it is readily absorbed in the small intestine and quickly transported by the watery bloodstream to all cells.[1] On the other hand, because it is water-soluble, vitamin C is not easily stored in the body's fat and fresh vitamin C must be consumed daily to meet the body's needs. Couple its water solubility with the unfortunate fact that the human body cannot synthesize vitamin C, while almost all other animals can, and the nutrient becomes an *essential* vitamin.[1] Vitamin C can be synthesized from ordinary glucose (sugar) through a four-step enzymatic process by most species. But sometime during evolution man lost the enzyme that catalyzes the fourth step[a,1] So, we must rely on exogenous vitamin C, sometimes called ascorbic acid, for our needs. Only apes and guinea pigs are also unable to make vitamin C.[2] Mice, curiously enough, generate vitamin C exuberantly.[3]

---

[a] Linus Pauling speculates in the foreword to Irwin Stone's 1972 book *The Healing Factor: "Vitamin C" Against Disease* that perhaps "an evolutionary accident. . .occurred many millions of years ago," at which time ancestors of human beings. . .were living in an area where the natural foods available provided very large amounts of ascorbic acid." Then, "a mutation occurred that removed from the mutant the ability to manufacture ascorbic acid within his own body. Circumstances were such that the mutant had an evolutionary advantage over the other members of the population, who were burdened with the (enzymatic) machinery for manufacturing additional ascorbic acid. The result was that the part of the population burdened with the machinery gradually died out, leaving the mutants, who depended upon their food for an adequate supply of ascorbic acid."[4]

History of Vitamin C

The ever-alert Hippocrates is credited with being the first person to recognize scurvy, in ancient Greece in about 400 B.C. During the Crusades (1095–1291), "scurvy took a far greater toll. . .than all of the weapons of the Saracens," reported Stone.[4] Later, in the Middle Ages (5th–16th centuries), epidemics decimated — in the nonhistorical sense — Europe. In the 14th century, the Black Death — the bubonic plague — killed millions of people. The plague, which is a bacterial infection, typically struck people previously enfeebled by scurvy. The disease killed over 25 million people in Europe alone.

While ancient folklore had taught mankind that scurvy was related to a lack of fresh fruits and vegetables, in the absence of a printing press that knowledge was "forgotten and had to be rediscovered again and again at great cost in lives and suffering."[4] Then, with the advent of oceanworthy ships enabling long voyages, scurvy flourished. Sailors soon succumbed to the disease. "In a few short months, out of what started out as a seemingly healthy crew, only a few remained fit for duty and were able to stand watch."[4] Scurvy ended up destroying "more sailors than all other causes, including the extremely high toll of naval warfare."[4] First, in 1497, Vasco da Gama lost 100 of his 160-man crew to scurvy during his voyage around the Cape of Good Hope.[b] Next, Ferdinand Magellan set out in 1519 with a fleet of five ships to circumnavigate Earth. Three years later, the expedition returned to Spain with only one ship and 18 men, scurvy having claimed the lives of the rest, except Magellan, who was killed in a battle with natives in the Battle of Mactan, an island in the Philippines (the navigator became captain upon Magellan's death).[4] And, finally, between 1772 and 1775, Captain James Cook suspected that a lack of fresh fruits and vegetables was the cause of scurvy and made it a point to touch land often and to replenish his stores of fruits and vegetables. He lost but one man in three years. When he and his healthy crew returned to England, Cook was rewarded by the British Royal Navy Society, which handed him the Copley Medal[c] for his singular voyage.[4]

---

[b] Curiously, Christopher Columbus' crews did not develop scurvy. This may have been due to the fact that some of his voyages were shorter than those of other explorers and he unwittingly avoided the disease, or to his ship's physician having learned from Carib Indians that yams and turnips were very nutritious and kept crews healthy.[5] Carib Indians live on an island in the Caribbean Sea.
[c] The Copley Medal is awarded annually by the Royal Society of London for an "outstanding achievement" in scientific research. The first such award was made in 1731, 44 years before Cook received his medal.[6]

Fig. 23.1.

While Cook's suspicions were correct, it wasn't until 1747 that James Lind (Table 23.1), a ship's surgeon (or a ship's doctor),[d] proved that scurvy was a preventable disease. In what was later hailed as the first controlled medical experiment, Lind took 12 seamen suffering from the same degree of scurvy and divided them into six groups. One group received a quart of cider daily, a second received 25 drops of dilute sulfuric acid three times a day, the third was given two spoonfuls of vinegar three times a day, the fourth drank a half pint of seawater three times a day, the fifth ate a concoction of garlic, mustard seed, gum myrrh, and balsam of Peru, and the last group ate two oranges and one lemon daily for six days. (All of these peculiar potions were believed, at the time, to have some therapeutic effect on scurvy.) There was slight improvement in the cider group but no benefit in the others. . .except in the citrus fruit group. The men in the citrus group "improved with such astonishing rapidity that they were used as nurses to care for the others."[4] Here, at last, was clear-cut evidence that citrus fruits can cure scurvy.[e,f]

But the British Navy wasn't as quick to catch on to the citrus fruit cure of scurvy. Not until 1795, one year after Lind's death, did the Navy adopt Lind's simple regimen of consuming one ounce of lemon juice a day to prevent

---

[d] Typically, an 18th century ship's surgeon had some medical training, but was not a physician or surgeon as we consider physicians and surgeons today. They cared for the ship's crew, dealing mostly with battle wounds and some minor ailments.

[e] The citrus-fruit–scurvy link is remindful of the chance observation that thalidomide quickly reverses the painful skin lesions of leprosy (Chap. 22).

[f] After leaving the navy, Lind went on to medical school at Edinburgh University. In 1753, he published a classic monograph, *A Treatise of the Scurvy*.[8]

scurvy.[g,h] As Stone put it, "It takes much more than logic and clear-cut demonstrations to overcome the inertia and dogma of established thought."[4] Yet, the U.S. Army shattered the British Navy's 48-year record to endorse citrus fruits as a simple way to prevent and cure scurvy. More than 30,000 cases of scurvy were seen during the American Civil War, but still the U.S. Army didn't adopt issuing citrus fruit portions to its soldiers until 1895.[4] And then when WWI broke out in 1914, so did another wave of scurvy. This edition was on land, not sea. Ships carried fruits, but it was harder to get those rations to ground soldiers in "distant areas," leading to a "reappearance [of scurvy] on a large scale."[9]

Vitamin C in the 20th Century

The inertia of the 19th century vanished during the early 20th century. In 1907, it was learned that guinea pigs are also susceptible to scurvy, providing a ready means to perform experimental work on the disease. (It still was not known that vitamin C was the key.) The disease could be prevented and cured without knowing its biochemical cause. In 1928, Albert Szent-Gyorgyi (from Hungary), while working in Cambridge, England on a "biochemical problem unrelated to scurvy or vitamin C," came across a compound with a sugar-like quality in the adrenal gland of oxen.[4] He also found a similar substance in oranges. It took several years to connect the dots, but in 1931 he did. He determined that the elusive molecule that could prevent and cure scurvy was ascorbic acid — vitamin C. The mystery was solved. The search was over. In 1937, Szent-Gyorgyi's accomplishments were recognized in Stockholm, where he was awarded the Nobel Prize for biochemistry.

Sources of Vitamin C

Fruits and vegetables are excellent sources of vitamin C. Guava and kiwi are two (reasonably) available if somewhat exotic fruits with an outstanding

---

[g] It is estimated that the 42-year delay led to about 100,000 casualties from scurvy in the British Navy.[4]

[h] While scurvy vanished on British Navy ships in 1795, it remained rampant aboard British merchant ships on long voyages to the Far East. Finally, in 1844, through an Act of Parliament, merchant ships, too, were required to stock lemon or lime juice on long sea voyages. With that, scurvy disappeared on Britain's high seas.[9]

Table 23.1 Vitamin C timeline.

400 B.C.	Hippocrates first recognizes scurvy.
15th–18th centuries	Scurvy kills thousands of sailors on long oceanic voyages.
1747	James Lind shows that citrus fruits prevent scurvy in the first controlled medical experiment at sea.
1795	British Navy begins to stock citrus fruits on its ships at sea.
1928	Vitamin C is discovered by Albert Szent-Gvorgyi in Hungary.
1934	Vitamin C is mass produced for retail sale by Hoffman-La Roche.
1937	Szent-Gvorgyi receives the Nobel Prize for biochemistry for discovering vitamin C.
1965	Irwin Stone discovers that the lack of a final enzyme in humans makes man susceptible to scurvy.
1970	Linus Pauling publishes the best-selling book *Vitamin C and the Common Cold.*
1972	Stone publishes *The Healing Factor: Vitamin C Against Disease.*
1989	The Food and Nutrition Board sets the RDA for vitamin C at 60 mg/day for adults.
2002	The Food and Nutrition Board increases the RDA to 90 mg/day for men and, 75 mg/day for women.

amount of vitamin C (Table 23.2). Since most fruits and some vegetables are eaten raw, they are the best sources because cooking can diminish or destroy vitamin C.[10] Liver and kidney also have a good amount of vitamin C, but given the choice of eating a slice of liver or drinking a glass of OJ, many of us might opt for the latter.

Absorption of vitamin C

Because vitamin C is water-soluble, it is rapidly absorbed in the small intestine. Importantly, however, as the amount of vitamin C ingested *increases,* its absorption *decreases.*[1] For example, with a low vitamin intake (10–60 mg/day) absorption is 98%, while with a high intake (>100 mg/day) absorption is only 16%.[12] What's more, in depletion/repletion studies, the blood vitamin C concentration falls quickly when a person eats a vitamin C–deficient diet, reaching a low value within 30 days. When vitamin C is restarted, little accumulation occurs in taking 50 mg/day, but with 100 mg/ day, the blood concentration rises sharply (Fig. 23.2). With a daily

Table 23.2   Vitamin C content of selected fruits.

Fruit	Milligrams of vitamin C per average size fruit/slice*
Guava	165
Kiwi	108–162
Orange	70
Mango	57
Papaya	47
Grapefruit	44
Tangerine	26
Honeydew melon	20
Pineapple	13
Apricot	10
Apple	8
Strawberry	7
Pear	7
Plum	6
Peach	6
Apricot	4
Blackberry	0.6

*A "slice" is roughly 1/8th of a medium-sized fruit, or 1/4th of a smaller fruit.
Source: Reference 11.

consumption of 200 mg, the concentration continues to rise, but with a consumption of 400–2,500 mg/day, the concentration plateaus. Consequently, megadoses of the vitamin (>500 mg), while safe, may not offer any advantage over standard doses. Also, as Anita Carr at the Linus Pauling Institute pointed out, ". . .there is little or no additional benefit from vitamin C intakes greater than 90–100 mg/day, likely because of tissue saturation at this level." Vitamin C is certainly cheap and harmless, other than having some undesirable GI effects at doses of 2000 mg/day and beyond, but because the blood level rises sharply as the daily dose is increased from 30 to 100 mg and then plateaus as the dose is further increased from 100 mg to 2500 mg (Fig. 23.3), it seems unnecessary to take megadoses with the hope of being as healthy as possible.

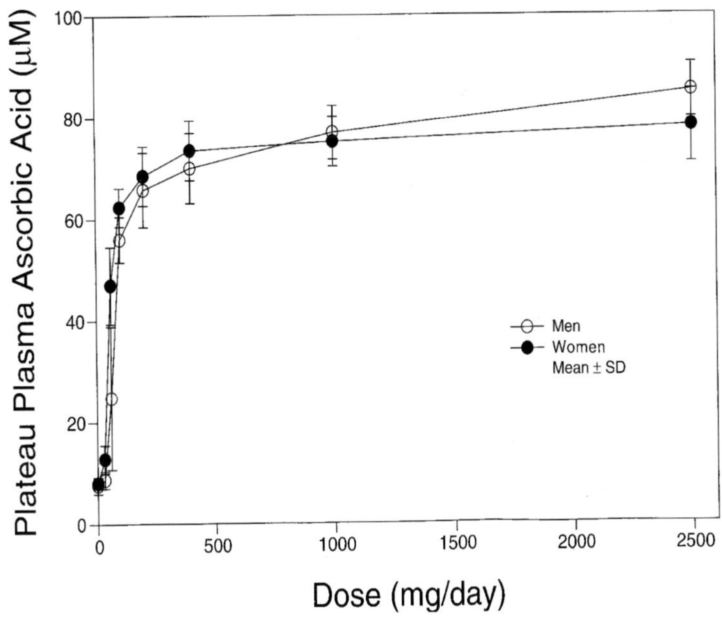

Fig. 23.2. The blood vitamin C concentration as a function of the dose in 7 healthy men and 15 healthy women. Following vitamin C depletion, the vitamin was given by mouth in doses of 30, 60, 100, 200, 400, 1000, and 2500 mg. The blood vitamin C concentration rose sharply between 60 and 100 mg/day. Two hundred to 400 mg lied to a less sharp further rise in the vitamin C concentration, followed by a plateau when the dose was further increased to 1000 and 2500 mg/day, indicating that little benefit is likely to accrue from doses above 100–400 mg/day.

(*Source*: Reproduced with permission from the publisher). Reference 12.

Storage of Vitamin C

Yet, even when body stores are fully saturated, the human body holds just enough vitamin C to fit on the head of a pin — 5 g (0.176 ounces). For that reason, the body requires a continuous supply of vitamin C to replenish that used by its cells and tissues. The vitamin is concentrated in organs with the greatest metabolic activity — the adrenal cortex, (where steroids are made), the pituitary gland (where growth hormone, endorphins, and thyroid-stimulating hormone are made), the brain, the eyes, and the reproductive organs.[4] In the absence of vitamin C, healthy persons become vitamin C–deficient within 30 days, even if vitamin C stores were fully saturated at first.[13]

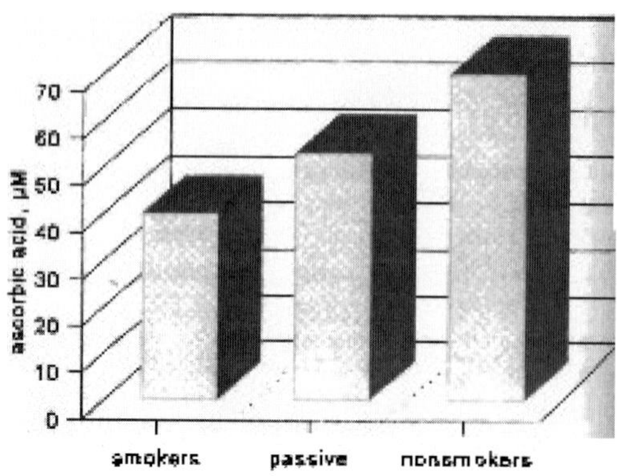

Fig. 23.3. Cigarette smoke is a rich source of free radicals, which oxidize vitamin C. Consequently, smokers require about 40% more vitamin C than do nonsmokers. Nonsmokers exposed passively to cigarette smoke have vitamin C levels in between those of smokers and nonsmokers not passively exposed to smoke.

(*Source*: Reproduced with permission from the publisher.) Reference 10.

Scurvy

The first *symptom* of scurvy is pronounced fatigue and "a desire to sleep."[4] (A symptom is *felt* by the patient. A sign is a *visible* abnormality discoverable upon physical examination of the patient.) The first *sign* of scurvy is a change in the complexion. The skin becomes "sallow or muddy."[4] Next, there are aches in the muscles and joints, particularly in the legs. Then the gums become sore, spongy, and friable — bleeding readily. The breath becomes fetid and teeth begin to fall out. The body is visibly rotting away. Soon, tiny hemorrhages appear at the base of hair follicles. The eyelids become puffy and purplish, and nosebleeds begin for no apparent reason. Fresh wounds don't heal and old wounds break open. The limbs become so painful that the patient is immobilized. The bones become brittle and break with minimal activity such as rolling over in bed. The joints crackle. All this happens over just a few months. Death often follows an infection — such as pneumonia — that ordinarily would be treatable.

Biological Significance of Vitamin C

The creation of collagen is the principal biological function of vitamin C. This protein constitutes about one-third of the body's entire protein content.[4]

Collagen is the main "structural element" in the cardiovascular system and, as such, is responsible for the formation of the arterial system into which the heart pumps its blood and the venous system which later collects the blood and returns it to the heart. It is collagen that imparts the "strength, elasticity, and ruggedness" that characterize all flexible arteries.[10] Collagen also acts as the "glue" that holds other tissues and many organs together. It is the principal protein in bones, teeth, cartilage, spinal disks, and the lens.[10] Without vitamin C we develop scurvy, which is fundamentally a rotting process where tissues ranging from the soft skin to hard bones fall apart. And because collagen in the skin is sensitive to free radicals and ultraviolet light, smokers and beach lovers develop wrinkles faster than nonsmokers and shade seekers.

Antioxidant Function of Vitamin C

Vitamin C is an antioxidant owing to its ability to *reduce* free radicals — the superoxide anion, the peroxyl radical, the alkoxyl radical, and the hydroxyl radical (Chap. 9). Cells must defend themselves against damage by free radicals, which are normal byproducts of all cells. Ironically, cells must guard against harm from the waste products of the very element — oxygen — that sustains life, as mentioned earlier. When free radicals are *reduced* by vitamin C — or any other antioxidant — they *gain* an electron while the antioxidant *loses* an electron. (Recall the mnemonic "oil rig": oxidation is loss, reduction is gain.) As free radicals are reduced, vitamin C is oxidized. This see-saw between the oxidized and reduced forms of vitamin C accounts for most, if not all, of vitamin C's physiologic functions.[10] Tobacco smoke is a rich source of free radicals. Consequently, cigarette smokers require about 40% more vitamin C than do nonsmokers.[10] And those of us exposed passively to tobacco smoke have lower circulating vitamin C levels than do those of us not (significantly) exposed to smoke (Fig. 23.3).

Vitamin C as a Pro-oxidant

For every good thing there is potential for a bad thing. And so it is that vitamin C *can* be a pro-oxidant. But it can't misbehave all by it self. It needs the help of iron (or copper) ions to oxidize other compounds. Vitamin C oxidizes by donating one electron to an iron ion, which is now reduced and can react with oxygen to form free radicals.[10] The crucial question is: To what extent does vitamin C act as a pro-oxidant in the human body? This has been the focus of considerable research and debate. Wrote Carr, "We found compelling evidence for antioxidant protection of lipids by vitamin C. . .in humans, both with and without iron supplementation."[14] Wikipedia agrees, noting that while "vitamin C can reduce metal ions, which leads to the generation of free radicals," "in the human body, vitamin C acts mostly [as an] antioxidant."[15]

Popularity of Vitamin C

Vitamin C is now the "most widely used vitamin in the world."[2] And well it should be, since its principal job is to protect important molecules such as carbohydrates, proteins, and fats — even DNA — from damage by free radicals.[2] On the side, it promotes the absorption of iron and regenerates both vitamin E and β–carotene.[2] So it is not a surprise that in our quest to remain healthy about 50% of adult Americans take a vitamin–mineral supplement at an estimated cost of >1.5 billion dollars each year.[16]

RDA and EAR

It takes just 10 mg a day of vitamin C[13] to ward off the most significant consequence of too little of that vitamin — scurvy. The RDA for a vitamin, according to the Food and Nutrition Institute of Medicine, is roughly 20% above the EAR (estimated average requirement), which is based on "scientifically determined values."[17] In 2002, based on new reports, the Food and Nutrition Board boosted the RDA from 60 mg a day in both men and women to 90 mg a day for men and 75 mg a day for women (Table 23.3). The RDA is meant to provide enough of a given nutrient "to satisfy the requirements of almost all healthy persons."[17] The recommended upper limit (UL) of vitamin C is 2,000 mg/day to avoid bothersome GI side effects.

Although the Food and Nutrition Institute recommends those RDAs, that doesn't mean everyone will follow the agency's recommendations. In fact, a report from the Linus Pauling Institute indicates that 25% of men and women in the U.S. consume less than 60 mg per day of vitamin C and 10% less than 30 mg per day.[13] On top of that, the American Cancer Society recommends the consumption of "at least five fruits and vegetables a day,"[20] but in as much as 25% American adults consume less than 60 mg a day of vitamin C, one out of four of us isn't eating the recommended amount of fruits and vegetables.

Table 23.3   Dietary intake values for vitamin C.

Life stage	Gender	Age (years)	EAR	RDA	UL
Infants (months)	Boys and girls	0–6			
		7–12			
Children	Boys and girls	1–3	13	15	400
		4–8		25	650
	Boys	9–13	39	45	1,200
		14–18	63	75	1,800
	Girls	9–13	39	45	1,200
		14–18	56	65	1,800
Adults	Men	19–>70	75	90	2,000
	Women	19–>70	60	75	2,000
Pregnancy		14–18	66	80	1,800
		19–30	70	85	2,000
		31–50	70	85	2,000
Lactation		14–18	96	115	2,000
		19–30	100	120	2,000
		31–50	100	120	2,000
Smokers					
	Men	>19	110	130	
	Women	>19	95	115	

EAR = estimated average requirement; RDA = recommended daily allowance; UL = tolerable upper limit.

(*Source*: Modified and reproduced with permission from the publisher.) Reference 18.

High-Dose Vitamin C

In the 1970s, large doses of vitamin C were advocated by Irwin Stone, a biochemist at the Wallerstein Company in New York City. The firm manufactures industrial enzymes, making it a good fit with Stone's background. In his book, Stone does a fine job of marshaling the evidence that indicates most of us ingest less vitamin C than required to maintain good health. He makes an eloquent, passionate plea to expand the use of high-dose vitamin C (3–5 g a day) to prevent a slew of disorders. With specific regard to the ravages of atherosclerosis, he says, "In spite of the due need to do something effective in the prevention and therapy of this terrible plague of cardiovascular disease. . . .all this provocative and suggestive research has been glossed over and ignored and none of the critical large-scale tests have been made."[4]

Now, in the early 21st century those studies have finally been performed. Here's what's unfolded.

Clinical Trials of Vitamin C to Prevent Coronary Heart Disease

While there are many, many studies of vitamin C's effect on a number of noncardiac conditions ranging from the common cold to cancer, there are but two reports concerning the vitamin's impact on coronary heart disease. One is in the *Journal of the American Medical Association* (*JAMA*) and the other is in the *Archives of Internal Medicine*, both popular forums for large clinical trials on popular topics. Phase II of the Physician's Health Study, mentioned in Chap. 18, was done in doctors illustrating that physicians are willing to be "guinea pigs" themselves — at least for non-invasive research. In phase II, 14,642 men aged 50 or more were assigned randomly to take 500 mg of vitamin C daily or a placebo. The end points were a nonfatal heart attack or stroke or death due to cardiovascular disease. As you might imagine, doctors tend to take good care of themselves, so only 5% had evident coronary heart disease at the trial's beginning,[21] largely due to skirting the usual risk factors for coronary heart disease — cigarettes, elevated cholesterol, and high blood pressure — and they exercise a lot. (It always moves me at the annual meetings of the American Heart Association and the American College of Cardiology, attended by some 15,000 cardiologists, to see several thousand physicians running in the streets of the host city starting at the crack of dawn to finish, shower, and dress before the first sessions begin at 8 a.m.)

After taking vitamin C for an average of eight years, the bottom line was that vitamin C did not reduce the risk of "major cardiovascular events."[22] The men who took the vitamin had just as many heart attacks and strokes and just as much congestive heart failure as those who took a placebo. The mortality rate and need for procedures such as coronary artery bypass surgery and angioplasty or a stent were not affected by the vitamin either. The authors wrote, "[The] data provided no support for the use of [vitamin C] supplements for the prevention of cardiac disease in middle-aged and older men."

It's not only doctors who submit to clinical trials, — nurses do too. At Boston's Brigham and Women's Hospital, 8171 female healthcare professionals, all 40 or more and all at "high risk for cardiovascular disease," were treated with 500 mg of vitamin C daily (or with beta-carotene or vitamin E) or placebo and followed from 1995 to 2005. Each woman had either a prior confirmed heart attack (36%) or three or more cardiovascular disease risk factors (64%). The risk factors were high blood pressure, an elevated cholesterol level, diabetes mellitus, a premature myocardial infarction (≤ 60 years), obesity, and smoking. The women were followed for 10 years to determine if there was any difference between the two groups in the primary end point of a heart attack, a stroke, or the need for a coronary bypass operation or an angioplasty.

After 10 years, there was absolutely no difference in mortality between the two groups (or in the incidence of an acute myocardial infarction, a stroke, or the need for a bypass operation or an angioplasty. The authors wrote, "The [negative] results for the primary end point seen in the Women's Antioxidant Cardiovascular Study further limits enthusiasm for the role of ascorbic acid in cardiovascular risk protection."[21]

And so vitamin C's dream of preventing or treating coronary heart disease came to an end. Two antioxidants down, one to go.

References

1. http://en.wikipedia.org/wiki/Vitamin_C
2. Blake S. (2008) *Vitamins and Minerals Demystified.* McGraw Hill Medical, New York, Chicago, San Francisco, Lisbon, London, Madrid, Mexico City, Milan, New Delhi, San Juan, Seoul, Singapore, Sydney, Toronto.
3. Vitamin C. In: *The Vitamins: Fundamental Aspects in Nutrition and Health*, 3rd ed. Combs GF, Jr (eds.), Elsevier, Amsterdam, Boston, Heidelberg, London, New York, Oxford, Paris, San Diego, San Francisco, Singapore, Sydney, Tokyo. Chap. 9, pp. 235–264.
4. Stone I. (1972) *The Healing Factor: Vitamin C Against Disease.* Grosset & Dunlap, New York.
5. Carpenter KJ. (1986) *The History of Scurvy and Vitamin C.* Cambridge University Press, London, New York, New Rochelle, Melbourne, Sydney.
6. http://en.wikipedia.org/wiki/Copley_Medal
7. http://www.ehow.com/about_451150_does-vitamin-c-do-body
8. Lind J. (1753) *A Treatise of Scurvy.* Printed by Murray & Cochran for Kincaid A, Donaldson A. Edinburgh.
9. Wilson LG. (1975) The clinical definition of scurvy and the discovery of vitamin C. *J Hist Med* **30**: 40–60.
10. Naidu KA. (2003) Vitamin C in human health and disease is still a mystery? An overview. *Nutr J* **2**: 7–10.

11. http://www.beta-glucan-info.com/vitaminchistory.htm

12. htttp://www.naturalhub.com/natural_food_guide_fruit_vitamin_c.htm

13. Carr AC, Frei B. (1999) Toward new recommended dietary allowances for vitamin C based on antioxidant and health effects in humans. *Am J Clin Nutr* **69**: 1086–1107.

14. Carr A, Frei B. (1999) does vitamin C act as a pro-oxidant under physiological conditions? *FASEB* **13**: 1007–1024.

15. http://en.wikipedia.org/wiki/Pro-oxidant

16. Balluz LS. Kieszak SM, Philen RM, Mulinaro J. (2000) Vitamin and mineral supplementation use in the United States: results from the third National Health and Nutrition Examination Survey. *Arch Fam Med* **9**: 258–262.

17. McCormack DB. (2006) The dubious use of vitamin–mineral supplements in relation to cardiovascular disease. *Am J Clin Nutr* **84**: 680–681.

18. Food and Nutrition Board, Institute of Medicine. (2002) *Dietary Reference Intakes for Vitamin C, Vitamin E, Selenium, and Carotenoids.* National Academy Press, Washington, DC.

19. www.hsph.harvard.edu/.../what.../vegetables-full.../index.html

20. Sesso HD, Buring JE, Christen WG, Kurth T, Belanger C, Mac Fayden J, Bubes V, Manson JE, Glynn RJ, Gaziano JM. (2008) Vitamin E and C in the prevention of cardiovascular disease in men. The Physicians' Health Study II. Randomized controlled trial. *JAMA* **300**: 2123–2133.

21. Cook NR, Albert CM, Gaziano JM, Zaharris R, Mac Fadyan J, Danielson E, Buring JE, Manson JE. (2007) A randomized factorial truial of vitamins C and E and beta carotene in the secondary prevention of cardiovascular disease in women. *Arch Intern Med* **167**: 1610–1618.

24

VITAMIN E

Akin to vitamin A, vitamin E is a collective name for a family of molecules. Vitamin A has four molecules and vitamin E twice as many, four called tocopherols and four others called tocotrienols.[1] Within these two groups, the four molecules are structurally related to one another.[2] In humans, α-tocopherol is the predominant and most active form. Unlike vitamin C, vitamin E is not water-soluble (it is fat-soluble), but like vitamin C, vitamin E cannot be synthesized by the human body; therefore it, too, is essential.

Curiously, vitamin E is shuttled through the bloodstream by LDL, the very same carrier that transports cholesterol to all peripheral tissues.[2] Being fat-soluble, vitamin E becomes embedded in LDL particles for transport purposes. That "intimacy" with LDL should give vitamin E — the emperor of the antioxidant vitamins — a better chance than vitamin C to protect LDL against oxidation and to prevent its deposition in atherosclerotic plaques. In the best of all possible worlds, vitamin E would remove plaque, converting a coronary artery that looks like the one on the right (Fig. 24.1) to one that is plaque-free, on the left — clean as a whistle. Can vitamin E do it?

We'll soon find out.

History of Vitamin E

It wasn't until 1922 that Herbert Evans and Scott Bishop, working at the University of California, Berkeley, isolated a substance from wheat germ with biological activity.[4] They called it tocopherol. The name comes from the Greek words "*tokos*" (meaning "offspring") and "*pherein*" (meaning "to bear," with "-ol" tacked on the end to remind us that it is an alcohol). It was coined to highlight the important role of vitamin E in reproduction. The following year, two other tocopherols were found in vegetable oils, and they were called

333

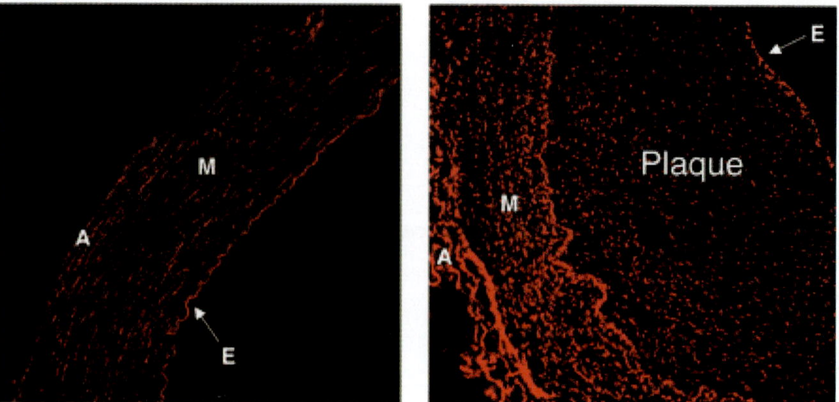

Fig. 24.1.   Microscopic sections of two human coronary arteries stained with a compound that allows easy visualization of the vessel's wall under a microscope. The coronary artery on the left is from a 42-year-old woman with cardiomyopathy, a disorder in which the heart becomes enlarged and very weakened (but it's free of coronary heart disease), leading to congestive heart failure. This heart was obtained at the time of cardiac transplantation. There is no plaque formation. The one-cell-thick endothelium (E) lies directly atop the thick, muscular media (M), making this a normal coronary artery. The coronary arterial endothelium, media, and adventitia (A) exhibit uniformly low levels of red fluorescence. See Fig. 2.1 for a photomicrograph of a normal coronary artery. The coronary artery on the right is from a 69-year-old man with coronary heart disease. A large plaque lies between the media (M) and the endothelium (E). This plaque, like all other plaques, was filled with cholesterol. It markedly narrowed the vessel, leading to a heart attack.

(*Source:* Reproduced with permission from the publisher.) Reference 3.

β-tocopherol and γ-tocopherol, while the first tocopherol became α-tocopherol, in keeping with the practice of using the Greek alphabet to name relatives of a vitamin in their order of discovery while using the English alphabet to name the parent vitamins in their order of discovery.[5] Some of the more significant moments in the history of vitamin E from the time of its discovery in 1922 to the large clinical trials in 2005 are listed in Table 24.1.

Structure of Vitamin E

The structure of vitamin E (for the few readers with an interest in such matters) is shown in Fig. 24.2. The eight forms of the vitamin fall into one of two classes, the tocopherols and the tocotrienols. It takes close scrutiny to see that the four tocotrienols differ from the tocopherols only in the number of double bonds in the side chain attached to the chromanol nucleus on the left.

Table 24.1 Vitamin E timeline.

Year	Event
1922	Vitamin E is discovered by Evans and Bishop.
1938	Swiss chemist Paul Karrer synthesizes vitamin E in its α-tocopherol form and wins the 1937 Nobel Prize in Chemistry for also synthesizing vitamin A in 1931 and vitamin B2 in 1935 a year before synthesizing vitamin E in 1938.
1945	The first antioxidant theory of vitamin E is proposed.
1962	Biochemist Al Tappel at the University of California, Davis suggests that vitamin E acts in human beings to protect cell membrane lipids from free radical oxidation.
1968	The U.S. Food and Nutrition Board recognizes vitamin E as an essential nutrient for humans.
1980	P.B. McKay and M.M. King propose that vitamin E functions as an antioxidant in cell membranes.
1980	Vitamin E is shown to be the major fat-soluble vitamin antioxidant protecting cell membranes from oxidation by free radicals.
1990	Vitamin E is demonstrated to be effectively inhibit the oxidation of LDL in humans.
2004	Vitamin E is demonstrated to have a function in addition to its antioxidant function of regulating the gene that governs scavenger receptor formation.
2005	Large clinical trials testing vitamin E's ability to prevent coronary heart disease are published.

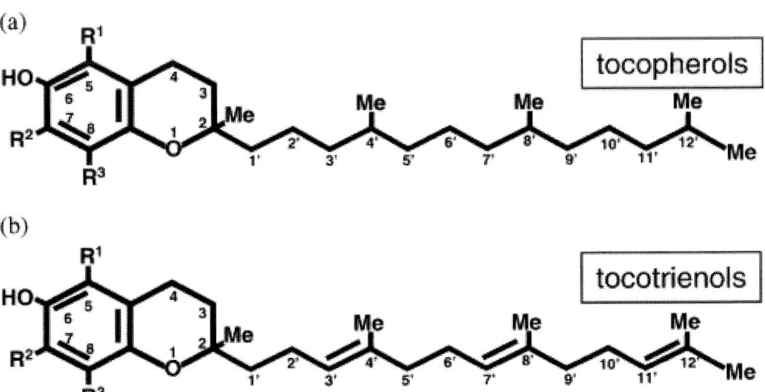

Fig. 24.2. The structure of tocopherols and tocotrienols. Three double bonds in the side chain attached to the chromanol nucleus are the only difference. The double bonds render tocotrienols more unsaturated than tocopherols, making them more easily oxidized by free radicals.

Fig. 24.3.   Liver (rear, right) is a good source of vitamin E...but lobster, olives, and nuts may be a bit more palatable choices.

(*Source*: Reproduced with permission from the publisher.) Reference 9.

Each of the four forms of tocopherols and tocotrienols, in turn, differ only in the number of methyl groups on the chromanol nucleus — α has three, β and γ have two, and δ has one.[6] But that subtle difference has a powerful effect on the antioxidant activity of the four foms, with α-tocopherol being about twice as potent as β-tocopherol and four times as potent as γ- and δ-tocopherol.[6] This makes α-tocopherol a more effective scavenger of free radicals, vitamin E's principal biological function, than its mates. The human body recognizes α-tocopherol's virtue and "preferentially retains" it.[7] α-tocopherol is the sole form of vitamin E in natural vitamin E preparations, while synthetic vitamin E preparations contain all eight forms.[8]

Sources of Vitamin E

Vitamin E is synthesized only by plants, with green-leaved plants making more of the vitamin than yellow-leaved ones.[a] The richest sources (Fig. 24.3) are the oils made by both types of plants. The F.D.A's Office of Dietary

---

[a]While plants alone synthesize vitamin E, many animals eat plants and most humans eat (primarily) domestic animals … hence the liver in the above photograph.

Table 24.2   Selected food sources of vitamin E (α-tocopherol).

Food	IU (mg) per serving	Percent DV
Wheat germ oil, 1 tablespoon	30.2 (20.3)	100
Almonds, dry-roasted, 1 ounce	11.0 (7.4)	40
Sunflower seeds, dry-roasted, 1 ounce	8.9 (6.0)	30
Sunflower oil, 1 tablespoon	8.3 (5.6)	28
Safflower oil, 1 tablespoon	6.8 (4.6)	25
Peanut butter, 2 tablespoons	4.3 (2.9)	15
Peanuts, dry-roasted, 1 ounce	3.3 (2.2)	11
Corn oil, 1 tablespoon	2.8 (1.9)	10
Spinach, boiled, ½ cup	2.8 (1.9)	10
Broccoli, chopped, boiled, ½ cup	1.8 (1.2)	6
Soybean oil, 1 tablespoon	1.6 (1.1)	6
Kiwi, 1 medium	1.6 (1.1)	6
Mango, sliced, ½ cup	1.3 (0.9)	5
Tomato, raw, 1 medium	1.2 (0.8)	4
Spinach, raw, 1 cup	0.9 (0.6)	4

DV = daily value. DVs were developed by U.S. Food and Drug Administration (F.D.A) to help consumers compare the nutrient content of different foods within the context of a total diet. The DV for vitamin E is 30 IU (approximately 20 mg) of natural α-tocopherol for adults and children aged 4 and older. However, the F.D.A does not require food labels to list vitamin E content unless a food has been fortified with this nutrient. Foods providing lower percentages of the DV also contribute to a healthful diet. The U.S. Department of Agriculture's Nutrient Database Website (http://www.nal. usda.gov/fnic/foodcomp/search) lists the nutrient content of many foods, including, in some cases, the amount of α-, β-, γ-, and δ-tocopherol. The U.S.D.A also provides a comprehensive list of foods containing vitamin E.

Source: Reference 10.

Supplements ranks wheat germ oil and almonds No. 1 and 2 in the amount of vitamin E (Table 24.2). Sunflower seeds aren't far behind in vitamin E content and all three are (reasonably) low in price. In America, γ-tocopherol is the dominant form of dietary vitamin E, due to the generous use of soybean, canola, and corn oils for cooking purposes.[10]

Approximately 70% of the vitamin E, consumed in the typical American diet is γ-tocopherol.[11] Importantly, α-tocopherol, the most potent form of vitamin E, decreases blood and tissue concentrations of γ-tocopherol, and γ-tocopherol is inversely associated with cardiovascular disease. The higher the γ-tocopherol level, the lower the risk of coronary heart disease. In experimental animals rats a γ-tocopherol-rich diet is "much more potent than α-tocopherol … in reducing the oxidizability of LDL."[12]

Absorption of Vitamin E

Like vitamin A, the other fat-soluble antioxidant vitamin, vitamin E is absorbed in the small intestine and requires the presence of fat in the gut for full absorption.[13] After absorption, both are transported through the lymphatic system to the liver. In the liver, the α- tocopherol form of vitamin E is preferentially incorporated into LDL for further transportation and distribution to the various body tissues. Both vitamins then leave the liver through the thoracic duct and finally enter the circulatory system in the upper chest (Fig. 21.5).

Storage of Vitamin E

The bulk of vitamin E is stored in fat (adipose tissue), which releases it at a snail's pace, as needed. So slow is mobilization that, following a change in vitamin E intake, the concentration of the vitamin in adipose tissue "may not reach a new steady state for two or more years",[9] — just the opposite of vitamin C's depletion rate. It's gone within a month.

Biological Function of Vitamin E

You'd think that everything that could be known is known about a substance discovered in 1922[4] and first synthesized 16 years later. It isn't. Even in the early 21st century, its biological function remains unsettled. As recently as 2009, Regina Brigelius-Flohé, from Germany, said, "Vitamin E is the last of all vitamins whose [biological function] is not yet [fully] understood." She likened vitamin E to "an Elizabethan virgin immortalized by William Shakespeare in [*The Taming of the Shrew*] — [who] persistently guards the secrets that might explain its odd behavior."[14] And yes, it's an antioxidant — but is that all it is? Does it have any other purpose(s)?[b] That became the focus of a "debate"

---

[b]Vitamin E does have at least two other biological functions. It inhibits several enzymes, including lipoxygenase (Chap. 10), and it exerts a modifying influence on several genes,

in the pages of *Free Radical Biology and Medicine* in July 2007. Maret Traber and Jeffrey Atkinson, both biochemists at the Linus Pauling Institute in Oregon, took the position that vitamin E is an antioxidant *only* and that any other effects it has are consequences of its antioxidant activity. Angelo Azzi at Tufts took the position that vitamin E is *not* an antioxidant in the human body and cites two other esoteric functions, "cell signaling" and "gene transcription," as its *raison d'être*. (It's not necessary to know what cell signaling and gene transcription are to appreciate the debate.)

Here are a few excerpts from these two papers. Opened Traber and Atkinson, "We propose...that all of the observations concerning the *in vivo* mechanism of action of α-tocopherol result from its role as a potent antioxidant."[15] They believed that "...α-tocopherol's major function, if not only function, is that of a peroxyl radical scavenger" (Chap. 9). "The importance of this function is to maintain the integrity of long-chain polyunsaturated fatty acids in the membranes of cells and thus maintain their bioactivity.... Thus, virtually all of the variation and scope of vitamin E's biological activity can be seen and understood in the light of protection of polyunsaturated fatty acids and the membrane qualities...that they bring about."[15]

Azzi shot back, That's a bunch of mush, in effect. He wrote, "...α-tocopherol has been shown...not to protect *in vivo* against oxidative damage or to prevent diseases which have as their basis an oxidative insult." (These diseases include coronary heart disease, cancer, and Alzheimer's disease.) "Altogether, the conclusion can be drawn that α-tocopherol is not physiologically acting as an antioxidant."[16]

Now come German Regina Brigelius-Flohé and Kevin Davie, from the University of Southern California, in a commentary attempting to reconcile these two very different viewpoints. They note that in this issue of *Free Radical Biology and Medicine* two groups offer views about "the biological role of vitamin E that could not be more controversial.... The article by Traber and Atkinson with the provocative title 'Vitamin E, an Antioxidant and Nothing More' reiterates the 'classical' view that α-tocopherol acts as an antioxidant" and any other effects "result from preventing lipid oxidation."[17] The article by Azzi challenges that statement and claims, "... α-tocopherol is not able, at physiological concentrations, to protect against oxidation of lipids or to

---

including the gene that governs scavenger receptor formation.[8] Some believe that these nonantioxidant activities of vitamin E "represent the main biological reasons for the selective retention of α-tocopherol in the body."[8]

prevent diseases allegedly caused by [such oxidation]. . . . The controversy reflects our state of knowledge on vitamin E, which 85 years after its discovery, is still rudimentary." (That may be the take-home message for users and would-be users of vitamin E supplements. Is it a good idea to take vitamin E beyond that provided through the diet, even if it is "only" a nutrient, when knowledge about it is "rudimentary" nearly a century after its discovery?) "The seemingly unbridgeable discrepancy between the views. . .is further exacerbated by different uses of fashionable terms. . . ." (The terms referred to are "antioxidants" and "α-tocopherol.") "There is not even agreement about whether pharmacological doses are physiologically or pathologically relevant" — meaning, does a "standard" dose of, say, 400 IU of vitamin E a day have any impact on the well-being of healthy or unhealthy people or should we "save our money" as Robert Greenberg at Dartmouth wondered.[18] They wind up reminding us that the question of how and whether vitamin E works is "one that has been waiting for an answer since the beginning of the last century!" — begging the rhetorical question: Should people take supplemental vitamin E? Robert Blendon at the Harvard School of Public Health told us that vitamin supplements — not just vitamin E, but vitamin A and vitamin C as well — are so popular in the U.S. that many Americans who take supplements do not tell their physician that they do so because "they believe that physicians know little or nothing about these products and may be biased against them. Many users felt so strongly about the potential health benefits of some of these products that they reported that they would continue to take them even if they were shown to be ineffective in scientifically conducted clinical studies."[19]

Vitamin E Deficiency

Other than a rare genetic defect and in very premature babies[c] (<3.5 pounds; 1500 g), there are only two ways to become vitamin E–deficient — not eating

---

[c]An inherited defect in the gene that controls the synthesis of the protein that transports vitamin E — LDL — can lead to vitamin E deficiency.[21] The vitamin must be carried through the bloodstream to every cell, where it is used to make the cell's membrane. The inherited deficiency disorder is known as ataxia and vitamin E deficiency syndrome (AVED). Huge, lifelong doses of vitamin E — 800–3,600 IU/day — are needed to treat adults with this disorder. Very premature babies who are not yet able to absorb vitamin E are also prone to vitamin E deficiency. These infants are given intramuscular injections of vitamin E to tide them over for a few weeks or months. The signs of vitamin E deficiency are primarily neurological, ocular,

enough of the vitamin or not absorbing enough from foodstuffs ingested. It's not easy to become vitamin E–deficient through malabsorption. You have to whack out a lot of small intestine — about 8–10 feet of the full 21 feet — to become deficient in vitamin E.[d,20] (There is no significant absorption of the vitamin in the stomach.) Replacement therapy with vitamin E depends on how much of the small intestine is resected, the range being 800–3,600 IU per day.[20] But inadequate dietary *intake* can lead to vitamin E deficiency even in the U.S. and Canada.[23] Older Americans and Canadians living in nursing homes — and some in the community — are most vulnerable. With ageing come poor dentition, depression, a reduced appetite, loss of the husband or wife, and reliance by many on limited Social Security funds to buy food. Yet, aside from inference, there is no way to diagnose vitamin E deficiency — other than measuring its concentration in the blood — because there are no clear-cut signs or symptoms of vitamin E deficiency.[24] That's because the vitamin's principal function is forming protective membranes around all cells. As a result, a clinically significant deficiency leads to diverse signs and symptoms as many organs suffer. The protective myelin sheath surrounding long axons of nerve cells (to be covered in Chap. 31) are especially vulnerable, leading to muscle weakness and impaired vision.[24] But these symptoms are not specific enough to permit the diagnosis of vitamin E deficiency. Pragmatically, it may be simplest for concerned children in their 30s, 40s, and 50s to see to it that Mom and/or Dad take a multivitamin each day — and to visit with them often so as to lift their spirits.

---

and hematological, reflecting vitamin E's critical role in forming cell membranes that protect nerves, the retina, and red blood cells, respectively. In the case of red blood cells, the outer membrane guards the hemoglobin within to prevent hemolysis (breakdown of the membrane, with loss of the hemoglobin leading to anemia). In the case of the eye, very premature infants develop retrolental fibroplasia (retinopathy of prematurity) (Chaps. 5–7) as a result of damage to the outer membrane (sheath) which ordinarily protects delicate, immature nerves.

[d]Another important cause of malabsorption is pancreatic insufficiency in children and adults with cystic fibrosis. Cystic fibrosis is an uncommon, inherited disorder affecting mostly people of northwestern European extraction.[22] It is far less common in blacks and is very rare in Orientals. The defective gene was isolated in 1989. Absence of that gene has grave consequences. It leads to the lack of a single amino acid — phenylalanine. The flawed protein distorts the normal movement of salt and water across the membranes (linings) of the gut and lungs, leading to easy dehydration and the development of thick, sticky mucus plugs which clog the airways and intestinal tract. In the GI tract, the mucus plugs interfere with the absorption of essential nutrients, such as vitamin E, causing malnutrition.

Popularity of Vitamin E

Vitamin E battles with vitamin C for public popularity even though they work side by side in the body as antioxidants. Greenberg noted that "a third of all [American] adults and half of those older than 55 years of age take at least one vitamin supplement daily."[18] Vitamin E is the most widely used specific vitamin supplement, taken by 22% of U.S. adults 55 or more years of age.[18] The nutrient accounts for a whopping US$18.8 billion in sales each year.[18]

Safety/Toxicity of Vitamin E

There's little doubt that vitamin E ought to be safe. It is, after all, a vitamin that is taken by millions and millions of American and nets billions and billions of dollars each year. No disputing the numbers; yet, is it safe? We thought it was…until 2009, when a report from The Netherlands emphasized that women who took vitamin E supplements in addition to consuming a vitamin E–rich diet gave birth to infants with a "ninefold" greater incidence of congenital heart disease than did those women who did not take supplemental vitamin E.[25] The congenital heart diseases included unusual defects such as transposition of the great vessels, tetralogy of Fallot, and hypoplastic left heart syndrome — all complex heart problems — the first two correctable by surgery, the third untreatable. The authors concluded, "High vitamin E by diet and supplements is associated with an increased risk of congenital heart defects."[25] Thus, supplemental vitamin E, like supplemental vitamin A, can cause serious birth defects. The message may be, once again, that fertile women — really all of us — should consume a diet, such as the Mediterranean diet, (Chap. 17) with abundant fruits, vegetables, fish, and nuts, and eschew vitamin supplements.

In addition to more congenital heart disease at the start of life in infants borne by mothers who take too much vitamin E, there's a totally different consequence of taking too much vitamin E near the end of life — more strokes. These strokes are hemorrhagic strokes, not the far-more-common thrombotic strokes, which are due to a thrombus (a blood clot) forming over an atherosclerotic plaque in the brain, just as a thrombus forms over an atherosclerotic plaque in the heart at the time of a heart attack. Strokes may be caused by too much blood flow (hemorrhagic strokes) or too little blood flow (thrombotic strokes). At the root of these hemorrhagic strokes is vitamin E's anticoagulant activity. It interferes with a fellow vitamin — vitamin K — which is needed to stem bleeding from any cause. Vitamin K is a major player in the coagulation cascade that allows a clot to form at a bleeding site. (In the soon-to-be-mentioned HOPE-TOO trial, there were slightly more strokes in those participants who took vitamin E (5.7%) than in those who took placebo (5.1%).[26] The difference did not reach statistical significance. But some say it may not make sense to take a vitamin if it has even a small chance of a serious side effect.) The drug Coumadin warfarin, mentioned in Chap. 21, is taken by a number of people at high risk of forming a clot, not at a site of bleeding but in the center of the vascular system (such as on an artificial heart valve); it works by interrupting the coagulation cascade. It is vitamin E's interaction with vitamin K that may lead to a hemorrhagic stroke. Just another example of one vitamin accentuating the effect of another or acting synergistically with an unwanted effect. Just another reason many physicians believe it is best to avoid supplemental vitamin E and to take a multivitamin once a day … and only one at a time.

RDA for Vitamin E

The RDA for vitamin E at all ages is shown in Table 24.3.

Table 24.3   Recommended dietary allowance (Rda) for vitamin E.

Life stage	Age	IU/day	Males (mg/day)	IU/day	Females (mg/day)
Infants	0–6 months	6	(4)	6	(4)
Infants	7–12 months	7.5	(5)	7.5	(5)
Children	1–3 years	9	(6)	9	(6)
Children	4–8 years	10.5	(7)	10.5	(7)
Children	9–13 years	16.5	(11)	16.5	(11)
Adolescents	14–18 years	22.5	(15)	22.5	(15)
Adults	19 years and older	22.5	(15)	22.5	(15)
Pregnancy	All ages			22.5	(15)
Lactation	All ages			28.5	(19)

(*Source*: Reproduced with permission from the publisher.) Reference 27.

Clinical Trials of Vitamin E

As in the case with supplemental vitamin A and vitamin C, frequent, consistent epidemiological studies documenting a lower prevalance of heart attacks among people who consume lots of fruits and vegetables formed the basis for believing supplemental vitamin E should lead to fewer heart attacks. So, in the early years of the 21st century, Edgar Miller and his colleagues at Johns Hopkins performed a meta-analysis of all English-language-published trials regarding the affect of high-dosage vitamin E supplements on the risk of cardiovascular disease and on mortality. [Recall that the RDA for vitamin E is 22.5 IU/day (15 mg day) for both men and women.] When the numbers were crunched, 135,967 men and women who took 16.5–2000 IU/day of vitamin E (400 IU/day mean dosage) for 1.4–8.2 years were identified for the analysis. The results were surprising, if not alarming. While low-dose vitamin E (≤150 IU/day) cut all-cause mortality, "progressively increased for dosages...greater than 150 IU/day" (Fig. 24.4). This dosage is substantially lower than the tolerable upper intake level for vitamin E, currently set at 1100 IU/day.[27] At the time of publication in the *Annals of Internal Medicine* on January 4, 2005, the research team had little choice but to say frankly that "vitamin E supplements may increase all-cause mortality and should be avoided."[28]

Large-scale studies such as this often prompt an accompanying editorial. In this case, Robert Greenberg at Dartmouth pointed out that since one-third of all adults and one-half of adults older than 55 take "at least one vitamin supplement daily" — at a cost of about $18.8 billion dollars each year — the "finding of possible harm with higher doses of vitamin E is surprising and has serious implications for the tens of millions of people who regularly use vitamin E supplements."[18] But he offered an "out" to all those who do take supplemental vitamin E, saying that although the meta-analysis indicated that "higher doses of vitamin E [can] cause death, the case is not iron-clad"[18] because vitamin E was not given alone in many of the trials. Vitamin A or vitamin C was given along with vitamin E in a number of the trials included in the meta-analysis. (It's always easier to point an incriminating finger at an isolated person than one in a crowd.) Although he acknowledged the "case is not iron-clad," he was quick to add that he ". . .fully agree[d] with the authors' conclusion that high-dose vitamin E supplementation is unjustified."[18] Large doses of vitamin E have "no favourable health effects and may cause harm. Thus, our message to the public must be clear on this point. Vitamin E supplements won't help, and might harm, so save your money."[18]

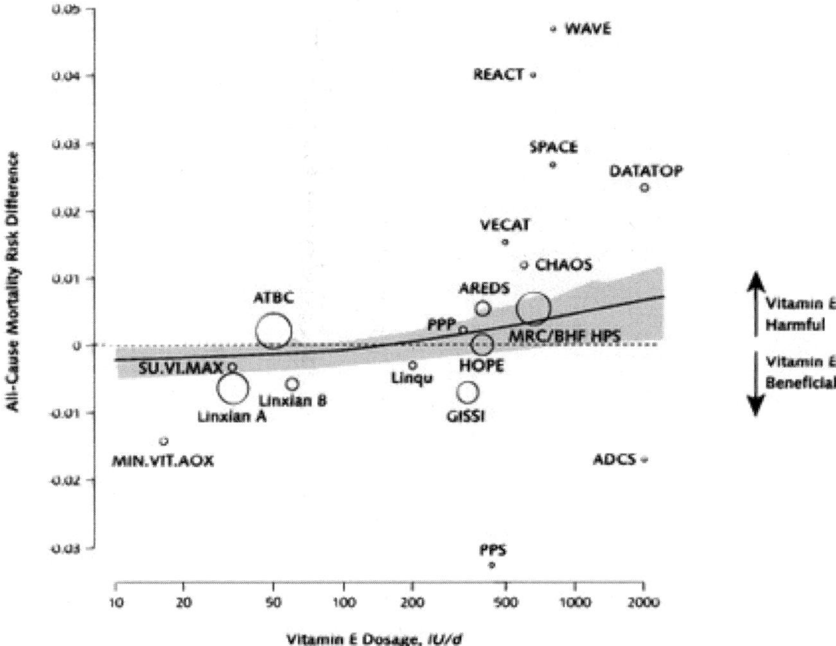

Fig. 24.4. This plot of the relationship between vitamin E dosage and all-cause mortality in Miller's meta-analysis shows that lower doses of vitamin E (<150 IU/day) reduce mortality, while dosages ≥150 IU/day progressively increase mortality. The data is derived from 18 trials of varying sizes, the size of the circles reflecting the size of the trial.

Abbr: ADCS = Alzheimer's Disease Cooperative Study; AREDS = Age-Related Eye Diseases Study; ATBC = Alpha-Tocopherol, Beta Carotene Cancer Prevention Study Group; CHAOS = Cambridge Heart Antioxidant Study; DATATOP = Deprenyl and Tocopherol Antioxidative Therapy of Parkinsonism; GISSI-Prevenzione = Gruppo Italiano per lo Studio della Sopravvivenza nell'Infarcto Miocardio Prevenzione; HOPE = Heart Outcomes Prevention Evaluation; MIN.VIT.AOX = The Geriatrie/MINéraux, VITamines, et AntiOXydants Network; MRC/BHF HPS = Medical Research Council/British Heart Foundation Heart Protection Study; PPP = Primary Prevention Project; PPS = Polyp Prevention Study; REACT = Roche European American Cataract Trial; SPACE = Secondary Prevention with Antioxidants of Cardiovascular disease in Endstage renal disease; SU.VI.MAX = SUpplementation en VItamines et Minéraux AntioXydants; VECAT = Vitamin E, Cataracts, and Age-Related Maculopathy; WAVE = Women's Angiographic Vitamin and Estrogen Study.

(*Source*: Reproduced with permission from the publisher.) Reference 28.

That was in January 2005. Vitamin E knew it would *really* be in hot water if another disappointing clinical trial was published. The expectations and high hopes of the three vitamin antioxidants 10 years earlier now rested entirely on its sagging shoulders. Knowing it was the best and strongest of all three, vitamin E may have been feeling like The Little Engine That Could — saying

Fig. 24.5.   The popular children's story used to teach children the value of optimism and hard work.

(*Source*: Reproduced with permission from the publisher.) Reference 19.

sotto voce, "I think I can, I think I can" as it huffed and puffed up the steep hill before the finish line (Fig. 24.5). But could it?

Hope Springs Eternal

Two months later, investigators in Canada weighed in with a multicenter (174 hospitals) study carried out from 1999 to 2003 of 9541 men and women, all of whom either had an earlier acute myocardial infarction or had diabetes. All participants took 400 IU of vitamin E daily or a placebo for seven years. The study, ironically dubbed the Heart Outcomes Prevention Evaluation — The Ongoing Outcomes (HOPE-TOO) trial, contrasted the number of heart attacks, strokes, hospitalizations for congestive heart failure, and deaths over seven years in a group of men and women taking vitamin E[e]

---

[e] One reason vitamin E was ineffective in the HOPE-TOO trial — and others — may be that the vitamin was given in its α-tocopherol form. Increasing the blood and tissue α-tocopherol concentration decreases the γ-tocopherol concentration, and the γ-tocopherol concentration is inversely associated with coronary heart disease. Thus, clinical trials testing the ability of the α-tocopherol form of vitamin E to prevent heart attacks and strokes may have been barking up the wrong tree. A group of clinicians and physiologists at the University of Arkansas found that in experimental animals a "tocopherol mixture is superior to α-tocopherol alone in blocking

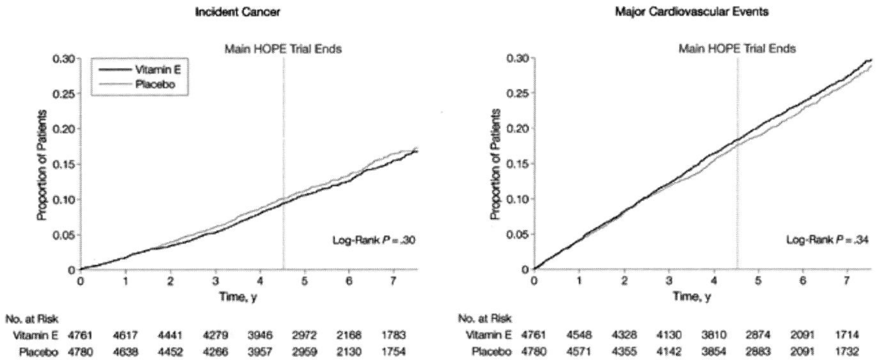

Fig. 24.6. The incidence of cancer and major cardiovascular events over seven years in the HOPE-TOO trial. There were slightly fewer persons who developed cancer and slightly more who had a major cardiovascular event, neither reaching statistical significance.

(*Source*: Reproduced with permission from the publisher.) Reference 26.

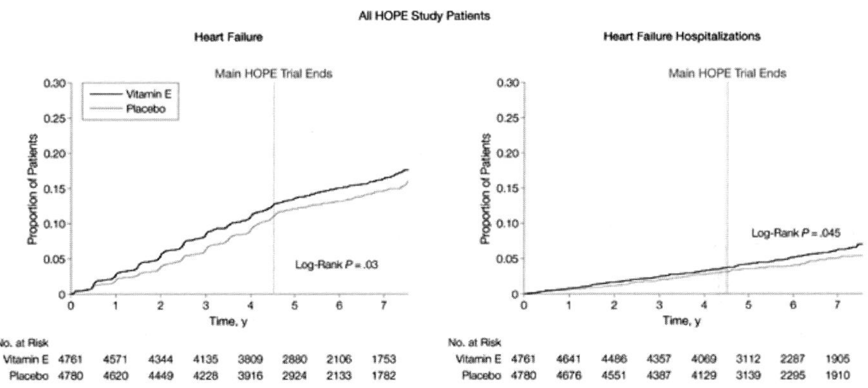

Fig. 24.7. The proportion of patients with congestive heart failure is consistently greater in those individuals receiving vitamin E over seven years than in those taking a placebo, as is the frequency of hospitalizations for treatment of more severe heart failure.

(*Source*: Reproduced with permission from the publisher). Reference 26.

the molecular and cellular responses that may predispose and facilitate the development of atherosclerosis."[29] So, they said, "…it seems logical that vitamin E supplements combining α-and γ-tocopherol may provide better protection against [coronary heart disease] than α-tocopherol alone. Such mixtures are expected to replicate more faithfully the cardioprotective effects observed with vitamin E–rich diets."[29]

with the number in a placebo group. (The frequency of cancer was also assessed in the two groups.) At the end of the trial, published in *JAMA*, there was absolutely no difference in the frequency of any of those awful cardiovascular outcomes between the two treatment groups (Fig. 24.6) with one exception — congestive heart failure (Fig. 24.7). Curiously and surprisingly, there were more instances of congestive heart failure in the vitamin E group than in the placebo group! The investigators reached the conclusion that the HOPE-TOO trial did not determine the outcome it hoped to show. Au contraire, it found "a lack of [any] benefit for vitamin E in preventing. . .major cardiovascular events after a prolonged period [seven years] of treatment and observation. Furthermore, our study raises concern about an increased risk of heart failure related to vitamin E."[26] They closed with a quasi-cautionary, quasi-philosophical message: "There is a tendency to accept 'natural products' (*e.g.* vitamins) as being safe, even if they have not been proven to be effective. However, our findings emphasize the need to thoroughly evaluate all vitamins, other natural products, and complementary medicines in appropriately designed trials before they are widely used for presumed health benefits."[f,26]

A skeptical editorial entitled "Is There Any Hope for Vitamin E?", written by cardiologist Greg Brown and biostatistician John Crowley at the University of Washington, came with the Canadian study. It would become yet another thorn in vitamin E's side. They said that not only did vitamin E alone increase mortality in healthy women in the Women's Health Study (discussed shortly), but a combination of all three antioxidant vitamins — vitamins A, C, and E — appeared to interfere with the proven beneficial (cholesterol-lowering) effects of the statins (drugs like Mevacor, Lipitor, and Zocor) and niacin in men and women alike. They pointed out that in the British Heart Protection Study of patients with established coronary heart disease, men and women taking antioxidants in addition to statins and niacin alone (statins lower the "bad cholesterol" level, while niacin raises the "good cholesterol" level) had more "cardiovascular events" over seven years than those who took statins and niacin alone. And those who took antioxidants had more "cardiovascular events" than those who did not.[30] (Because of the significance of that study, it will be reviewed more thoroughly in Chap. 25.)

---

[f]It might not have been so damaging to vitamin E's reputation if these 2005 reports were published in off-the-beaten path journals, but *JAMA* and the *Annals of Internal Medicine* are decidedly mainstream.

Returning to the HOPE-TOO trial, Brown and Crowley went on to note that vitamin E *did* decrease the risk of lung cancer (compared to placebo) at 4.5 years, but, alas, the reduction didn't last long. By seven years the scores were tied. And they said, "…of greater concern…[vitamin E] was associated [with an] increased risk of heart failure…[at] 4.5 years…[and that risk] persisted in the 7-year extended follow-up." They ended on a downbeat note saying, "…this report effectively closes the door on the prospect of a major protective effect of long-term exposure to this supplement…on atherosclerosis and overall cancer incidence."[g,30] You could sense the end was near for vitamin E.

The Last Gasp

But there would be one last gasp for the troubled vitamin, one last opportunity to shore up its sagging image, one last chance to see if vitamin E's hallowed position embedded deep in the LDL particle could save a sinking ship.[h] That gasp would come from Boston and be heard from coast to coast. A report from the Division of Epidemiology in the Department of Medicine at Harvard was printed yet again in *JAMA*, on July 6, 2005. The studies cited earlier were done on folks who already a serious cardiac event making them "secondary prevention studies." All participants had either known coronary heart disease or two or more risk factors for the disease. Vitamin E didn't help them one bit. While there is no shortage of people who have had a heart

---

[g] In the midst of all the negative press regarding vitamin E and coronary heart disease came a dollop of good news for America's 50 million or so smokers. Because cigarette smoke is such a potent free radical stimulant — a single puff of smoke generates 1014–1015 free radicals[i] — smokers *do* benefit from vitamin E supplementation, even though nonsmokers do not. That blessing is because body tissues of smokers are spewing out such huge numbers of free radicals. In the absence of an outside freeradical stimulus, the body's elaborate and diverse antioxidant mechanisms protect its tissues from freeradical damage, but the enormous efflux of free radicals from tissues while one is puffing on a cigarette quickly overwhelms the body's antioxidant defenses.[31] Nonsmokers, on the other hand, don't benefit from extra vitamin E — even when the blood vitamin E level is increased fivefold — because diet-derived vitamin E is plenty to keep free radicals in check.[32]

[h] On the side, vitamin E was being tested in patients with Alzheimer's disease. But the embattled vitamin just couldn't get a break. It did not slow — let alone reverse — the progression of the disease in patients with either an early mild cognitive deficit or advanced disease. It simply had no effect.[33]

attack, there are a still greater number who have not and don't want one. So you might wonder, *Well, what about vitamin E in healthy people who want to stay healthy?* Maybe it *will* work in that setting. And since an estimated 50 million Americans spend US$18.8 million dollars each year on vitamin supplements,[18] there's gotta be a substantial number of perfectly healthy people taking vitamin E supplements hoping to stay healthy.

Clinicians at Boston's Brigham and Women's Hospital wondered the same thing. So they tested whether or not vitamin E supplementation might reduce the risk of coronary heart disease (and cancer) in healthy women — a "primary prevention study." Called the Women's Health Study, it enrolled 39,876 (apparently) healthy women — all health care professionals — 45 or more years of age. The trial was carried out between 1992 and 2000. One group of women was assigned randomly to receive vitamin E (600 IU of natural vitamin E every other day) or placebo, while another group received aspirin or placebo. The two groups of women were then tracked to see how many developed a nonfatal myocardial infarction, a nonfatal stroke, or died from cardiovascular disease over the next 10 years.

Here's what happened. Ten years later, the women who took vitamin E had slightly fewer "cardiovascular events," which did not reach statistical significance. Specifically, the incidences of acute myocardial infarction and stroke were remarkably similar (Fig. 24.8.).[34] The incidence of cancer in the two groups was even more similar, the coordinates being almost identical (Fig. 24.9).[i] There was, however, a vitamin E advantage in women 65 years of age or older. Elderly women who took vitamin E had fewer cardiovascularly related deaths than those who took placebo. The Women's Health Study, by the way, is the largest and longest randomized trial of vitamin E yet.

The researchers wrote, "In this trial, 600 IU of natural-source vitamin E every other day for 10 years did not provide any statistically significant benefits on the primary endpoints of major cardiovascular events or cancer in almost 40,000 healthy women. We [also] observed no significant effect of vitamin E on total mortality. [Thus], "… theWHS does not support recommending vitamin E supplementation for cardiovascular disease or cancer

---

[i]Cancer is not at all the focus of this book, nor does the author have more than a thimbleful of knowledge about this disease. But since some of the larger vitamin E trials include the effect of the vitamin on cancer and some figures include cancer data, the cancer bar graphs are included (rather than cropped out). Many readers may have as much interest in cancer as incoronary heart disease.

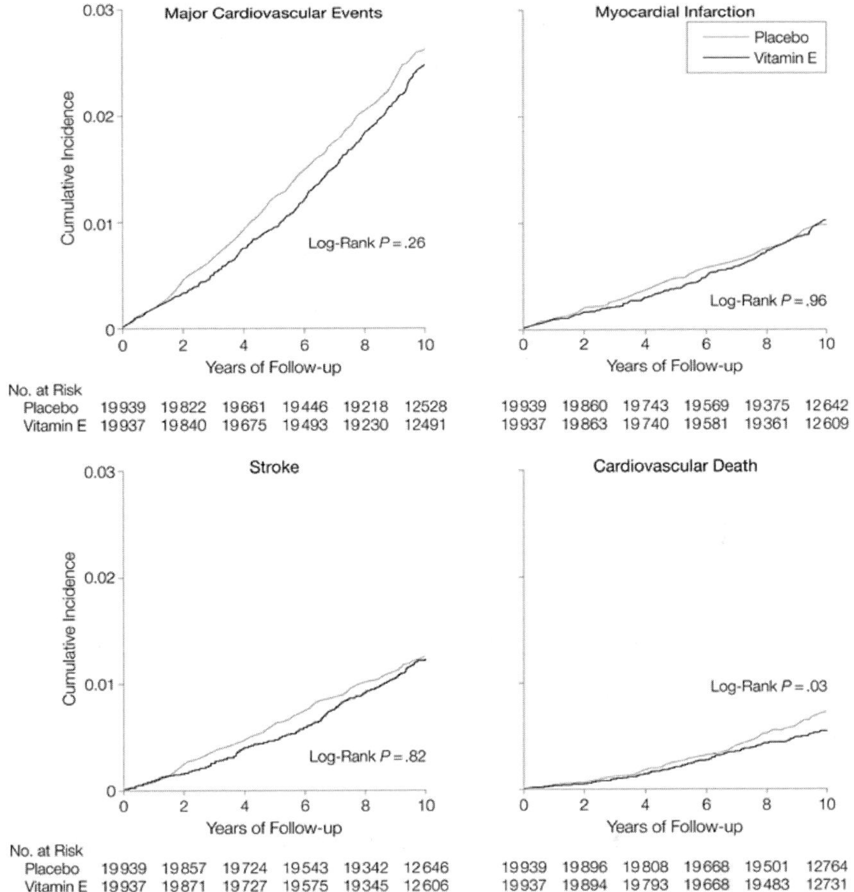

Fig. 24.8. There were slightly fewer major cardiac events in women taking vitamin E in the Women's Health Study, which did not reach statistical significance. The difference was even less when the incidences of acute myocardial infarctions, strokes, and cardiovascular deaths were considered separately.

(*Source*: Reproduced with permission from the publisher.) Reference 34.

prevention among healthy women. This large trial [does] support current guidelines stating that use of antioxidant vitamins is not justified [for cardiovascular disease] risk reduction."[34]

Because of the magnitude of this study and its largely negative findings regarding the effect of vitamin E on cardiovascular disease and cancer, not one but two editorials accompanied it in *JAMA*. One editorial was written by Rita

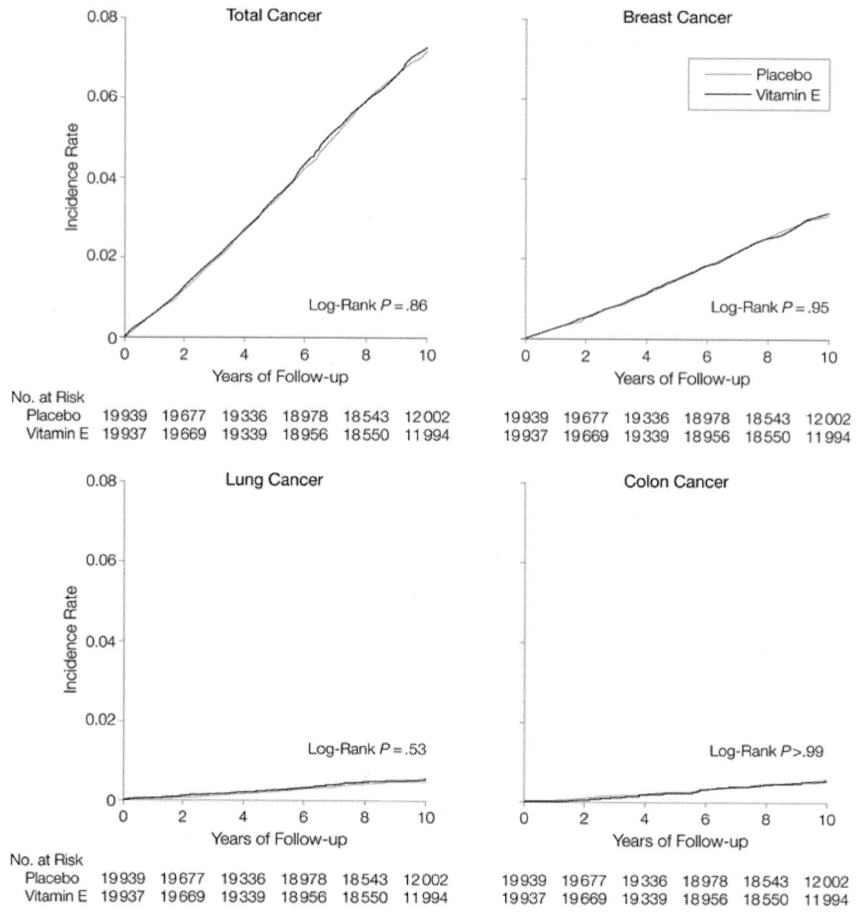

Fig. 24.9. The incidences of specific cancers and all cancers in the Women's Health Study. The coordinates are almost identical.

(*Source*: Reproduced with permission of the publisher.) Reference 34.

Redberg at the University of California, San Francisco. She poignantly pointed out that in the Women's Health Study mortality increased among women taking 600 IU of vitamin E every other day — akin to the findings of Miller *et al.* several years earlier, who noted an increased "all-cause mortality" among both men and women as the dosage of vitamin E increased from 150 IU per day a day to 2000 IU per day. She wisely wrote, "The study by Lee *et al.* also noted increased mortality for women taking 600 IU of vitamin E every other day. Although the increase was not statistically significant, in a healthy low-risk population it is difficult to justify an intervention with no proven benefit and a suggestion of harm"[35] She finished on a downbeat note, saying, "Thus, vitamin E enters the category of therapies that were promising in epidemiological and observational studies but failed to deliver in…randomized controlled clinical trials. As in other studies, the 'healthy user' bias must be considered, i.e., the healthy lifestyle behaviors that characterize individuals who care enough about their health to take various supplements are actually responsible for the better health, but this is minimized with the rigorous trial design."[35]

Two oncologists with the American Cancer Society contributed the second editorial. They noted, "The null results for vitamin E [in the form of α–tocopherol] and cancer from the WHS add to the evidence from 2 previous trials that even relatively long-term α-tocopherol supplementation is unlikely to have any substantial effect on cancer risk, at least in women." They added that "… no agent has been shown to do for cancer what statins do for cardiovascular disease, namely substantially and relatively safely reduce disease occurrence in individuals at especially high risk."[36]

As if that rebuke weren't enough, there came yet another verbal volley — this one not anticipated — from Hopkins. It appeared in the very journal that had published the sensational meta-analysis six months earlier on January 4, 2005, in the *Annals of Internal Medicine*. (It's not common in medicine for an editorial to be printed in a journal other than the journal publishing the original work. In this case, because of the enormity of the U.S. Women's Health Study and the Canadian HOPE-TOO study and the vast impact they would inevitably have on the practice of cardiovascular medicine, an exception was made.) Three epidemiologists with the Johns Hopkins School of Public Health noted that 2005 had been an *annus horribilis* for vitamin E, referring to the negative January and March reports followed by the huge, again-negative Women's Health Study (on healthy women) in July. They felt that because high-dose vitamin E supplements may increase all-cause mortality, both "high-risk and healthy people should avoid this vitamin at high dosages."[33] They added, "We think…that current evidence does not justify using it to reduce the risk or cancer, cardiovascular disease, or Alzheimers disease."[33] They then offered specific advice to the public.

So significant is this advice to all Americans, Canadians — all of us — it will be reproduced in its entirety. It reads, "Vitamin E enjoyed superstar status among dietary supplements. Because of perceived health benefits, vitamin E supplements are consumed by many people. As this editorial shows, recent trials have further weakened the evidence for benefit, while the evidence for harm has accumulated. We did not fully appreciate just how many people may be putting themselves at risk by using high-dosage vitamin E supplementation until this issue of *Annals*, in which Ford and colleagues use data from the 1999–2000 National Health and Nutrition Examination Survey to estimate that about 12% of U.S. adults (24 million people) consumed 400 IU or more of vitamin E daily from supplements. Furthermore, older adults were more likely to use high-dosage vitamin E supplementation. We are disturbed that a presumption of benefit by so many may increase risk for mortality. We call on health professionals to warn the public against the use of ineffective or even harmful interventions such as vitamin E that may compete with well-established preventive measures. High-dosage vitamin E is a prime example of misplaced priorities."[33] Amen!

And so, with those words, vitamin E fell from its lofty pedestal. The emperor was dead.

References

1. Murphy RT. (2007) Smoking and tocopherol status. In: Preedy VR, Watson RR (eds.), *The Encyclopedia of Vitamin E.* Cromwell, Trowbridge, UK pp. 683–689.

2. Keaney JF, Jr, Simon DI, Freedman JE. (1999) Vitamin E and vascular homeostasis: implications for atherosclerosis. *FASEB* **13**: 965–975.

3. Shihabi A, Wei-gen LI, Miller FJ, Jr, Weintraub NL. (2002) Antioxidant therapy for atherosclerotic vascular disease: the promise and the pitfalls. *Am J Physiol Heart Circ Physiol* **282**: 797–H802.

4. Evans H, Bishop KS. (1922) On the existence of a hitherto unrecognized dietary factor essential for reproduction. *Science* **56**: 649–651.

5. Traber MG. (2006) Vitamin E. In: Shils ME, Shike M, Ross AC, Cabellaro B, Cousins RJ, (eds.), *Modern Nutrition in Health and Disease.* Lippincott, Williams & Wilkins, Philadelphia, Baltimore, New York, London, Buenos Aires, Hong Kong, Sydney, Tokyo, Chap. 21, pp. 396–411.

6. Traber MG, Packer L. (1990) Vitamin E: beyond antioxidant function. *Am J Clin Nutr* **22**: 1501S.

7. Zingg JM, Azzi A. (2004) Non-antioxidant activities of vitamin E. *Curr Mod Chem* **11**: 1113–1133.

8. Combs GF, Jr. (ed.). (2008). In: Vitamin E, *The Vitamins. Fundamental Aspects in Nutrition and Health,* 3rd ed. Elsevier, Amsterdam, Boston, Heidelberg, London, New York, Oxford, Paris, San Diego, San Francisco, Singapore, Sydney, Tokyo, Chap. 7, pp. 181–212.

9. http://en.wikipedia.org/wiki/Vitamin_E

10. http://www.nal.usda.gov.fppdcomp/Data/SR20/mutrlist/sf20w323.pd

11. http://ods.od.nih.gov/factsheets/VitaminE.asp. Vitamin E Fact Sheet. Office of Dietary Supplements. National Institutes of Health.

12. Devaraj S, Jialal I. (2001) Failure of vitamin E in clinical trials. Is gamma-tocopherol the answer? *Arch Intern Med* **63**: 290–293.

13. In: Vitamin E. *The Vitamins. Fundamental Aspects in Nutrition and Health*, 3rd ed. Combs GF, Jr. (ed.). 2008. Elseiver. Amsterdam, Boston, Heidelberg, London,New York, Oxford, Paris, San Diego, San Francisco, Singapore, Sydney, Tokyo. Chapter7. pp: 181–212.

14. Brigelius-Flohé R. (2009) Vitamin E: the shrew waiting to be tamed. *Free Rad Biol Med* **46**: 543–544.

15. Traber MG, Atkinson J. (2007) Vitamin E, an antioxidant and nothing more. *Free Rad Biol Med* **43**: 4–15.

16. Azzi A. (2007) Molecular mechanisms of α–tocopherol action. *Free Rad Biol Med* **43**: 16–21.

17. Brigelius-Flohe R, Davies KT. (2007) Is vitamin E an antioxidant, a regulator of signal transduction and gene expression, or a "junk" food? *Free Rad Biol Med* **43**: 2–3.

18. Greenberg ER. (2005) Vitamin E supplements: Good in theory, but is the theory good? *Ann Intern Med* **142**: 75–76.

18.5 http://www.ebay.com/sch/Books-/267/i.html?_nkw=the+little+engine+that+could

19. Blendon RJ, Des Roches CM, Benson JM, Brodie M, Altman DE. (2007) Americans' views on the use and regulation of dietary supplements. *Arch Intern Med* **161**: 805–810.

20. Cagir B, Sawyer MAJ. Short-bowel syndrome. http://emedicine.medscape.com/article/193391-overview (accessed May 28, 2010).

21. Schuellke M. Ataxia with vitamin E deficiency. http://www.ncbi.nlm.nih.gov/bookshelf/br.fcgi?book=gene&part=aved

22. *The World Book*. (2009) World Book, Vol. 4, pp. 1208.

23. Dehghan M, Merchant AT. (2007) Vitamin status in the elderly. In: Preedy VR, Watson DA (eds.), *The Encyclopedia of Vitamin E*. Cromwell, Trowbridge, UK, pp. 285–296.

24. Williams SR. (1999) *Essentials of Nutrition and Diet Therapy*. 7th ed. Mosby, St Louis, Baltimore, Boston, Carlsbad, Chicago, Chap. 7 (Vitamins), pp. 111–146.

25. Smedts HPM, De Vries JH, Rakshandehroo M, Wildhagen MG, Verkleig-Hagroort AC, Steegers EA, Steeger-Thunison RPM. (2009) High maternal vitamin E intake by diet or supplements is associated with congenital heart defects in the offspring. *BJOG: An International Journal of Obstetrics and Gynecology* **116**: 416–423.

26. HOPE and HOPE-TOO trial investigators. (2005) Effects of long-term vitamin E supplementation on cardiovascular events and cancer: a randomized controlled trial. *JAMA* **293**: 1338–1347.

27. http://lpi.oregonstate.edu/infocenter/vitamins/vitaminE

28. Miller ER III, Pastor-Barriuso R, Dalal D, Riesmersma RA, Appel AJ, Gullan E. (2005) Meta-analysis: high-dosage vitamin E may increase all-cause mortality. *Ann Int Med* **142**: 37–46.

29. Sarkar K, Parthasarthy S, Saldeen TGP, Mehta JL. (2007) Efficacy of tocopherol mixtures compared with alpha-tocopherol in cardioprotection. In: Predy VR, Watson RR (eds.), *The Encyclopedia of Vitamin E.* Cromwell, Trowbridge, UK, pp. 299–306.

30. Brown BG, Crowley J. (2005) Is there any hope for vitamin E? *JAMA* **293**: 1387–1390.

31. Morrow JD, Frei B, Longmier AW, Gaziano JH, Lynch SM, Shyr Y, Strauss WE, Oates JA, Roberts LJ. (1995) Increase in circulating products of lipid peroxidation (F-2-isoprostones) in smokers: smoking as a cause of oxidative damage. *N Engl J Med* **332**: 1198–1203.

32. Meagher EA, Barry OP, Lawson JA, Rokach J, Fitzgerald GA. (2001) Effects of vitamin E on lipid peroxidation in healthy persons. *JAMA* **285**: 1178–1182.

33. Guallar E, Hanley DF, Miller ER III. (2005) *Annus horribilis* for vitamin E. *Ann Intern Med* **143**: 143–145.

34. Lee IM, Cook NR, Gaziano JM, Gordon D, Ridken PM, Manson JE, Hennekens, CH, Buring JE. (2005) A randomized trial of vitamin E in the primary prevention of cardiovascular disease in 39,876 women: The Women's Health Study. *JAMA* **294**: 56–65.

35. Redberg R. (2005) Does sex matter? *JAMA* **294**: 107–109.

36. Jacobs EJ, Thun MJ. (2005) Low-dose aspirin and vitamin E. Challenges and opportunities in cancer prevention. *JAMA* **294**: 105–106.

25

DO COMBINATIONS OF ANTIOXIDANT VITAMINS WORK BETTER THAN INDIVIDUAL VITAMINS IN PATIENTS WITH CORONARY HEART DISEASE?

Individual antioxidant vitamins didn't work, but what happens when *all three* are taken over a period of time? That might be all that needs to be done to put a dent in coronary heart disease's armor.

That important and very practical question was tackled in England by the British Heart Protection Study in 2002. The study directors chose to give vitamins A, C, and E–a pretty intimidating opponent. Rather than a traditional drug versus placebo, it would, in effect, be the three antioxidant vitamins versus a powerful cholesterol-lowering agent known as "simvastatin" (brand name Zocor). Simvastatin belongs to a class of drugs called "statins," referring to the suffix "-statin," common to all drugs in the family (*e.g.* lovastatin, atorvastatin, and pravastatin, sold commercially as Mevacor, Lipitor, and Pravachol, respectively). All statins lower cholesterol by blocking the rate-limiting enzyme in the LDL cholesterol ("bad" cholesterol) synthetic pathway. The cholesterol level consequently falls, often within one week, maximally within four to six weeks. Statins may well be the most significant new class of cardiac drugs developed in the 20th century.

In the well-designed and -conducted British Heart Protection Trial,[1] more than 20,000 men and women of all ages at "high risk" for a major cardiovascular event (defined as an acute myocardial infarction, a stroke, the need for coronary artery bypass surgery or angioplasty, or death), one-half of the participants were randomly assigned to take simvastatin and niacin, while the other half took a three-vitamin "antioxidant cocktail" and selenium, a mineral antioxidant. The antioxidant dosages were β-carotene 20 mg per day, vitamin C 250 mg per day, and vitamin E 600 mg per day. Then the participants were followed for five years to determine how many in each group had one of the predefined end points listed above.

Simvastatin

The results were *very* impressive. Simvastatin cut all-cause mortality from 14.7% to 12.9%, due to a "highly significant" 18% reduction in the coronary heart disease death rate, irrespective of age (Fig. 25.1).[1] The medication also reduced the incidence of nonfatal myocardial infarctions, strokes, and the need for coronary artery bypass surgery or angioplasty by 25%. And, perhaps most notably, ". . .even [patients] with an LDL cholesterol below 116 mg/dL or a total cholesterol below 193 mg/dL" had fewer cardiovascular events when taking simvastatin. In other words, the drug offered cardiovascular protection regardless of the cholesterol level — even when normal.

It was calculated that the statin prevented 70–100 people per 1000 population from at least one of three major cardiovascular events — an acute myocardial infarction, a stroke, and a revascularization procedure, i.e. coronary artery bypass surgery, a stent or coronary angioplasty. For each reduction in the blood cholesterol level by 1 mmol/L (38.67 mg/dL) over five years, the risk of a nonfatal acute myocardial infarction was reduced by about 25%.[1] And the good results were not due to picking "weller" participants. They all had advanced atherosclerosis and/or diabetes and the study "deliberately included large numbers of older individuals." Nonetheless, simvastatin improved survival and lowered the cholesterol level across the board. And the reduction in heart attacks and strokes by simvastatin was *in addition to* reductions from aspirin, β-blockers, and ACE inhibitors. The broad reduction in cholesterol level and lower coronary heart disease mortality allowed the conclusion that ". . .the. . .study demonstrated unequivocally that lowering LDL cholesterol (not *total* cholesterol) from 116 mg/dL to below 77 mg/dL reduces vascular disease risk" by about one quarter.[a,1]

---

[a] The value of simvastatin was also shown through a separate study that looked at the rate of progression of atherosclerosis as measured by carotid artery ultrasound. (We can't do coronary artery ultrasound noninvasively, but we can do carotid ultrasound since the carotid arteries are

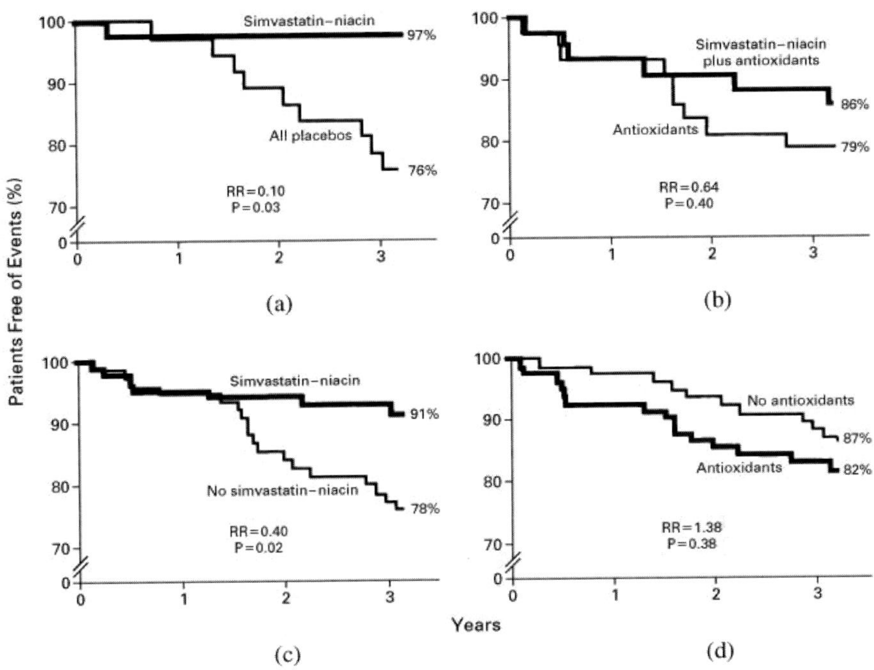

Fig. 25.1.   This figure is from the British Heart Protection Study, which assessed the effect of simvastatin (a "bad cholesterol"-lowering drug), niacin (a "good cholesterol"-raising drug), and a combination of all three vitamin antioxidants on coronary events (defined in the next chapter) in men and women with advanced coronary heart disease over three years. Panel a shows that 97% of patients taking simvastatin and niacin had no coronary event over three years, while 24% of those taking placebo did have such an event. Panel b shows that those taking simvastatin and niacin plus antioxidants had more coronary events than those taking simvastatin and niacin alone, and that those taking antioxidants alone fared no better than those taking placebo in panel a. Panel c shows that those patients taking simvastatin and niacin had fewer coronary events than those who took nothing, while panel D shows that those taking antioxidants had more coronary events than those taking no antioxidants.

(*Source*: Reproduced with permission from the publisher.) Reference 38.

easily accessible in the neck.[2] And because atherosclerosis is a diffuse disease affecting blood vessels throughout the body, the extent of carotid artery atherosclerosis can be used to infer the extent of coronary atherosclerosis.) We can do coronary *angiography*, however. And serial coronary angiograms have consistently shown that "...treatment with cholesterol-lowering agents is associated with the slowing of [the progression of] coronary atherosclerosis."[4]

Three Antioxidant Vitamins

It was a very different matter with the three antioxidant vitamins. Much the opposite was the case.

To contrast the clinical response to simvastatin with that to antioxidants, another arm of the Heart Protection Study randomized men and women of the same age at "high risk" for a major cardiovascular event to *all three* antioxidants and selenium or placebo for five years. The dosages of the antioxidants were β-carotene 20 mg per day, vitamin C 250 mg per day, and vitamin E 600 mg per day. Then everyone was followed for five years to see how many in each group had one — or more — of the predefined end points mentioned above.

Fascinatingly, ". . .although (the vitamin antioxidant) regimen increased blood vitamin concentrations substantially, it did not produce any significant reduction in the five-year mortality from, or the incidence of any type of vascular disease, cancer, or other major outcomes" (Fig. 25.2).[4] On top of that, the three antioxidant vitamin preparations[5] produced "a small, but definite, increase in the measured values of plasma triglycerides and LDL cholesterol concentrations," while lowering the HDL cholesterol ("good cholesterol") level.[4] Of even greater concern, ". . .although this regimen increased blood

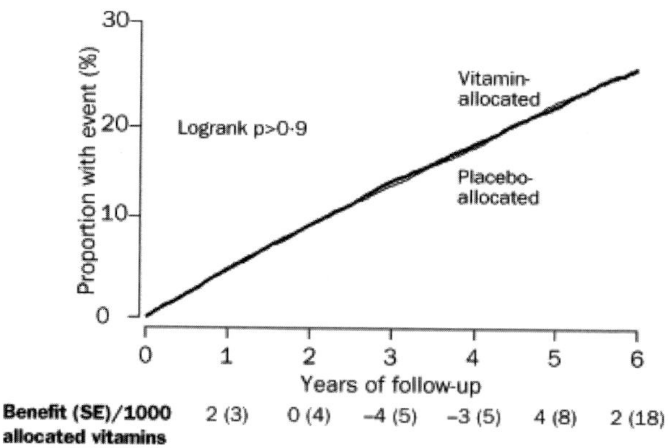

Fig. 25.2.   The proportion of patients at high risk for a heart attack who actually had a heart attack, a stroke, or needed coronary artery bypass surgery or coronary angioplasty while taking vitamins A, C, and E or a placebo daily for six years in the British Heart Protection Study. The lines are virtually superimposable. Antioxidant vitamins didn't make one bit of difference.

(*Source*: Reproduced with permission from the publisher.) Reference 5.

Table 25.1   Drawbacks of antioxidants for treating cardiovascular disease.

May lower the "good cholesterol" concentration.

When combined with a cholesterol-lowering statin, antioxidants blunt the favorable rise of the HDL level produced by the statin.

Do not prevent heart attacks or strokes.

---

vitamin concentrations substantially, it did not produce any significant reduction in the 5-year mortality from, or incidence of any type of vascular disease, cancer, or other major outcomes" (Table 25.1).[3] Triple antioxidant therapy did manage to "blunt the increase in cardioprotective HDL that was seen with. . .lipid-lowering therapy," however.[3] Seems like vitamin antioxidants just couldn't get a break. They not only didn't help, but they caused some harm — violating a major principle of medicine, *Primum non nocere*, meaning "First, do no harm."[4] (That maxim is vowed in the Hippocratic oath by all medical students at the time they receive their M.D. degree.) Or, put another way, the maxim says, "Given an existing problem, it may be better to do nothing than to do something that risks causing more harm than good." This is particularly significant when "debating the use of an intervention that carries an obvious risk of harm but also a less certain chance of benefit."[5] The results of triple antioxidant vitamin therapy in the Heart Protection Study prompted the opinion that the report ". . .effectively rules out any substantive reduction. . .in heart attacks, strokes, or other major adverse events," the result of a three-vitamin antioxidant mixture.[5]

Editorial Opinion

The accompanying editorial pointed out that a 1 mmol/L (38.67 mg/dL) reduction in the blood cholesterol level "led to a 25% reduction in risk" of an acute myocardial infarction or stroke.[3] And the reduction applied to all age groups (including people over 75 years of age) and all cholesterol concentrations (even people with a normal or low LDL cholesterol level). That means ". . .practically all patients with vascular disease will benefit from statins."[3]

Yet another encouraging point made by the Heart Protection Study was that because the benefits of the four medications — statins, aspirin, β blockers, and angiotensin-converting enzyme (ACE) inhibitors to treat high blood pressure (if present) — prescribed routinely following an acute myocardial infarction ". . .appear to be largely independent [of each other] . . .when used together in appropriate patients it is reasonable to expect that about two-thirds to three-quarters of future [acute myocardial infarctions and strokes] can be prevented."[6] And if someone with coronary heart disease were to choose to stop smoking and lower his or her blood pressure by just 10 mmHg (if elevated), "it may be possible to lower the risk of future events by more than four-fifths in high-risk individuals."[6] So, there's good reason for optimism among the millions of Americans with coronary heart disease today.

Why Did the Clinical Trials of Antioxidant Vitamins Fail?

The failure of clinical trials to demonstrate a clear antioxidant benefit could be due to the lack of an antioxidant deficiency in healthy subjects. Or, restated, in normal people or those with only a mild disease, there may be no need for high antioxidant concentrations. In an analogous way, aspirin can effectively lower fever, but if the body temperature is already normal it will not further lower the fever. An antioxidant may be effective only in the presence excessive free radicals.[7]

But for those who do n't want to give up hope on antioxidant supplements as a nonpharmacologic means of lessening atherosclerosis, a report from Denmark and Finland offers hope. It ain't much hope, and taken at face value it applies only to smokers with a high cholesterol level, but the study may lighten the dark medical cloud that seems to hover above smokers. In 520 smoking (and some nonsmoking) men and postmenopausal women, all with an elevated cholesterol level (>193 mg/dL), a combination of slow-release vitamin C and vitamin E slowed down the progression of atherosclerosis over six years.[8] It didn't reverse the damage that had been done, but it did slow the pace of progression. And it was in only 520 people, a small number in a world where we are used to seeing trials of thousands and thousands of participants.

But it offers hope!

References

1. MRC/BHF Heart Protection Study Collaborative Group. (2002) MRC/BHF Heart Protection Study of cholesterol lowering with simvastatin in 20,536 high-risk individuals: a randomized placebo-controlled trial. *Lancet* **360**: 7–22.

2. Salonen RM. Six-year effect of vitamin C and E supplementation on atherosclerosis progression: the Antioxidant Supplementation in Atherosclerosis Prevention (ASAP).

3. Brown BG, Crowley J. (2005) Is there any hope for vitamin E? *JAMA* **293**: 1387–1390.

4. http://www.wikipedia.org/wiki/Primum_non_nocere

5. MRC/BHF Heart Protection Study Collaborative Group. (2002) MRC/BHF Heart Protection Study of antioxidant vitamin supplementation in 20,536 high-risk individuals: a randomized placebo-controlled trial. *Lancet* **360**: 23–34.

6. Yusef S. (2002) Two decades of progress in preventing vascular disease. *Lancet* **360**: 2–3.

7. Padaytty SJ, Levine M. (2000) Vitamin C and myocardial infarction: the heart of the matter. *Am J Clin Nutr* **71**: 1027–1028.

8. Salonen RM, Nyyssonen, Kaikonen J *et al.* (2003) Six-year effect of combined vitamin C and E supplementation on atherosclerosis progession. The antioxidant supplementation in atherosclerosis prevention (ASAP) Study. *Circulation* **107**: 947–953.

26

A REQUIEM

Why didn't antioxidant vitamin supplements work? Why didn't they improve health and longevity? What went wrong? As a trio of Johns Hopkins epidemiologists wrote in a 2005 editorial what would amount to a requiem for vitamin E, "We call on health professionals to warn the public against the use of ineffective or even harmful interventions such as vitamin E that may compete with well-established preventive measures. High-dosage vitamin E is a prime example of misplaced priorities."[1]

There are several possible explanations for what went wrong. One is that fruits and vegetables contain many, many nutrients in addition to vitamins A, C, and E. So, any single or any of the nearly infinite number of permutations and combinations of all the *other* nutrients in fruits and vegetables could account for their cardioprotective effects. Indeed, the epidemiological studies documented a 31% lower risk, on average, of cardiovascular death among persons who eat generous amounts of fruits and vegetables (Fig. 26.1). We chose to test vitamins A, C, and E because of the epidemiological studies *and* a plausible biological explanation. Turns out vitamin supplements didn't translate into better survival in the clinical trials, but that doesn't invalidate the epidemiological findings. They're real. *So you can't use the results of the antioxidant trials as an excuse not to eat your fruits and vegetables!*

A second reason antioxidant vitamins failed may be that participants in the megatrials consumed many different diets. The amount of an antioxidant vitamin in a given supplement can be rather easily controlled,[a] but it's not so

___

[a] Following a fierce struggle between vitamin supplement manufacturers and health food stores versus the F.D.A., the Dietary Supplement Act of 1994 was passed. The major force behind the bill leading to the Act was Senator Orrin Hatch, from Utah. The Act affected the quality assurance of vitamin supplements and other nonprescription entities. It "removed the F.D.A.'s authority to test dietary supplements"[3] and the F.D.A. was not permitted to oversee "vitamins,

365

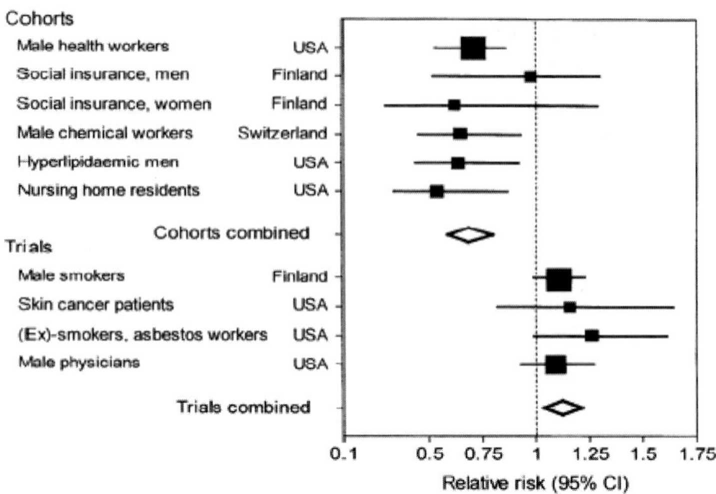

Fig. 26.1. Epidemiological studies (to the left of the vertical line) showed a significantly lower — 31% lower — risk of cardiovascular death among those who ate a diet high in β-carotene. Yet randomized clinical trials (to the right of the vertical line) failed to show any benefit. Instead, they showed a significantly higher — 12% higher — risk of cardiovascular death among those who took β-carotene supplements. This graph speaks for β-carotene alone, but a similar dichotomy was found for vitamins C and E.

(*Source*: Reproduced with permission from the publisher.) Reference 2.

easy to control the (natural) vitamin content in the diet of a mobile, multiethnic society of Americans whose members have an unrestricted choice of foods.

A third reason why antioxidant vitamin supplements didn't work could be that "by eliminating free radicals from our organism [by taking antioxidant vitamin supplements], we interfere with some essential defensive mechanisms like apoptosis [programmed cell death], phagocytosis [removal of bacteria, dead cells, and other 'debris' by phagocytes, which are white blood cells that

---

minerals, herbs, botanical, and amino acids."[3] Plus, manufacturers were "allowed to make unproven claims about the effect of the supplement on the structure or function of the body."[3]

For their part, manufacturers needed only to put a disclaimer on the supplement label saying, "This statement [referring to unproven claims] has not been evaluated by the F.D.A. The product is not intended to diagnose, treat, cure, or prevent any disease."[3] The result of all this is that ". . .no government agency is responsible for ensuring [the] safety and efficacy [of vitamin supplements] prior to [their] reaching store shelves."[3]

act like a garbage disposal], and 'detoxification' [the removal of toxic substances].[b,3] We need *some* free radicals. To be sure, free radicals[c] can cause and/or contribute to disease, but we need some. We need enough to stay healthy![4] As Steinberg said, when trying to account for why we have genes that make "bad" cholesterol (and even "bad" cholesterol has a few "good" qualities), "Teleological thinking is out of place when discussing atherogenesis [the development of atherosclerosis]. . . .There are 'good' genes and there are 'bad' genes regulating lipoprotein metabolism. However, those genes are 'good' or 'bad' by chance, not by selection."[5] So, in a nutshell, significantly reducing free radicals through exogenous (supplemental) antioxidants may have impaired the body's ability to ward off infectious agents and the removal of other toxic substances. Nature has done a fine job in creating the human body. Some say we may best take care of it for 70 or 80 years through weight control, a nutritious diet, regular exercise, and avoiding cigarettes as well as too much alcohol (one or two glasses of red wine a day actually provide cardioprotection, particularly in individuals with one or more risk factors for coronary heart disease; see the Nurses' Health Study in Chap. 19) while taking as little exogenous vitamins as possible — and certainly as few prescription medications as possible.[d,8]

---

[b] Other explanations for why no protective effect of antioxidant vitamins was found may be that the "wrong" dosages were used (maybe more or less would favor survival) and that the duration of the trials was too short (maybe 10- or 20-year trials would have had different outcomes.[2])

[c] Being more into instant gratification than delayed satisfaction, free radicals oxidize as much LDL as possible, as quickly as possible. They "inhale," then "swallow" LDL particles, allowing little or no time to savor the meal they just ate. Not only do they gulp down huge LDL particles, risking a distressing case of indigestion, they do it with not even a hint of (table) manners in the dive — otherwise known as the subendothelial space — they call home. No elegant food, no five-star dining here. It's every antioxidant vitamin, and every bayonet-toting free radical on its own in the hostile environment that is the subendothelial space. Yet, even in this war zone, LDL oxidation slogs ahead at a snail's pace, taking 24 h or more for full oxidation (see Chap. 9, footnote 8) — more like Union General William Sherman's March to the Sea (the 300-mile march from Atlanta to the seaport of Savannah took his army five weeks, a pace of about eight miles per day[6] or *El Führer's* drive toward Moscow during the brutal, hostile Russian winter of 1941–42. Hitler's troops got within 19 miles of the Kremlin, a pace of about 3.75 miles (6.0 km) per day, before they were stopped by the Soviet Army and the harsh winter weather during the Battle of Moscow — with enormous casualties on both sides.[7]

[d] There are certainly special groups, such as pregnant women, heavy drinkers, the elderly, and some poorer inner-city residents, who may benefit from supplemental vitamins.

Timing

All those reasons are valid speculations as to why the antioxidant trials failed. But perhaps the major reason antioxidants didn't work is that they were started *too late in life* to have a favorable effect on atherosclerosis. While adults are, certainly, the ones at risk for an acute myocardial infarction, with the risk increasing each year, they may be a bit too old for antioxidants to have a beneficial effect on a disease that began in childhood. And coronary heart disease *does* begin in childhood. Fatty streaks — the first visible lesions of atherosclerosis — are present in many children and young adults (see Fig. 12.10). Heart attacks — the ostensible sudden expression of atherosclerosis — don't occur until we are in our 40s, 50s, 60s, or beyond. But the atherosclerotic process, from a fatty streak on to a soft, cholesterol-rich plaque, started when we were small children and adolescents. [Recall the astonishing 77.3% prevalence of atherosclerosis in young soldiers — average age 22 years — who died in action during the Korean War (Chap. 12)]. It is during childhood and adolescence that LDL is actively being oxidized and the plaque is growing steadily. If antioxidant vitamin supplements were started at the time LDL was being actively oxidized, they might stand a better chance of reducing the number of acute myocardial infarctions later in life.[e] In the fallen soldiers, the extent of the disease ranged from early fatty streaks, through advanced lesions, to total occlusions of coronary arteries. (Such total occlusions are the vascular hallmark of an acute myocardial infarction.) The recruits with totally occluded coronary arteries had had a heart attack and

---

[e] Later in life, plaques become hardened and brittle, giving rise to the phrase "hardening of the arteries" (see Fig. 12.12). The hardening is due to the deposition of calcium within plaques as they age. Such calcifications can be detected by a "heart scan." These scans have been promoted in the 1990s and the first decade of the 21st century as a noninvasive way to diagnose the presence or absence of coronary heart disease on an outpatient basis. But they have limitations. One is that a "negative" scan does not at all mean that a person is free of coronary heart disease. A scan cannot "see" fatty streaks and soft, early plaques. Nor does a "positive" scan mean that someone should begin taking antioxidants with the expectation of improvement or resolution. By the time plaques become calcified and brittle, the active oxidation of LDL cholesterol that initiated the plaque long ago has largely stopped. Neither the American Heart Association, the American College of Cardiology, nor the Mayo Clinic recommends the routine use of heart scans in people with no cardiac symptoms.[9] For more information about heart scans, go to http://www.mayoclinic/health/heart-scan/MY00327

likely mistook the chest pain caused by it to something less ominous, such as "indigestion." Fatty streaks and soft plaques probably are reversible. Had those GIs lived, many fatty streaks and advanced lesions may well have progressed to even advanced lesions by age 35, leading to a heart attack in their 40s, 50s, 60s, or beyond. Antioxidant vitamins would be less likely to have had a cardioprotective effect beyond age 40 or so, when hardened calcified plaques prevail. While atherosclerosis starts out casually enough, causing a few innocent-looking fatty streaks in children (see Fig. 12.10), it too often ends up viciously in adults (see Fig. 12.11).

Yet, it is almost inconceivable that a large-scale, randomized, placebo-controlled clinical trial of antioxidants in children and adolescents will be performed. Many of us may be willing to be participants in clinical trials. But our children? Many parents may be unwilling to enroll their children in an antioxidant trial. And many investigators may be hesitant to put time and money into such a trial because of anticipated parental reluctance, *and*, while younger children may do what their parents tell them to do, will teenagers do so? How can you be certain that busy, sometimes rebellious teenagers will take their vitamins? Investigators may not want to begin an antioxidant trial on kids aged 5–12 that may become flawed by teenage noncompliance. For the same reason, agencies that fund large clinical trials — like the N.I.H., and pharmaceutical companies — may have a dim view of putting big, big bucks into such a study. And then there's the very practical issue of conducting a large and very long antioxidant vitamin clinical trial, beginning in childhood but measuring the effect of antioxidants on the incidence of acute myocardial infarctions, strokes, and cardiovascular deaths many years later. The followup logistics would be overwhelming. It might be more practical to figure out a way to measure the extent of fatty streaks during the early phases of plaque formation — perhaps through high-resolution coronary ultrasound or MRI — than to wait 40 or 50 years for an answer.

Teach Kids Healthy Living by Example

Rather than enlist our children into a lifelong antioxidant vitamin clinical trial — which is extremely unlikely to happen — we *can* introduce them to healthy lifestyle habits. If we do that when they're kids, it is more likely that healthy living may become a habit lasting a lifetime. We all know habits are easier to form than to break. And we all know that the best way to teach kids is *by example*, not words. (Girls mimic moms; boys mimic dads.) Children tend to follow in their parents' footsteps. If kids see us doing the "right thing," they are likely to mimic our good examples. And if they see us not living healthfully, they are apt to copy that lifestyle, too. We owe it to ourselves to take care of our bodies and we owe it to our children to teach them, *by example*, how to live healthfully — day after day, month after month, year after year. *And if kids learn to limit their cholesterol intake, avoid cigarettes, exercise regularly, and watch their weight, there may be no need for antioxidant vitamins later in life!*

References

1. Guallar E, Hanley DF, Miller ER III. (2005) *Annus horribilis* for vitamin E. *Ann Intern Med* **143**: 143–145.
2. Ness AR. (2001) Commentary: Beyond beta-carotene — antioxidants and cardiovascular disease. *Int J Epidemiol* **30**: 143–145.
3. http://www.fda.gov/food/dietarysupplements/consumerinformation/ucm110417.htm
4. Salganik RI. (2001) The benefits and hazards of antioxidants: controlling apoptosis and other protective mechanisms in cancer patients and the human population. *J Am Coll Nutr* **20** (Suppl): 464S–472S.
5. Tsimakis S. Glass CK, Steinberg D, Witzum JK. (2004) Lipoprotein oxidation, macrophages, immunity and atherosclerosis. Chein KT. Saunders, Philadelphia, pp. 385–413.
6. http://en.wikipedia.org/wiki/Sherman%27s_March_to_the_Sea
7. http://en.wikipedia.org/wiki/Battle_of_Moscow
8. Willett WC, Stampfer MJ. (2001) What vitamins should I take, doctor? *N Engl J Med* **345**: 1819–1824.
9. http://www.mayoclinic.co/health/heart-scan/MY00327

27

MOLECULAR MARKERS, REPORTER GENES AND SUICIDE GENES

Coronary heart disease is the leading cause of death in the United States and the financial burden of its treatment — about US$350 billion (that's 10 zeros after the 35)[a] per year — is "staggering."[1] In light of these imposing figures and the fact that some — though not all — of the stem cell transplantation trials have shown improved cardiac function following such a transplantation after a heart attack (see Table 28.1), the talented, multidisciplinary stem-cell-biology/tissue-regeneration team at Stanford decided to take on the challenge of tracking stem cells after their transplantation into the heart. They reasoned that because "some transplanted stem cells may misbehave [unpredictably] over a lifetime,"[1] it is important to know their whereabouts over a lifetime. So in 2007, the Stanford team devised "molecular markers" to follow the fate of stem cells during life after their transplantation, rather than waiting years to determine their destiny by getting a single snapshot of the transplanted cells under a microscope after an autopsied death. Any naughty stem cells could be identified by the marker and brought back into line.

The marker consisted of a versatile triple-function "reporter gene" that could enter embryonic stem cells to carry out undercover investigation. (A reporter gene is simply a gene attached to another gene being studied to permit its easy detection — through its marker — after its introduction into a stem cell.) A reporter gene, they figured, should allow day-by-day

---

[a] By comparison, the national debt is US10 trillion — that's 12 zeros after the 10 — and climbing, according to the national "Debt Clock."[2]

Fig. 27.1.   A female firefly with a green-glowing flasher.

(*Source*: Reproduced with permission from the publisher.) Reference 5.

Fig. 27.2.   A sea anemone.

(*Source*: Reproduced with permission from the publisher.) Reference 6.

tracking of labeled embryonic stem cells without adversely affecting the cells' ability to differentiate into specialized cells. They designed the reporter gene to also serve as a "suicide gene" (to be discussed soon), thus "providing an extra safety mechanism against cellular misbehavior."[1] To track embryonic stem cells, they employed not one, not two, but three reporter genes to follow (rat) embryonic stem cells after their transplantation into the heart. Once in the heart, one reporter gene bioluminesced,[b] another fluoresced,[c] and a third made the cells visible using positron emission tomography (PET) imaging. [Reporter genes have been used in the past to track tumor metastases (the spread of a tumor to other organs), but not to follow the fate of transplanted stem cells.] The transplanted embryonic stem cells had no place to hide from such a sophisticated surveillance system. The new "triple threat" technique is superior to standard radioisotope imaging (such as a thyroid or lung scan), which uses an isotope with a short half-life limiting its diagnostic potential to one day after cell transplantation. It is also better than magnetic resonance imaging (MRI), which cannot distinguish between living and dead cells because macrophages (the "big eaters") near living cells take up the labeled radioisotope after donor cell death. While the primary purpose of the 2007 report was to demonstrate the utility of its novel three-pronged imaging technique for tracking embryonic stem cells, an ancillary aim was to call attention to the occurrence of teratomas in all mice that had embryonic stem cells injected into their hearts and the elimination of the tumors by an antiviral drug that kills cancer cells (Fig. 27.3).[3,4]

Fluorescent Proteins and Bioluminescence

In the summer of 1961, a young man arrived at a small fishing village on an island in Puget Sound, 85 miles from Seattle, just off the coast of Victoria,

---

[b] Bioluminescence refers to the production of light by a living organism as the result of a chemical reaction. Fireflies (lightning bugs) are common insects in southern U.S. noncoastal and coastal states with this fascinating talent (Fig. 27.1).[5] Sea anemones are small marine animals (Fig. 27.2) armed with a potent toxin that can immobilize larger ocean prey. Their poison features a neurotoxin quite capable of paralyzing predators, and making ocean swimmers feel very uncomfortable as well.
[c] Fluorescence is a form of luminescence. It is characterized by the absorption of ultraviolet (invisible) light and its immediate conversion to visible light.[7,8]

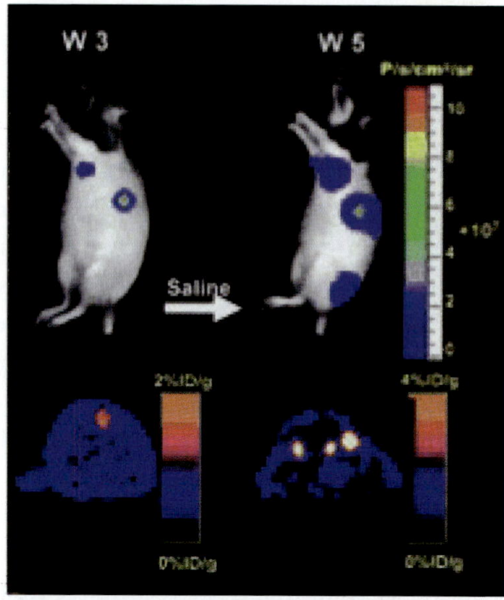

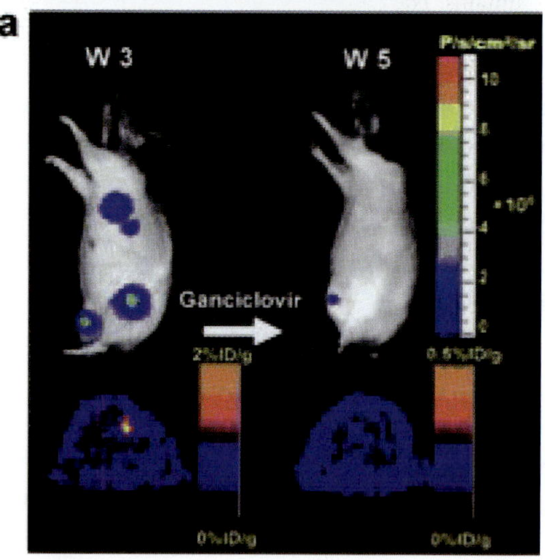

Fig. 27.3. Two mice that each had embryonic stem cells injected directly into their hearts three weeks before the images on the left were obtained (W3). Both developed bioluminescence images (explained below) represented by blue circles detectable by positron emission tomography three weeks later. The images were later shown to be due to teratomas. The upper mouse received saline (salt water) injections beginning at week 3, while the lower mouse received ganciclovir, an antiviral drug that kills cancer cells caused by a virus, during weeks 3–5. The tumors disappeared both by bioluminescence imaging and later postmortem examination.

(*Source*: Reproduced with permission from the publisher.) Reference 1.

Fig. 27.4.   A map of the scenic northwest United States and southeast Canada region. The 172 islands making up the San Juan Archipelagos were formed many centuries ago by the exposed tops of submerged mountains. The Canadian and American islands are colored differently. San Juan Island, the largest, is the commercial and cultural center of the islands. It is colored blue, while the red circle identifies Friday Harbor, a tiny town of just 2000 people on the eastern shore of the island, 85 miles from Seattle and 90 miles from Victoria, British Columbia. It is accessible by passenger ferry — no cars — and seaplane only. On the island, bicycles rule. Always-snow-capped Mount Rainier soars 14,410 feet (4392 m) above sea level, 120 aeronautical miles away.

(*Source*: Modified and reproduced with permission from the publisher.) Reference 9.

British Columbia. This is as far northwest as you can go in the U.S. before either falling into the Pacific Ocean or traveling, by ferry, to Victoria (Fig. 27.4). From the village, Mount Rainier towers 60 aeronautical miles away in the background (Fig. 27.5). He brought with him his wife and two small children. But this was not a family vacation and fishing wasn't his reason for being there, 3000 miles from home in Princeton, New Jersey. He didn't even bring a fishing pole. It really didn't matter if he caught any fish. . .at least not any edible ones. He had his eye on jellyfish. . .lots of them. . .and a net on the tip of a long pole caught a lot more jellyfish than a hook, line, and sinker. Day after day, he tossed his catch into buckets on the pier (Fig. 27.6) and saved each one. Then, at summer's end, he carted his booty back to Princeton for analysis.

Fig. 27.5.   Mount Rainier as reflected in a lake at dawn.

(*Source*: Reproduced with permission from the publisher.) Reference 10.

Fig. 27.6.   The harbor — called Friday Harbor — near Victoria, British Columbia, where Osamu Shimomura gathered nearly one million jellyfish in the 1960s. They were then used to isolate and purify the compound responsible for their brilliant fluorescence.

(*Source*: Modified and reproduced with permission from the publisher.) Reference 11.

Japanese-born Osamu Shimomura came to Princeton University on a Fulbright Scholarship to do "research [on] the luminescent properties of marine organisms."[12] What makes them glow brightly, not in the water but when exposed to ultraviolet light in a laboratory? It didn't take long to get

Fig. 27.7.   A number of the jellyfish with the technical name of *Aequorea victoria* which glow green under ultraviolet light. When the light strikes a protein in the jellyfish, the protein re-emits — as opposed to reflecting — it at a longer wavelength. The jellyfish now glow green.

(*Source*: Reproduced with permission from the publisher.) Reference 11.

an answer. In 1962, the then 33-year-old scientist published his first of many papers on the bioluminescent (light-emitting) jellyfish that glowed bright green when put under ultraviolet (UV) light[d] (Fig. 27.7).[13] The protein in jellyfish that made them glow green "absorbed UV light and emitted a green glow, which unlike. . .the light-emitting chemicals in fireflies, did not require the addition of any chemicals."[14] (For the few, if any, physical chemists reading this book, the "green protein" in the jellyfish "converted blue light to green light through an energy transfer process that did not involve radiation."[5] And that's a watered-down explanation of the process that touts activated electrons and electron transfer.) The newly discovered compound was at first called just that — "green protein." In 1969, two biologists at Harvard referred to it as "green fluorescent protein."[15] The name stuck.[e]

---

[d] Shimomura isolated the "juice" of jellyfish by "squeezing the excised bioluminescent tissues of the jellyfish through a cotton bag."[13,17] Nothing fancy, yet highly effective.

[e] Green fluorescent protein in jellyfish evolved over many millennia to function most efficiently at the cool temperatures of the ocean. When the protein was found useful in human imaging techniques such as PET and MRI, biological engineers adapted green fluorescent protein to function at the physiological temperature of 36.8 degrees Celsius (98.2 degrees Fahrenheit) in humans.[17]

Coral and Coral Reefs

Far beneath the sea's surface where Shimomura skimmed thousands and thousands of jellyfish[f] that glowed green (under UV light) lie coral reefs on the sea's floor, or at least on the floor of seas lying between 30° north and 30° south of the Equator (Fig. 27.8). There warm, tropical waters bathe coral reefs that corals call home. The stony-hard reefs are where soft-bodied corals spend their lives, lasting typically 2–3 years.[19] And, yes, corals are marine animals — polyps — that live in inanimate coral reefs made of limestone. Living corals — generally only 1–3 inches in length — long ago figured out a way to protect themselves from larger ocean predators by forming an external skeleton — an "exoskeleton" — that acts as a shell. The armor is durable and is a characteristic that links corals to other marine mollusks, like clams, snails, or oysters. And, as a backup defense, coral polyps are armed with tentacles around a central "mouth," tentacles equipped with cells that sting and toxins capable of immobilizing or killing unfriendly marine life.[21] If such a dual defense system isn't a clear sign that corals are animals, proof positive that they rightfully belong to the animal kingdom is that they have a sex life — a

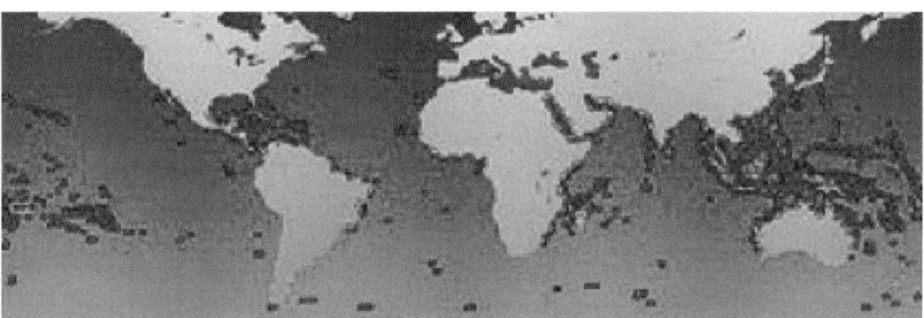

Fig. 27.8. The famous coral reefs of the world include the Great Barrier Reef, the largest coral reef system in the world, near Queensland, Australia; the Belize Barrier Reef, the second-largest coral reef, stretching from southern Quintano Roo, Mexico, and all along the coast of Belize, down to the Bay Islands of Honduras; and the Andros, Bahamas Barrier Reef, the third-largest coral reef, along the east coast of Andros Island between Andros and Nassau.

(*Source*: Reproduced with permission from the publisher.) Reference 18.

---

[f] The jellyfish in Puget Sound where Shimomura caught his many, many jellyfish have a diameter of five to ten centimeters (two to four inches) during their brief life of six months, but they reproduce prolifically during that short time.[20]

rather wide-ranging sex life, to boot. They like to mate at night, around the time of a full moon. They can reproduce sexually or asexually and can choose either reproductive mode during their lifetime. They can be male or female, or both (a hermaphrodite), depending upon their mood and the ambience.[g] They can even change their sex along the way![23]

Once a coral polyp's busy reproductive life[h] is over, its exoskeleton falls to the ocean floor, where many (inanimate) exoskeletons, over time, form large — sometimes massive — reefs that support the animate coral. Coral reefs are simply tropical water graveyards for coral skeletons. They may well be the most beautiful graveyards in the world. Corals themselves are tiny and live in almost-as-tiny, cup-shaped structures within the reefs. Individual corals take the shape of a polyp but are far more engaging and colorful than their human colonic counterparts (Fig. 27.9). Ironically, the beauty of corals is linked to wretched algae — tiny, one-celled plants responsible for unwanted aquarium slime.[i,25] Corals — animals — live in a symbiotic relationship with (saltwater) algae — plants. The corals "harbor [the] single-celled algae [within] their bodies," providing them with needed "shelter and minerals" and carbon dioxide.[28] The algae, for their part, give off oxygen and synthesize food from sunlight — accounting for the proximity of coral reefs to the sea's surface where it is sunniest[28] — for the corals, as "rent." This mutually nurturing partnership works well.

Akin to jellyfish, to which they are related, corals are translucent and have a soft body. But the similarities don't stop there. Like jellyfish, reef corals are "brightly colored and intensely fluorescent under UV light,"[29] making them popular among scuba divers and snorkelers in the Caribbean and South

---

[g] Many flowers have both male and female reproductive organs, perhaps shooting down the notion that corals' bisexuality places them in the animal kingdom.

[h] At a July 2008 conclave of coral scientists in Fort Lauderdale, it was pointed out that corals "spawn almost around the world at the same time in what. . .may be the world's biggest [orgy]. But. . .global warming could leave corals confused. . .and start spawning all over the place at different times, reducing fertilization success and leading to less effective replenishment of coral populations. If the corals fail to regenerate properly. . .they become less able to withstand human impacts such as overfishing and [water] pollution or climatic impacts such as hurricanes."[24]

[i] Not everyone would label algae as "wretched." Some see the aquatic organisms as "an everyday miracle" as they apply algae-based beauty creams to their skin and shampoo their hair with a similar product.[26] Others add "a sachet to (their) bath (water) . . . to help relieve aches . . . (and) relax . . .."[27]

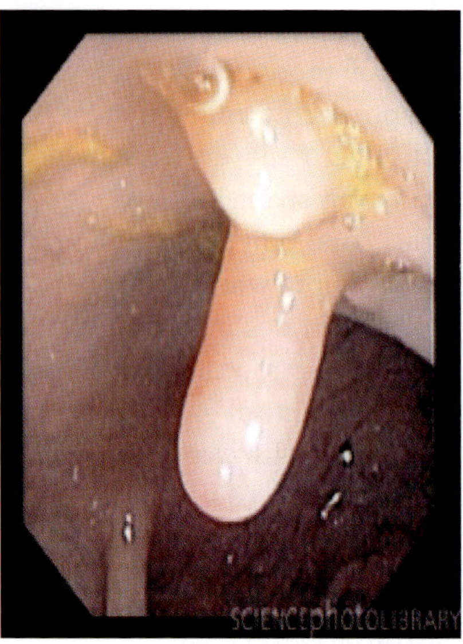

Fig. 27.9.   A (pillar) coral polyp and a colonic polyp in a stalagmite and stalactite formation, respectively.

(*Source*: Reproduced with permission from the publishers.) References 21 and 22.

Pacific. Their fluorescence acts as a "sunscreen" shielding these attractive creatures from harmful UV light able to penetrate into shallow, tropical waters.[28] Their bright colors — imparted by natural pigments — vary from red, orange, and yellow hues to green, blue, and purple shades (Figs. 27.10 and 27.11).

At the electron-biochemical level, when fluorescent, electrons within the coral's pigment are "excited," leading to colors in the blue range of the color spectrum. Later, as electron activity dies down, the colors change from blues to greens, then yellows and, lastly, reds. During this process, the coral pigment converts solar radiation into energy "wavelengths seen in the colors of the rainbow and by absorbing higher energy — UV and blue light — and emitting it in green to red fluorescent colors, they transform damaging light into less. . .damaging wavelengths,"[28] endearing corals to us.

In addition to blocking harmful UV light, bioluminescence may serve as a physiologic "burglar alarm."[17,32] Approaching predators may be startled by the corals' "sudden flash of bright light," distracting or even scaring off other marine organisms intent on harm, and allowing the corals to escape.

Pea Crab in Cup Coral Polyp **Coral Polyps** **Marine Red Coral Polyp**

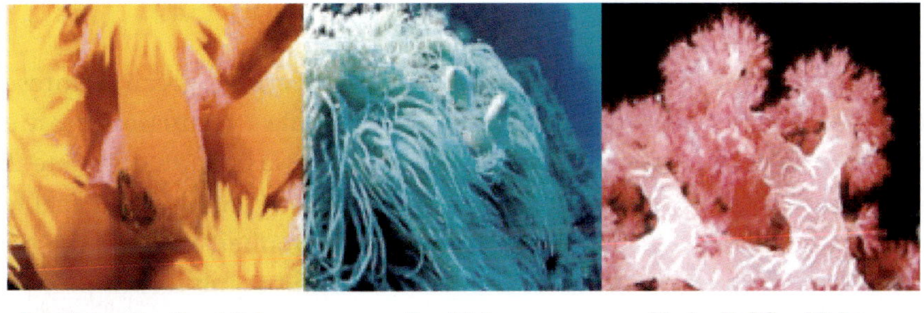

Fig. 27.10. A sampling of coral polyps.

(*Source*: Reproduced with permission from the publisher.) Reference 30.

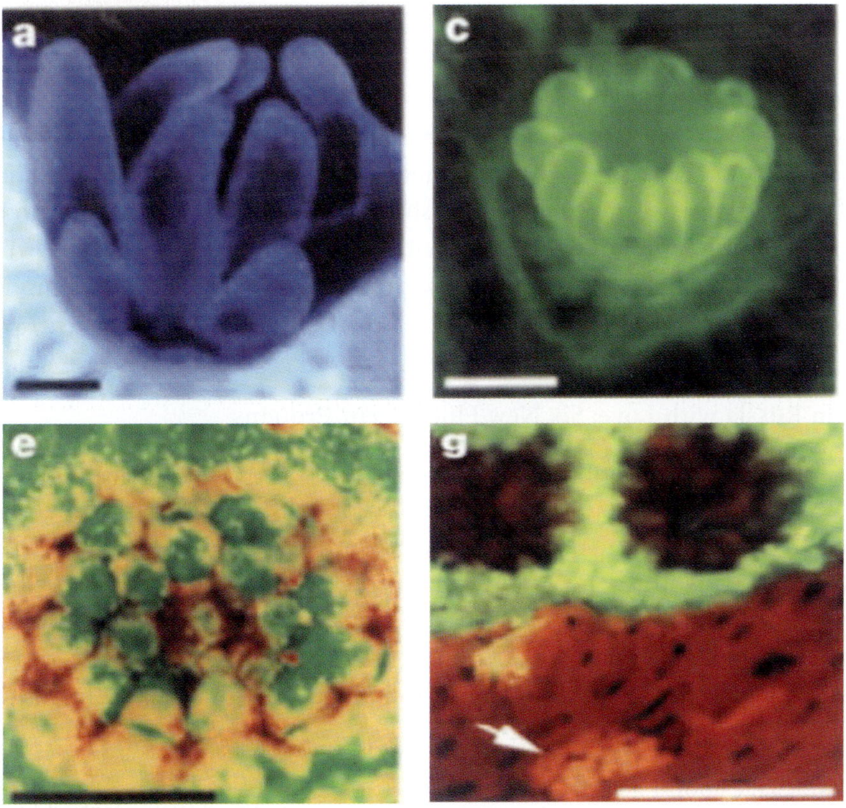

Fig. 27.11. Coral pigments may be blue, green, yellow, or red alone or in multiple combinations.

(*Source*: Reproduced with permission from the publisher.) Reference 31.

Gene Shuffling

In a marriage of nature and biological engineering, green fluorescent protein — which had evolved over many millennia to function most efficiently at the cool temperatures of the ocean — was adapted by late 20th century biological (protein) engineers to function at the physiological temperature of 37.0 degrees Celsius (98.2 degrees Fahrenheit[j]) so that the pleasantly bright compound could be used in human imaging techniques (such as PET and MRI).[35] That goal was reached in 1996 by chemists working at the Affymax Research Institute in Palo Alto, through the use of "DNA shuffling." This new technique involves the shuffling of genes "within a family of related genes," enabling the creation of a being with mixed genes — a chimera.[34] The methodology is a bit daunting but the end result is worth the effort. It leads to the rapid propagation of "beneficial mutations" and, as such, is now a popular way to "rapidly increase [the size of] a DNA library."[34] The "new and improved" green fluorescent protein sends a "stronger fluorescence signal" — 42 times stronger than the original, natural green fluorescent protein.[35] And, in a tip of their hat to Mother Nature, the Silicon Valley biochemists said their results showed how "molecular evolution can solve a complex practical problem without needing to first identify which process is limiting."[35] In other words, sometimes Nature still bests modern biotechnology.

From Russia with Love: Six New Fluorescent Proteins

On top of gene shuffling, marine biologists fascinated by green fluorescent protein began searching for more fluorescent colors in other marine organisms. In October 1999, a team of Russian scientists hit the jackpot. They found not one or two, but six new fluorescent proteins in corals living in coral reefs in the warm Indian Ocean and western Pacific waters bathing the archipelago between India and Australia.[37] And just as impressive as the discoveries was the Russian scientists' out-of-the-box thinking leading to the findings.[k] They

---

[j]For years, the normal human body temperature was said to be 98.6 degrees Fahrenheit. In April 2008, the *Harvard Health Letter* announced, after reviewing previously published data, that a value of 98.2 degrees Fahrenheit should now be considered normal.[33]

[k]An amusing vignette related to the discovery of the six new fluorescent proteins is told by Robert Campbell, now a chemistry professor in Canada, then a graduate student in the

reasoned that "green fluorescent protein–like proteins [may not] necessarily be linked to bioluminescence." They speculated that "bioluminescence [could have] evolved independently and relatively recently in different [evolutionary] groups" of organisms. And, "if so, such ancestral domains could still be present in non-bioluminescent organisms."[39] They were right.

In an accompanying editorial, the author praised the researchers from the Russian Academy of Science for even considering the possibility that green fluorescent protein "could have cousins elsewhere in nature."[38] And, the

---

laboratory of Yulii Labas, an evolutionary biologist at the Russian Academy of Sciences in Moscow. Campbell writes that by the mid-1990s "coral and sea anemone fluorescence [had] been known for decades, and the need for more fluorescent proteins besides GFP [green fluorescent protein] was widely recognized by the biotech community. So why did it take so long to make a connection? The reason was in the mindset: everyone (including us in Lukyanov's lab in the beginning of the project) assumed that FP, [fluorescent proteins] just like…GFP… must be found in bioluminescent animals, and they should be green. It took the intervention of Yulii A. Labas…to break out of the box. Being interested in the evolution of bioluminescence for most of his career, Labas realized that GFPs, like all other components of bioluminescent systems, must not have evolved from scratch for the purpose of bioluminescence, but must have been recruited from other places and functions. So it made perfect sense for him to look for GFPs in non-luminous, but fluorescent animals — such as corals and sea anemones. Now, Labas was arguably the closest living approximation of a 'mad scientist' — old, hunched, grey-bearded, chain-smoking, completely fascinated with science, and shooting all sorts of ideas (ranging from improbable to completely crazy) at anybody (literally, anybody) who would listen. And I was a young and very snobby graduate student. So when one day Labas called me and said that a friend of his — a professional aquarium keeper, Andrei Romanko — had some amazing fluorescent creatures in his tank in his tiny flat on the outskirts of Moscow, and I should come to check them out, my feelings were rather mixed. Then, imagine us staring into the tank, with Labas stabbing his nicotine-stained finger into the glass at the bright green tentacle tips of a sea anemone (that was *Anemonia majano*) shouting, 'See?? See??? GFP!!!' GFP my ass, thought I (I remember this exact thought with an exceptional clarity). It is just bloody too bright. But I took advantage of Andrei Romanko's generosity and snipped off a few tentacles, along with a yellowish piece of *Zoanthus* (why am I doing this? GFP must be green for heaven's sake! But I cannot disappoint the old looney), and a *Clavularia* polyp which was also unnaturally brightly fluorescent. Then, a couple of weeks later, Anemonia cDNA library yielded the first snippet of an obvious GFP-like sequence. The world became a whirlpool and everybody got so excited. Then Andrei Romanko called me from work, saying, 'Look, I just got a live rock that I have to deliver to a customer tonight, with three tiny mushroom anemones on it, and they are red! If you come right now, we can steal one of them!' Needless to say, I was there as soon as Moscow's subway would carry me. That's how we got the sample that yielded DsRed. By the way, since I took only a part of the mushroom anemone, the rest happily regenerated and until this day lives in Romanko's tank."[17]

editorialist pointed out, the marine biologists used methods to "isolate novel genes" from corals living in some of the world's oldest coral reefs in a "non-destructive" fashion, in deference to a number of environmental protection groups. In the well-balanced commentary addressing the need for the biotechnology industry and various environmental protection groups to get along with each other for the benefit of humanity, he or she said, "alarmingly…almost 60% of the world's reefs are under assault due to human activity, including overexploitation, coastal development, and runoff of pesticides and sediment from intensive inland farming activities." After citing a report from the World Resources Institute (based in Washington, D.C.) estimating the extent of the threat of reef destruction, the (unidentified) author emphasized that [on the other hand] "a significant part of the search for new drugs against cancer and antibiotic-resistant bacteria now focuses on compounds from marine organisms such as these, [corals]." He or she ended up saying that it is crucial for reef preservation to establish a "partnership between the proponents of biotechnology and [reef] conservationists" and the time to form this alliance is "now".[38]

A Green Fluorescent Lantern

In 1988, Martin Chalfie, a scientist at Columbia University in Manhattan, first learned about green fluorescent protein at a conference being held at Columbia.[39] He realized quickly that if he were to put the green protein gene next to — not merely nearby — any other gene in a transparent roundworm, which was the focus of his research at the time, the worm would glow green.[l,39] (A classic example of one scientist with a prepared — yet open — mind building on another's observations to the benefit of both and to science.) More specifically, he realized the cells in the worm that actually were making the fluorescent protein would glow green. They would then function as a "lantern," allowing him to peer inside the worm. The only problem was that the gene that makes the protein — one gene, one protein — had not yet been found. [See footnote c of Chap. 9 regarding the "one-gene-one-protein (enzyme) hypothesis."] In 1992, Douglas Prasher, who also worked at the Woods Oceanic Institute in Massachusetts (like Osamu Shimomura), came to

---

[l]The green gene would be controlled by the same genetic "switch" as the protein. That way, whenever the protein was made, GFP would be made, too.

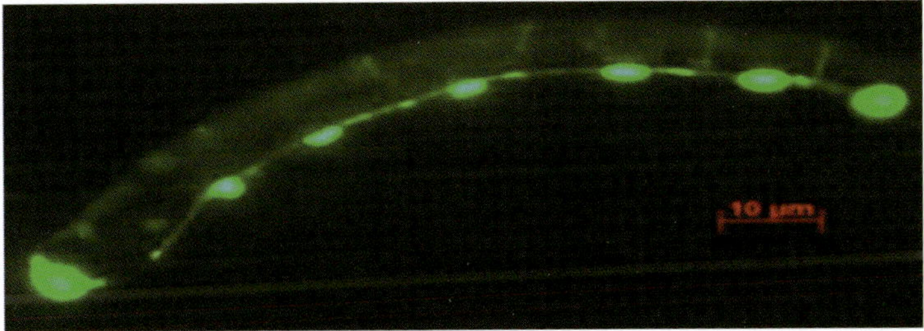

Fig. 27.12. A photograph of a roundworm labeled with green fluorescent protein. The six bright green dots are neurons (also called nerve cells), containing green fluorescent protein. Long, thin axons wrapped in a myelin sheath extend from one neuron to the next, forming the cable-like structure shown above. Neurons are highly specialized cells in the brain, spinal cord, and all nerves that transmit electrical impulses to muscles. The electrical impulses themselves are transmitted by chemicals, particularly norepinephrine, dopamine and acetylcholine. The human brain has about 100 billion neurons. In contrast, the roundworm has just 302 neurons, making it an ideal organism for studying nerve impulses. A roundworm neuron is about 1 mm long, the calibration bar being 10 μm or *x* mm.

(*Source*: Reproduced with permission from the publisher.) Reference 41.

the rescue. He found the "green gene"[40] and passed it on to Chalfie. In 1994, Chalfie put the protein into six cells of the *C. elegans* roundworm. Later, when he peeked at those six cells under UV light, they "shined green, revealing their location" (Fig. 27.12), allowing him to visualize the worm's inner machinery.[39] Without the fluorescent lighting, the worm would look like the one in (Fig. 27.13).

Then, three years later, Roger Tsien (pronounced "Chen"), a chemist and Howard Hughes Medical investigator at the University of California, San Diego, also contacted Prasher. He asked Prasher if he could use his "green gene" to develop other-than-green mutants.[39] Prasher said yes, and Tsein went to work. Within five years he created red and yellow strains.[43] Since then, he has "expanded the color palette of fluorescent proteins — adding a magnificent array of hues, including cherry, strawberry, tangerine, tomato, orange, banana and honeydew" (Fig. 27.14), according to the Howard Hughes Medical Institute's website.[44]

Remember those names — Shimomura, Chalfie and Tsien. They'll come up again.

Fig. 27.13. A roundworm *sans* green fluorescent protein. Even though its skin is transparent, it's far easier to see its innards with the fluorescent protein than without it.

(*Source*: Reproduced with permission from the publisher.) Reference 42.

Fig. 27.14. Seven new fluorescent proteins viewed in visible light (panel c) and in ultraviolet light (panel d). The colors from left to right are: honeydew, banana, orange, tangerine, tomato, strawberry, and cherry. They look delicious!

(*Source*: Reproduced with permission of the publisher.) Reference 45.

References

1. Swijneburg RJ, van der Bogt KEF, Sheikh AY, Cao F, Wu JC. (2007) Clinical hurdles for the transplantation of cardiomyocytes derived from human embryonic stem cells: role of nuclear imaging. *Curr Opin Biotechnol* **18**: 38–45.
2. http://www.brillig.com/debt_clock
3. Cao F, Lin S, Xie X, Ray P, Patel M, Zhang X, Drukker M, Dylla SJ, Connolly AJ, Chen X, Weissman IL, Gambir SS, Wu JC. (2006) *In vivo* visualization of embryonic stem cell survival, proliferation, and migration after cardiac delivery. *Circulation* **113**: 1005–1014.
4. Rubsam LZ, Davidson BL, Shewach DS. (1998) Superior cytotoxicity with ganciclovir compared with acyclovir and 1-B-D-arabinofuranosyl thymine in herpes simplex virus-thymidine kinase–expressing cells: a novel paradigm for cell killing. *Canc Res* **58**: 3873–3882.
5. http://en.wikipedia.org/wiki/Bioluminescenc
6. http://en.wikipedia.org/wiki/Sea_anemone
7. http://en.wikipedia.org/wiki/Fluorescence
8. www.universityscience.ie/pages/glossary.php
9. http://www.andersenhouse.com/areamap.html
10. http://www.terragalleria.com/parks/np-mount-rainier.html Accessed June 21, 2010.
11. Miyawaki A. (2008) Green fluorescent protein glows gold. *Cell* **135**: 987–990.
12. http://www3.interscience.wiley.com/egi-bin/fulltext/118735599/HTML_START
13. Shimomura O, Johnson FH, Saiga Y. (1962) Extraction, purification and properties of *Aequorin*, a bioluminescent protein from luminous hydromedusan *Aequorea*. *J Cell Comp Physiol* **59**: 223–239.
14. Coleman R. (2008) Jellyfish fluorescent proteins. Nobel Prizes and pioneers in biochemistry. *Acta Histochemica* **110**.
15. Hastings JW, Morin JG. (1969) Comparative biochemistry of calcium-activated photoproteins from the ctenophere *Mnemiopsis* and the coelenterates *Aequorea, Obelia, Pelagia and Renilla. Biol Bull* **137**: 402.
16. Reference needed.
17. http://www.scholarpedia.org/article/Fluorescent_proteins
18. http://en.wikipedia.org/wiki/Coral_reef
19. http://www.floridamarine.org/support/view_faqs.asp?id=10
20. http://faculty.washington.edu/cemills/Aequorea.html
21. http://en.wikipedia.org/wiki/Coralhttp://en.wikipedia.org/wiki/Coral
22. http://stockindexonline.com/index.php?q=keyword_search&edtKeyword=colon%20polyps&rdbKeyword=1§ion=keywordsearch
23. http://en.wikipedia.org/wiki/Coral_reef
24. http://earthdive.com/site/news/newsdetail.asp?id=2656
25. http://freshaquarium.about.com/cs/maintenance1/p/algaeslime.htm

26. http://www.bohemia-style.com/Algae-Seaweed-Cosmetics-orderby0-p-1-c-49.html

27. http://www.amazon.com/Thalgo-Micronized-Marine-Algae/dp/B0002VHIEI

28. http://wildshores.blogspot.com/2008/11/why-do-some-corals-glow-in-dark.html

29. Cox G, Salih A. (2007) Fluorescence lifetime imaging of coral fluorescent proteins. *Microsc Res Tech* **70**: 243–251.

30. Salih A, Larkum A, Cox G, Kuhl M, Hoegh-Guldberg O. (2000) Fluorescent pigments in corals are photoprotective. *Nature* **408**: 850–853.

31. http://www.fotosearch.com/photos-images/coral-polyps.html

32. Abrahams MV, Townsend LD. (1993) Bioluminescence in dinoflagellates: a test of the burglar alarm hypothesis. *Ecology* **74**: 258–260.

33. https://www.health.harvard.edu/press_releases/normal_body_temperature.htm

34. Reference needed.

35. http://en.wikipedia.org/wiki/DNA_shuffling

36. Crameri A, Whitehorn EA, Tate E, Stemmer WP. (1996) Improved green fluorescent protein by molecular evolution using DNA shuffling. *Nat Biotechnol* **14**: 315–319.

37. Matz MV, Fradkov AF, Labas YA, Savitsky AP, Zaraisky AG, Markelov ML, Lukanov SA. (1999) Fluorescent proteins from nonbioluminescent *Anthozoa* species. *Nat Biotehnol* **17**: 969–973.

38. A common green ground. *Nat Biotechnol* **17**: 933.

39. http://www.nytimes.com/2008/10/09/science/09nobel.html

40. Prasher DC, Eckenrode VK, Ward WW, Prendergast FG, Cromier MJ. (1992) Primary structure of the *Aequorea victoria* green-fluorescent protein. *Gene* **111**: 228–233.

41. http://pets.webshots.com/photo/104084275702926014tqizAO

42. http://image02.webshots.com/2/4/27/57/40842757tqizAO_ph.jpg

43. Heikal AA, Hess ST, Baird GS, Tsien RY, and Webb WW. (2000) Molecular spectroscopy and dynamics of intrinsically fluorescent proteins: coral red (DsRed) and yellow (citrine). *Proc Natl Acad Sci* **97**: 11996–12001.

44. http://www.hhmi.org.researcgh/investigators/tsein_bio

45. Shaner NC, Campbell RE, Steinbach PA, Glepmans BNE, Palmer A, Tsein RY. (2004) Improved monomeric red, orange and yellow fluorescent proteins derived from *Discoma sp.* red fluorescent protein. *Nat Biotechnol* **22**: 1567–1572.

28

STEM CELLS FOR THE HEART: HYPE OR HOPE?

A basic question regarding the treatment of coronary heart disease with stem cells is whether the heart has stem cells and, if it does, can the heart regenerate cardiac tissue after birth?[1] Certainly, the normal, healthy heart can get bigger (thicker). By so doing, the "hypertrophied" (in medical parlance) vital pump becomes more efficient, ejecting more blood with each heartbeat at a slower heart rate. For example, aerobic exercise — such as serious running, bicycling, and swimming — can make the heart stronger by causing individual cardiac cells to contract more powerfully, at a slower heart rate. (A resting heart rate of 30–40 beats per minute is frequently recorded in well-conditioned athletes. The effects of regular, rigorous training appear quickly — within 3–6 months — and vanish just as fast upon ending intense, aerobic exercise.[2]) Exercise does so by making cardiac muscle cells — called cardiomyocytes — bigger, just as weightlifting makes the biceps bigger, but it does not make cardiomyocytes proliferate. Or, put another way, exercise strengthens existing heart muscle, but it does not make cardiomyocytes multiply. However, strengthening heart muscle through exercise does not imply that exhausted or dead heart muscle can be replaced by healthy tissue following, say, a heart attack.

Yet, a number of studies since 1995 have shown, contrary to conventional teaching, that heart muscle cells *can* be made to proliferate, as opposed to individual cardiac muscle cells being made stronger through exercise or by heart-strengthening drugs such as digitalis.[2] These investigations have established that the heart — like the blood, liver, skin, and even brain — can regenerate new, functional muscle cells, leading to a life of 100 years or more, if death from another cause doesn't cut life shorter.[1]

Stem Cell Niches

Nestled in sheltered spaces known as "niches," stem cells are sequestered and nurtured until needed to repair damaged tissue. Niches are small, three-dimensional spaces where adult stem cells are stored during life *and* a place where the dormant cells are influenced by local and remote signals that maintain the precious cells until they are released to repair injured or dying tissue. In essence, a niche is a safe haven. Adult stem cells need to be treated respectfully and gently given their rarity in any given organ. As pointed out in Chap. 14, even in the bone marrow, where adult stem cells are most abundant, only 1 in 100,000 to 1 in 10 billion bone marrow cells is an adult stem cell. Couple their rarity with their inability to be seen using standard histological methods, and it's easy to understand why stem cell biologists from the University of Michigan and the Carnegie Institution in Baltimore said, "Accurately identifying stem cells *in vivo* [in a living person] remains the biggest obstacle to progress in understanding stem cell biology."[3]

Each niche houses one or two stem cells and, more importantly, regulates their self-renewal and the release of their progeny during life.[4] Asymmetric cell division is the key to a mother stem cell's unique ability to produce two daughter cells with different cell fates even though they live in the same environment. One of the daughter cells is identical to the mother cell and has the capacity for unlimited self-renewal to ensure propagation of the cell. The other daughter cell is able to differentiate into each and every one of the 210 specialized cell types present in the human body.[5,6] By housing stem cells, a niche lives up to the *Merriam–Webster* definition of "niche" as "a place… suitable for the capabilities or merits of a person or qualities of a thing," not excluding a stem cell as a "thing."

Regeneration of the Heart by Stem Cells

Before 2001, basic scientists and physicians believed that were all dealt the number of cardiac stem cells we would play the game of life with. We thought that it was a set number and that we could not discard (lose) cardiac stem cells and draw (replace) the lost cells from somewhere else deep in the heart. It looks like we may have been wrong, according to Piero Anversa and his colleagues at New York Medical College (Valhalla, NY) and Donald Orlic from the N.I.H. (Bethesda, MD), who first proposed and provided evidence for regeneration of lost heart muscle cells from "resident" stem cells lurking in "niches" somewhere deep in the heart. They said, No, the hand we're dealt at birth is *not* one we are destined to play until death with regard to the number of heart muscle cells. Cardiac regeneration is possible.[7]

The myocardial regeneration door officially opened in April 2001. At that time, Anversa and Orlic first reported in a landmark study, published in the highly respected scientific journal *Nature*, regeneration of "infarcted [dead] myocardium" by bone marrow cells injected into the viable edge of a myocardial infarct [dead or nearly dead heart muscle, the result of a heart attack] 3–5 h after tying off a coronary artery in 12 out of 30 mice (40%).[7] Inability to restore cardiac function in the other 60% was attributed to the technical difficulties encountered while trying to inject bone marrow cells into the very thin (1 mm) wall of a tiny mouse heart beating 600 times a minute. (The normal human heart, in comparison, is 10–12 mm thick and beats 60–90 times a minute, at rest.) When sacrificed nine days after transplanting the bone marrow cells into the heart, "newly formed myocardium occupied 68% of the infarcted portion of the ventricle." In mice with such a partially repaired heart, cardiac function measured just before sacrifice was 30%–40% better than in untreated mice that also had a coronary artery tied off leading to an acute myocardial infarction — a heart attack.

In a commentary by **Mark** Sussman from Cincinnati's Children's Hospital and Research Foundation in the same issue of the journal *Nature*, the author pointed out that "replacing...dead heart cells with living ones would be an attractive alternative [to replacing the whole heart through cardiac transplantation, in man], but has been stymied by biological, technical and ethical issues. Many of these problems [would be] circumvented by [using] a specific population of cells derived from bone marrow...to produce functional cardiac cells in damaged hearts."[8] Just how bone marrow stem cells may repair an ailing heart is depicted in Fig. 28.1.

Using bone marrow cells to repair the damaged heart makes sense because bone marrow contains a large number of stem cells (while the heart has very few) *and* they are able to transdifferentiate (see Chap. 14) into healthy heart muscle cells when put into the damaged area of the heart shortly after a heart attack, or so it seemed in 2001. Also, bone marrow cells can be (relatively) easily obtained by a biopsy of the sternum or hip bone. And — importantly, in man — injecting s person's bone marrow cells into his or her own heart eliminates the concern about rejection, likely to occur if another person's bone marrow is used.

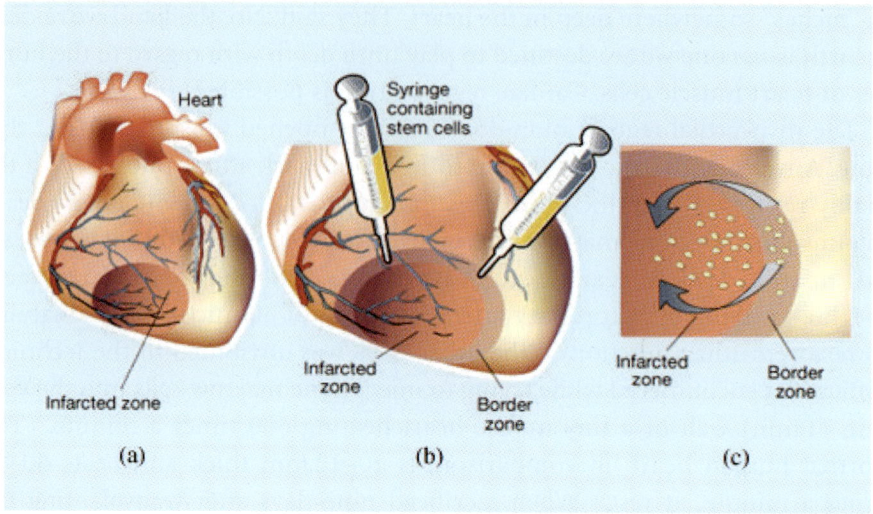

Fig. 28.1.   Panel a shows a heart soon after a heart attack, with its damaged area (infarcted zone) colored tan. Panel b shows two syringes holding bone marrow stem cells with their needles in the adjacent (viable) border zone, colored light purple. Panel c shows the stem cells migrating into the damaged region, where they form new heart muscle cells.

(*Source*: Reproduced with permission from the publisher.) Reference 8.

The Counterpunch

On April 8, 2004, three years following publication of Anversa's paper claiming to show regeneration of 68% of the dead heart muscle in mice, two new reports from different centers appeared in the same journal that published the original landmark paper. Neither study came close to corroborating the findings of the initial study from New York Medical College.

A collaborative investigation between Charles Murry at the University of Washington and Loren Field at Indiana University and their respective colleagues was the basis of one of the two new reports. These two groups took "considerable care to reproduce the methods for stem cell isolation and transplantation [in mice] used by Orlic and colleagues," and despite doing so were unable to replicate the results of Orlic and Anversa. Murry and Field wrote that their data "suggest that direct injection of [bone marrow cells] into the mouse heart does not result in *de novo* cardiomyogenic [giving rise to cardiac muscle] events or tissue regeneration….Indeed, not a single cardiomyogenic event was detected in the 145 [bone marrow stem cell] transplants that were analyzed. These results contrast with the work of Orlic *et al.*, who reported extensive cardiac regeneration after direct injection of [bone marrow] stem cells into infarcts."[9] From regeneration of 68.8% of dead heart muscle to 0% regeneration in three years!

In the second new report in the April 2004 issue of *Nature*, Irving Weissman and his colleagues at Stanford stated that in contrast to Orlic and Anversa "we demonstrate…that bone marrow cells do not differentiate into cardiac muscle…when injected directly into ischaemic[a] myocardium."[10] They went on to add, "Our data are of particular importance given the fact that a variety of groups worldwide have initiated clinical trials of bone marrow transplantation into ischaemic myocardium. Without additional preclinical experimental data, these studies are premature and may in fact place a group of sick patients at risk. At the very least, many more preclinical experimental data should be collected before such phenomena can be fully understood or clinically exploited."[10]

An editorial and a commentary accompanied the two new reports in this power-packed issue of *Nature*. The editorialist lamented the "two reports of failures to replicate the findings" of Orlic and Anversa three years earlier and suggested that the groups "exchange samples [of their heart tissues] to try to understand why their results are at odds."[11] In a commentary by Kenneth Chien from the Institute of Molecular Medicine in La Jolla, he said, "The new reports raise serious concerns regarding the feasibility of using stem cells derived from the bone marrow to drive cardiac regeneration. We should be wary of prematurely pushing laboratory research into clinical practice."[12] Or, restated, don't rush into clinical trials hoping to regenerate heart muscle from bone marrow — as exciting as that possibility sounds.

---

[a]"Ischemic" is an alternative spelling of "ischaemic." The former is used far more often in the U.S., whereas the latter prevails in the U.K. Ischemia is caused by a local diminution of blood supply to heart muscle due to a narrowed — but not completely occluded — coronary artery. Ischemic heart muscle has at least a chance of remaining alive if blood flow can promptly be restored by an angioplasty, a stent, or cardiac bypass surgery. Ischemia is less serious than infarction, where there is no viable heart muscle beyond the completely occluded coronary artery. An acute myocardial infarction (a heart attack) is due to a fresh blood clot forming at the site of a previously narrowed — but still open — coronary artery. Because of the freshness the of blood clot, a clot-dissolving medication has the ability to dissolve it *if* the medication is given within the first six hours after the beginning of a heart attack.

The Performance of a Clinical Trial

A clinical trial is a carefully designed investigation to assess the efficacy of a new medical treatment in a group of people.[b] The treatment may be a new drug, a new device, or a new use of an established treatment. In the U.S., clinical trials are regulated by the F.D.A., while in Canada and Europe, Health Canada and the European Agency for the Evaluation of Medical Products (EMEA) oversee trials of new drugs and devices, respectively.[13]

A clinical trial is conducted in three distinct phases (Fig. 28.2). In phase I, a small group of healthy, paid volunteers is given the new treatment to evaluate its *safety* and possible side effects. In phase II, a larger group of 30–300 people is given the new treatment to assess its *efficacy*.[c] A phase II trial is ran-

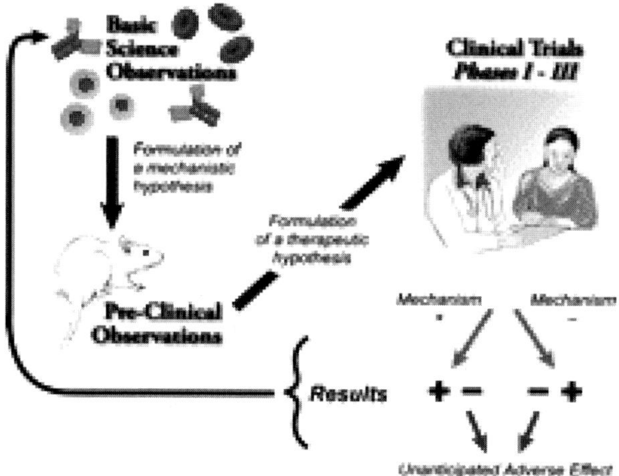

Fig. 28.2. This figure depicts the interrelationship of basic science observations and clinical trials. Observations by basic scientists in the laboratory lead to studies on animals (often mice), which in turn lead to clinical trials by physicians. The trials are performed in three phases, with phases I and II being to establish the feasibility and safety of a new treatment and phase III being large-scale testing in a randomized, double-blind, placebo-controlled manner. Phase III trials are usually done at multiple centers, coordinated by a single center.

(*Source*: Reproduced with permission from the publisher.) Reference 17.

---

[b] Some say, whimsically, the definition of a "clinical trial" is a "malpractice suit."

[c] Establishing the "safety" and "efficacy" of a drug and what constitutes "substantial evidence" of that claim is a requirement of state and federal law. "Substantial evidence" is defined as "evidence consisting of effectiveness of the drug involved...[allowing] experts qualified by scientific training and experience...[to conclude] that the drug will have the effect it purports or is represented to have...." The statutory "substantial evidence" requirement is "a rigorous one."[14]

domized, i.e. some people receive the new drug or device, and others get a placebo or the standard treatment for the condition at the time. A phase II trial may or may not be blinded and includes far fewer patients than a phase III trial. A phase III trial typically enrolls 300–3000 or more patients to compare the new drug or device with the current "gold standard" treatment. Size and duration make a phase III trial costly and challenging to run, requiring the coordination of many centers and registering a large number of patients.[13]

In the 1980s, when the new clot-busting drug Activase, known generically as tissue plasminogen activator (t-PA), was introduced by the Genentech Corporation in San Francisco, the Thrombolysis in Myocardial Infarction trial (TIMI) was created and included 2952 patients with a fresh heart attack enrolled at 50 U.S. centers.[15] Yet that number pales in comparison with a similar trial to measure the effect on survival after a heart attack of another clot-dissolving drug, called streptokinase. The *Gruopo Italiano per lo Studio della Strptochinasi, nell'Infarto Miocardico* (GISSI) (medicine is fond of catchy acronyms) trial enrolled 11,712 patients with an acute myocardial infarction (heart attack) in Italy. It found a very significant lowering of mortality in the streptokinase group compared with the control group.[16] And even the nearly 12,000 patients in the GISSI trial was later dwarfed by the 41,021 patients with a heart attack who took part in the Global Utilization of Streptokinase and Tissue Plasminogen Activator to Treat Occluded Arteries (GUSTO-1) Study, which compared the two clot-dissolving medications head to head in an international study coordinated by Duke University and conducted in 15 countries at 1084 hospitals in just 26 months.[17] So cardiology is used to really big phase III trials to test new treatments for the leading cause of death in the Western world.

Should We or Should We Not Perform Large-Scale Clinical Trials?

The debate regarding whether bone marrow stem cells transdifferentiate into cardiac muscle after injection into the heart and the related extent of cardiac regeneration certainly tempered enthusiasm about continuing clinical trials in the latter half of 2004 and all of 2005. In July 2006, *Circulation*, the American Heart Association's official journal, put the controversy front and center, publishing articles by protagonists and an antagonist.

In one corner were Andrew Boyle, Steven Schulman, and Joshua Hare from Johns Hopkins, and in the other was Peter Oettgen from Boston's Beth Israel

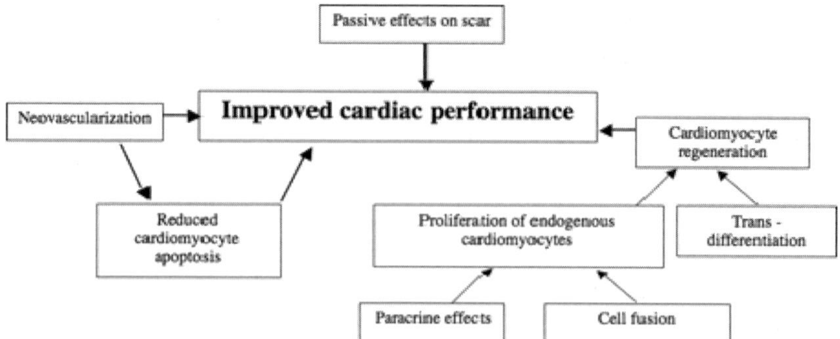

Fig. 28.3. This figure depicts several ways that stem cell therapy may improve cardiac performance after an acute myocardial infarction. Stem cells may promote the growth of new blood vessels into the damaged area (neovascularization), reduce cardiomyocyte death (apoptosis), and stimulate cardiomyocyte regeneration.

(*Source*: Reproduced with permission from the publisher.) Reference 18.

Deaconess Center. The Hopkins group was pro larger clinical trials, saying that stem cell therapy had "captured the imagination of both the medical and popular communities."[17,18] "Early studies of more than 400 patients," they continued, "have demonstrated the safety, feasibility and efficacy" of stem cell therapy in the setting of an acute myocardial infarction.[18] They acknowledged that while transdifferentiation turned out not to be how stem cells generate new heart muscle, nonetheless stem cell therapy works. Its efficacy could be due to stimulating the growth of new blood vessels into the oxygen-starved area or to diminished cardiomyocte apoptosis[d] (Fig. 28.3). They concluded, "Although at this time no data support…stem cell therapies as standard clinical practice…there is a wealth of preclinical and early clinical data showing safety, feasibility, and efficacy of adult cell–based therapy. Adult stem cell therapy should therefore proceed into randomized, placebo-controlled, double blind

---

[d] Apoptosis (a-pop-to'-sis) refers to the genetically controlled death of a cell. It is characterized by shrinkage of the cell as it ages, followed by fragmentation of its nuclear DNA. About 10 billion cells die each day. This is balanced by the number of cells arising from the body's stem cell population. Thus, the body's homeostasis is regulated. Homeostasis refers to the normal physiological processes that keep the body's metabolism, temperature, and blood pressure in equilibrium with the outside environment. For example, the body temperature is maintained at 98.4 degrees Fahrenheit in hot and cold external environments, although not at extreme temperatures.

clinical trials. The ongoing rigorously designed trials will contribute greatly to this emerging and exciting new therapeutic approach for diseases of the cardiovascular system."[18]

In the other corner, Peter Oettgen argued that the clinical trials done so far had shown mixed results. For example, in one study from Germany, the heart beat stronger six months after a heart attack in those patients who *received* an intracoronary infusion of bone marrow stem cells within hours of the attack than it did in those patients who did *not* receive such an infusion.[19] In another trial, from Belgium, there was *no difference* in cardiac function between the stem cell treatment group and the control group six months after a heart attack.[20] And, in a third study from Norway, the heart was actually stronger six months after a heart attack in those patients who did *not* receive stem cells than in those who did receive a stem cell infusion within several hours of a heart attack.[21] Therefore, Bettgen felt, "We should err on the side of caution before committing precious resources to conduct additional large clinical trials."[22] Before we go on, he added we "need to identify which stem cell types to use…to determine the mechanisms by which stem cells promote myocardial function [and] repair…and [determine] if stem cells can be retained efficiently within the heart" [rather than most being washed out]. He concluded that, because of the expense of large clinical trials, "it is particularly important that we address the aforementioned questions before proceeding with larger clinical trials."[22]

Clinical Trials of Stem Cell Transplantation in Acute Myocardial Infarction

The first clinical trial of stem cell transplantation after an acute myocardial infarction was published in October 2002, just 18 months after the groundbreaking report from Valhalla, NY. (Valhalla is a suburb of New York City, and Piero Anversa, the senior author of the paper, holds appointments at both the New York Medical College in Valhalla and Albert Einstein University in the nearby Bronx.) Such a short turnaround from bench to bedside was possible because bone marrow transplantations had been done for 30 years for conditions such as leukemia and aplastic anemia (a lack of red blood cell production). It had become the gold standard for tissue transplantation[23] and physicians were comfortable doing the procedure. (Additionally, bone marrow cells are delivered easily to all other tissues in the body because the delivery boy — the bloodstream — passes by the marrow's front door 24 h a day with complimentary service to all destinations.) Similarly, cardiac catheterization and coronary artery bypass surgery, the two procedures used to deliver the cells to the heart, had also been around for more than 30 years. Thus, transplanting bone marrow cells into the heart was simply a new twist to two time-honored procedures. The only two questions left were: Was the new twist safe, and would it work, would it help patients?

Cardiologists and stem cell biologists in Dusseldorf, Germany, chose 10 patients who had just had a heart attack for the world's initial procedures.[24] All were men whose average was 50. All underwent a cardiac catheterization within a few hours of arriving at the hospital. During the procedure, blood flow to the injured area was restored by a balloon angioplasty, followed by the insertion of a stent to keep the artery open. Then, on day 7 after their heart attack, stem cells obtained from each person's hip bone—called *autologous* stem cells, as opposed to *allogeneic* stem cells, which come from another person — were infused into the coronary artery that had become blocked by a small blood clot leading to the heart attack. Because the artery had been opened several days earlier by a balloon angioplasty and stent, another balloon catheter could now be inserted just beyond the point of previous occlusion to deliver the stem cells directly into the damaged area (Fig. 28.4).

The 10 men then went home. Four months later, they returned to have a third cardiac catheterization done. Each person had a smaller damaged area than he had had three months earlier and a smaller damaged area compared with a control who didn't receive stem cell transplantation. The authors felt that stem cell transplantation soon after a heart attack is "safe and seems to be effective."[24]

In the next early report, also a phase I safety and feasibility study, 37 men and 3 women who received intracoronary bone-marrow-derived stem cells had a significant improvement of their cardiac function four months later.[25] These authors concluded that the administration of intracoronary stem cells was "feasible and safe" in patients with an acute myocardial infarction.[25]

Since 2002, six more phase I or phase II trials have evaluated the benefit or lack of benefit of infusing bone-marrow-derived stem cells for cardiac function in the setting of an acute myocardial infarction (Table 28.1). At this time (summer 2008), no phase III trials have been reported, although several are underway in Europe (Germany, Finland, Poland, Belgium) and in Brazil. Most, though not all, of the phase I and phase II trials have shown an improvement in the "ejection fraction" of the heart in individuals who receive

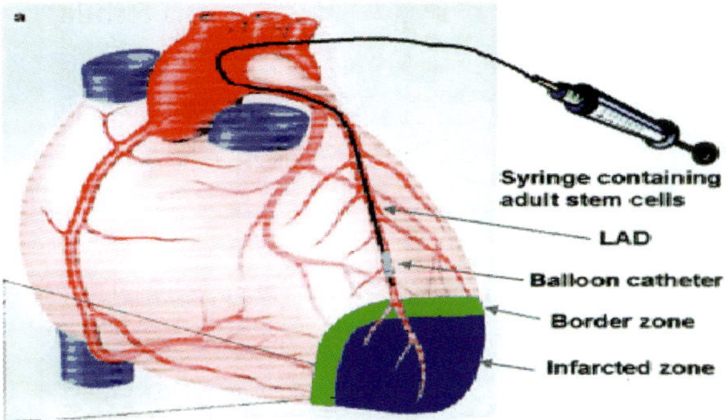

Fig. 28.4. The transplantation of bone-marrow-derived stem cells into the human heart. A standard balloon catheter attached to a syringe holding stem cells derived from the patient's own bone marrow is inserted into the femoral artery (groin) of the patient and advanced under an X-ray machine (fluoroscope) into the coronary artery that had become blocked and caused the heart attack. The tip of the balloon catheter is placed just above the border zone lying between healthy heart muscle and infarcted (dead) muscle. The balloon (light blue) is inflated for 2–4 min to prevent backflow of the injected cells out of the heart. After inflation, a suspension of stem cells is infused under high pressure. The cells then migrate into the border zone and the infarcted zone.

(*Source*: Reproduced with permission from the publisher.) Reference 24.

such infusions within a few days of a heart attack. The ejection fraction is simply the amount — the fraction — of blood, expressed as a percentage, the heart pumps (ejects) with each heartbeat. Normally, the heart ejects 60%–65% (not all) of the blood within it with each heartbeat at rest, and a bit more during exercise. But after a heart attack the damaged area is not able to contract with its former force. The ejection fraction falls as a result; the bigger the heart attack, the greater the drop. The two largest trials appeared in the September 21, 2006, issue of *The New England Journal of Medicine*. The first trial, done at 16 centers in Germany and one in Switzerland, found a significantly better ejection fraction among those men and women who received an infusion of bone-marrow-derived stem cells into the affected region of the heart four days after the attack.[30]

But the second, at two university hospitals in Oslo, Norway, did not.[31] It found "no effect" of a similar infusion of bone marrow stem cells on the ability of the heart to pump. The ejection fraction was no better in the stem cell group than in the control group (Fig. 28.5).[31] In fact, there was "an excellent clinical outcome in both the treatment and the control group," with a 0%

Table 28.1　Summary of stem cell transplantaion trials in acute myocardial infarction.

Study name or lead author name (year) (reference)	Number of treated subjects/control	Day of bone marrow transplantation after onset of AMI	Number of stem cells infused	Result
Strauer (2002) (24)	10/10	7 ± 2	1.5×10^6	Smaller infarct size
TOPCARE-AMI (2002) (25)	20/11	4.3 ± 1.5	$7.35\pm7.3\times10^6$	Improved EF in stem cell group
BOOST (2004) (18), (26)	30/30	6	25×10^9	Improved EF at 6 months in stem cell group. At 18 months, no difference in EF between the 2 groups
Chen (2004) (27)	34/35	18	48×10^{10}	Improved EF in stem cell group
Fernandez-Avila (2004) (28)	20/13	13.5 ± 5.5	$78\pm41\times10^6$	Improved EF in stem cell group
Bartunek (2005) (29)	19/16	11.6 ± 1.4	1.5 to 33.6×10^6	No difference in EF between the 2 groups
REPAIR-AMI (2006) (30)	102/102	4	2.4×10^8	Improved EF in stem cell group
Janssens (2006) (20)	33/34	2	3×10^8	No difference in EF between the 2 groups
ASTAMI (2006) (31)	50/50	6	8.7×10^7	No difference in EF between the 2 groups

EF = ejection fraction (explained in text).

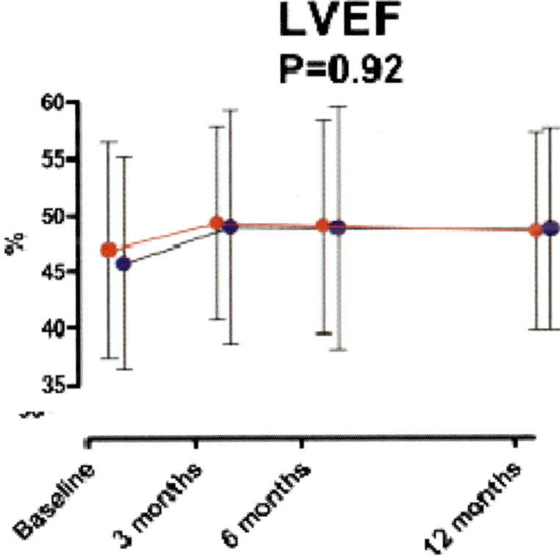

Fig. 28.5. The left ventricular ejection fraction (LVEF) at three points in time over the first year after stem cell transplantation in the group that received stem cells (red) and the control group (blue). The curves are almost superimposable.

(*Source*: Reproduced with permission from the publisher.) Reference 32.

mortality in both groups 12 months after the acute infarction. The failure of stem cells to improve cardiac function in this study may be related to the smaller number of cells infused in comparison with the number infused in the BOOST and REPAIR-AMI trials (Table 28.1). These "mixed results" led an editorialist from the Stem Cell Institute at Harvard to say of stem cell therapy, "We should guard against both premature declarations of victory and premature abandonment of a promising therapeutic strategy."[33]

In yet another editorial in the same issue of the *Journal*, Robert Schwartz took then-president George Bush to task for vetoing a bill just two months earlier that would have enabled "the derivation of embryonic stem cell lines from fertilized eggs stored in freezers and already tagged for destruction."[34] He also pointed out that Karl Rove's assertion that adult stem cells had "far more promise" than embryonic stem cells was "patently false." (Karl Rove was Bush's head of the White House's Office of Political Affairs.) Schwartz went on to say that Singapore had recently attracted "gifted American investigators fed up with political restrictions on their research" in the U.S. Singapore is a vibrant, ultramodern, forward-thinking city-state which scientists have flocked to since 2002,[35] drawn by "the government's financial largesse and relatively permissive regulations on stem cell research," *Time Magazine* reported in October 2006.[36] Schwartz wound up saying, "The delay of medical advances by theological disputes is not in the best interests of the sick and disabled."

Physicians doing clinical research at university hospitals are fond of randomized, double-blind, placebo-controlled trials of new drugs or new devices. Studies regarding the intracoronary infusion of stem cells soon after a heart attack are not exempt from this preference. In the setting of administering stem cells soon after an acute myocardial infarction, a placebo group is preferred to ensure that any improvement of cardiac function seen after an intracoronary infusion of stem cells is due to the stem cells, not a "side effect" of the catheterization procedure used to deliver the cells to the injured heart zone.[e] This concern pops up because a tiny balloon on the tip of the catheter used to infuse the cells into the damaged area is inflated three times for 3 min (each inflation separated by a 3 min deflation) to temporarily block the artery, thereby preventing the cells from being washed out by the flow of blood through the vessel. But it doesn't work very well, as will be explained two paragraphs hence. Nonetheless, the balloon is fully inflated during each of the 3 min infusions and this could lead to "preconditioning" of the heart muscle beyond the balloon. "Preconditioning" refers to the better ability — condition — of the heart to withstand a total cessation of blood flow during the early hours of a heart attack if it has experienced brief, repeated periods of no blood flow before total occlusion of the vessel. These temporary disruptions of blood flow render the myocardium beyond the ultimate total occlusion more resistant to the deleterious effects of the prolonged occlusion when no oxygen reaches the heart muscle. The protection afforded by preconditioning is due to the release of nitric oxide, free radicals, and adenosine during the periods of balloon occlusion — all powerful vasodilators that allow more oxygen-rich blood flow into the damaged area.[37] Misattributing the benefit of a stem cell infusion to the protective effect of brief, repeated coronary occlusions during

---

[e] A standard coronary artery balloon angioplasty is done immediately before the intracoronary infusion of stem cells. This procedure (usually) breaks up the small blood clot in the coronary artery that caused the heart attack. The clot formed a few hours earlier, when the patient's chest pain started, cutting off the delivery of oxygen-rich blood to the infarcted (dead or severely damaged) zone beyond the now-occluded artery. Continued complete occlusion of the artery by the blood clot would prevent the soon-to-be-infused stem cells from reaching the dead or dying myocardium downstream.

the infusion of the cells is avoided by including a control group that does not receive the cells (saline is infused instead) but has a similar number of brief, repeated occlusions.

At this time, only two of the reported trials of infusing stem cells directly into the heart soon after an acute myocardial infarction have been randomized, double-blind, placebo-controlled trials. Janssens and his colleagues in Belgium[20] found no difference in cardiac function four months after a heart attack between the group that received a stem cell infusion and the group that received a placebo infusion. On the other hand, Schachinger and his colleagues in Germany found significantly better cardiac function in the stem cell treatment group than in the placebo infusion group.[29] And those patients with bigger heart attacks benefited the most from stem cell treatment. One year after the heart attack, there were fewer deaths and fewer recurrent heart attacks in the stem cell treatment group than in the group that received a placebo infusion.

The Matter of Stem Cell Washout

One reason the Belgian study did not find any strengthening of the heart after an infusion of stem cells may be that there weren't enough cells staying in the heart long enough to do any good. In a 2005 report from Germany, investigators discovered that less than 3% of stem cells infused directly into the coronary artery nourishing the damaged area remained in the heart for an hour.[38] The rest were washed out in spite of the balloon being fully inflated. They ended up in nearby neighbors — the liver and spleen — but not near enough to help the heart (Fig. 28.6). A second reason may be that the patients in this study had their blocked artery reopened by a stent about 4 h after the heart attack began, thereby restoring oxygen delivery to the damaged region and resulting in a smaller heart attack. Thirty-eight percent of the patients in this study had an ejection fraction of 55% (normally it is 60% or more).[40]

It's harder to show an improvement (from stem cells) in a mildly damaged heart than in a heart with major damage. One might even wonder if it's "right" to do a bone marrow biopsy and infuse stem cells into the heart in patients with such small heart attacks. In the BOOST trial, in which stem cells did strengthen the heart, the blocked artery was reopened twice as late (8.0–9.8 h after the heart attack began).[19,26] Consequently, the size of the myocardial infarction was larger and the chances of more improvement following stem cell therapy greater. Figure 28.7 shows, in color, the improvement of the

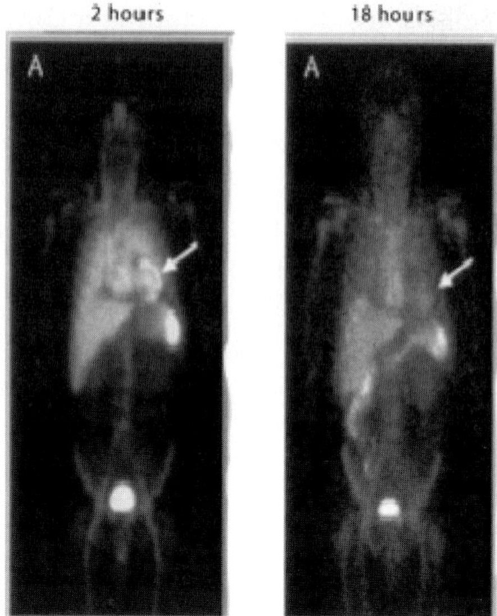

Fig. 28.6.   A whole-body MRI scan 2 h (left) and 18 h (right) after an intracoronary infusion of stem cells labeled with a radioactive tracer. Less than 5% of the cells remain in the heart (arrow) 2 h after the infusion. The great majority are in the spleen to the right and below the heart (arrow) and the bladder. After 18 h there is almost no cardiac activity left.

(*Source*: Modified and reproduced with permission from the publisher.) Reference 39.

heart (in two patients) after a stem cell transplantation and bypass surgery.[f] Clearly, these color images show more live and less damaged myocardium, but the effectiveness of the stem cell transplantation *per se* cannot be assessed because of the simultaneous bypass surgery. And a third reason why the heart wasn't helped in this study may be that the relatively thick mixture of stem cells infused into the large (3–3.5 mm[g]) coronary artery clogged up smaller vessels downstream, causing — believe it or not — the very problem the cells

------

[f]Why bypass surgery? If the goal of a stem cell transplantation is to regenerate heart muscle, wouldn't a successful stem cell transplantation obviate the need for simultaneous coronary bypass surgery? Bypass surgery can improve the strength of viable myocardium but not necrotic (dead) heart muscle. These two patients were part of an early phase 1 trial (in 2001) to assess the "feasibility and safety" — but not the effectiveness — of transplanting autologous (one's own) stem cells into severely damaged heart muscle.

[g]The coronary artery diameter is related to the size (height) of the individual. In general, women have smaller coronary arteries than men, which may contribute to the overall lower success rate of coronary artery bypass surgery in women than in men.

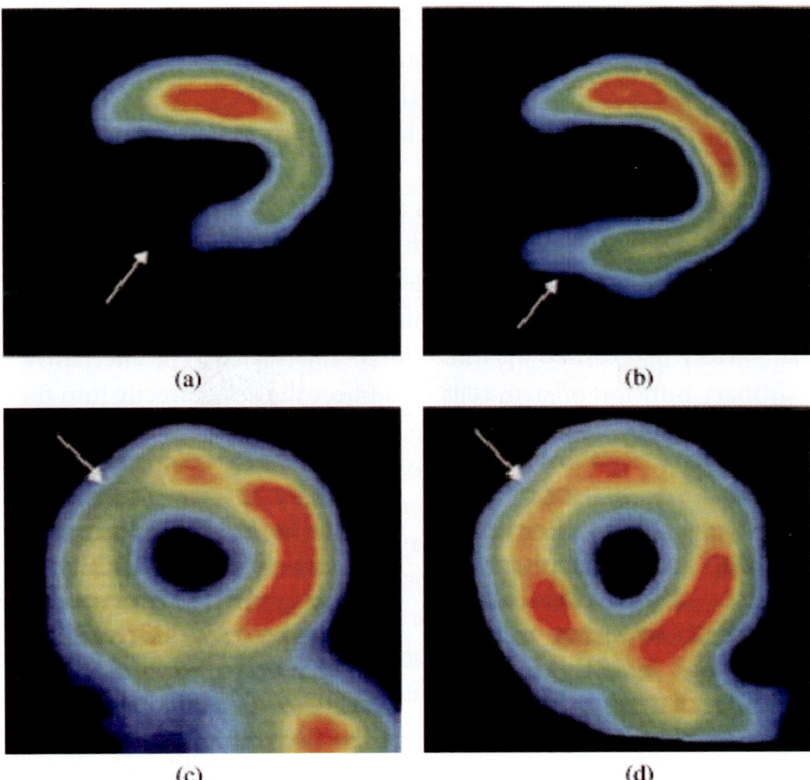

Fig. 28.7. Heart scans from two patients before and after stem cell transplantation at the time of coronary artery bypass surgery. The scans at the top are from one patient, while the ones at the bottom are from another. Both a and c are before stem cell transplantation, while b and d are after the procedure. The arrows indicate where the stem cells were injected into the heart. The upper scan images should have a donut configuration, like the lower ones. The hole in the donut represents blood in the heart's cavity. The black void from 6 to 9 o'clock represents severely damaged heart muscle unable to take up the radioisotope. The red region is normal heart muscle, while the yellow, green, blue, and black regions are progressively damaged regions, with black representing dead or nearly dead muscle. After stem cell transplantation and heart surgery, in scans b and d, there is more red muscle and less black, with varying amounts of yellow, green, and blue.

(*Source*: Reproduced with permission from the publisher.) Reference 41.

were designed to treat: a myocardial infarction, albeit a microscopic one. In healthy dogs, multiple injections of stem cells — just as is done in humans with not-so-healthy hearts — caused transient changes of myocardial injury on the EKG and, one week later, microscopic evidence of myocardial infarctions.[42] Ditto in pigs.[43] On the one hand, numerous clinical studies have established, using readily available techniques such as routine blood tests and the EKG and the not-so-routinely-available technique of MRI, that stem cell transplantation is safe — at least for the short haul (six months).[44] But what about down the road, or as assessed by a technique no one is eager to have done soon — an autopsy!

Because stem cell transplantation is a safe procedure, few patients have died after stem cell therapy for an acute myocardial infarction. And then, even if someone were to die, how could any stem cell therapy–induced areas of infarction be distinguished from areas of infarction caused by the blood clot in the coronary artery that caused the infarction being treated? The alternative to the intracoronary infusion of stem cells is to inject the cells directly into the heart muscle. But this, of course, requires open-heart surgery with its list of possible complications, not to mention pain, cost, and loss of time from work.

In addition to early washout of stem cells from the heart, there may be a different fate awaiting those that stay put. They may not turn into heart muscle at all. They may turn into bone! In a fascinating study from Germany in the journal *Blood,* laboratory mice whose hearts were injected with stem cells shortly after an acute myocardial infarction developed patches of bone (Fig. 28.8).[43] This "unexpected and disturbing" finding caused the investigators to wonder if clinical trials involving the injection/infusion of stem cells

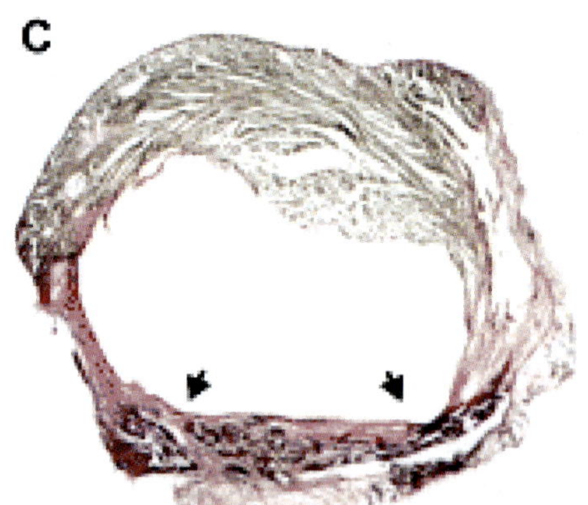

Fig. 28.8.   The left ventricle of a mouse sliced open in a "bread loaf" manner 28 days after a stem cell transplantation for an acute myocardial infarction produced simply by ligating a coronary artery. The upper half of the ventricle is largely normal myocardium, measuring 12 mm in thickness. The lower half is thinned, infarcted myocardium with striking calcification, stained black and marked by two arrows.

(*Source*: Reproduced with permission from the publisher.) Reference 45.

into the heart after an acute myocardial infarction should be continued. It was not the first time such a strange transformation has taken place. The versatile embryonic stem cell from which all other cells are derived can also give rise to a teratoma — a bizarre tumor that contains cells from all three embryonic layers (Chap. 14), including teeth, hair, and bone.[46] This is a major issue, quite apart from ethical concerns, about the use of embryonic stem cells. Couple the formation of teeth, hair, and bone in the heart after stem cell transplantation with the development of patchy areas of myocardial damage in healthy dogs, and there may understandably be reluctance to infuse/inject bone marrow stem cells into human hearts.

Bone Marrow-derived Intracardiac Stem cell Transplantation versus Subentaneous Injection of a Cytokine

But, nonetheless, suppose the notion of replacing dead heart tissue with fresh tissue still sounds good, but you simply don't like the idea of a bone marrow biopsy under either a local or a general anesthetic. In that case there is a nearly painless way to do the trick. All it takes is a single subcutaneous needle stick in the abdominal wall, just as diabetics who need insulin do every day. The stem cells are left in the bone marrow and, rather than drilling a hole in the bone and aspirating them into a syringe, they are coaxed out with a cytokine (Chap. 8). This goes by the cumbersome name (and don't let its name discourage you from reading on) of "granulocyte-colony-stimulating factor," or G-CSF, and perhaps its ease of administration and sparing of a bone marrow biopsy make up for its name.[h] It "mobilizes" marrow stem cells into the bloodstream, where they travel to the damaged heart muscle, which then releases a compound that guides the cells exactly to where they are needed.

---

[h] Human granulocyte-colony-stimulating factor was first isolated in Japan in 1986.[47] It is a cytokine (Chap. 8) that stimulates the release of granulocytes from the bone marrow. Granulocytes are one of the five types of white blood cells circulating in the blood. They are the most abundant type, constituting 60%–65% of the 5000–10,000 white blood cells (WBCs) in the peripheral (circulating) blood. Granulocytes are characterized by granules in their cytoplasm and are better known as polymorphonuclear leukocytes, or "polys" in the medical vernacular. The term "colony" refers to the collection of cells derived from a single cell on the laboratory culture medium used to grow them. Following a series of G-CSF injections, the WBC count can rise to 50,000–75,000, close to the leukemia range, yet with no adverse health effects. In fact, the compound is often given during treatment of cancer since chemotherapy can cause dangerously low levels of WBCs, making patients prone to infections.

But Does it Work?

The first clinical trial of this new compound was in 2004 in Seoul, South Korea. In it, 10 patients were treated with a standard intracoronary infusion of stem cells plus G-CSF, while 10 others received subcutaneous injections of G-CSF for four days.[48] Although the ejection fraction increased with the combination of stem cells and G-CSF, it did not increase with G-CSF alone. And 7 of 10 patients who received a stent insertion on day 1 of their heart attack developed stenosis (narrowing) of its diameter within the next six months — an inauspicious beginning for G-CSF. The next trial appeared in April 2005, from Italy. It, too, showed no effect of G-CSF on left ventricular function (Table 28.2).[49] Two trials, two strikes. Then came three small reports, one from Spain and two from Germany. Collectively they showed an improved ejection fraction following serial subcutaneous injections of G-CSF.[50–52] But they would be the last breath of life for the cytokine in the seting of a heart attack. Two randomized, double-blind, placebo-controlled trials were lurking right around the corner. In the spring of 2006, they appeared in *JAMA* and *Circulation*, the American Medical Association's and the American Heart Association's official journals, respectively. The trial in *JAMA* was done in Germany. In it, 56 patients received G-CSF once daily for five days soon after their heart attack, while 58 others received placebo.[53] After six months there was no difference in the ejection fraction between the two groups, leading the authors to conclude the "use of G-CSF therapy to mobilize bone marrow–derived stem cells does not improve left ventricular recovery in patients with AMI" (acute myocardial infarction). This, in turn, led the editorialist to quip that it provided a "dose of reality" to the earlier high hopes for G-CSF.[56]

The trial in *Circulation* was done in Denmark. Different country, same results. (The many reports in the cardiology literature from Europe simply reflect the Bush ban[i] on embryonic stem cell research and, yet, the investigators' preference to publish their cutting-edge research in American scientific journals.) The compound was "well tolerated and seemed safe," but didn't affect the ejection fraction any more than did placebo.[54] The editorialist commenting on this study pointed out that using cardiac MRI, the "'gold standard' for measuring serial changes of ejection fraction," the investigators of the

---

[i]On August 7, 2001, Bush banned the creation of future lines of embryonic stem cells through federal funds. Only the 21 lines developed before August 2001 would be eligible for federal money.

Table 28.2   Summary of G-CSF trials in acute myocardial infarction.

Study name or lead author name (year) (reference)	Number of treated subjects/control	Days of G-CSF injection after onset of AMI	Dose of G-CSF (μg/kg of body weight) and frequency	Result
MAGIC (2004) (48)	10/10/7	0–4	10 μg/kg once daily	Improved EF with stem cells, no change in EF with G-CSF; stent stenosis in 7 of 10
Valgimigli (2005) (49)	10/10	0–4	5 μg/kg once daily	No difference in EF between the 2 groups
Suarez de Lezo (2005) (50)	13	5–15	10 μg/kg once daily	Improved EF in G-CSF group; No control
Kuethe (2005) (51)	14/9	2–7	10 μg/kg once daily	Improved EF in G-CSF group
FIRSTLINE-AMI (2005) (52)	25/25	0–6	10 μg//kg twice daily	Improved EF in G-CSF group
REVIVAL (2006) (53)	56/58	0–5	10 μg/kg once daily	No difference in EF between the 2 groups
STEMMI (2006) (54)	39/39	0–6	10 μg/kg once daily	No difference in EF between the 2 groups
G-CSF–STEMI (2006) (55)	23/21	5 days after delayed hospital admission	10 μg/kg once daily	No difference in EF between the 2 groups

EF = ejection fraction.

STEMMI trial detected no difference in the ejection fraction between the treatment and placebo groups.[57] He wondered, "Will G-CSF simply join the vast spectrum of other experimental agents designed to aid myocardial infarction healing that did not live up to the preclinical and early-phase clinical hype?"

It took just five more months to get the answer. In September 2006, another group from Germany, aware that preceding trials of G-CSP given after the immediate reopening of a blocked coronary artery with a stent reported mixed results from the cytokine, decided to evaluate the effect of G-CSP in patients who have a blocked artery reopened later because they are admitted to the hospital beyond the optimal 6 h window of opportunity to restore blood flow with a stent. The usual reason that patients arrive at the hospital too late for the timely placement of a stent is living in a rural area in which the nearest hospital does not have a cardiac catheterization laboratory to do the procedure in. Those people must then be transferred to an urban hospital that does have such a laboratory. Less often, the reason for late arrival is that the patient misattributes his or her chest pain to "indigestion" and takes antacids for several days before going to the ER. In this study, patients with chest pain of 6 h to 7 days in duration were included. They were randomly given either a G-CSP or a placebo infusion after the delayed insertion of a stent. Six months later, there was no difference in the ejection fraction between the two groups. The authors concluded, "Patients do not benefit from G-CSP when [the stent is inserted] late."[55]

In the accompanying editorial, James Forrester at Cedars-Sinai Hospital in Los Angeles lamented that this "study seems to signal the end to an era that began just 5 years ago."[58] Forrester, one of the finest cardiologists in the U.S. later added that now we need "inspiration for a new generation of clinical testing." And, in a postmortem review of all clinical reports of G-CSP in the setting of an acute myocardial infarction, Kastrup from Copenhagen concluded, "...presently there is no indication for the use of G-CSF mobilized stem cells for clinical regenerative treatment of infarcted or degenerated ischaemic myocardium."[59] Although G-CSF–induced stem cells failed to pass muster in the treatment of coronary heart disease, G-CSF is still widely used to raise the white blood cell count in patients undergoing chemotherapy for leukemia and other malignancies, which was, after all, its original purpose. By so doing, G-CSF affords protection against otherwise potentially lethal infections by increasing the number of polys in the bloodstream.

The reason for even mentioning this five-year experience with a drug that will never find its way into a hospital pharmacy is that it typifies what happens to so many other promising drugs in the U.S. Sure, new drugs are very expensive. Many physicians feel ambivalent about prescribing them because of their cost. On the one hand, they want needy patients without insurance to benefit from them. On the other hand, pharmaceutical companies have to recoup their huge R&D expenses during the 17 years they have until their patent expires. And pharmaceutical companies are the major source of funding for physician continuing-medical-education expenses to keep practicing physicians informed about new treatments and recent developments in their specialty. It's drugs like G-CSF that will never put a dime in the pocket of a pharmaceutical company after the company has spent millions of dollars to do the basic research to develop the drug and to conduct largely negative clinical trials that successful drugs must pay for during their 17 profit-making years. The solution for the consumer is, of course, medical insurance through the individual's employer. Unfortunately, the employer takes the hit now, but large employers are generally better prepared to pay the bill than are their employees. Bottom line is, I think, pharmaceutical companies should recoup their R&D expenses through high prices for successful drugs and offer the medications to indigent patients at a discounted cost, which many companies do.

Sources of Stem Cells That Can Regenerate the Heart

There are five sources of stem cells with the capability of forming cardiac tissue. They are depicted in Fig. 28.9 and will be discussed in the order shown, from left to right. All except embryonic stem cells can be grown from a person's own tissue, obviating the problem of rejection. The rationale behind using stem cell therapy in the setting of an acute myocardial infarction is, simply, that repopulating damaged heart muscle with a "new pool of functional cells" will help the heart.[61]

Circulating blood contains stem cells using the "stem cell highway" shown in Fig. 14.2. These cells can be isolated in the laboratory, cultured for three days, and then injected into the heart. In practice, 250 cc (8 ounces) of venous blood is removed. After centrifugation, the cells in the sediment are plated on a culture dish and grown for three days in the lab. Then the carpet of cells on the surface of the culture dish is removed and put into solution for reinjection into the self-donor.

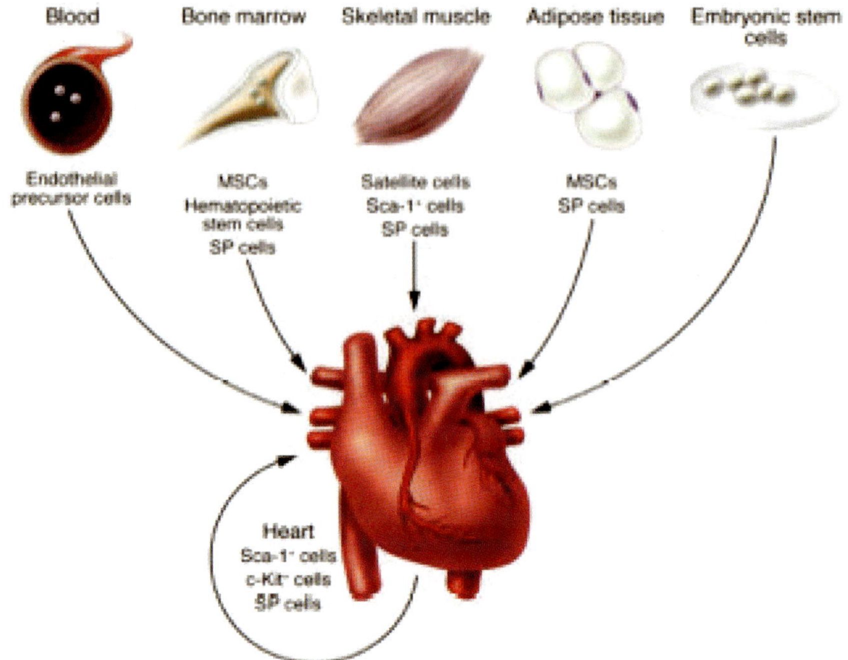

Fig. 28.9. Sources of stem cells for cardiac regeneration. The pros and cons of each are discussed in text.

(*Source*: Reproduced with permission from the publisher.) Reference 60.

Bone marrow contains a rich variety of stem cells and is the source of stem cells used for cardiac regeneration clinically. The marrow is aspirated from the iliac crest (hip bone) under a local or a brief general anesthetic. Approximately 20–50 cc (about 1–2 ounces) of marrow is withdrawn into a syringe. After centrifugation, the sediment is infused (using a catheter inserted in the groin and advanced under fluoroscopic vision) into the coronary artery that caused the heart attack or (less often) injected directly into the left ventricle at the time of bypass surgery.[61]

Ordinary skeletal muscle cells, such as the biceps, were the first cell type to be used for cardiac regeneration, in 1992.[j,62] In practice, the vastas lateralis

____

[j] The specialty of "regenerative medicine" was born in 1998. And no, you won't find this specialty listed in the Yellow Pages. If you want to find such a species in the near future, the best bet is to Google the term "regenerative medicine" or to see a cardiologist (or the appropriate specialist for liver, lung, brain, etc. disorders) at your local university hospital. He or she will

muscle — a muscle in the thigh[k] — is biopsied to obtain the tissue needed to build heart muscle. Literally millions of stem cells can be grown from a small muscle biopsy, making skeletal muscle a practical source of cells for treating coronary heart disease. And since a person's own skeletal muscle is used to create fresh cardiac muscle, the problem of rejection is avoided. But, alas, there's a catch. When transplanted into the heart seven days after an acute myocardial infarction, they contract totally out of sync with the heart, even when in direct contact with heart muscle, in rats and mice.[63,64] They are, in effect, "electrically insulated" from the rest of the heart.[65] On top of that, they do not transdifferentiate into cardiac muscle after transplantation and no amount of coaxing changes their mind.[66]

Even *fat (adipose tissue)* can be used to grow new blood vessels into damaged heart muscle! Fat can transdifferentiate into blood vessels.[67] How this strange conversion occurs is not clear. But who cares as long as it works and it's safe? Isn't it nice to know that the stuff we try to limit by diet and exercise can be cut out and transplanted into the heart to grow new blood vessels into oxygen-deprived myocardium after a heart attack?

And, finally, the gifted *embryonic stem cell* can also be used to regenerate cardiac tissue. Embryonic stem cells offer several advantages over all other sources. They have the unique capacity to divide infinitely, providing a (potentially) unlimited number of stem cells for cardiac transplantation — as well as many other organs (Table 28.3).[68] They are also relatively easy to grow in the laboratory. Yet, their tendency to form teratomas is a problem. Teratomas are, in effect, a "side effect" of embryonic stem cells, a manifestation of the cells' vast regenerative capacity run amok by the random formation of bizarre (benign) tumors from all three embryonic layers (Fig. 28.10). Teratomas can arise in any organ, but when in the heart they may behave in a malignant manner, causing skipped beats and even cardiac arrest.[69] When mouse embryonic stem cells were transplanted into the hearts of rats following an induced myocardial infarction (by ligating one of the three coronary arteries), they improved cardiac function within three weeks — an improvement that did not diminish over the next nine weeks.[74]

---

refer you to the university's "regenerationist," as they soon will be called, or, lacking one, will refer you to another university hospital that does have such a rare species (now).

[k]There are three vastus muscles; the vastus lateralis, the vastus medislis, and the vastus intermedius. Together they form the quadriceps muscle, so well-developed in runners and bicyclists, that extends the knee joint.

Table 28.3 Advantages and disadvantages of embryonic and adult stem cells.

Embryonic		Adult	
Advantages	**Disadvantages**	**Advantages**	**Disadvantages**
• Can be grown for a long time in the laboratory, possibly infinitely.	• Teratomas.	• No ethical issues.	• Restricted ability to differentiate into diverse tissues.
• Relatively easy to grow in the laboratory.	• Possibility of immune rejection.	• No or minimal risk of immune rejection.	• Restricted lifespan in culture.
• Potentially unlimited supply of stem cells.	• Limited supply due to ethical concerns about their use and current U.S. ban on new lines since August 7, 2001.[1]	• Only a small sample of tissue is required to grow stem cells in the lab.	• Very small number in each tissue.
• Availability — large number of "unused" embryos discarded by *in vitro* fertilization clinics.	• Destruction of a potential human life.	• Easy to harvest from skin, muscle, bone marrow, and fat. Umbilical cord and placenta at birth.	• More difficult and expensive to grow in the lab.
• Potential to make any of the 210 cell types.		• Do not form tumors.	• Difficult to obtain in the large numbers that will be needed for stem cell.

(*Source:* References 70–72).

---

[1]The Bush ban on federal funding of embryonic stem cell research was lifted by President Barack Obama on March 9, 2009, less than two months after taking office. In so doing, it ended "a long, bleak period in which the moral objections of religious conservatives were allowed to constrain the progress of medically important science," according to *The New York Times*.[73]

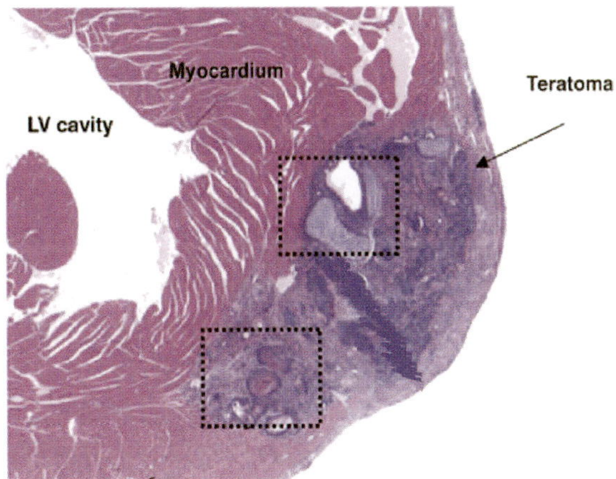

Fig. 28.10. A microscopic section of the left ventricle (from a mouse) about three months after an embryonic stem cell transplantation for an acute myocardial infarction. The teratoma occupies a significant chunk of myocardium. The several shades of purple reflect teratoma cells derived from the skin (lower box), cartilage (upper box), and intestine. This photomicrograph is from a mouse. Teratomas also formed after human embryonic stem cells were transplanted into rats. (*Source*: Modified and reproduced with permission from the publisher.) Reference 75.

What then is the "ideal" stem cell for cardiac repair? In theory, the ideal stem cell would migrate from the blood or from an undamaged area of the heart into the damaged region and build new cardiac tissue without giving rise to unwanted cell types, forming a teratoma. It should also have no, or at least a low, probability of causing rejection. Such a cell might also be grown easily in the lab, making it and its progeny rapidly available for clinical application (Table 28.4). And, finally, the ideal stem cell should be free of both rejection and teratoma formation, as well as being relatively inexpensive to grow in the laboratory because a large number of stem cells will be needed for therapeutic purposes in the future. The heart on the right in Fig. 28.11 represents an example of a heart that could be helped by a stem cell transplantation. Certainly, whole-heart cardiac transplantation isn't the answer to long-standing coronary heart disease because of the steady shortage of donors and problems with rejection, mandating drugs that predispose the recipient to unusual, even exotic, infectious diseases.[76] Skeletal muscle cells seemed to be the ideal cell type for cardiac repair since they can be collected ("harvested") easily from an autologous source (a person's own vastus lateralis muscle), grown rapidly in the lab to yield "large numbers of

Table 28.4 Characteristics of an embryonic, an adult, and an "ideal" stem cell.

Type of stem cell	Origin	Amount	Ability to replicate	Ease of identification and growth in laboratory	Chances of rejection	Chances of teratoma
Embryonic	Only in inner cell mass of days-old embryos	Abundant	Can double up to 300 times	Relatively easy	High	Low, but possible
Adult	Fully formed "adult" tissues throughout life	Rare	Limited replication	More difficult	None	None
Ideal	Circulating blood	Abundant	Unlimited	Easy	None	None

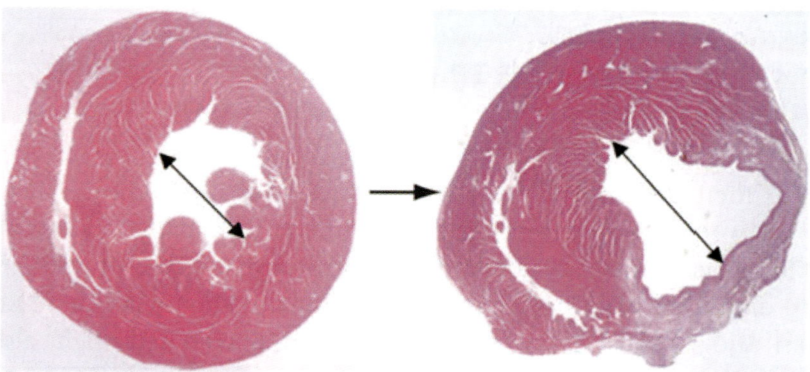

Fig. 28.11. Photographs of two human hearts sliced in a "bread loaf" fashion. The heart on the left is a normal heart. The heart on the right is from someone who died several months after an acute myocardial infarction. The thickness of the ventricle near the tip of the arrow pointing in the 10 o'clock direction is normal (15 mm). The myocardium near the tip of the arrow pointing in the 4 o'clock direction is thinned significantly, measuring only 6 mm. The thinning is due to the loss of about one billion cardiomyocytes following the infarction. The region adjacent to the 4 o'clock tip, stained pale pink, is composed of noncontracting scar tissue. It is this region that might be helped by a stem cell transplantation leading to fresh, contracting cardiomyocytes. Ironically, skeletal muscle transplantation can generate the billion or more new cardiomyocytes needed to replace the scar by muscle — but the new muscle beats independently of the host's heart.

(*Source*: Reproduced with permission from the publisher.) Reference 75.

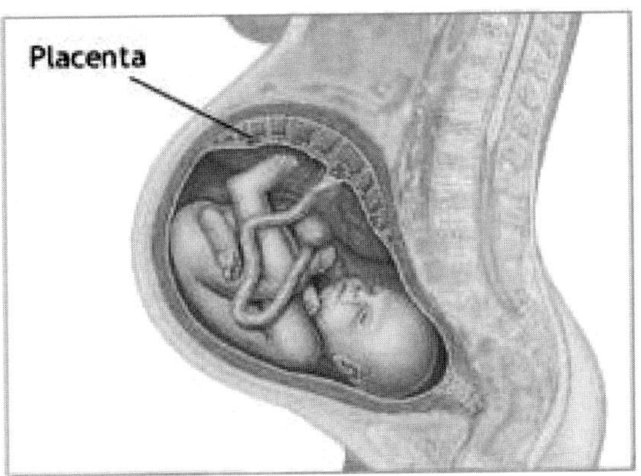

Fig. 28.12.   A near-term baby, its umbilical cord, and the placenta. (*Source*: Reproduced with permission from the publisher.) Reference 78.

stem cells" for autotransplantation, and they do not cause tumors (teratomas) after transplantation.[76] But, unfortunately, they also don't couple electrically and mechanically with host heart tissue, leading to inefficient pumping of blood.

There is one other source of stem cells for regenerating the heart, which first became available in 1993. Hematologists at the New York Blood Center introduced the use of *umbilical cord blood* (Fig. 28.12) (also called "placental blood") rather than bone marrow transplantation to shorten the time between when the decision to perform a bone transplantation to shorten the time between when the decision to perform a bone marrow transplant is made and the time of actual transplantation, when no related donor with a human leukocyte antigen (HLA) type identical to that of the prospective recipient is available.[74,77] (That is the scenario 70% of the time.) The term "human leukocyte antigen" refers to a group of six major antigens — together called "transplantation antigens" — on the surface of leukocytes (white blood cells) that regulate the human immune system's response to foreign proteins introduced into the body. The six HLA types are inherited — akin to the three ABO blood types familiar to most people. The closer the match, the less the likelihood of a rejection, or in the case of ABO matching, the less the chances of a "transfusion reaction." As it turns out, HLA mismatches are better tolerated by the body than are ABO mismatches. Up to three (out of six) HLA mismatches may not lead to rejection because of the "immaturity" of an infant's immune system,[79] its naiveté being an advantage in this setting because one, two, or even three mismatches are less likely to trigger an immune reaction.[80] In a large, multicenter, international study, while the success rate of cord blood or bone marrow transplantation among unrelated adults with leukemia was greatest in those who received fully matched transplants, the success rate was nearly as good in patients who had one or two HLA mismatches.[81] The similar favorable results led to the conclusion that "HLA-mismatched cord blood should be considered an acceptable source of [bone marrow] stem-cell grafts for adults in the absence of an HLA-matched donor."[81,82]

In a recent Policy Statement issued by the American Academy of Pediatrics, practicing pediatricians were advised that the one-year survival rate "may be as high as 75% to 90% after sibling HLA-matched cord blood stem cell transplantation and 40% to 80% after unrelated cord blood stem cell transplantation."[82] The Academy pointed out that "the use of less-than-completely matched HLA cord blood stem cells may incur less risk of [rejection] than mismatched cells from either a related or unrelated...donor."[m] Autologous (derived from the patient) cells are certainly best, but if not available the next-best option is cells from an "ALA-matched sibling." However, even if autologous cells are available the likelihood a child may need his or her own stored cord blood is estimated to be "1 in 1,000" to "more than 1 in 200,000" — in other words, it is an open-ended issue. On top of that, the Academy advises its members that "most conditions that might be helped by cord blood stem cells already exist in the infant's cord blood," i.e. there may be precancerous changes in the child's stem cells at birth. The position paper concludes that "cord blood donation should be encouraged when the cord blood is stored in a bank for public use" while the "private storage of cord blood as 'biological insurance' should be discouraged."[82]

Finding unrelated individuals who are HLA-identical to a potential bone marrow transplant recipient is "very difficult,"[77] because of the many (36) permutations and combinations (according to the "*n* choose *k*" formula[83])

---

[m] On the other side of the Atlantic, physicians in Europe organized in 1995 a group called Eurocord to standardize methods of "collecting, testing and cryopreserving cord blood from both related and unrelated donors."[80]

that arise when trying to match six HLA and three ABO types between a donor and a recipient. Computer databases greatly facilitate finding a match. But even when a computer search and compatibility testing go smoothly, 4–5 months may go by before a bone marrow transplant can be done — too long for many patients with a rapidly deteriorating condition.[n]

To hasten the process, the New York hematologists proposed turning to what had been considered as obstetrical waste — the umbilical cord blood and placenta. Cord blood can quickly be collected, typed, and frozen for future use when needed. The advantages of such a system are the abundance of umbilical cord blood, the ease of its collection, the absence of any risk to the mother or the child, and the low incidence of serious infectious agents in newborns compared with adults (Table 28.5).[77,88] Umbilical cord blood comes entirely from the infant, by the way.[79] Its chief disadvantages are the low number of stem cells in cord blood[81] and the potential for future problems currently unrecognized because of the newness of the procedure.

Collecting cord blood after delivery of the infant (Fig. 28.13), but before delivery of the placenta, takes advantage of continued uterine contractions leading to the collection of 80–160 cc (3–5 ounces) of blood (Fig. 28.14).[75] Following delivery, the placenta is most often incinerated (at an

---

[n] Sadly, the majority of individuals who need a cord blood transplant are children and young adults with leukemia. About 600 cord blood transplants are now done each year in the U.S. Once collected, cord blood can be stored in a public or private facility. Public storage banks are operated in the same way as public blood banks and are a source of cord blood from unrelated donors. They do not charge for collection or storage. They do charge U.S. 15,000–25,000 when a unit of cord blood is used for transplantation — a fee usually covered by health insurance. Private storage banks charge U.S. 1500–2500 for collection and U.S. 100 a year for storage. Most of the cord blood stored in private facilities is used to treat a sibling of the varied donors.[84] Private facilities are used primarily by couples expecting a child. They store cord blood for the future use of the donor or the donor's family.

An editorialist for *The New England Journal of Medicine*, not happy about the idea of, in effect, selling "a baby's placental blood [the blood is technically the baby's]...to the newborn's mother by charging her for collecting and storing it," likened the practice to a custom in Russia wherein you can "sell your testicles to a firm...that will give you [the equivalent of] four thousand (U.S.) dollars and then remove the items surgically...mash them up and extract the vital substances and market the resulting syrupy stuff as rejuvenating beauty cream, for a profit that is awesome."[85]

Table 28.5   Advantages of umbilical cord blood for allogeneic transplantation.

- Extremely abundant.

- Absence of risk to mother or child.

- No ethical concerns.

- Low incidence of infectious diseases.

- No risk of teratoma.

- Can be frozen and stored to be used when needed.

(*Sources*: References 77 and 86.)

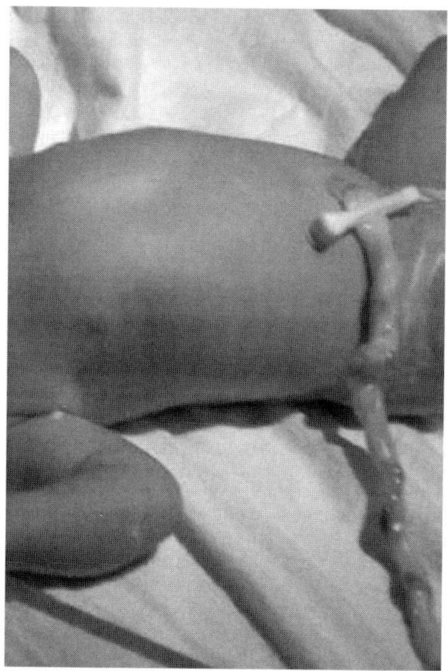

Fig. 28.13.   A 3-min-old baby, seconds after the cord was clamped.
(*Source*: Reproduced with permission from the publisher.) Reference 87.

off-premises, biological waste incinerator) in the Western world. However, in some Eastern cultures it is buried. In Cambodia, for example, it is buried to ensure "the health of the baby and the mother."[88] In Turkey, the placenta is wrapped in a clean cloth and then buried, often in the courtyard of a mosque.[88]

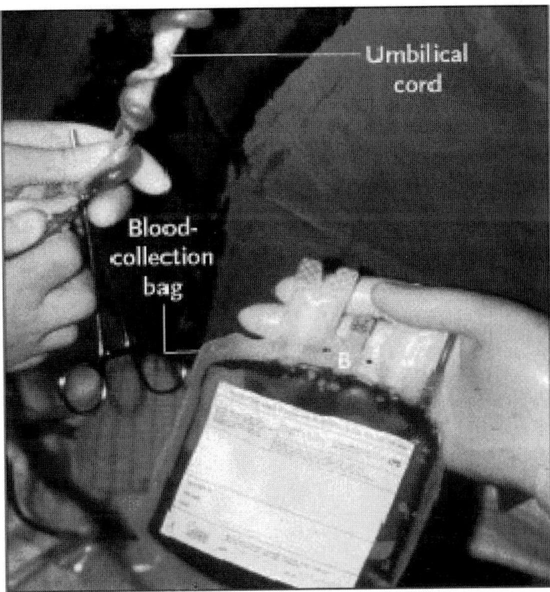

Fig. 28.14.   After birth, blood is collected — by an obstetrician, a nurse–midwife, or a designated collector — either from the placenta *in utero* or from the delivered placenta and drained by gravity into a sterile blood-collection bag. The blood is then sent to a central facility, where it is processed to remove excess cells and plasma. The nucleated cell content of the cord blood, a surrogate for the stem cell content, is determined and the mother's blood and sometimes the cord blood are tested for relevant infectious diseases. The unit is then frozen and stored in liquid nitrogen.

(*Source*: Reproduced with permission from the publisher.) Reference 84.

Evidence that the Human Heart Can Regenerate Healthy Muscle After an Acute Myocardial Infarction

In June 2001, the cardiology world was rocked by a paper in *The New England Journal of Medicine* bearing the title "Evidence That Human Cardiomyocytes Divide After Myocardial Infarction."[89] For many years, it had been thought that the loss of heart muscle after an acute myocardial infarction was irreversible. Unlike newts and salamanders, which are able to regenerate limbs after an amputation[90] (Fig. 28.15), we cannot regenerate even a little heart muscle after a heart attack to replace heart muscle lost. (Newts and salamanders aren't limited to regenerating a limb, by the way. They can also regenerate the intestine, the lens, the retina, segments of the brain and spinal cord, heart

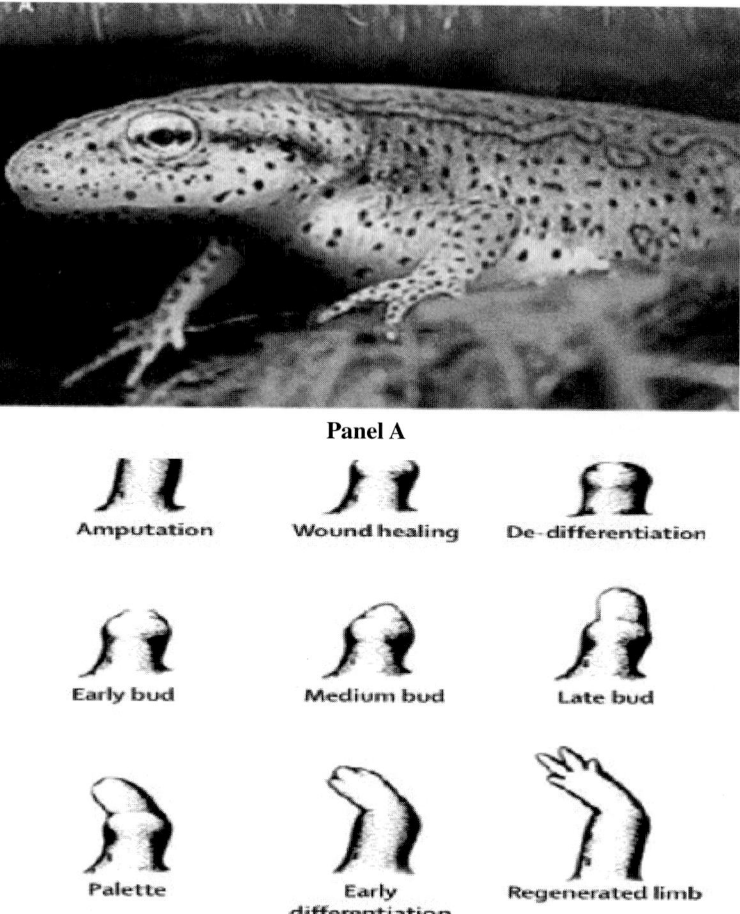

Fig. 28.15.   Panel A. A regenerated left forelimb ("arm") in a newt. The North American red spotted newt, a brilliantly colored salamander, can regenerate a lost limb — complete with skin, muscle and nerves — in about two months, allowing it to resume its dual life swimming in water and climbing on land and rocks. Panel B. A diagrammatic representation of stages from amputation to complete regeneration of a limb.

(*Source*: Reproduced with permission from the publisher.) Reference 90.

muscle — and, yes, even a whole eye![90]) The damaged heart muscle is replaced, instead, by scar tissue, which cannot contract, and hence the damaged region is no longer able to contribute to the pumping of blood to the rest of the body. The concept that the heart cannot renew itself was considered sacrosanct. It just didn't happen.

Then, on June 7, 2001, Italian-born Piero Anversa and his colleagues at New York Medical College in Valhalla shattered that shibboleth in a landmark publication maintaining that, in fact, the heart *can* form new, healthy muscle after a heart attack. To establish the heart's ability to repair itself, the authors painstakingly examined more than 100,000 cardiomyocyte nuclei from 13 patients who died 4–12 days following a heart attack. Their diligence paid off. There was clear evidence of fresh myocardium (heart muscle) in the border zone between dead and healthy myocardium, though no new muscle in the center of the damaged region — which, alas, was always far larger in size than the border zone.[89] (The border zone was kept alive by blood flowing through nearby patent blood vessels.) Thus, this study challenged the long-held view that heart muscle can hypertrophy (see first paragraph of this chapter), but not proliferate. And — overnight — the heart joined the skin, the red blood cells, the cornea, and, to a lesser extent, the liver as organs capable of regeneration. In an accompanying editorial by Nadia Rosenthal from the Massachusetts General Hospital she pointed out that, "Ironically, this most critical organ is also the most mortal" one.[91] While evidence of cardiac regeneration was certainly welcomed, the extent of cardiomyocyte proliferation would be much less than that needed to repair the damage wrought by a heart attack. Also, to help the acutely damaged heart, new cardiomyocyte formation must be fast *and* fresh blood vessels must simultaneously nourish the new heart muscle cells with oxygen to sustain them." Not dwelling on the limitations of the discovery, she concludes optimistically, "Sidestepping the need for banks of embryonic stem cells and potential immune rejection, the successful incorporation of a patient's own banked stem cells into multiple cardiac tissues might induce repair of a damaged heart."[91]

Chimerism of the Transplanted Human Heart

On January 3, 2002, the cardiology world was stunned by yet another paper from Valhalla, again in *The New England Journal of Medicine*, bearing the title "Chimerism of the Transplanted Human Heart." (It is unusual for physicians and scientists to have their research published twice in a lifetime in the *Journal*, and rarer still to do so in short order — twice in six months.) Some *Journal* readers grabbed a dictionary to check the meaning of the

Fig. 28.16. A mythical chimera. The monster was composed of segments from several animals. A lioness formed its body, its tail was a snake, and a goat sprang from the middle of its back. To make matters scarier, the grotesque "being" breathed fire on all who dared to near her. For unclear reasons, the monster was thought to be female.

(*Source*: Reproduced with permission from the publisher.) Reference 93.

word "chimerism."° Others soon learned through an accompanying editorial that, in ancient Greece, a chimera was a "mythical monster — part lion, part goat, and part snake" (Fig. 28.16).[95] In Greek mythology, the centaurs, the sirens, and the Minotaur are well-known examples of chimeras.[95] Fantasy became reality in the 1970s and 1980s, when chimeric mice (Fig. 28.17) and sheep–goat chimeras were developed.[95] Chimeric mice are a far cry from the ferocious-looking mythical characters. They're tame, nearly "pettable," almost cuddly — a totally different pedigree. (And these genetically engineered creatures eat far less than the fictional chimeras might have devoured.)

°Chimerism refers to the "presence of two cell types in a given individual each of which is derived from a different zygote" (fertilized egg)[92] and therefore the individual has two genetically distinct cell types. Chimeras can arise through a variety of spontaneous and experimental processes, including when the blood of fraternal twins (twins developed from two fertilized eggs, as opposed to identical twins, developed from a single fertilized egg) mixes *in utero*.[91] Chimerism also may occur long after birth as a result of "any organ transplantation other than between identical twins."[91] The several paths to chimerism share a common ending: two genetically different cell lines derived from two different zygotes in one individual.

Fig. 28.17.   Chimeric mice.

(*Source*: Reproduced with permission from the publisher.) Reference 94.

In addition to animal–animal chimeras, there are human–human chimeras (basically hybrids[P].) Human chimerism was documented in the 1990s through the detection of fetal cells in the mother's blood during pregnancy and the bidirectional "migration of cells between transplanted organs and their recipients."[95] Then, in early January 2002, Anversa and his colleagues first verified the occurrence of chimerism in hearts from females transplanted into male recipients.[96] Among 16,834 cardiomyocytes examined microscopically from eight men who had received a heart from a female donor, 18% of the heart muscle cells had the Y chromosome.[96] The chimerism under these circumstances was manifested by cells that (probably) migrated from the (male) recipient heart to the (female) donor heart. In these sex-mismatched transplants, confirmation of chimerism was relatively easy — the presence of

---

[P] Broadly speaking, a hybrid is "the offspring of two animals (or plants) of different races, varieties, breeds, species, etc."[92] Strictly speaking, a hybrid is "any offspring of two parents with different genotypes" genetic makeups. "In this sense, all humans are hybrids."[91] Speaking genetically, a hybrid is "a cell formed by the fusion of two genetically distinct cells."[91] For example, a hybrid develops following the mating of a female horse and a male donkey. The product is a mule — a true mule, as opposed to an obstinate person, who may be as stubborn as a....

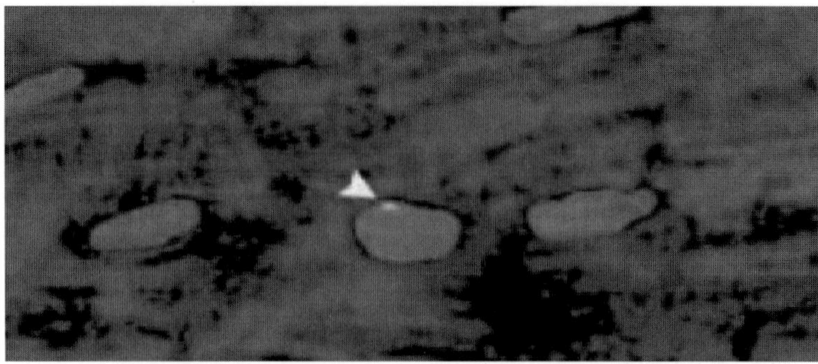

Fig. 28.18. A microscopic section of a human heart from a female donor transplanted into a male recipient. The arrowhead indicates a Y chromosome (with a fluorescent label) from the recipient in a nucleus of the donor.

(*Source*: Reproduced with permission from the publisher.) Reference 96.

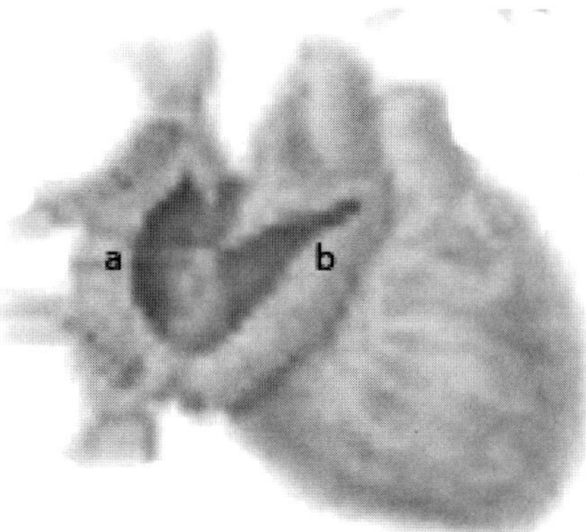

Fig. 28.19. A portion of the recipient right atrium (a) is sewn to the donor right atrium (b) to anchor the "new" heart during a cardiac transplant. Stem cells from the recipient right atrium may migrate across the suture line into the donor heart.

(*Source*: Reproduced with permission from the publisher.) Reference 97.

a Y chromosome, the definitive marker of a male cell, or chromosomes in a female heart where it/they didn't belong (Fig. 28.18). Just how a Y chromosome ended up in a female donor heart isn't hard to understand. When a cardiac transplant is performed, a portion of the recipient's right atrium (the blood-receiving chamber) is intentionally left behind to anchor the donor heart soon to be sewn to the recipient heart (Fig 28.19). Stem cells holding a Y chromosome in their nuclei may simply migrate across the suture line into the female donor heart, or Y-chromosome-containing stem cells in the

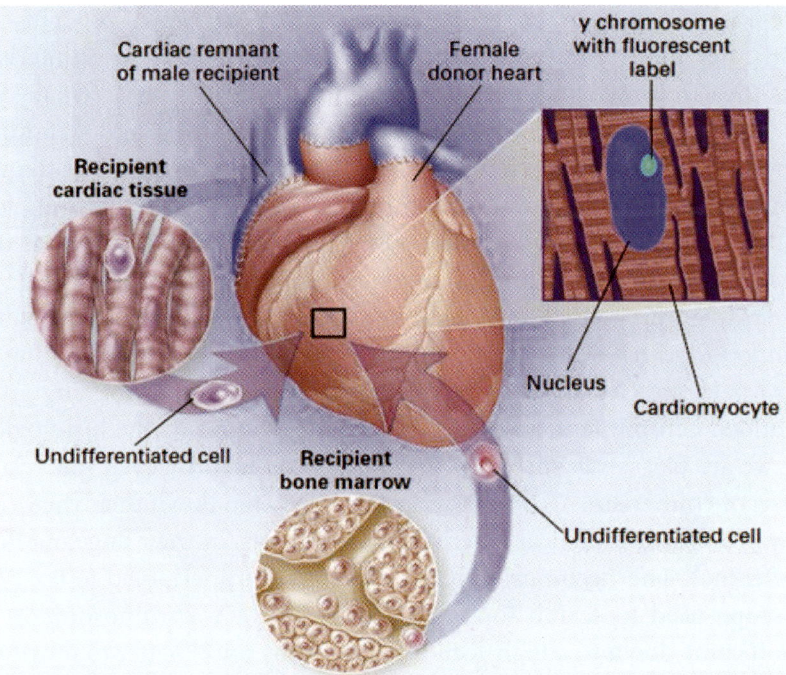

Fig. 28.20.   Two ways a heart from a female donor may incorporate cells with a Y chromosome from a male recipient. An undifferentiated stem cell from the recipient's *cardiac tissue* may migrate across the suture line between the cardiac remnant (a portion of the two atria) of the male recipient directly into the donor's heart, or a stem cell may circulate from the recipient's *bone marrow* into the donor's heart. A Y chromosome with a fluorescent label (as in Fig. 27.15) is shown in the inset.

(*Source*: Reproduced with permission from the publisher.) Reference 98.

recipient's bone marrow may circulate through the blood stream before reaching the heart, and then differentiate into heart muscle cells (Fig. 28.20). Irrespective of how a Y chromosome got from point A to point B, its presence provides "compelling evidence" that cells from the recipient heart migrated into the transplanted heart, according to a perspective by the editors of the *Journal*[98] — or does it?

The Great Chimerism Debate

Remarkably discrepant data about the frequency of cardiac chimerism quickly came to light. The disparity focused attention on the equipment and technique used by different investigators at different institutions to determine the presence (or absence) of cardiac chimerism. In early April 2002, just three months after the initial report from Valhalla finding cardiac chimerism in 18% of the cardiomyocytes in male recipients of a heart from a female donor, researchers at the University of Washington (Seattle) and Washington University (St. Louis) found a much lower rate of chimerism.[99] The frequency was only 0.04% or four instances of chimerism per 10,000 cardiac muscle cells visualized microscopically! The authors speculated that the "enormous difference" could be due to "technical differences" between the two studies.[99]

Anversa's group fired back quickly. In a May 17, 2002, letter to the editor of *Circulation Research*, which had published the "0.04% report," the gang from Valhalla paid the University of Washington–Washington University group a backhanded compliment, saying, "Given the resolution of the histologic sections...we are pleasantly surprised that they could identify even this relatively low level of chimerism."[100] Both warring sides watered down their rhetoric just a bit by proffering "technical differences" as the basis for their huge mathematical difference. The "technical differences" referred to pertained to the type of microscope used to search for cardiac cells with a Y chromosome in their nuclei. It came down to a high-tech version of "my toy is better than your toy." Anversa said his more modern scope had better resolution, allowing him to see things better (in more detail) than the University of Washington–Washington University group. The University of Washington–Washington University group fired back, "Conventional microscopy [which they used] offers much better detection of cell borders than does fluorescence microscopy" (which the Anversa group used).[99] And that, they felt, allowed them to tell heart muscle cells from white blood cells, of all things, that may have been "sandwiched between" the two cell types. Such "sandwiching" could have led to white blood cells being "misidentified as cardiomyocyte nuclei" by the fluorescent microscopy technique used by Anversa and his colleagues. Anversa's comeback? "While we commend their thoroughness, we are surprised that they seem to doubt our ability to do the same."[100] Anversa ended on an upbeat note, saying that his opponents "document beyond doubt that new myocytes are generated in the adult heart from nonmyocte precursors,"[100] something the University of Washington researcher had previously disclaimed.

There the debate would die down...until early July 2002. Then, on July 2, everything exploded, almost as if the controversy was somehow linked to the upcoming Fourth of July fireworks. Yet another group joined the fracas. Cell biologists from the University of Pennsylvania further fueled the dispute by reporting that they had looked carefully at the microscopic sections obtained from the hearts of six men who died six months to ten years after receiving a heart from a female donor. Upon doing so, the U.P. scientists wrote, "After examination of >6,000 myocyte nuclei...we were unable to detect any nuclei with the clear presence of the Y chromosome,"[102] implying that cardiac chimerism did not occur. Not a single instance of chimerism, in contrast to an 18% frequency reported six months earlier by Anversa *et al.*

Such antithetical findings served as fodder for an editorial. Cardiologists and pathologists from Duke University and Johns Hopkins merged to write that their combined experiences (with regard to the frequency of chimerism) were similar to the results from the University of Pennsylvania.[103] They found "very low levels" of chimerism in the human heart "very consistent" with the reports from both the University of Pennsylvania and the University of Washington.[103] Being questioned by an influential academic quadrumvirate such as Duke, Johns Hopkins, the University of Pennsylvania, and the University of Washington must have shaken even the prolific Valhalla group. (Figure 28.21 shows they consistently publish 10–14 papers per year on stem cells and myocardial regeneration.) In fact, it was more than "just" their frequency of cardiac chimerism that came under fire, it was also their belief that the human heart could repair itself that was being challenged. If you're going to believe myocardial chimerism is real, you're also likely to believe the heart can repair itself. Cardiac chimerism and cardiac regeneration go hand in hand. The skeptical, four-university alliance seriously questioned the novel Jack-and-Jill, cardiac–chimerism–cardiac self-repair notion. The coalition claimed that while the fluorescent images published by the Valhalla group

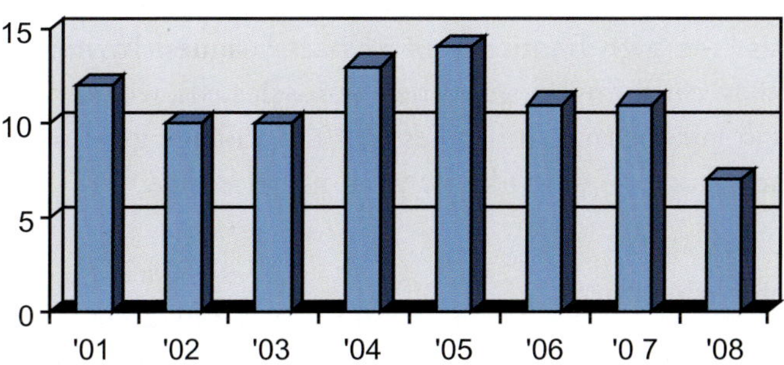

Fig. 28.21. The average number of papers published by Anversa and his colleagues from 2001 through June 2008.

were "aesthetically pleasing" they did not "definitively demonstrate" cardiac chimerism.[103] A "sandwiching" effect could easily have created the illusion of chimerism.[103] The east coast–west coast connection summed up, "The completion of this paradigm shift, from the dogma of nonregenerating myocardium to that of a self-repairing heart, requires unquestionable proof and a reproduction of the data from a number of investigators. The reports [from the University of Pennsylvania and the University of Washington] do not support such extraordinary regenerative capacity for the myocardium. Although we recognize the importance of challenging dogmas for the progress of science and the welfare of humankind, we also urge all involved in this field to rigorously evaluate the data."[103]

The next salvo came from Valhalla on October 29, 2002, in the form of a letter to the editor of *Circulation*, which had published the July edition of the spat. After reiterating that the "published values [of cardiac chimerism] range from as high as 18% [his group's value] to as low as 0.04% [the University of Washington–Washington University group's value] to totally absent" (the University of Pennsylvania group's value), Anversa continued the spring dialog on "technical differences" as the basis for the dissimilar numbers. Coming down hard on the groups from Duke, Hopkins, and the University of Washington, which used a conventional — though 50-year-old — microscope, he said "these investigators [should] improve their technical approach to the higher standard of the year 2002."[104] He wound up pointedly, "…it is imperative that the challenger…use techniques…that are at least as sensitive and discriminating as those used to obtain the observations being challenged."[104]

In a brief rebuttal, the Duke arm of the squabble wrote that the issue was not the age or the price of the microscope used by different groups. Rather, the issue was "the high frequency of artifacts" caused by mistaking nearby cells, other than cardiomyocytes, as heart muscle cells, when they were not.[104] (This is tough talk, because it implies that the Valhalla group — while using a fancier microscope — didn't know what it was seeing.)

The Lull Before the Storm

The "great chimerism debate" simmered down for the next four-and-one-half years — in part because during that time the gang from Valhalla was designing and conducting a study, in conjunction with three other universities, that they expected would greatly impact the disagreement — but one (unchallenged) paper worthy of mention did appear in print during the lull. The three other universities were Temple University in Pennsylvania, the University of Louisville in Kentucky, and San Diego State University in California. If nothing else, this new four-university alliance balanced the scale of academicians weighing in on cardiac chimerism and myocardial regeneration. (The results of that four-university investigation were published in November 2007 in the *Proceedings of the National Academy of Sciences*.[105]) In October 2005, Anversa and his colleagues published a review paper on the topic of cardiac stem cells and myocardial regeneration.[106] And what a review it was! The majority of original scientific reports are 5–7 pages in length, with 20–30 references, while review papers tend to be 15 to 20 pages long, with 100–250 references — but *this* one ran into 43 pages and included more than 500 references!

The paper opened provocatively, saying the discovery that bone–marrow-derived stem cells can form other tissues — such as heart muscle — had been "perceived with excitement and mistrust, which [has] equally divided the scientific and clinical communit[ies]. The prevailing reaction on the part of the skeptics was disbelief for this alternative view of stem cell biology. Studies reporting negative findings were published, and questions raised about the validity of the shift in paradigm advanced in early work. This controversy is feeding a fire that may disrupt the stream of scientific consciousness and delay the clinical application of a prospectively revolutionary therapeutic tool."[106] Pressing on, it said that "if the goal of medicine has been to treat symptoms of disease and possibly remove their causes, scientists and physicians point now to a more ambitious target. The primary role of regenerative medicine is the complete structural and functional recovery of the damaged organ."[106] The specialty of regenerative medicine was born in 1998.[q]

Concerning the controversy over transdifferentiation and cardiac repair by stem cells, Anversa continued, "Because of the compelling need to seek new treatments for severely ill patients, clinicians are leading the field of [stem] cell therapy and myocardial regeneration."[106] But, because some investigators

---

[q] A new journal, *Regenerative Medicine*, also sprang up eight years later. Published in London, the inaugural issue in January 2006 said that the magazine was "devoted to the emerging field of regenerative medicine and stem cell biology."[107] By then, the "field of regenerative medicine and stem cell biology [had] become one of the hottest areas in contemporary science."[107]

involved deeply in the field have urged caution before proceeding with clinical trials of bone-marrow-derived stem cells to regenerate cardiac muscle (see pp. 279–281), Anversa once again went on the defensive, saying that the "revolutionary work on the brain and the heart that has led to the identification of neural stem cells[r] and cardiac stem cells has been attacked as inconclusive, methodologically incorrect, and, more recently, a collection of artifacts."[106] And he pointed out, "The most relevant observation that dramatically challenged the old paradigm of the heart as an...organ (incapable of self-repair) was the identification of male cells in female hearts transplanted in male recipients. In these cases of sex-mismatched cardiac transplants in humans, the female heart in a male host had a significant number of Y-chromosome positive myocytes," implying that these male cells "colonized" the female heart and later formed functional cardiomyocytes.[105] (There are at least four other reports from independent institutions indicating the presence of the Y chromosome in sex-mismatched cardiac transplantations.[108–111] You gotta admit that's pretty convincing.

References

1. Anversa P, Leri A, Nadal-Ginnard B, Kajtura J. (2004) Stem cells and heart disaese. In: Lanza R, Blauu H, Melton D, Moore, West M (eds.), *Handbook of Stem Cells*. Elsevier, Amsterdam, Boston, Heidleberg, London, New York, Oxford, Paris, San Diego, San Francisco, Singapore, Sydney, Tokyo, pp: 715–719.
2. Nadal-Ginnard B, Kajstura J, Leri A, Anversa P. (2003) Myocyte death, growth, and regeneration in cardiac hypertrophy and failure. *Circ Res* **92**: 139–150.
3. Morrison SJ, Spadling AC. (2008) Stem cells and niches: mechanisms that promote stem cell maintenance throughout life. *Cell* **132**: 598–611.
4. Spadling A, Drummond-Barbosa D, Kai T. (2001) Stem cells find their niche. *Nature* **414**: 98–104.
5. Watt FM, Hogan BLM. (2000) Out of Eden: stem cells and their niches. *Science* **287**: 1427–1430.
6. Smith O. (2008) In this issue. *Cell* **132**: 501–503.
7. Orlic D, Kajstura J, Chimenti S, Jakoniuk I, Anderson SM, Li B, Pickel J, McKay R, Nadal-Ginard B, Bodline DM, Leri A, Anversa P. (2001) Bone marrow cells regenerate infarcted myocardium. *Nature* **410**: 701–705.

---

[r]Although a few studies suggest that the human brain can form a limited number of new cells, brain stem cells have not yet been identified.[112]

8. Sussman M. (2001) Hearts and bones. *Nature* **410**: 640–641.

9. Murry CE, Soonpaa MH, Reinecke H, Nakajima H, Nakajima HO, Rubart M, Pasumarthi KBS, Virag KI, Bartelmez SH, Poppa V, Bradford G, Dowell JD, Williams DA, Field LJ. (2004) Haematopoietic stem cells do not transdifferentiate into cardiac myocytes in infarcted hearts. *Nature* **428**: 664–668.

10. Balsam LB, Wagers AJ, Christensen JL, Kofidis T, Weissman I, Robbins RC. (2004) Haemtopoietic stem cells adopt mature haematopoietic fates in ischaemic myocardium. *Nature* **428**: 668–673.

11. Editorial. (2004) No conensus on stem cells. *Nature* **428**: 587.

12. Chien KR. (2004) Lost in translation. *Nature* **428**: 607–608.

13. http://en.wikipedia.org/wiki/Clinical_trial

14. *American Jurisprudence,* 2d (2004).

15. Gore JM, Sloan M, Price TR, Ranbdall AMY, Bowell E, Collen D, Forman S, Knatterud GL, Sopko G, Terrin ML, and the TIMI investigators. (1991) Intracranial hemmorhage, cerebral infarction, and subdural hematoma after acute myocardial infarction and thrombolytic therapy in the thrombolysis in myocardial infarction study. Thrombolysis in Myocardial Infarction, Phase II, Pilot and Clinical Trial. *Circulation* **83**: 448–459.

16. Rovelli F, De Vita C, Feruglio GA, Lotto A, Selvini A, Tognoni G. (1986) GISSI trial: early results and late follow-up. *Gruppo Italiano per lo Sperimentazione della Streptochinasi nell'Infarto Miocardico. J Am Coll Cardiol* **10** (**Suppl B**): 33B–39B.

17. Gusto Investigators. (1993) An international randomized trial comparing four thrombolytic strategies for acute myocardial infarction. *N Engl J Med* **329**: 678–682.

18. Boyle AJ, Schulman SP, Hare JM. (2006) Stem cell therapy for cardiac repair: ready for the next step. *Circulation* **114**: 339–352.

19. Wollert KC, Meyer GP, Lotz J, Ringes-Lichtenberg S, Lippolt P, Breidenbach C, Fichtner S, Korte T, Hornig B, Messinger D, Arseniev LL, Hertenstein B, Ganser A, Drexler H, Hertenstein B, Ganser A, Drexler H. (2004) Intracoronary autologous bone-marrow cell transfer after myocardial infarction: the BOOST randomized controlled clinical trial. *Lancet* **364**: 141–148.

20. Janssens S, Dubois C, Bogart J, Theunissen K, Deroose C, Desmet W, Kalantzi M, Herbots L, Sinnaeve P, Dens J, Maertens J, Rademakers F, Dymarkowski S, Ghetsens O, Van Cleemput J, Bormans G, Nuyts J, Belmans A, Mortelmans L, Boogaerts M, Van de Werf F. (2006) Autologous bone marrow–derived stem-cell transfer in patients with ST-segment elevation myocardial infarction: double-blind, randomized controlled trial. *Lancet* **367**: 113–121.

21. Lunde K, Solheim S, Aakuhus *et al.* (2005) Autologous stem cell transpantation in acute myocardial infarction: the ASTAMI randomized controlled trial.

Intracoronary transplantation of autologous mononuclear bone marrow cells, study design and safety aspects. *Scand Cardiovasc J* **39**: 150–158.

22. Oettgen P. (2006) Cardiac stem cell therapy. Need for optimization of efficacy and safety monitoring. *Circulation* **114**: 353–358.

23. Daley GQ, Scadden D. (2008) Prospects for stem cell–based therapy. *Cell* **132**: 544–548.

24. Strauer BE, Brehn M, Zeus T, Kostering M, Hernandez A, Sorh RV, Kogler G, Wernet P. (2002) Repair of infarcted myocardium by autologous intracoronary mononuclear bone marrow transplantation in humans. *Circulation* **106**: 1913–1918.

25. Assmus, Schachinger V, Teupe M, Britten M, Lehmann, Dobert N, Grunwald F, Aicher A, Urbich C, Nartin H, Hoelzer D, Dimmeler S, Zeiher AM. (2002) Transplantation of progenitor cells and regeneration enhancement in acute myocardial infarction. (TOPCARE-AMI). *Circulation* **106**: 3009–3017.

26. Meyer GP, Wollert KC, Lotz J, Steffens J, Lippolt P, Fichtner S, Hecker H, Schaefer A, Arsenier L, Hertenstein B, Ganser A, Drexler H. (2006) Intracoronary bone marrow cell transfer after acute myocardial infarction. Eighteen months' follow-up data from the randomized controlled BOOST [BOne marrOw transfer to enhance ST-elevation infarct regeneration Trial]. *Circulation* **113**: 1287–1294.

27. Chen SI, Fang HW, Olan J, Ye F, Liu YH, Shan SJ, Lin S, Liao LM, Zhao RCH. (2004) Improvement of cardiac function after transplantation of autologous bone marrow mesenchymal cells in patients with acute myocardial infarction. *Chin Med J (Engl)* **117**: 1443–1448.

28. Fernandez-Avila F, San Roman JA, Garci-Frade J, Fernandez ME, Penarrubia MJ, de la Fuente L, Gomez-Bueno M, Cantalapi A, Fernandez J, Guitterez O, Sanchez PL. (2004) Experimental and clinical regenerative capability of human bone marrow cells after myocardial infarction. *Circ Res* **95**: 742–748.

29. Bartunek J, Vanderheyden M, Vanderekhove B, Mansour S, De Bruyne B, De Bondt P, Van Haute I, Lootens N, Heyndtrickx G, Wijns W. (2005) Intracoronary injection of CD133-positive enriched bone marrow progenitor cells promotes recovery after recent myocardial infarction. Feasibility and safety. *Circulation* **112**: 178–183.

30. Schachinger V, Erbs S, Elsasser A, Haberbesch WM, Hambrecht R, Holschermann H, Yu J, Corti R, Mathey DG, Hahhm CM, Suselbeck T, Assmus B, Tonn T, Dimmeler S, Zeiher AM, for the PEPAIR-AMI investigators. (2006) Intracoronary bone marrow–derived progenitor cells in acute myocardial infarction. *N Engl J Med* **355**: 1210–1221.

31. Lunde K, Solheim S, Aakhus S, Arnesen H, Abdelnoon M, Egeland T, Endresen K, Ilebekk A, Mangschau A, Field JG, Smith HJ, Taraldsrud E, Grogaard HK,

Bjornerheim R, Brekke M, Muller C, Hopp E, Ragmarsson A, Brinchmann JE, Forfang K. (2006) Intracoronary injection of mononuclear bone marrrow cells in acute myocardial infarction. *N Engl J Med* **355**: 1199–1209.

32. Lunde K, Solheim S, Forfang K, Arnesen H, Brinch L. Bjornerheim R, Ragnarsson A, Egeland T, Endresen K, Ilebekk A, Mangaschau A, Aakhus S. (2008) Anterior myocardial infarction with acute percutaneous intervention and intracoronary injection of autologous mononuclear bone marrow cells safety, clinical outcome, and serial changes in left ventricular function during 12 months' follow-up. *J Am Coll Card* **51**: 674–676.

33. Rosenzweig A. (2006) Cardiac cell therapy — mixed results from mixed cells. *N Engl J Med* **355**: 1274–1277.

34. Schwartz RS. (2006) The politics and promise of stem cell research. *N Engl J Med* **355**: 1189–1191.

35. Coleman A. (2008) Stem cell research in Singapore. *Cell* **152**: 519–521.

36. Walsh B. (2006) Singapore: Asia's stem cell city. *Time Magazine*, Oct. 30.

37. Kalikki PC, Sachan RSGS. (2004) Ischemic and anesthetic preconditioning of the heart: an insight into the concepts and mechanisms. *Internet Anesthesiol* **8**(2).

38. Hofmann M, Wollert KC, Meyer GP, Menke A, Arseniev L, Hertenstein B, Ganse A, Knapp WH, Drexler H. (2005) Monitoring of bone marrow cell homing into the infarcted human myocardium. *Circulation* **111**: 2198–2201.

39. Penicka M, Widimsky P, Kobylka P, Kozart T, Long O. (2005) Early tissue distribution of bone marrow mononuclear cells after transcoronary transplantation in a patient with an acute myocardial infarction. *Circulation* **112**: e63–e65.

40. Penn MS. (2005) Stem-cell therapy after acute myocardial infarction: the focus should be on those at risk. *Lancet* **367**: 87.

41. Stamm C, Westphal B, Kleine HD, Petzsch M, Kittner M, Klinge H, Schumichen C, Nienaber CA, Steinhoff G. (2003) Autologous bone-marrow stem cell transplantation for myocardial infarction. *Lancet* **361**: 43–46.

42. Vulliet PR, Greeley M, Halloran M, Mac Donald A, Kittelson MD. (2004) Intracoronary arterial injection of mesenchymal stromal cells and microinfarction in dogs. *Lancet* **363**: 783–784.

43. Moelker AD, Spitskovsky D, Baks T, van Beutekom HMM, Muhs A, Duncker DJ, Wendt S. (2005) Intracoronary delivery of human umbilical cord derived stem cells does not improve ventricular function after four weeks in swine with a myocardial infarction. [Abstr.] *Eur Heart J* **26** (**Suppl**): 122.

44. Ben-Dor I, Fuchs S, Kornowski R. (2006) Potential hazards and technical considerations associated with myocardial cell transplantaion protocols for ischemic myocardial syndrome. *J Am Coll Cardiol* **48**: 1519–1526.

45. Breitbach M, Bostami T, Rosell W, Xia Y, Dewald O, Nygren JM, Fries JWW, Tiemann K, Bohlen H, Hescheler J, Welz A, Bloch W, Jacobaern FW,

Fleischmann BK. (2007) Potential risk of bone marrow cell transplantation into infarcted hearts. *Blood* **110**: 1362–9.44.

46. http://en.wikipedia.org/wiki/Teratoma

47. Nagata S, Tsuchiya M, Asano S, Yamamoto O, Hirata Y, Kubota N, Oheda M, Nomura H, Yamazaki Y. (1986) The chromosomal gene structure and two mRNAs for human granulocyte colony–stimulating factor. *EMBO J* **5**: 871–876.

48. Kang H-J, Kim H-S, Zhang S-Y, Park K-W, Cho H-J, Koo B-K, Kim Y-J, Lee D, Sohn D-W, Han K-S, Oh B-H, Lee M-M, Park Y-B. (2004) Effects of intracoronary infusion of peripheral blood stem cells mobilized with granulocyte-colony-stimulating factor on left ventricular systolic function and restenosis after coronary stenting in myocardial infarction: the MAGIC cell randomized clinical trial. *Lancet* **363**: 751–756.

49. Valgimigli H, Rigolini GM, Cittanti C, Walagutti P, Curello S, Peccoorco G, Bugli AM, Della Porta M, Bragotti MM, Feggi L, Lan Franchi A, Gigantoi R, Feggi L, Castoldi G, Ferrari R. (2005) Use of granulocyte colony stimulating factor during acute myocardial infarction to enhance bone marrow stem cell mobilization in humans: clinical and angiographic safety profile. *Eur Heart J* **26**: 1838–1845.

50. Suarez de Lezo J, Torres A, Herrera I. Pan M, Romoro M, Paulovic D, Segura J, Ojeda S, Sanchez J, Lopez RF, Medina H. (2005) Effects of stem-cell mobilization with recombinant human granulocyte colony stimulating factor in patients with percutaneously revascularized acute anterior myocardial infarction. *Rev Esp Cardiol* **58**: 253–261.

51. Kuethe F, Figulla HR, Herzau M, Voth M, Fritzenwanger M, Operfermann T, Pachmann K, Kraicic A, Sayer HG, Gottschild D, Werner GS. (2005) Treatment with granulocyte colony-stimulating factor mobilization of bone marrow cells in patients with acute myocardial infarction. *Am Heart J* **150**: 115–121.

52. Ince H, Petzsch M, Klein HD, Schmidt H, Rehders T, Korber T, Schumichen C, Freund M, Wienaber CA. (2005) Preservation from left ventricular remodeling by front-integrated revascularization and stem cell liberation in evolving acute myocardial infarction by use of granulocyte-colony-stimulating factor (FIRSTLINE-AMI). *Circulation* **112**: 3097–3106.

53. Zolinhofer D, Ott I, Mehilli J, Schomig K, Michalk F, Ibrahim T, Meisetschlager G, von Wedel J, Bollwein H, Seyfarth M, Dirschinger J, Schmitt C, Schwaiger M, Kaasrati A, Schomig A for the REVIVAL-2 investigators. (2006) *JAMA* **295**: 1003–1010.

54. Ripa RS, Jergensen E, Yong-Hong W, Thune JJ, Nilsson JC, Sendergaard L, Johnsen HE, Korber L, Grande P, Kastrup J. (2006) Stem cell mobilization induced by subcutaneous granulocyte-colony stimulating factor to improve

cardiac regeneration after acute ST-elevation myocardial infarction (STEMMI) trial. *Circulation* **113**: 1983–1992.

55. Engelmann MG, Theiss HD, Hennig-Theiss C, Huber A, Winterpeerger BJ, Werle-Ruedinger AE, Schoenberg SO, Steunbock G, Franz WM. (2006) Autologous bone marrow stem cell mobilization induced by granulocyte colony–stimulating factor after subacute ST-segment elevation myocardial infarction indergoing late revascularization: final results from the G-CSF STEMI [Granulocyte Colony-Stimulating Factor ST-segment Elevation Myocardial Infarction] trial. *J Am Coll Cardiol* **48**: 1712–1721.

56. Kloner RA. (2006) Attempts to recruit stem cells for repair of acute myocardial infarction: a dose of reality. *JAMA* **295**: 1058–1060.

57. Hill JM, Bartinek J. (2006) The end of granulocytye colony–stimulating factor in acute myocardial infarction? *Circulation* **113**: 1926–1928.

58. Forrester JS. Shah PK, Makkar RR. (2006) Myocardial regeneration by stem cells. *J Am Coll Cardiol* **48**: 1722–1724.

59. Kastrup J, Ripa RS, Wang Y, Jorgensen K. (2006) Myocardial regeneration induced by granulocyte colony–stimulating factor mobilization of stem cells in patients with acute or chronic ischaemic heart disease: a non-invasive alternative for clinical stem cell therapy? *Eur Heart J* **27**: 2748–2754.

60. Dimmeler S, Zeiher AM, Schneider MD. (2005) Unchain my heart: the scientific foundation of cardiac repair. *J Clin Invest* **115**: 572–583.

61. Caspi O, Gepstein L. (2006) Myocardial regeneration using human embryonic stem cell–derived cardiomyocytes. *J Cont Rel* **116**: 211–218.

62. Marelli D, Desrosiers C, el-Alfy M, Kao RL, Chiu RC. (1992) Cell transplantation for myocardial repair: an experimental approach. *Cell Transplant* **1**: 383–390.

63. Leobon B, Garrin I, Menasche P, Vilquin JT, Audinat E, Gharpaic S. (2003) Myoblasts transplanted into rat infarcted myocardium are functionally isolated from the host heart. *Proc Natl Acad Sci* **100**: 7808–7811.

64. Rubart M, Soonpaa MH, Nakajuma H, Field LJ. (2004) Spontaneous and evoked intracellular calcium transients in donor-derived myocytes following intracardiac myoblast transplantation. *J Clin Invest* **114**: 775–783.

65. Menasche P. (2007) Skeletal myoblasts as a therapeutic agent. *Prog Cardiovasc Dis* **50**: 7–17.

66. Reinecke H, Poppa V, Murry CE. (2002) Skeletal muscle stem cells do not transdifferentiate into cardiomyocytes after cardiac grafting. *J Molec Cell Cardiol* **34**: 241–249.

67. Planat-Benard V, Silvestre JS, Cousin B, Andrea M, Nibbelink M, Tamarat R, Cleergue M, Manneville C, Saillan-Barreau C, Duriez M, Tedgui A, Levy B, Penicaud L, Casteilla L. (2004) Plasticity of human adipose lineage cells toward

endothelial cells: physiological and therapeutic perspectives. *Circulation* **109**: 656–663.

68. Kehat I, Khimovich L, Caspi O, Gepstein A, Shofti R, Arbol G, Huber I, Satin J, Itskovitz-Eldor J, Gepstein L. (2004) Electromechanical integration of cardiomyocytes derived from human embroyonic stem cells. *Nat Biotecnol* **22**: 1282–1289.

69. Leri A, Kajstura J, Anversa P. (2005) Cardiac stem cells and mechanisms of myocardial regeneration. *Physiol Rev* **85**: 1373–1416.

70. http://stemcellworld.tripod.com/page9.html

71. http://www.stemcellresearchfacts.com/pros_cons.html

72. http://stemcells.nih.gov/info/basics4.asp

73. Editorial. (2009) Science and stem cells. *The New York Times,* Mar 9.

74. Hodgson DM, Behar A, Zingham LV, Kane GC, Perez-Terzic C, Alekseev AE, Puceat M, Torzic A. (2004) Stable benefit of embryonic stem cell therapy in acute myocardial infarction. *Am J Physiol Heart Circ Physiol* **287**: H471–H479.

75. Laflamme MA, Murry CE. (2005) Regenerating the heart. *Nat Biotechnol* **23**: 845–856.

76. Caspi O, Gepstein L. (2006) Stem cells for myocardial repair. *Eur Heart J* **8** (Suppl E): E43–E54.

77. Rubinstein P, Rosenfeld RE, Adamsom JW, Stevens CE. (1993) Stored placental blood for unrelated bone marrow reconstitution. *Blood* **81**: 1679–1690.

78. http://en.wikipedia.org.wiki/Placenta

79. Silberstein E, Jeffries LC. (1996) Placental blood banking — a new frontier in transfuaion medicine. *N Engl J Med* **335**: 199–201.

80. Gluckman E, Rocha V, Boyer-Chammard A, Locatelli F, Arcese W, Pasquini R, Ortega J, Souilleet G, Ferreira E, Laporte JP, Fernandez M, Chastang C, for the European. (1997) Transplant Outcome of Cord-blood Transplantation from Related and Unrelated Donors. *N Engl J Med* **337**: 373–381.

81. Laughlin MJ, Eapen M, Rubinstein P, Wagner JE, Zhang M-J, Champlin RE, Stevens C, Barker JN, Gale RP, Lazarus H, Marks DI, van Rood JJ, Scardavou A, Horowitz MM. (2004) Outcomes after transplanttion of cord blood or bone marrow from unrelated donors in adults with leukemia. *N Engl J Med* **351**: 2265–2275.

82. Policy Statement by American College of Pediatrics. Section on Hematolgy/Oncology and Section on Allergy/Immunology. Cord blood banking for potential future transplantation. *Pediatrics* **119**: 165–170.

83. http://en.wikipedia.org/wiki/Binomial_coefficient

84. Steinbrook R. (2004) The cord-blood-bank controversies. *N Engl J Med* **351**: 2255–2257.

85. Annas GJ. (1999) Waste and longing — the legal status of placental-blood banking. *N Engl J Med* **340**: 1521–1524.

86. Tse W, Laughlin MJ (2005) Umbilical cord blood transpantation: a new alternative option. *Hematology* **1**: 377–383.

87. http://en.wikipedia.org.wiki/Umbilical_cord

88. Buckley SJ. Placenta rituals and folklore from around the world. Mothering. *Natural Family Living*, Issue No. 131.

89. Beltrami AP, Urbanek T, Kajstura J, Yan S-N, Finato N, Bassani R, Nadal-Ginard B, Silvestri F, Leri A, Beltrami B, Anversa P. (2001) Evidence that human cardiac myocytes divide after myocardial infarction. *N Engl J Med* **344**: 1750–1771.

90. Mathur A, Martin JF. (2004) Stem cells and repair of the heart. *Lancet* **364**: 183–192.

91. Rosenthal N. (2001) High hopes for the heart. *N Engl J Med* **344**: 1785–1787.

92. *Blakiston's Medical Dictionary*, 4th ed. (1979). New York, St Louis, San Francisco, Auckland, Bogota, Dusseldorf, Johannesburg, London, Madrid, Mexico, Montreal, New Delhi, Panama, Paris, Sao Paulo, Singapore, Sydney, Tokyo, Toronto.

93. https://www.google.com/search?q=chimera+ancient+greece&hl=en&client=fir efox-a&hs=kNJ&rls=org.mozilla:enUS:official&prmd=imvns&source=lnms&t bm=isch&ei=rdTLT_XRIqie2QX4p63ZCw&sa=X&oi=mode_link&ct=mode& cd=2&ved=0CEgQ_AUoAQ&biw=1440&bih=756

94. http://en.wikipedia.org/wiki/Chimera_(genetics)

95. Bolli R. (2002) Regeneration of the human heart — no chimera? *N Engl J Med* **346**: 55–56.

96. Quaini F, Urbanek K, Beltrami AP, Finato N, Beltrami CA, Nadal-Ginard B, Kajstura J, Leri A, Anversa P. (2002) Chimerism of the transplanted heart. *N Engl J Med* **346**: 5–15.

97. www.netterimages.com/image/2815.htm

98. Schwartz MA, Curfman GD. (2002) Can the heart repair itself? *N Engl J Med* **346**: 2–4.

99. Laflamme MA, Myerson D, Saffitz JE, Murry CE. (2002) Evidence for cardiomyocyte repopulation by extracardiac progenitors in transplanted human hearts. *Circ Res* **90**: 634–640.

100. Nadal-Ginard B, Anversa P. (2002) Chimera or not chimera? *Circ Res* **90**: e72.

101. Glaser R, LU MM, Narula N, Epstein JA. (2002) Smooth muscle cells, but not myocytes of host origin in transplanted human hearts. *Circulation* 17–19.

102. Taylor DA, Hruban R, Rodriguez ER, Goldschmidt-Chermont PJ. (2002) Cardiac chimerism a mechanism for self-repair: does it happen and if so to what degree? *Circulation* **106**: 2–4.

103. Anversa P, Nadal-Ginard B. (2002) Cardiac chimerism: methods matter. *Circulation* **106**: e129–e131.

104. Rota M, Kajstura J, Hosoda T, Beaazi C, Vitale S, Esposito G, Iaffaldano G, Padin-Irvegas ME, Gonzalez-A, Rizzi R, Smal N, Murski J, Alvarez R, Chen X,

Urbanek K, Bolli R, Chen X, Urbanek K, Bolli R, Heuser SR, Leri A, Sussman MA, Anversa P. (2007) Bone marrow cells adopt the cardiomyogenic fate *in vivo*. *Proc Natl Acad Sci* **104**: 17783–17788.

105. Leri A, Kajstura J, Anversa P. (2005) Cardiac stem cells and mechanisms of myocardial regeneration. *Physiol Rev* **85**: 1373–1416.

106. Minger SL. (2006) Regenerative medicine. *Regen Med* **1**: 1–2.

107. Hocht-Zeisberg EH, Kahnert H, Guan K, *et al.* (2004) Cellular repopulation of myocardial infarction in patients with sex-mismatched heart transplantation. *Eur Heart J* **25**: 749–758.

108. Laflamme MA, Myerson D, Saffitz JE, *et al.* (2002) Evidence for cardiomyocyte repopulation by extracardiac progenitors in transplanted human hearts. *Circ Res* **90**: 634–640.

109. Muller P, Pfeiffer P, Koglin J, *et al.* (2002) Cardiomyocytes of noncardiac origin in myocardial biopsies of human transplanted hearts. *Circulation* **106**: 31–35.

110. Deb A, Wang S, Skelding KA, *et al.* (2003) Bone marrow–derived cardiomyocytes are present in adult human heart: a study of gender-mismatched bone marrow transplantation patients. *Circulation* **107**: 1247–1249.

111. Eriksson PS, Perfilleva E, Bjork-Erikson T, Alborn A-M, Nordborg C, Peterson DA, Gage FH. (1998) Neurogenesis in the adult human hippocampus. *Nat Med* **4**: 1313–1317.

29

THE PRIZEFIGHT

In October 2007, the gloves came off. The Anversa cohort opened the round saying, "The reason for the [myocardial regeneration] controversy is unclear, but it seems to reflect unshakable positions based on preconceived beliefs more than on careful analysis and proper consideration of published results… The old dogma has profoundly conditioned basic and clinical research in cardiology for the last three decades. Based on this paradigm, cardiomyocytes [can] undergo cellular hypertrophy (see p. 000) but cannot be replaced" during life…"However, [recent] efforts made to introduce a highly dynamic perspective of the heart have led to the identification and characterization of a resident pool of stem cells that can generate [cardio]myocytes.…This discovery has created a new, heated debate concerning the implementation of cardiac stem cells in the treatment" of coronary heart disease.[1]

The old dogma also claimed that "the postnatal heart is composed of a fixed number of myocytes and that, if myocytes die, they are permanently lost and the myocardium must then maintain its vital role with a reduced number of cells. Remarkably, there is not a single piece of evidence that demonstrates the inability of the heart to replace its dying myocytes. It seems rather extravagant that cardiomyocytes can contract 70 times a minute for over 100 years (in some human beings) and continue to be functional. In a centenarian, they would have contracted 3.7 billion times and still be operative. If myocytes were not replaced, they would be essentially immortal cells."[1]

The fundamental question posed by Anversa and his two most distinguished colleagues, Jan Kajstura and Annarosa Leri (Figs. 29.1 and 29.2), who occasionally write papers alone,[3-5] rather than with the other 15–20 members

Fig. 29.1. Jan Kajstura, Ph.D., associate professor of medicine, has assisted Piero Anversa, M.D., since 1992, longer than anyone in the three laboratories that constitute the Cardiovascular Research Institute in Valhalla, NY. The Polish-born cell biologist has seen the Institute triple in size from 7 M.D.s or Ph.D.s to 22. The focus of his research is the relationship between stem cells and the aging heart. Anversa says of Kajstura and Leri, "I would be dead without Jan and Annarosa." Beyond the lab, he enjoys classical music, rollerblading, and spending precious minutes with his wife and children.

(*Source*: Reproduced with permission from the publisher.) Reference 2.

of the international Valhalla entourage[a,6–9] is whether the heart is a static or dynamic organ. Not static in the sense that it doesn't move. It does. We all know it beats 60 or more times a minute (when we're awake) every day we're alive. But static in the sense that it cannot generate new heart muscle cells to replace cells worn out by constantly contracting or damaged by a heart attack. We also all know that the skin can heal after being cut. But the heart? Time-honored thinking says it can't. Anversa believes it can. Traditional thinking

---

[a]A typical paper from the Anversa group has 15–20 authors with ethnic roots in at least 7 countries, with roughly equal numbers of Ph.D.s and M.D.sn, 25–30 references, and is 5–10 pages in length. Anversa and his colleagues hold five N.I.H. grants amounting to more than US$1.9 million.[10]

Fig. 29.2. Annarosa Leri, M.D., is a cell biologist who collaborates with Anversa on research related to cardiac stem cells. To attract Leri, Anversa contacted his alma mater, the University of Parma, in Italy. There he found Leri and, over the years, the Valhalla–Parma connection has led to 11 faculty (out of 22) members from Parma, turning the lab into "Little Italy." While Leri enjoys the European lifestyle, she also appreciates the quality of research done at the Valhalla laboratory. With a "thoughtful; and subdued persona," she is easy to work with. Cultural plays, movies, and the beach are among her beyond-the-lab interests.

(Source: Reproduced with permission from the publisher.) Reference 2.

says that the hand we're dealt at birth, with regard to the number of heart muscle cells, is the hand we must play the game of life with. It is a set, number of cells (cards) and we cannot discard (lose) any stem cells and draw the lost cells from a reserve deck of stem cells somewhere else in the heart. Anversa thinks we can. Emotionally, we all hope he's right. Intellectually, we want to know that what he believes is supported overwhelmingly by facts.

The controversy regarding the heart's ability or inability to replace worn-out or dead muscle cells with fresh ones intensified when studies on the ability to activate stem cells from one person (a donor) by another person (a recipient) suggested just that. Studies by Anversa[1] and others[11] showed that cardiac chimerism was real. But the *extent* of chimerism became the point of disagreement. Was it 18% — as seen by Anversa — or 0.04% as found by the University of Washington's Pathology Department?[11] In the October 2007 bare-knuckle attack on the University of Washington's pathologists, Anversa

$$P = (d_Y/l_N) \cdot F$$
$$F = 1 - (k/(k + t_S))$$
$$k = [w_N \cdot \sin(\alpha)] + [l_N \cdot \cos(\alpha)]$$

Fig. 29.3. The three equations determining the probability of finding the Y chromosome in a particular microscopic sample of heart muscle by conventional microscopy.

(*Source*: Reproduced with permission from the publisher.) Reference 1.

restates the "technical differences" related to the dissimilar microscopes[b] used by the warring factions and offers somewhat abstruse calculations of the volume of a cardiomyocyte nucleus and that of a Y chromosome (Fig. 29.4) supporting his contention that the older instruments used by investigators at other institutions simply had little chance of "hitting the Y chromosome"[1] and — not surprisingly to him — found a much lower frequency of chimerism. To make clear his ability to see the Y chromosome in cardiomyocytes of hearts in male recipients of hearts from female donors with his microscope, he showed a number of colorful microphotographs, one of which is reproduced in Fig. 29.4. Pretty pictures aside, he contended, "The promoters of this new source of technical error" reached a conclusion based on a false

---

[b]In May 2007, a female science reporter — one Helen Pearson — for the highly regarded scientific journal *Nature*, which had published papers by both Anversa and his critics, did some investigative reporting on the two types of microscopes used by the warring groups. Hopeful of finding a reconciliatory explanation for the opposing results of Anversa and his critics, she (her gender is mentioned only because women tend to be more compassionate than men), interviewed a number of experts on microscopy.

Referring to the attractive images seen under a newer-vintage microscope, she wrote, "A pretty image is not necessarily a good one."[2a] "It may be either the instrument itself or the person using the instrument, or both, at fault. The instrument is so sensitive that the thickness of the glass cover slip atop the tissue being examined and the "intermittent cooling of [the ambient air] by an air conditioner can cause a microscope to drift in and out of focus." "The list of potential mistakes…is long and complex." After speaking with several cell biologists from The Netherlands, she wrote, "To correctly capture images using a modern microscope, researchers must have a good grasp of optics, an awareness of the microscope's complexity and an obsession for detail. Such skills can take months or even years to master, and, yet, owing to inexperience or the rush to publish, are all too often squeezed into hours or days. Popular methods such as [the modern-day] microscope are particularly fraught with dangers." Continuing, she said, "…inept microscopy, and subsequent analysis, can easily generate results that are misleading or wrong." Lamented the director of the imaging center at Harvard, "It's easy to pick up any journal — even *Nature* — and see poor microscopy data."

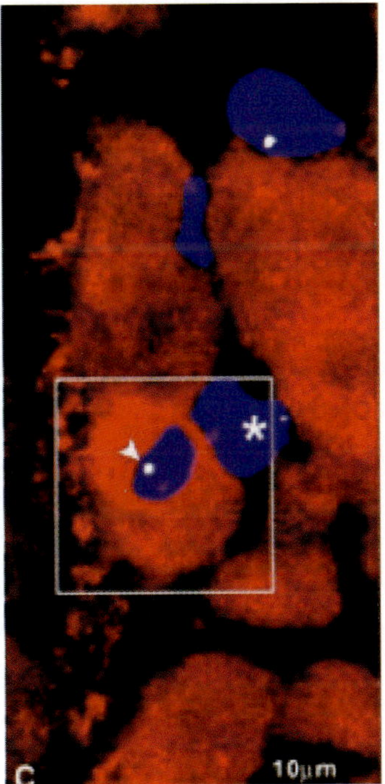

Fig. 29.4. A Y chromosome (the circular white dot) in the nucleus of a cardiomyocyte from the heart of a female transplanted into a male recipient is indicated by an arrowhead in the cell within the square. The X chromosome from the female donor is represented by the adjacent magenta dot. The other nucleus, labeled by an asterisk lying within the square, is in a different type of cell — not a cardiomyocyte — and does not have a Y chromosome.

(*Source:* Reproduced with permission from the publisher.) Reference 11.

syllogism — drawing a wrong conclusion from true major and minor premises. The major premise was that the heart had been known for years to be a nonregenerative organ. The minor premise was that the heart has a limited (or no) ability to activate stem cells from one person (a donor) in another person (a recipient). Therefore, the heart cannot regenerate itself. Not so fast, he said. He later retorted, "The critics decided to take the prerogative of reinterpreting data published by other groups, pointing to a number of crass morphological errors. Surprisingly these negative comments were published and, once again, these questionable criticisms became facts."[1]

Not only did questionable criticisms become facts, he went on, but "opponents of this emerging new paradigm in cardiology...continued to claim that these concepts are the product of artifacts. A recurrent unqualified statement is that they have introduced in their work 'rigorous' and 'stringent' criteria, which implies that others do not adhere to the same high scientific standards. The old paradigm [espoused by Anversa's opponents] has to survive, and the notion that [cardiac] myocytes cannot be generated postnatally, in adulthood or senescence, must be defended."[1]

He closed, "The recognition that stem cells reside in the myocardium had a dramatic impact on the field of cardiology from a biological and clinical perspective. The discovery has provided the missing link between the documentation of small dividing myocytes and the uncertainty concerning the origin of these repopulating cells, and it has laid the groundwork for the possibility of introducing the use of these cells in the treatment of the. . . human heart. A number of laboratories have invested resources and staff in exploring this novel and radically different perspective of cardiac pathophysiology [the abnormal physiology of a diseased organ, as opposed to its abnormal structure]. Because of these efforts, major advances are in progress with the objective of rapidly implementing this technology in humans."[1]

With a little luck and a lot of hard work by basic scientists and clinicians, that dream could become a reality by 2020.

Immune Rejection

Three biological obstacles must be overcome before the transplantation of tissues derived from stem cells becomes a reality in human beings. The first hurdle is the prevention of "immune rejection." When an organ or tissue from an unrelated donor is transplanted into a recipient, the transplantation process "almost invariably" results in rejection of the organ or tissue, unless immunosuppressive therapy[c] is given simultaneously.[12] [The susceptibility of an organ/tissue transplant between two unrelated individuals is so pronounced that a variation in the amino acid sequence of a single protein (in a person's DNA) between the donor and the recipient may be enough to "trigger an immune response that can result in the destruction or rejection of the transplant."[13]] In fact, "with the exception of an organ or tissue donated by an identical twin, virtually all transplant recipients require lifelong immunosuppressive

---

[c]Immunosuppressive therapy is the interference with or prevention of the natural immune response to a foreign agent, such as a transplanted heart, kidney, etc. Commonly used immunosuppressant drugs are Imuran (azathioprine), Cytoxin (cyclophosphamide), Rheumatrex (methotrexate), and Sandimmune (cyclosporine).

drugs to prevent rejection."[14] But "immunosuppressive treatment...[has] frequent side-effects that can be serious and even lethal."[12] While the five-year survival rate after organ transplantation is now about 70% while taking immunosuppressive drugs,[15] their frequent, unpleasant side effects [such as acne, alopecia (hair loss), anorexia (loss of appetite), nausea, vomiting, diarrhea, stomatitis (inflammation and painful ulcerations of the inner lining of the mouth), gingival hyperplasia (enlargement of the gums, sometimes requiring gingival resection),[16] myelosuppression (bone marrow suppression leading to too few red and white blood cells and platelets), and susceptibility to sometimes-exotic infections [such as the cytomegalovirus, which causes cytomegalic inclusion disease, and the herpes zoster virus, which causes always-painful herpes zoster (shingles)[17]] and certain types of cancer (such as melanoma and Hodgkin's disease)[18] leave daunting challenges for pharmaceutical manufacturers to develop better-tolerated drugs.

Perhaps the three most noteworthy diseases likely to benefit from embryonic stem cell therapy are coronary heart disease, diabetes, and neurodegenerative diseases such as Parkinson's disease, muscular dystrophy, and even Alzheimer's disease. There are two fundamental ways to put stem cells to therapeutic use: they may be used either to *replace* lost cells in tissues or to *protect* cells in tissues before they are lost. Human embryonic stem cells are the "most promising candidates" to *replace* lost cells and tissues.[12] Gene transfer (to be discussed later) is the principal *protective* technique. An army of bacteria, viruses, fungi, protozoa (such as *Giardia* and toxoplasma), and even worms (such as roundworms, tapeworms, and the flat worm in the *Trematodo* class which causes schistosomiasis) — collectively called pathogens — challenge human health endlessly. These pathogens rarely cause disease in people with a healthy immune system, but can do so if the immune system is impaired by disease or drugs. Given those circumstances, a so-called "opportunistic infection" may occur. The causative agents need an "opportunity" to infect people. The attempt of such opportunistic organisms — listed in Table 29.1 — to enter the body is the *raison d'être* of the immune system. [The three most common opportunistic infections are candidiasis (thrush), a fungal infection of the mouth, throat, or vagina; cytomegalovirus disease, a viral infection that can cause blindness; and genital herpes.[20]] Combating these infectious invaders is clearly a desirable consequence of the immune system. But the system may also have undesirable consequences, such as the rejection of transplanted tissues, hypersensitivity to common environmental allergens (like ragweed, which causes hay fever and aggravates asthma), and autoimmunity. The desirable consequences keep us healthy, while

Table 29.1   Opportunistic infections.

Bacterial and mycobacterial infections	Viral infections	Fungal infections	Protozoal infections
Syphilis	Cytomegalovirus	Aspergillosis	Cryptosporidiosis
Tuberculosis	Hepatitis	Candidiasis (thrush, yeast infection)	Isosporiaisis
Salmonellosis	Herpes simplex (genital herpes)	Coccidiodomycosis	Microsporidiosis
Bacillary angiomatosis (cat scratch disease)	Herpes zoster (shingles)	Cryptococcal meningitis	*Pneumocystis carinii* pneumonia
Mycobacterium avium complex	Human papilloma virus (genital warts, cervical cancer)	Histoplasmosis	Toxoplasmosis
	Molluscum contagiosum		
	Oral hairy leukoplakia		
	Progressive multifocal leukoencephalopathy		

Source: Reference 19.

the undesirable ones often lead to disease. Anaphylactic shock is the consequence of a full-blown immune reaction that can be lethal.[21] Autoimmunity occurs when the body reacts to its own tissue(s) as if it (they) were foreign. Even more alarming, there is no cure for most autoimmune diseases. Steroids can be used to suppress the autoimmune process, but the side effects of long-term steroid use are diverse and dangerous. Systemic lupus erythematosis and rheumatoid arthritis are two common expressions of an autoimmune disease.

Human Leukocyte Antigen

The six major antigens on the surface of human white blood cells (leukocytes), first mentioned on pages, are the root of immune rejection in humans.

Collectively, the six antigens form the "human leukocyte antigen system."[d] Drukker and his colleagues at the University of Jerusalem (now at Stanford University) found that although human embryonic stem cells have rather low levels of the human leukocyte antigen, the levels quadrupled as the cells differentiated into specialized cells,[22] such as cardiomyocytes and the even more remarkable cells that form the brain and spinal cord. Draper and his colleagues from the U.K. and the University of Wisconsin confirmed these results.[23] (In addition, the use of embryonic stem cells to treat acute myocardial infarctions, produced simply by tying off a coronary artery in laboratory mice and thereby depriving their heart muscle of oxygen, (unexpectedly) revealed that embryonic stem cells evoke an immune response.[24,25] These observations indicated that embryonic stem cells generate just enough human leukocyte antigen to "manifest their immune identity, thus creating [the need for] a strategy for immune matching in order to achieve graft acceptance following transplantation."[25] Or, restated, even a little bit of a bad medicine may make someone sicker, creating the need for yet another medicine — or no medicine at all. In the case of avoiding immune rejection, taking no medicine at all is not an option compatible with life for very long.

Coping with Immune Rejection

A sensible strategy for coping with immune rejection using immunosuppressive therapy was put forward by three Cambridge immunologists: J. Andrew Bradley, Eleanor M. Bolton, and Roger A. Pedersen. Recognizing both the power of the human leukocyte antigen (HLA) situated at the epicenter of the immune rejection quandary and the undesirable side effects of immunosuppressive therapy, they proffered the following words of wisdom — longer than any other quote in this book, but worth the publisher's ink and our reading: "Even if no attempt were made to match the HLA antigens of hES [human embryonic stem]-cell-derived tissue with its intended recipient there is no reason to doubt the ability of currently available agents effectively to control acute rejection.

---

[d] The human leukocyte antigen is found on the surface of all human cells other than cells in sperm and unfertilized eggs. As a result of its superficial location, the antigen is in an extremely good position to defend cells against foreign invaders, such as bacteria, viruses, and fungi. Any foreign pathogens hoping to enter a cell are recognized by the body's immune system and are promptly destroyed. This surveilanc-and-destroy system is reminiscent of a submarine recognizing an approaching foreign warship on the ocean's surface, and then quickly destroying it.

After all, living unrelated donor kidney transplants are usually poorly matched for HLA antigens, yet with modern immunosuppressive therapy, graft survival is as good or better than for well-matched cadaveric kidney transplants. Moreover, no attempt is made to match for HLA when undertaking heart or liver transplants, yet the results are generally very good.[e] The one-year graft survival after solid organ transplantation is currently around 85% and five-year graft survival is around 70% [as mentioned earlier]. Modern immunosuppressive therapy has also enabled the introduction of human pancreatic islet transplantation [in diabetics] and again no attempt is made to use rejection HLA-matched tissue."[28]

Continuing, the Cambridge trio said, "While immunosuppressive agents prevent acute graft rejection, this is achieved at a considerable price in terms of unwanted side effects. Such side effects are also a long-term problem, since immunosuppressive therapy must be continued indefinitely to prevent rejection. The non-specific immunosuppression that is an inevitable consequence of current immunosuppressive agents increases the risk of both infection and malignancy. In addition, immunosuppressive agents cause a wide range of potentially serious agent-specific side effects that may increase mortality and reduce the quality of life after transplantation....

"Whether the side effects of immunosuppressive therapy can be justified for hES-cell-derived tissue will depend on the likely benefit of the tissue transplanted. For example, if the transplant were a life-saving transplant of cardiac tissue, then justification for immunosuppressive therapy might be easy to argue. On the other hand, if the transplant were of insulin producing tissue to treat diabetes mellitus the benefit over [the patient's] existing insulin replacement therapy might be more marginal." End of quote!

Immune Tolerance

By July 2004, six years had passed since the historic isolation and culturing of human embryonic stem cells by Thomson and his colleagues at the University

---

[e] HLA matching is not done before a cardiac transplantation, because the oxygen-deprived donor heart cannot be preserved for more than 4 h after its removal — even when iced — from the brain-dead donor without substantial damage caused by the lack of oxygen.[26] Three to eight hours are needed to complete an HLA match between the donor and the recipient.[27] This time constraint also limits the distance that a donor heart can be transported, which in turn limits cardiac transplantation to larger urban areas.

of Wisconsin.[f,29] Despite a tsunami of subsequent research on human embryonic stem cells and the recognition that these remarkable cells had the potential to provide an unlimited source of stem cells to replace diseased or damaged tissues, the "immune responses to hESCs [had not] yet [been] evaluated."[32] Would human embryonic stem cells derived from the blastocyst of an embryo unrelated to a future recipient be accepted or rejected by the recipient? There was no reason to expect that such cells would not be rejected, since they would be complete foreigners in the body of anyone other than the two parents. Or. . .or was there a possibility that such cells might be accepted by an unrelated recipient? After all, a fetus is not rejected during pregnancy even though 50% of the genes in its cells/tissues come from the father!

With that question in mind, researchers in Ontario, Canada, injected human embryonic stem cells into mice with healthy immune systems perfectly capable of rejecting such foreign cells. And what happened? The mice accepted the human cells without a hint of rejection. They were "tolerant." The surprised investigators concluded that "hESCs [human embryonic stem cells] possess unique immune-privileged characteristics [that] provide an unprecedented opportunity to further investigate the mechanism of immune response."[32]

Definitions of Immune Tolerance

The definition differs between basic scientists doing research on immune tolerance and clinicians treating patients with transplanted organs and/or tissues who need every ounce of immune tolerance they can muster to hang on to their new organ/tissue. (They also need all the patience and will power they can muster to tolerate the unpleasant side effects of the immunosuppressive

---

[f]Thomson's embryonic stem cells were isolated in a laboratory on the University of Wisconsin campus, physically and financially independent of his N.I.H.-funded laboratory. A duplicate laboratory was necessary because of a U.S. law that "forbids the use of public funds for research on tissues derived from human embryos."[30] To comply with the law he turned to nongovernmental money from the privately owned Geron Corporation in Menlo Park, California, to equip and staff a matching lab on the University of Wisconsin campus. In return, Geron received exclusive licensing of Thomson's technology. Several years later, the University of Wisconsin Alumni Research Foundation (W.A.R.F.), which kept control of the patent, got in a spat with Geron over who had the commercial rights to the five "immortal" stem cell lines isolated by Thomson in 1998. The W.A.R.F. ended up with two lines, and Geron three.[31]

drugs they must take forever.) As a result, there is a "classic" and a "practical" definition of immune tolerance.

The *classic definition* is "a complete lack" of an immune response to any and all foreign antigens.[33] Or, restated, immune tolerance is a state in which the recipient "accepts" the foreign transplanted tissue as if it were his or her own without the need for chronic immunosuppression.[14] For example, if newborn mice are made immune — highly resistant to a foreign substance — very early in life to injected cells/tissue from an allogeneic (a genetically distinct) mouse, the animals will mature and accept the transplanted cells/tissue with "absolutely no immune response"— classic tolerance.[33] The proof of the pudding is most apparent in individuals who receive a bone marrow transplant from a closely matched HLA — but not HLA-identical — sibling for malignancies such as leukemia or multiple myeloma and can later receive a second transplant from the *same* donor to replace a different diseased organ (for example, a kidney) with "complete tolerance."[33]

The *practical definition* of immune tolerance is the "long-term. . .survival" of a transplanted tissue "without immunosuppression."[33] In other words, the practical definition requires that the tissue/organ recipient be tolerant of the *specific* tissue or organ being transplanted, as opposed to "any and all" tissues or organs.

Immune Tolerance Between Species

It turns out that not only mice[32] but also rats[34,35] and sheep[36] are tolerant of human embryonic stem cells. As a result, these extraordinary cells can add "immune-privileged" to their other two unique properties — the ability to "proliferate indefinitely" in a stem cell laboratory and to "differentiate into any type of cell in the body."[37] Figures 29.5–29.7 are photomicrographs of myocardium from a normal rat (Fig. 29.5) and two rats with an induced myocardial infarction (Figs. 29.6 and 29.7). Figures 29.5 and 29.6 illustrate in vivid fluorescent colors the tolerance of rats to the presence of human embryonic stem cell–derived cardiomyocytes. Looking at heart muscle under a microscope hasn't always been this esthetically pleasing. For 150 years the hematoxylin and eosin (H&E) stain was used; it stains tissues a mundane light pink (cytoplasm), darker pink (cytoplasm), and blue (nuclei), as shown in Fig. 29.7. If you had to peer into a microscope hour after hour five days a week, which stain would you pick? That's a no-brainer. But, in reality, the H&E stain remains the most widely

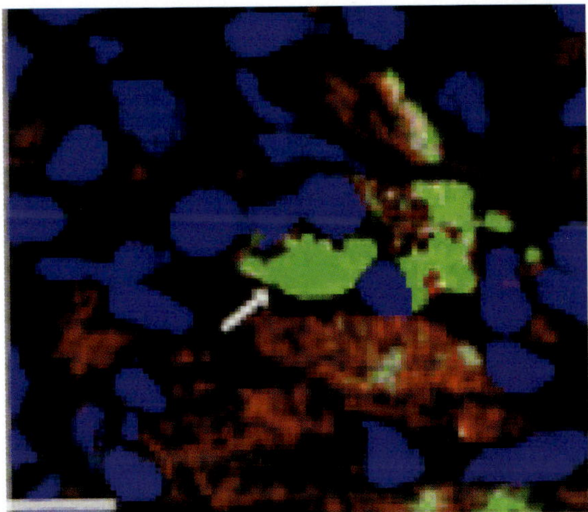

Fig. 29.5. A photomicrograph of heart muscle from a healthy rat 36 h after transplantation of human embryonic stem cell–derived cardiomyocytes. The grafted embryonic cardiomyocytes are stained green by green fluorescent protein. The nuclei of nearby cardiomyocytes are stained blue.

(*Source*: Reproduced with permission from the publisher.) Reference 116

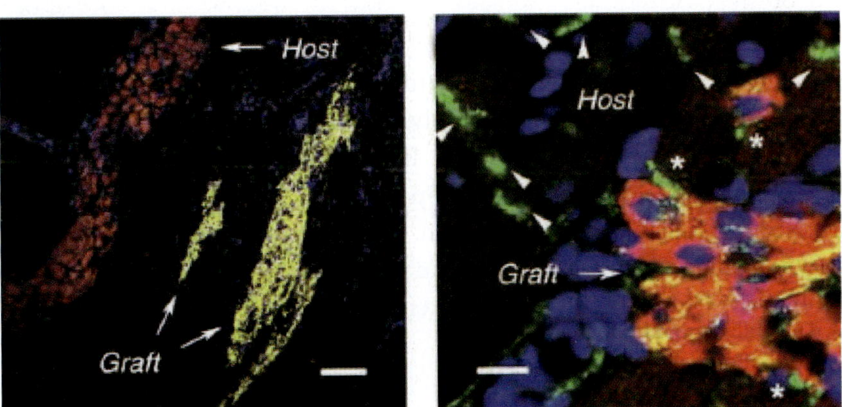

Fig. 29.6. Photomicrographs of heart muscle from two rats four weeks after transplantation of human embryonic stem cell–derived cardiomyocytes into infarcted myocardium. The host and grafted tissue are stained with different fluorescent proteins, creating a colorful and easy-to-look-at display. The surrounding scar tissue is unstained in these images, accounting for the black areas adjacent to the fluorescent regions.

(*Source*: Reproduced with permission from the publisher.) Reference 35.

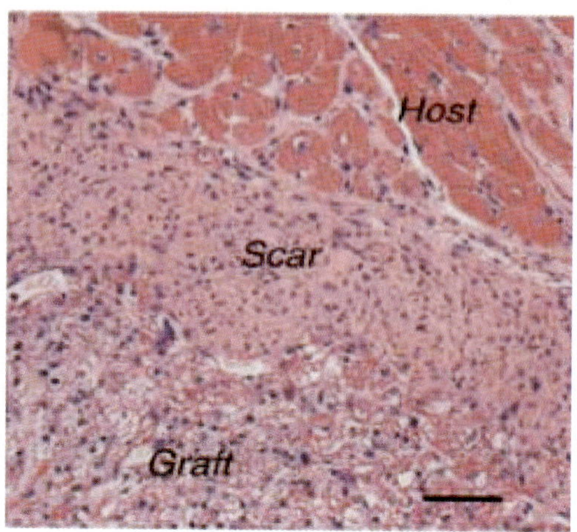

Fig. 29.7.   A photomicrograph of heart muscle from a rat four weeks after transplantation of human embryonic stem cell–derived cardiomyocytes into infarcted myocardium. The cardiac tissue in this figure is stained with a hematoxylin and eosin stain. The host tissue, graft tissue, and scar tissue are all stained in (boring) shades of pink. To distinguish scar tissue from graft tissue requires close examination to determine nonchromatic differences, such as the presence of vacuoles — the tiny clear cavities — in the graft tissue.

(*Source*: Reproduced with permission from the publisher.) Reference 35.

used stain for routine tissue examination in U,S, hospitals and laboratories because it is inexpensive, easy to apply, and tissues can be examined with a standard, also (relatively) inexpensive light microscope. Colorful fluorescent imaging using a fluorescence microscope is limited to university hospitals and outfits such as the Geron Corporation. (The background and current use of fluorescent imaging in science and the derivation of fluorescent colors from marine corals will be covered in Chapter 27.)

Maternal Tolerance of an Embryo

Survival of transplanted cells, irrespective of their cell type, correlates with the number of HLA mismatches between the donor and the recipient. Zero to six mismatches are possible; the fewer, the better. One significant exception to this rule is maternal tolerance to a fetus, half of whose genes come from the father and half from the mother.[32] While a fetus will soon become a baby profoundly nurtured by its mother, it nonetheless is a "foreign entity" to its mother's immune system, and yet it is not rejected.[38] Pregnancy is a unique symbiotic relationship between a mother and her fetus that "breaks" rules and usually gets away with it.

The rules transgressed are those of immunology. The maternal immune system "knows" there is a foreign tissue in the mother's body, but instead of triggering rejection — the normal response of the body to an outside entity, be it a bacterium or a splinter — it tolerates and fully supports the embryo-fetus for nine months. Pregnancy is also an "immunological balancing act" between the mother's immune system, which must be tolerant of the paternal antigens while retaining its immune defense against micro-organisms (viruses, bacteria) capable of causing pregnancy-ending infections.

Answers to questions about "this most unique example of human tolerance among nature's creations"[39] should help basic scientists and physicians better understand and avoid immune rejection of transplanted tissues and organs. Answers to such questions could also improve gestational outcomes by avoiding congenital defects, maternal infections, and maternal morbidity (i.e. any departure of maternal health during pregnancy, childbirth, or the postpartum period up to 90 days). Understanding the mechanism of the "immunological mystery" that enables a pregnancy to progress to childbirth without an immune-induced miscarriage also has implications *vis-à-vis* the immune-privileged status of human embryonic stem cells.

The immunologically privileged state of human embryonic stem cells is attributed to their similar unique ability to avoid immune rejection by the mother's immune system during pregnancy.[32] This fetal bliss is related to unusual "properties of embryonic human cells that inhibit local immune responses to the fetus."[32] This exceptional attribute of the human embryo was first recognized in 2004.[32] Just how the human embryo avoids immune-mediated rejection remains unclear. The immune-privileged status of human embryonic stem cells is, unfortunately, countered by their tendency to form teratomas (bizarre tumors) as will be discussed soon. This is tantamount to swapping graft acceptance for a neoplastic growth — a tumor.

A Brief History of Immune Privilege

A privilege is "a right or immunity granted as a peculiar benefit, advantage, or favor," according to Noah Webster's famed book.[40]

One hundred years ago, a scientist in a reflective mood at the Rockefeller Institute in New York wrote, "Gardeners and arborculturists have known for centuries that plants of one species could be grafted on to another without in any way affecting the essential characters of either. In more recent years, it has been shown that a like situation exists in certain of the lower animals."[41] He chose worms and tadpoles as examples of lower animals.

In contrast to those observations on "tissue transplantation" in plants and lower animals, James Murphy, the turn-of-the-20th-century Rockefeller Institute scientist, discovered that when cancerous tissue from mice was "transplanted" into rats, and then back into mice in so-called "zigzag" experiments," a "cellular reaction" occurred in most transplanted tissues, but not in the brain. For example, if a sarcoma[g] from a mouse was transplanted into a rat's brain it thrived, but not if transplanted beneath the skin (subcutaneously) or in a muscle (intramuscularly).[43] Curiously, the tumor's rapid growth in a rat brain happened only if the tumor was confined entirely within the brain. If any part of it came in contact with the cerebrospinal fluid within the ventricles of the brain,[h] a "cellular reaction [took] place with destruction of the graft."[43] Although the animals deteriorated and died within two weeks of transplantation with a "cellular reaction," the term "rejection" was not yet used. On occasion, the word "transplantability" was employed, implying the ability of a transplanted tissue to avoid what we now call "rejection."

In 1923, Murphy extended his earlier work by showing that if a "small bit" of the spleen from the recipient was transplanted (at the same time) into the brain along with the foreign tumor, the tumor also did not spread.[i,43] This suggested that something within the spleen destroyed the cancer. (Perhaps this feature meets one of Webster's definitions of a "privilege" — a "peculiar benefit.") Murphy's observations were the original demonstrations of what 30 years later would first be called "immune privilege."[45] And now, many who do research related to immune privilege can't remember a time when the term didn't exist.

---

[g] A *sarcoma* is a rare type of cancer that arises from connective tissues, such as bone, cartilage and muscle.[42] A *carcinoma*, in contrast, arises from tissues of epithelial (regarding the inner lining or outer surface of an organ) origin, such as the prostate, colon and breast. Sarcomas are often named according to the tissue from which they arise. For example, an osteosarcoma arises from bone and a chondrosarcoma arises from cartilage. Some sarcomas, such as Ewing's sarcoma, are more common in children and many sarcomas are more aggressive than carcinomas.

[h] Like the heart, the brain has ventricles. In fact, the brain has four ventricles, and the heart merely two. Unlike the heart, whose ventricles contain and pump blood, the ventricles of the brain contain cerebrospinal fluid and pump nothing. Nutrition and protection of the brain are the two principal functions of cerebrospinal fluid.

[i] Much later — in the 1990s — immune privilege also was found to be present when noncancerous tissues and even infectious tissues were implanted in the mouse brain.[44]

Immune-Privileged

The transplantation of another person's organ or tissue into an ailing recipient in need of a healthy organ/tissue provokes a "powerful immune response" that almost always results in rapid rejection — within 1–2 weeks — of the graft by a hostile immune system.[28] The human body has several "immune privileged sites" where transplanted cells escape detection by the immune system, even in the absence of immunosuppressive therapy.[28] The brain, anterior chamber of the eye, and testes — but not the ovaries — are the only three locales that afford such protection.[28] (Somehow, blackballing the ovaries doesn't seem "right." For an obscure reason, the testes trump egalitarianism.) In these sanctuaries, transplanted cells prosper. An anti-inflammatory cytokine (see Chap. 8) synthesized by these three tissues suppresses potentially damaging immune responses.[46] Site-specific protection is necessary because proteins synthesized in these three tissues do not travel in the largely unheralded lymphatic system — as opposed to the circulatory system — where specialized white blood cells called "T cells" are synthesized by the equally unheralded thymus gland.[j] Currently, corneal (regarding the anterior chamber of the eye) transplantation is the only procedure that takes advantage of an immunologically privileged tissue's special status.[40]

Teratomas

Teratomas (Greek. "*terat*"- meaning "monster", and "-*toma*" meaning "tumor)" are rare, (usually) benign tumors that affect mostly 20–40-year-old people but, unfortunately, may sometimes occur in children.[48] These disorganized growths "contain tissues of embryonic origin only"[49] and as a result reflect the three layers of an early embryo (the endoderm, mesoderm, and ectoderm), discussed in Chap. 14 (Table 14.1). Those three primitive layers allow the development of teeth, hair, and bone in these bizarre tumors.[43] Ninety percent of teratomas occur in the testes or ovaries, while 10% are in the upper chest (mediastinum) or lower-back region.[50]

The very trait that makes human embryonic stem cells so precious — their ability to form all other cells in the body even after birth — is also their greatest liability: the formation of teratomas.[49] So commonly do teratomas form after

---

[j] T cells are actually made in the bone marrow, like all other white (and red) blood cells. They then leave home and take up residence in the thymus the "T" in "T cells" stands for "thymus"), where they mature before being set free to begin their mission — ensuring immunity.[47]

transplantation of embryonic stem cells (in mice) that their presence is used as a parameter of a (technically) successful transplantation. And why couldn't an asset become a liability in a different setting? Embryonic stem cells, after all, are charged with replenishing diseased or damaged tissues during life from three layers that existed only in the early embryo. Their job description does not call for distinguishing desirable tissues from potentially undesirable ones, such as teratomas. In Thomson's milestone paper, the formation of teratomas was used as evidence that human embryonic stem cells could differentiate into a number of tissues.[29] Each of the original five lines of human embryonic stem cells produced teratomas within 7–8 weeks following injection of the stem cells into a leg muscle of newborn mice. Several years later, investigators at the Institute for Stem Cell and Regenerative Medicine in Seattle found that *un*differentiated embryonic stem cells "consistently formed cardiac teratomas" in immunocompromised[k] and syngeneic (genetically similar) mice (Fig. 29.8).[52] The tumors weren't picky about the quality of the myocardium they chose to call home. They developed in healthy as well as infarcted (damaged) heart muscle. The tumors also didn't waste any time growing to the size shown in Fig. 29.8. Just three days after transplantation, small clusters of teratoma cells were present in all animals which grew progressively to occupy a substantial amount of the heart wall within three weeks (Fig. 29.8). However, there was a concentration of undifferentiated embryonic stem cells that could be transplanted into the heart without causing teratomas. No tumor development was seen when fewer than 100,000 undifferentiated embryonic stem cells were injected, whereas all mice receiving 500,000 undifferentiated stem cells formed tumors (Fig. 29.9). But there was good news just around the corner for mice with teratomas induced by embryonic stem cells: the tumors were "immunologically rejected after several weeks."[52] They vanished.

So, rejection can be a good thing…when it cures cancer!

To Differentiate or Not to Differentiate? That Is the Question

Just as the injection of fewer undifferentiated stem cells leads to fewer teratomas, so does the injection of more undifferentiated stem cells favor teratoma formation. Stem cells fresh out of an embryo have all their options open.

---

[k] In mice, immunity is often compromised intentionally by the administration of cyclophosphamide, a potent anticancer drug used clinically during chemotherapy of human lymphomas and leukemias.[51]

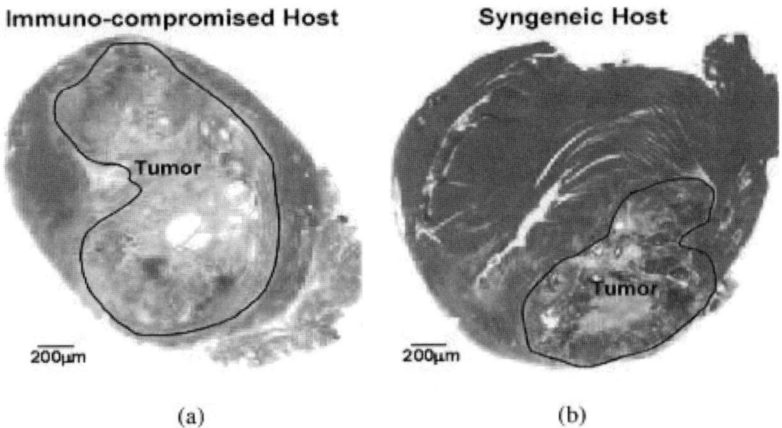

Fig. 29.8. This figure shows the hearts of two mice with tumors (teratomas) that developed within three weeks following the injection of *un*differentiated mouse embryonic stem cells into the heart of an immune-compromised and a syngeneic mouse. Syngeneic mice are obtained simply by inbreeding until their genetic similarity is sufficient to allow acceptance of a transplanted graft without rejection. The teratomas replace a substantial amount of the heart's wall under both circumstances.

(Source: Reproduced with permission from the publisher.) Reference 52.

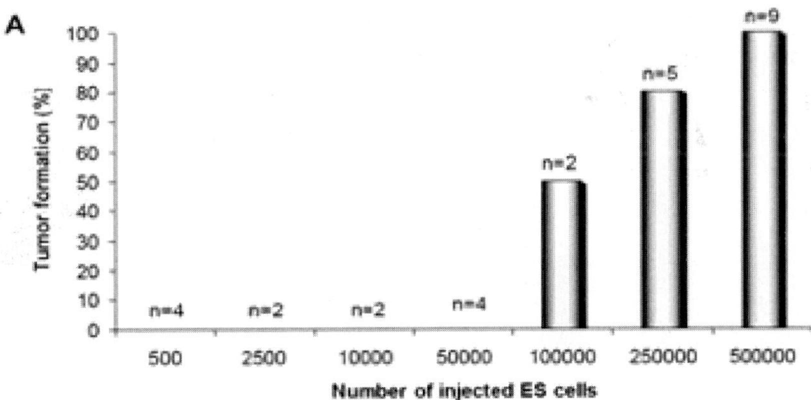

Fig. 29.9. The incidence of teratomas (ordinate) three weeks after the injection of various doses of undifferentiated embryonic stem cells (abscissa) into the heart (left ventricle) of syngeneic mice. Teratomas were seen in 50% of mice that received 100,000 embryonic stem cells and in 80% of mice that received 250,000 stem cells. All mice that received 500,000 stem cells developed teratomas. No tumor development was seen in mice that received fewer than 100,000 embryonic stem cells.

(Source: Reproduced with permission from the publisher.) Reference 52.

(In reality, there are no stem cells in a human embryo. There is an "inner cell mass" that can be nurtured in the laboratory to form stem cells, as described in Chap. 13.) They can form any of the 210 cell types in the human body; but they also can form undesirable tissues just as easily. Such cells are young and sometimes mischievous. Teratomas may be the unwanted result. But the more mature a stem cell becomes, the smaller the likelihood of its misbehavior is. Mature (differentiated) human embryonic stem cells transplanted into mice hearts underwent "acute donor cell death" followed by survival of the remaining cells over the next 2–3 weeks, while immature (*un*differentiated) stem cells "proliferated uncontrollably," leading to teratoma formation (Fig. 29.10).[53]

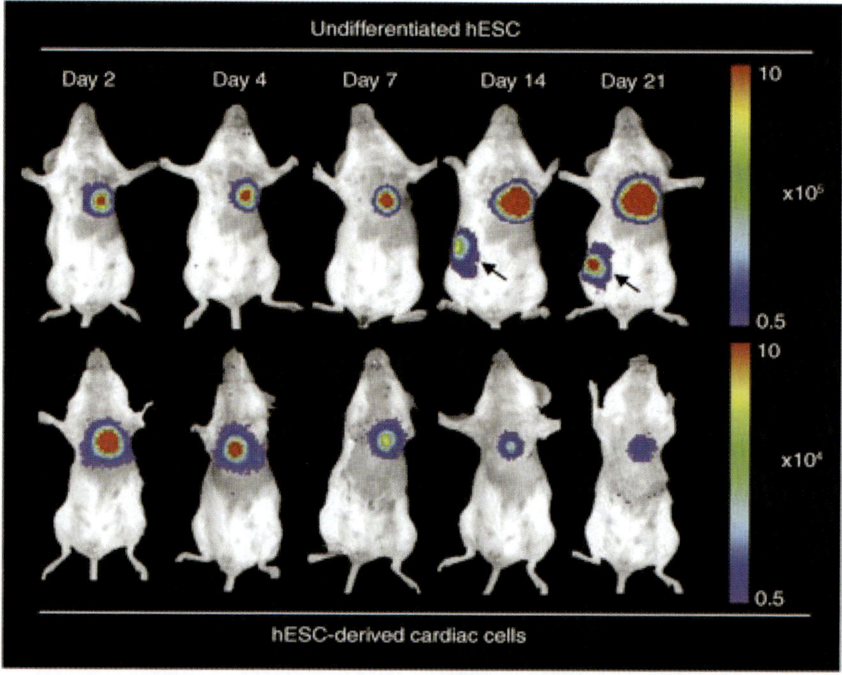

Fig. 29.10.   A comparison of *undifferentiated* (unspecialized) stem cells) top row) and human embryonic stem cells that have *differentiated* into contracting cardiac muscle cells — cardio-myocytes (bottom row) — on survival; following transplantation into the heart of mice. The red bullseye represents dead myocardium. The size of the bullseye grows progressively larger between day 2 and day 21 after *undifferentiated* cells are transplanted. Teratomas developed in all such mice. In contrast, after 4 days of acute cell death, the red zone shrinks progressively following the transplantation of differentiated (specialized) stem cells. No teratomas formed in these mice.
(*Source*: Reproduced with permission from the publisher.) Reference 51.

References

1. Anversa P, Leri A, Rota M, Hosada T, Bearzi C, Urbanek K, Kajstura J, Bolli R. (2007) Concise review: *Stem Cells* **25**: 589–601.

2. http://www.nymc.edu/pubs/chironian/fall2003/cardioresearch.asp Accessed March 17, 2006.

2a. Pearson H. (2007) The good, the bad and the ugly. *Nature* **447**: 138–140.

3. Anversa P, Leri A, Kajustra J. (2006) Cardiac regeneration. *J Am Coll Cardiol* **47**: 1769–1776.

4. Anversa P, Kajustra J, Leri A. (2007) If I can stop one heart from breaking. *Circulation* **115**: 829–832.

5. Leri A, Kajustra J, Anversa P. (2005) Cardiac stem cells and mechanisms of myocardial regeneration. *Physiol Rev* **85**: 1373–1416.

6. Kajstura J, Rota M, Whang B, Cascapera S, Hosoda T, Bearzi C, Nurzynska D, Kasahara H, Zias E, Bonaf M, Nadal-Ginard B, Torella D, Nascimbene A, Quaini F, Urbanek K, Leri A, Anversa P. (2005) Bone marrow cells differentiate in cardiac cell lineages after infarction independently of cell fusion. *Circ Res* **96**(1): 127–137.

7. Muraski JA, Rota M, Misao Y, Fransioli J, Cottage C, Gude N, Esposito G, Delucchi F, Arcarese M, Alvarez R, Siddiqi S, Emmanuel GN, Wu W, Fischer K, Martindale JJ, Glembotski CC, Leri A, Kajstura J, Magnuson N, Berns A, Beretta RM, Houser SR, Schaefer EM, Anversa P, Sussman MA. (2007) Pim-1 regulates cardiomyocyte survival downstream of Akt. *Nat Med* **13**(12): 1467–1475.

8. Rota M, Kajstura J, Hosoda T, Bearzi C, Vitale S, Esposito G, Iaffaldano G, Padin-Iruegas ME, Gonzalez A, Rizzi R, Small N, Muraski J, Alvarez R, Chen X, Urbanek K, Bolli R, Houser SR, Leri A, Sussman MA, Anversa P. (2007) Bone marrow cells adopt the cardiomyogenic fate *in vivo*. *Proc Natl Acad Sci USA* **104**(45): 17783–17788.

9. Bearzi C, Rota M, Hosoda T, Tillmanns J, Nascimbene A, De Angelis A, Yasuzawa-Amano S, Trofimova I, Siggins RW, Lecapitaine N, Cascapera S, Beltrami A, D'Alessandro DA, Zias E, Quaini F, Urbanek K, Michler RE, Bolli R, Kajstura J, Leri A, Anversa P. (2007) Human cardiac stem cells. *Proc Natl Acad Sci USA* **104**(35): 14068–14073.

10. http://www.nymc.edu/pubs/Chironian/Fall2000/anversa.htm

11. Laflamme MA, Myerson D, Saffitz JE, Murry CE. (2002) Evidence for cardiomyocyte repopulation by extracardiac progenitors in transplanted human hearts. *Circ Res* **90**: 634–640.

12. Mitjavila-Garcia MT, Simonin C, Peschanski M. (2005) Embryonic stem cells: meeting the needs for cell therapy. *Adv Drug Deliv Rev* **57**: 1891–1897.

13. Boyd AS, Tigashi Y, Wood KJ. (2005) Transplanting stem cells: potential targets for immune attacks. Modulating the immune response against embryonic stem cell transplantation. *Adv Drug Deliv Rev* **57**: 1944–1969.

14. Adler SH, Bensinger SJ, Turka LA. (2002) Stemming the tide of rejection. *Nat Med* **8**: 107–108.

15. Bradley JA, Bolton EM, Pedersen RA. (2002) Stem cell medicine encounters the immune system. *Nat Rev Immunol* **2**: 859–871.

16. Slapak CA, Kufe DW. (1994) Principles of cancer therapy. In: Isselbacher KJ, Fauci AS, Kasper DL (eds.), *Harrison's Principles of Internal Medicine*, 13th ed. McGraw-Hill, New York, St. Louis, Colorado Springs, Auckland, Bogotá, Caracas, Hamburg, Lisbon, London, Madrid, Mexico City, Milan, Montreal, New Delhi, Paris, San Juan, Singapore, Sydney, Tokyo, Toronto, pp. 1826–1840.

17. Simon DM, Levin S. (2001) Infectious complications of solid organ transplantation. *Infect Dis Clin North Am* **15**: 521–549.

18. Penn I. (2000) Post-transplantation malignancy: the role of immunosuppression. *Drug Safety* **23**: 101–113.

19. http://www.aegis.com/topics/oi/

20. http://www.aids.org/factSheets/500-Opportunistic-Infections.html#anchor 51213

21. Playfair JHL, Chain BM. (2005) *Immunology at a Glance.* Blackwell, London, pp. 8–18.

22. Drukker M, Katz G, Urbach A, Schulder M, Markel G, Itskovitz-Eldor J, Reubinoff B, Mandelbolm O, Benvenisty N. (2002) Characterization of the expression of MHC proteins in human embryonic stem cells. *Proc Natl Acad Sci* **99**: 9864–9869.

23. Draper JS, Pigott C, Thomson JA, Andrews PW. (2002) Surface antigens of human embryonic stem cells: changes upon differentiation in culture. *J Anat* **200**: 249–258.

24. Swijnenburg RJ, Tanaka M, Vogel H, Baker J, Kofidis T, Gunawan F, Lebl DR, Caffarelli AD, de Bruin JL, Fedoseyeva DR, Robbins RC. (2005) Embryonic stem cell immunogenicity increases uoon differentiation after transplantation into ischemic myocardium. *Circulation* **112**(9 Suppl): I166–I172.

25. Kofidis T, de Bruin JL, Tanaka M, Zwieczchoniewska M, Weissman I, Fedoseyeva A, Robbins RC. (2005) They are not stealthy in the heart: embryonic stem cells trigger cell infiltration, humoral and T-lymphocyte-based host immune response. *Eur J Cardiothorac Surg* **28**: 461–466.

26. Morris PJ. (1994) HLA matching and cardiac transplantation. *N Engl J Med* **330**: 857–858.

27. Thompson JS, Thacker LR, Takemoto S. (2000) The influence of conventional and cross-reactive group HLA matching on cardiac transplant outcome: an analysis from the United Network of organ sharing scientific registry. *Transplantation* **69**: 2178–2186.

28. Bradley JA, Bolton EM, Pedersen RH. (2004) ES cells for transplantation: coping with immunity. In: Odorico J, Zhang S, Pedersen S (eds.), *Human Embryonic Stem Cells.* Garland Science, BIOS Scientific, Oxon, New York, pp. 231–256.

29. Thomson JA, Iskovitz-Eldor J, Shapiro SS, Waknitz MA, Swiergiel JJ, Marshall VS, Jones JM. (1998) Embryonic stem cell lines derived from human blastocysts. *Science* **282**(5391): 1145–1147.

30. Marshall E. (1998) A versatile cell line raises scientific hope, legal questions. *Science* **282**: 1014–1015.

31. http://en/wikipedia.org/wiki/Stem_cell_controversy

32. Li L, Baroga ML, Majumdar A, Chadwick K, Rouleau A, Gallacher L, Ferbee I, Lebkowski J, Martin T, Madrenas S, Bham M. (2004) Human embryonic stem cells possess immune-privileged properties. *Stem Cells* **22**: 448–456.

33. Hildebrand P, Salomon DR. (2004) Molecular biology of transplantation and xenotransplantation. In: Chien KR (eds.), *Molecular Basis of Cardiovascular Disease.* Saunders, Philadelphia, pp. 649–666.

34. Caspi O, Huber I, Kehat, Habib M, Arbel G, Gepstein A, Yankelson L, Aronson D, Beyar R, Gepstein L. (2007) Transplantation of human embryonic stem cell–derived cardiomyocytes improves myocardial performance in infarcted rat hearts. *J Am Coll Cardiol* **50**: 1884–1893.

35. Laflamme MA, Chen KY, Naumova AV, Muskheli V, Fugate JA, Dupras SK, Reinecke H, Xu C, Hassanipour M, Police S, O'Sullivan C, Collins L, Chen Y, Minami1 E, Gill EA, Ueno S, Yuan CL, Gold J, Murry C. (2007) Cardiomyocytes derived from human embryonic stem cells in pro-survival factors enhance function of infarcted rat hearts. *Nat Biotechnol* **25**: 1015–1024.

36. Airey JA, Almei da-Porada A, Colletti EJ, Porada CD, Chamberlain J, Movsersian M, Sutko HL, Zanjani ED. (2004) Human mesenchymal stem cells form Purkinje fibers in fetal sheep heart. *Circulation* **109**: 1401–1407.

37. Koch CA, Geraldes P, Platt JL. (2008) Immunosuppression by embryonic stem cells. *Stem Cells* **26**: 89–98.

38. Jiang S-P, Vacchio MS. (1998) Cutting edge: multiple mechanisms of peripheral T cells tolerance to the fetal "allograft." *J Immunol* **160**: 3086–3090.

39. Guleria I, Sayegh MH. (2007) Maternal acceptance of the fetus: true human tolerance. *J Immunol* **178**: 3345–3351.

40. *The Random House Dictionary of the English Language,* unabridged ed. Stein J, Urdang L (eds.), (1967) Random House, New York.

41. Murphy JB. (1913) Transplantability of tissues to the embryo of foreign species. Its bearings on questions of tissue specificity and tumor immunity. *J Exp Med* **17**: 482–493.

42. http://en.wikipedia.org/wiki/Sarcoma

43. Murphy JB, Sturm E. (1923) Conditions determining the transplantability of tissues in the brain. *J Exp Med* **38**: 183–197.

44. Stevenson PG, Hawke S, Sloan Dj, Bangham CRM. (1997) The immunogenicity of intracerebral virus infection depends on anatomical site. *J Virol* **71**: 145–151.

45. Billingham RE, Boswell T. (1953) Studies on the problem of corneal homografts. *Proc R Soc Lond Biol Sci* **141**: 392–406.

46. http://en.wikipedia.org/wiki/immunnologically_privileged_site

47. http://en.wikipedia.org/wiki/T_cell

48. Blossom GB, Steiger Z, Stephenson LW. (1997) Neoplasms of the mediastinum. In: DeVita VT, Jr, Hellman S, Rosenberg SA (eds),*Cancer: Principles and Practice of Oncology,* 5th ed. Lippincott-Raven, Philadelphia, pp. 951–969.

49. Blum B, Benvenisty N. (2005) Differentiation *in vivo* and *in vitro* of human embryonic stem cells. In: Bongso A, Lee EH (eds.), *Stem Cells: From Bench to Bedside.* World Scientfic, New Jersey, London, Singapore, Beijing, Shanghai, Hong Kong, Taipei, Chennai, pp. 123–143.

50. Johnson PA. (2002) *The Gale Encyclopedia of Cancer.* In: Thackery E (ed.), Gale Group, Detroit, New York, San Diego, San Francisco, Boston, New Haven, Waterville, London, Munich, pp. 405–408.

51. http://en.wikipedia.org/wiki/Cyclophosphamide

52. Nussbaum J, Minami E, Laflamme MA, Vrag JAI, Ware CB, Masino A, Muskheli V, Reiniecke A, Murry CE. (2007) Transplantation of undifferentiated murine embryonic stem cells in the heart: teratoma formation and immune response. *FASEB* **21**: 1345–1357.

53. Swijneburg RJ, van der Bogt KEF, Sheikh AY, Cao F, Wu JC. (2007) Clinical hurdles for the transplantation of cardiomyocytes derived from human embryonic stem cells: role of nuclear imaging. *Curr Opin Biotechnol* **18**: 38–45.

30

BRAINBOW

Many, if not all, of us probably have ever become frustrated trying to untangle a ball of string or a knot of cords lying amid several appliances and the nearest wall electrical outlet. Many of us may also have thought we were oh-so-smart to buy a so-called "outlet adapter" that would accommodate up to six plugs. . . and triple the chances of knotting the cords. Yet the challenge of untangling up to six cords is nothing compared to that of separating the several hundred thousand individual neurons that constitute a nerve. The sciatic nerve, the longest and widest nerve in the human body (Fig. 30.1), is three-quarters of an inch (20 mm) in diameter,[1] comparable to the diameter of a person's index finger, at its beginning in the low-back (lumbar) region and is easily visible to the naked eye when exposed during low-back surgery. As it travels down the leg, it tapers gradually, becoming 6.4 mm in the upper thigh and 4.8 mm at the point where it divides behind the knee into the tibial nerve and the peroneal nerve.[3]

Neural Circuit Mapping and Connectomics

A ball of string? A knot of cords? Sort out neurons in the sciatic nerve? They're child's play when compared with trying to untangle individual neurons within the brain.[a] Now, just imagine the scientific world's utter astonishment when, on November 1, 2007, a paper doing exactly that — untangling individual neurons within the brain *in color* — was published in the prestig-

---

[a] The human brain is estimated to have about 100 billion neurons.[5]

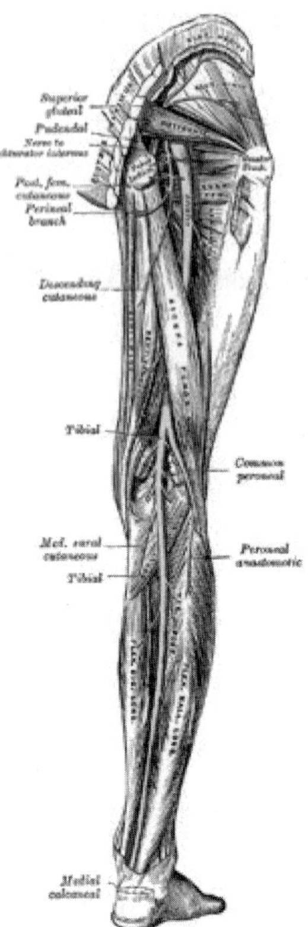

Fig. 30.1. The distribution of the sciatic nerve and its tapering branches in the right leg viewed from the back.

(Source: Reproduced with permission from the publisher.) Reference 2.

ious scientific journal *Nature*. Written by a team of seven molecular and cellular biologists at Harvard, the article introduced the term "Brainbow."[b,4]

---

[b] You may wonder, *What's the connection between Brainbow mice and antioxidant vitamins and stem cells?* The answer is, there really is none. Then, you might reasonably ask, *Well, why include a seemingly irrelevant chapter at the very end of an already lengthy book? Have you no mercy?*

The answer to that question is that I chose to include the exciting, exhilarating, breathtaking new Brainbow method of "staining" the central nervous system — the brain and larger peripheral nerves — as a "reward" for stoically trudging through some tough earlier chapters, such as the biochemistry of LDL oxidation (Chap. 9) chock-full of equations and foreign biochemical terms, and the many, negative clinical trials of antioxidants testing the ability of supplemental vitamin A, vitamin C, and vitamin E to prevent or treat coronary disease. [Chapters 21, 23, and

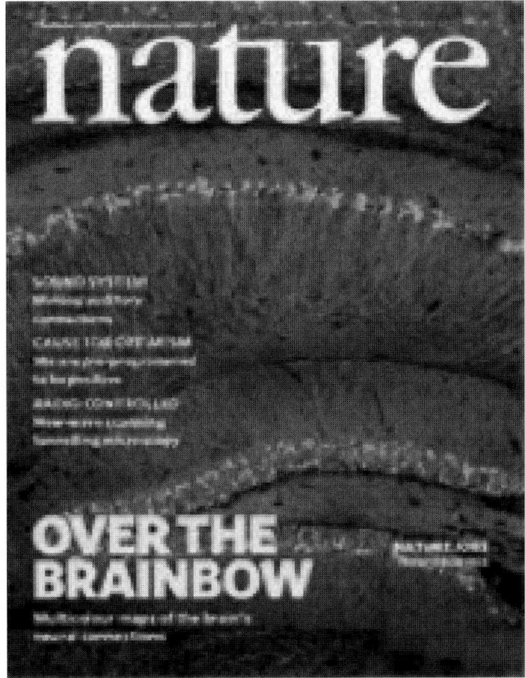

Fig. 30.2. The cover of the November 1, 2007, issue of *Nature,* reporting the historic Brainbow technique of coloring individual neurons in some 90 luminescent colors.

(Source: Reproduced with permission from the publisher.) Reference 6.

From the moment the story broke, basic scientists and physicians alike knew unmistakably this article would become a historic paper. The editor of the journal recognized science history was in his hands and devoted the cover to Brainbow (Fig. 30.2). Then, he wrote on page vii, "This issue features neuroscience on the cover in the form of Brainbow, a remarkable new technique that visualizes the track of hundreds of neurons as they wend their way through the brain."[7] And with those few words and a dazzling display of color photographs of the mouse brain in the pages that followed, a new era in neuroscience began.

---

24 — not to mention three chapters on retrolental fibroplasias (retinopathy of prematurity) that you may have found valueless other than from a historical standpoint.] You were likely hoping for — expecting — just a ray, a glimmer of good news in those early chapters. And you found none. So, I decided to finish this book on a picturesque, upbeat note, blending science and beauty — unforgettably.

We'll return to the fascinating Brainbow technique after a brief detour into the structure and function of a neuron and one of neurology's most devastating diseases — multiple sclerosis, a disease that one day could benefit from stem cell transplantation. [It's not only heart disease that may witness improvement following a stem cell transplantation (see Chap. 28, Table 28.1).]

Structure and Function of a Neuron

Neurons, Aka nerve cells, are the fundamental cells that conduct electrochemical impulses in the brain, spinal cord, and peripheral nerves. They characteristically have a cell body (soma) protecting their precious nuclei and (usually) multiple dendrites and a long, single axon.[8,10] The impulses themselves are produced and transmitted through the brain, spinal cord, and peripheral nerves by one of six neurotransmitter chemicals (such as acetylcholine)[c] that produce a small electrical charge that moves along the axon until it reaches a "synapse" — the site where an axon ends and delivers its electrochemical message to a postsynaptic nerve cell, using one of the six neurotransmitters [the other five being epinephrine (adrenaline), norepinephrine (noradrenaline), dopamine, serotonin, and histamine] without making physical contact with the cell. The messenger — the neurotransmitter — simply diffuses across a tiny "synaptic gap" to reach the outer membrane of the postsynaptic (receiving) nerve cell. A synapse, then, has three components: an axon, a narrow synaptic cleft (gap), and receptor sites along the dendrites of the receiving neuron, as shown diagrammatically in Fig. 30.3. Figure 30.4 depicts, in contrast, a *real* synapse in vivid fluorescent

---

[c] There is a thought-provoking vignette regarding the discovery of acetylcholine (recorded by the University of Washington on its website, "*Neuroscience for* Kids") that is worth repeating. "In 1921, an Austrian scientist named Otto Loewi discovered the first neurotransmitter. In his experiment (which came to him in a dream), he used two frog hearts. One heart (heart #1) was still connected to the vagus nerve. Heart #1 was placed in a chamber that was filled with saline (salt water). This chamber was connected to a second chamber that contained heart #2. Fluid from chamber #1 was allowed to flow into chamber #2. Electrical stimulation of the vagus nerve (which was attached to heart #1) caused heart #1 to slow down. Loewi also observed that after a delay, heart #2 also slowed down. From this experiment, Loewi hypothesized that electrical stimulation of the vagus nerve released a chemical into the fluid of chamber #1 that flowed into chamber #2. He called this chemical "*Vagusstoff*" (obviously German for the 'stuff' released by the vagus nerve). We now know this chemical as the neurotransmitter called acetylcholine."[9]

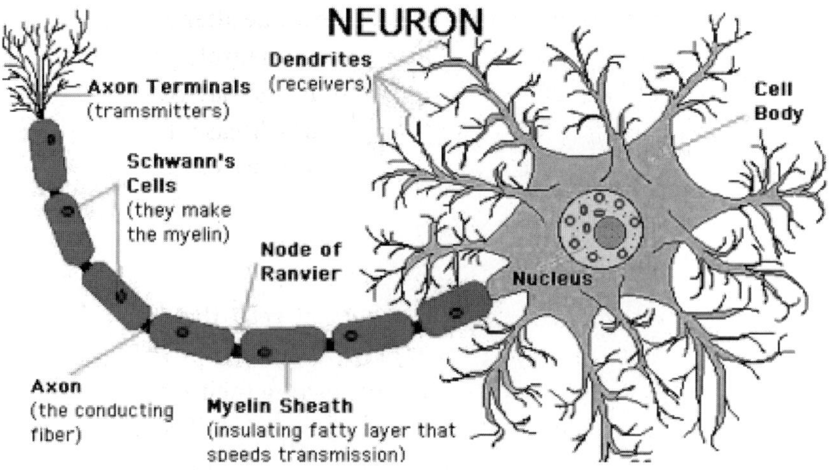

Fig. 30.3. This diagram depicts the structure of a neuron. The expanded bulblike end of a neuron containing the cell's important nucleus holding its DNA is known as the cell body (light purple). It is also called its "soma" (meaning "body" in Greek). The size of the soma ranges from 10 to 25 mm in humans. The neuron is a "conducting cell" that transmits electrical impulses along the nerve. The dendrites (receivers) of a neuron are simply branched extensions of the cell body that collect electrochemical signals from other neurons and transfer them *into* the cell body. Next in line is the neuron's axon, a thin cable-like structure (shown in black) that carries received signals *away from* the cell body. Most neurons have but one axon and many dendrites. The axon ends (upper left) the same way dendrites began — as a branched structure.[10]

This leaves nonconducting supporting structures — the myelin sheaths, Schwann cells, and nodes of Ranvier — to account for. A *myelin sheath* (red) is an insulating structure. It is composed of about 80% lipid (fat) and 20% protein. Like subcutaneous fat, it keeps what's beneath it warm. But myelin's main function is to increase the speed of electrical impulses (signals) plodding through an axon. Its function is best illustrated by its absence. Axons without a myelin sheath conduct impulses in a continuous-wave-like manner. Axons with a myelin sheath conduct impulses in a hop like manner, a faster mode of transmission than a continuous wave. Myelin increases the electrical resistance encountered by impulses traveling within the axon by about 5000 times, thus preventing the impulse from leaving the axon too soon.

The myelin sheath also has a secondary, yet critical, function. If a nerve is cut, the myelin sheath offers a tunnel-like structure for regrowth of the damaged nerve. Nerves without a myelin sheath do not regenerate. Loss of the myelin sheath occurs in multiple sclerosis and several less common neurological disorders. Nerves without a myelin sheath eventually wither away. *Schwann cells*, named after the German physiologist Theodore Schwann (1810–82), are the cells that make the myelin needed to insulate axons in peripheral nerves.[11] There are no Schwann cells in the central nervous system — the brain and spinal cord. Each Schwann cell is shaped like a rolled-up sheet of paper, with layers of Schwann cells wrapping themselves

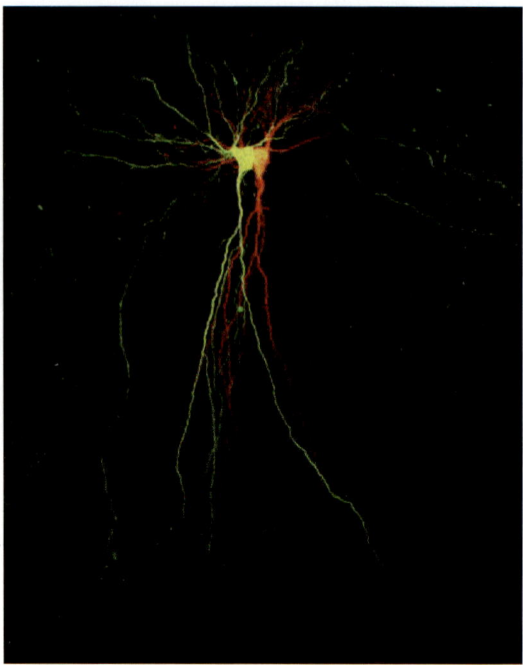

Fig. 30.4. A presynaptic neuron (right) "stained" with red fluorescent protein and a post-synaptic neuron (left) "stained" with green fluorescent protein forming a synapse. The cell body soma) is colored yellow in both neurons. It gives rise to a sprout of dendrites akin to antennae and a single, long, less-branched axon curving gracefully downward.

(Source: Reproduced with permission from the publisher.) Reference 14.

Fig. 30.3. (*Continued*)   around a central nerve fiber in a spiral fashion forming a tunnel, as shown in Fig. 30.5, and going about making the protective myelin coat surrounding peripheral nerves.[12] Each Schwann cell makes just enough myelin insulation for one axon. As a result of their stinginess, hundreds to thousands of Schwann cells are needed to cover an axon, which can stretch literally the length of the body. The stump of a damaged nerve is then able to sprout through the Schwann cell tunnel at an accelerated rate. Since 2001, Schwann cells have been implanted (autografted) — as opposed to transplanted — in patients with multiple sclerosis to promote myelin formation, thereby avoiding immune rejection.[13] Thinning or loss of the protective myelin sheath around nerves in the brain and spinal cord is the root cause of multiple sclerosis. The term *nodes of Ranvier* refers to periodic constrictions of the myelin sheath surrounding an axon. The nodes demarcate the junction between adjacent Schwann cells. At a node of Ranvier an axon is not insulated. Without a "jacket," an axon must generate some electrical activity for warmth.

(Source: Reproduced with permission from the publisher.) Reference 10.

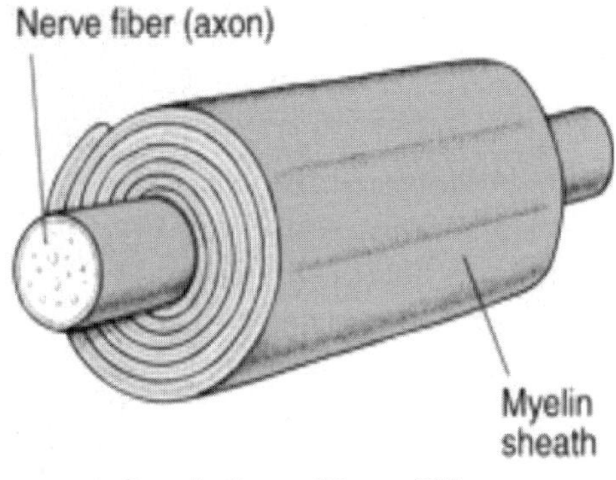

Insulating a Nerve Fiber

Fig. 30.5. Schwann cells make the myelin sheath that wraps around a nerve fiber (axon) in a spiral fashion. The insulating myelin sheath, like the insulation around an electrical wire, improves conduction.

(*Source*: Reproduced with permission from the publisher.) Reference 15.

colors, as elegantly applied using the Brainbow technique, introduced in November 2007. The contrasting images each have their virtues: more informational detail in the schematic version; visual appeal — beauty — in the fluorescent protein rendition. On the outer membrane, the postsynaptic cell has receptors so configured as to accept only a similarly configured neurotransmitter, *à la* lock and key. Figure 9.6 depicts how LDL — employing the lock-and-key principle — is transferred from the bloodstream into the subendothelial space using such similarly configured cell surface receptors that pinch off to form vesicles, which then ferry LDL from point A (the bloodstream) to point B (the subendothelial space). Neurotransmitters are transferred similarly. (A static and an animated illustration of a synapse may be found at http://facstaff.gpc.edu/~brown/psyc1501/brain/synapses.htm.) While acetylcholine was the first neurotransmitter discovered, five more have now been found. They are the chemicals mentioned earlier — dopamine, epinephrine (adrenalin), norepinephrine (noradrenalin), serotonin, and histamine. Both axons and dendrites, by the way, are called "nerve fibers."

Multiple Sclerosis

Multiple sclerosis is a neurological disease characterized by thinning or complete loss of the protective myelin sheath surrounding nerves. It is also an autoimmune disease, in which a person's immune system turns against the protective wrapping of nerves.[16] The term "sclerosis" refers to scars within the white matter (explained momentarily) of the brain and the spinal cord, both of which are formed mainly by myelin. Without insulating myelin, axons fail to conduct electrochemical signals (impulses) as they should.[17] White matter is composed of "bundles" of myelin-covered axons (an axon is depicted as a black cable within a red myelin sheath in Fig. 30.3), which connect gray matter areas (composed of light-purple cell bodies in Fig. 30.3) to each other.[18]

The function of white matter is to "carry signals between grey matter areas," where neural computing is done, and then to relay that data to "the rest of the body."[19] Extending the computer analogy, white matter represents computer cables linking gray matter areas — the computers. The whiteness of white matter is due chiefly to the color of myelin sheaths. In living tissue, gray matter has a gray hue that arises from its unmyelinated cell bodies.[20] The good news regarding white matter is that "unlike grey matter, which peaks in [its] development [during] a person's twenties, white matter continues to develop, and peaks in late-middle age"[17]. . .at the cost of less gray matter, which *really* counts.

The disturbing news about multiple sclerosis being an autoimmune disease — particularly disturbing for those with the disorder — emphasized a pair of neurobiologists, Jeffrey Kocsis and Stephen Waxman, at Yale, in a commentary[21] accompanying a paper in *Brain* reporting the "favorable" use of Schwann cells to repair the brain and spinal cord in multiple sclerosis[13] — is that though autologous (derived from a person's own body) stem cell transplants have shown "profound suppression of inflammatory activity in many patients, [this intervention] was developed in [an] attempt to mute the immune attack on the nervous system in MS, and not with the goal of repairing demyelination. Thus, even if the immune assault on the nervous system in MS could be halted by a new immuno-modulatory therapy and the subsequent cascade of tissue damage thereby stalled, hundreds of thousands of people harboring MS lesions would still be left with neurological deficits."[21] With this realization, "myelin repair has become an area of major interest in MS research."[21]

Back to Brainbow

Think of Brainbow as a clever way to visualize individual neurons within a nerve in Technicolor, just as insulating individual wires within a cable in housings of multiple colors is a clever way to allow easy visualization of

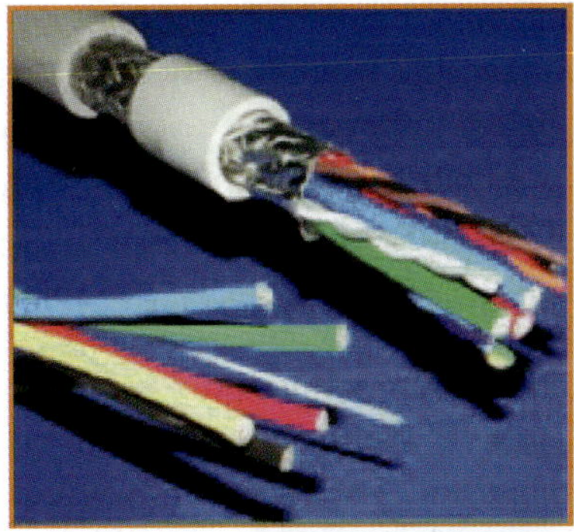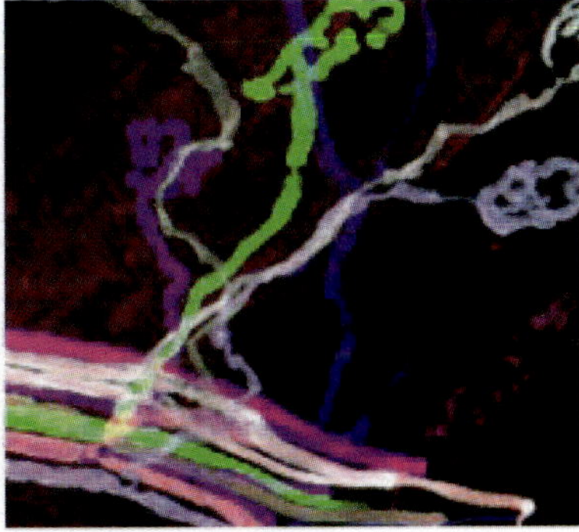

Fig. 30.6. *Left*: multicolored Video graphics array (VGA) and super VGA electronic cables used in computers. *Right*: Multi-colored axons, the thin cable-like neurons that carry signals received by dendrites away from the cell body, i.e. "soma" in a Brainbow mouse.

(*Source*: Reproduced with permission from the publisher.) References 22.

individual wires within a cable (Fig. 30.6). Think of Brainbow as a method to harness the enormous intellectual and emotional powers of the human brain described eloquently by Hippocrates in the fourth century B.C. (see box below). Think of Brainbow as a technique that randomly shuffles the genes in each neuron (nerve cell) responsible for synthesizing various fluorescent proteins. Think of Brainbow — as Jean Livet, one of the seven Brainbow investigators, put it — as being like "a molecular slot machine. [If] each cell [were to] play the slot machine, [each cell] would [randomly] be attributed a different [fluorescent] color."[24] Think of "Brainbow" as a method to "paint" individual neurons within a nerve using roughly 90 distinctive hues, producing a detailed map of neuronal circuits within the brain that looks like (Fig. 30.7),[d]

---

[d] Before Brainbow and since 1878, neuropathologists worldwide have visualized the brain and its peripheral (distant) nerves by staining them black using a compound developed by the Italian pathologist Camillo Golgi (1843–1926) — a compound which bears his name . . . the Golgi stain.[26] (We shall return to Golgi and two other early-20th -century pioneers in neurobiology later in this chapter. Their personal lives and professional rivalry are a story worth telling.) The Golgi compound stains some — not all — nerve cells (neurons) within the central nervous system (CNS) it comes in contact with. Yet, to this day, it is one of the mysteries of neuroscience just how it does what it does.[27] The chief limitation of the Golgi technique is that because it stains black *multiple* neurons within the brain or a peripheral nerve, *individual* neurons cannot be distinguished from one another (Fig. 30.7), making it impossible to follow the course of a single neuron from its origin.

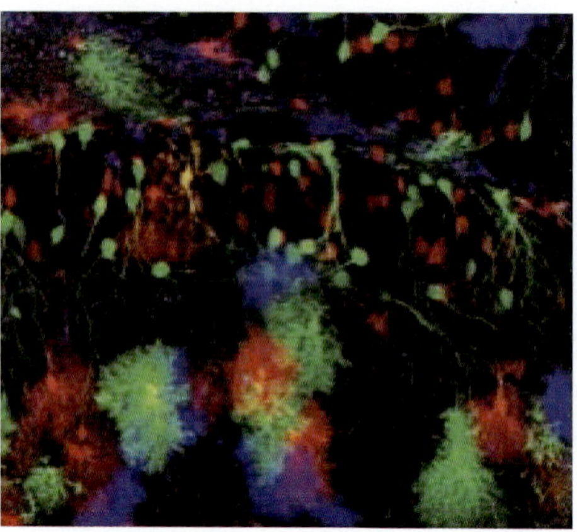

Fig. 30.7.   A segment of the brain from a Brainbow mouse. The cell bodies of neurons, distinguishable by their green, circular-to-elliptical shape, give rise to slender, green axons that connect with other neurons. The larger, more diffuse structures colored green, blue, and red are astrocytes — cells that provide structural and biochemical support for neurons.

(*Source*: Reproduced with permission from the publisher.) Reference 3.

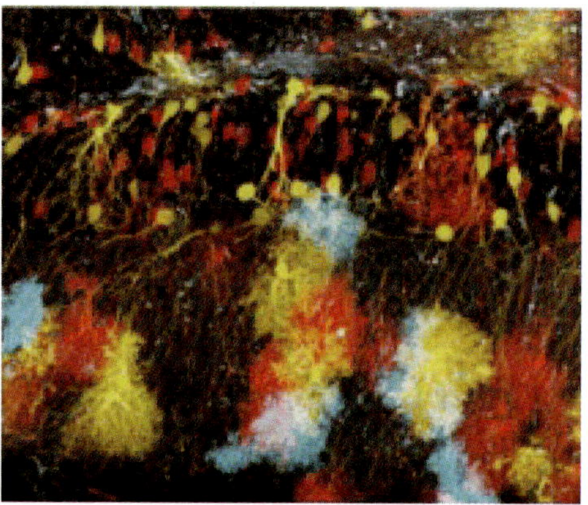

Fig. 30.8.   Neurons in the brain of a transgenic Brainbow mouse displayed in a kaleidoscope of colors.

(*Source*: Reproduced with permission from the publisher.) Reference 3.

Or Fig. 30.8, or even Fig. 30.9 As Marc Zimmer, a chemist from Johannesburg now Connecticut College, gushed, "the photgraphsw of the mouse brains. . . could be housed in the Museum of Modern Art or could be used to decorate Joseph's Technicolored dream coat. But it is not their colorful splendor that

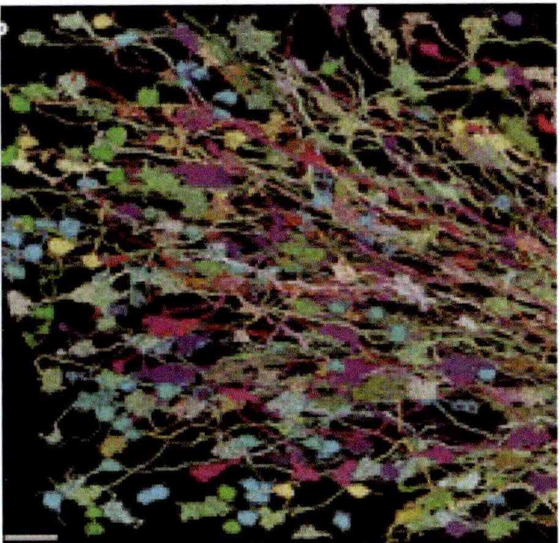

Fig. 30.9. A number of neurons in a mouse brain as seen through a confocal microscope using the Brainbow method of coloring individual neurons. Long, slender axons with their cell bodies (the circular structures) colored in several fluorescent hues can be distinguished from neighnoring axons.

(*Source*: Reproduced with permission from the publisher.) Reference 3.

makes these genetically modified mice so amazing. It is their potential to revolutionize neurobiology that excites scientists like myself and has our neurons firing away, creating oodles of endorphins."[25]

> Men ought to know that from nothing else but the brain come joys, delights, laughter and sports, and sorrows, griefs, despondency, and lamentations. And by this, in an especial manner, we acquire wisdom and knowledge, and see and hear and know what are foul and fair, what are bad and what are good, what are sweet and what are unsavory. . . .And by the same organ, we become mad and delirious, and fears and terrors assail us. . . .All these things we endure from the brain when it is not healthy. . . .In these ways I am of the opinion that the brain exercises the greatest power in the man."
>
> (*Source*: Hippocrates, *On the Sacred Disease*, fourth century B.C.)

While the beauty of Brainbow may release a torrent of endorphins and encephalins from the estimated 100 billion neurons in the human brain,[4] it takes more than the mere presence of fluorescent color genes to create a Brainbow mouse. As Jean Livet made clear, "In order for the color genes to be

Fig. 30.10. Jean Livet, from Paris, during a postdoctoral fellowship at Harvard in 2007. He came up with the Brainbow concept. Post-doctoral research is considered "essential to the scholarly mission of the host institution and is expected to produce relevant publications."[28] In this instance, Livet excelled.

(*Source*: Reproduced with permission from the publisher.) Reference 29.

expressed [made manifest], the mice cells must also contain another gene, called Cre.[e]

Derived from bacteria, "Cre activates the color genes inside the cell"[4] randomly — almost whimsically. Using a Las Vegas staple as an analogy, Livet explained that "if the color genes are the slot machine, then Cre is the hand [of the person playing the slot machine] pulling the lever over and over again."[24] Working together, color genes and Cre can create whatever [color] you want," he indicated.[24] Livet (Fig. 30.10), the youngest of the Brainbow investigators, is credited by his mentor, Jeff Lichtman, as the person who came up with the idea of randomly "painting" brain neurons using genes of distinctive hues.[30] For so doing, he was awarded the lead-author position in the now-classic Brainbow paper in *Nature*. (Remarkably, that groundbreaking report was Livet's first scientific publication.[31] He has since returned to France, where he works at the Institut de la Vision in Paris.[31]

---

[e]"Cre" is a shortened form of the term "*cyclization recombination*" protein. The protein is used in science to quickly and efficiently cut both strands of DNA, allowing the insertion or deletion of genes. The strands are then reconnected (recombined); hence the term "recombinant DNA." Because of the compound's enormous versatility — it will work with virtually any type of cell — Cre will be discussed more thoroughly within the next 10 pages or so.

A Fluorescence Microscope

There is one more item needed to visualize a brain or a peripheral nerve — it makes no difference — in Technicolor (Fig. 30.11). You gotta have a "fluorescence microscope." Not a "fluorescent microscope," but a "fluorescence microscope." The microscope itself isn't fluorescent, but specimens "stained" with fluorescent proteins are, when viewed through a florescence microscope.[f] And, yes, they are pricey. According to Ker Than, a writer for *LiveScience*, interviewing Jean Livet on the eve of the publication landmark paper of the, they can cost "several hundred thousand dollars."[33] (Yet, you might shell out US$250,000, or more, for such an instrument if it assured you'd be gazing at such stunning imagery eight hours a day.) A fluorescence microscope also differs from other scopes in that the specimen being examined is itself the light source.[34] In other words, the light source of the microscope — an ordinary incandescent bulb housed beneath the viewing stage — is not the source of the light making the specimen visible. The light you see is from the fluorescent specimen being examined.[35]

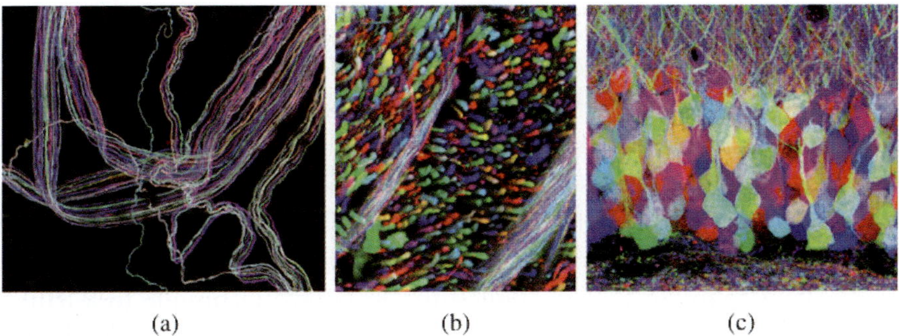

(a)   (b)   (c)

Fig. 30.11. Three images of mouse neurons colored by the Brainbow method. It makes no difference whether the neurons form a peripheral nerve (panel a) or synapses in the brain (panels b and c) — the tissue can be transformed into colorful splotches of this and that hue. (*Source*: Reproduced with permission from the publisher.) Reference 32.

---

[f]When viewed under ordinary incandescent light (from a tungsten filament bulb), using a standard "light microscope," Brainbow-mice brains have a drab-gray hue. . .like the brains of all other mice — and men.[32] Yet under UV light they glow in brilliant Technicolor (Fig. 30.11).[32]

Limitations of the Brainbow Technique

Still, even Brainbow is not perfect. For one thing, the microscope required to view the pretty specimens is expensive, as already indicated. Also, as Jean Livet explained, ". . .[Brainbow] only works with genetically modified, or transgenic, animals which at the moment only include mice [human brain tissue cannot be colored through the Brainbow technique.]. With the [century-plus-old] Golgi stain you can do everything, including humans."[33] On top of that, the mice must be specially bred from embryonic stem cells to create a useful animal — a process that is "time-consuming and complex."[36] And then, even if two such transgenic mice are successfully created, there is no guarantee all the progeny will be the same as their parents.

Ideally, the Brainbow method could be applied to mammals. But the "incredible [number and] diversity. . .[and] density of neurons [in the central nervous system of mammals] coupled with the presence of long tracts [of] axons [such as in the sciatic nerve shown in Fig. 30.1] make viewing lager regions of the CNS with high resolution difficult. While Brainbow is useful [when] examining single nerve fibers within a complex muticellular environment, the resolution limits of optical microscopy [render] identification of. . . connections between neurons challenging."[36]

A "Flybow"

Limitations aside, Brainbow is a spectacular advance in neuroscience. Impressed by the esthetics of the technique in transgenic mice, fruit fly (*Drosophila*) researchers at the Howard Hughes Medical Institute's branch in Ashburn, Virginia, decided to determine what a fruit fly's brain might look like in living color. So. . .they applied the Brainbow technique to a fruit fly brain in 2011 and found it looks like the one in Fig. 30.12. The investigators named it a "flybow." And should one day it be possible to view a human brain with fluorescent colors, I suppose it may be called a "humbow."

A Tug of War

About then, along came a gifted (science) writer, one Helen Pearson, a developmental biologist with a Ph.D. in genetics from the University of Edinburgh. After she served as a reporter for the Nature Publishing Group (London), her writing talents earned her a position preparing frequent feature articles for a

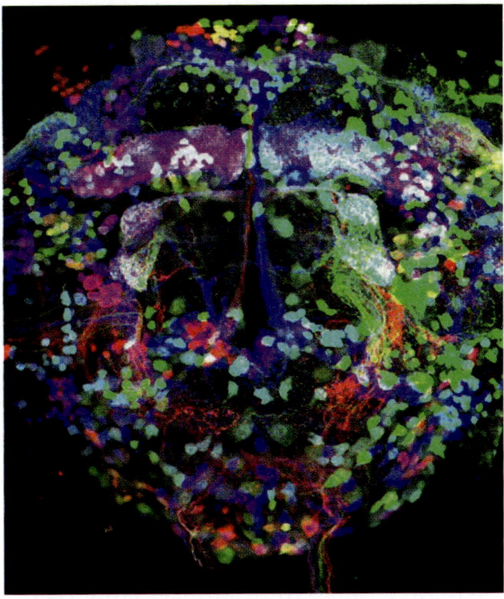

Fig. 30.12. A fruit fly brain set aglow in living color through the Brainbow method of "staining" neurons.

(*Source*: Reproduced with permission from the publisher.) Reference 37.

Fig. 30.13. Helen Pearson, Chief Features Editor of *Nature* in London.

(*Source*: Reproduced with permission from the publisher.) Reference 40.

couple of years as a member of a team of other skilled authors, before being catapulted in 2009 to her current position as the Chief Features Editor of the Nature Publishing Group (Fig. 30.13).[38] (Along with that promotion, she ended up with the British Science Writers Award in her lap the next year.[39] The group is a print and online publishing company with international tentacles that explore "the physical, chemical and applied sciences [as well as] clinical medicine."[41] It boasts several premier academic journals and just happens to publish *Scientific American* (in 16 languages), a magazine often read by college-educated but not necessarily scientifically trained people.

While Pearson was employed by the Nature Publishing Group, the Brainbow paper hit the streets in the November 1, 2007, issue of *Nature*. And she, naturally, as a feature editor, wrote an exemplary commentary. She also had penned a commentary just six months earlier in the May 10, 2007, issue of *Nature*,[42] spawned by a cautionary letter-to-the-editor of *Nature Biotechnology* by a group of cell biologists at the Microscope Imaging Centre in The Netherlands, titled "ATP and FRET — A Cautionary Note."[43] The pertinence of her commentary to the landmark article made it appear as if it were written in anticipation of the November paper. But Helen Pearson, while a talented science writer, is (presumably) not prescient. The (landmark) article's title page indicates it was received for publication on July 17, 2007[4] — two months after Pearson's May commentary. How could she possibly have anticipated the Brainbow bombshell?

Pearson's May 2007 story, titled "The Good, the Bad and the Ugly," spotlighted the reliability — the possible inaccuracies — of observations made by cell biologists using the new fluorescence microscope.[42] The instrument created a tug of war between beauty and science — although the two certainly can coexist. Viewing the skies with a powerful telescope on a clear night is sheer beauty — sheer bliss — to an astronomer, even though it's basically a black-and-white image. [On the other hand, the Hubble Space Telescope, functioning 24 h a day, transmits spectacular, color images of outer space (Fig. 30.14)]. Yet the galaxy of spectacular photographs in the November 1, 2007, issue of *Nature* and other publications that soon followed raised concerns that journal reviewers and journal editors might be swept off their feet by the jaw-dropping esthetics of the central nervous system when viewed in Technicolor under a fluorescence microscope and not critical enough of the pitfalls associated with the use of such a sensitive instrument by insufficiently trained individuals feeling the pressure of "publish or perish." Footnote 100, earlier in this chapter, is a synopsis of Pearson's article in the May 10, 2007, issue of *Nature*. Given the extraordinary and unprecedented impact of the November 1, 2007, publication from Harvard on the fluorescent imaging technique and the remarkable photographs of what's going on inside nerve cells, the technique will now be promoted from a footnote into prime time.

Fig. 30.14. A color photograph of outer space taken by the Hubble Space Telescope. (*Source:* Reproduced with permission from the publisher.) Reference 44.

Pearson began, "The satellite imagery of Google Earth offers homeowners the chance to zoom in from outer space and hover above their rooftops."[42] (Just who hasn't zoomed in on his or her neighborhood, indeed his or her own home, using Google Earth as his or her outer-space vehicle?) "For biologists, a microscope gives a similarly exhilarating view of a cell's innards."[42] But, she warned, ". . .looks can be deceiving." Plus, it's not easy to dismiss the adrenaline rush that comes with using a fluorescence microscope with all its "accoutrements and [a] price tag of a high-speed racing car [that] offers an exhilarating ride. It can boast numerous knobs, a foot pedal, winking lights and touch-control climate. Such microscopes can cost anything from US$50,000 to $1 million. But not everyone should be allowed behind the eyepiece."[42] Pearson quoted Simon Watkins, "who runs a biological-imaging center at the University of Pittsburgh in Pennsylvania," as saying that using a fluorescence microscope is "much more complicated than sitting down and pressing buttons. If you got into a fast car but didn't know how to drive it, you'd crash very quickly."[42]

And that's just what happened, Pearson lamented. "Many biologists' ability to handle the instruments has not kept pace with the technology, and the road to results is becoming littered with scrapes, prangs and outright wrecks." Couple that with the tendency of "journals and reviewers [to be] too often impressed by pretty images" and it becomes a "big problem," according to Alison North, a microscopist with Rockefeller University in New York. "When the reviewers [of scientific journals]," chimed in Pearson, "are more concerned with how aesthetically pleasing an image is than whether the scientific content is clear,"[42] then science is skating on thin ice.

Continuing, the developmental-biologist-turned-science-reporter-writer-and-editor and 2010 Winner of the Association of British Science Writers Award[42] cautioned, "The list of potential mistakes[42] in fluorescence microscopy is long and complex." After speaking with several cell biologists from The Netherlands, she wrote, "To correctly capture images using a modern [fluorescence] microscope, researchers must have a good grasp of optics, an awareness of the microscope's complexity and an obsession for detail. Such skills can take months or even years to master, and yet, owing to inexperience or the rush to publish, are all too often squeezed into hours or days. Popular methods, such as fluorescence microscopy, are particularly fraught with dangers."[42] Complementing the latest technology with its skilled usage and balancing the threat of "publish or perish" with meticulous research are art forms mastered by many — but not all — science investigators. We all want to think the neurobiologists at Harvard are among those investigators who are experts in using the fancy fluorescence microscope to look inside nerve cells, and that the colorful photographs are free of artifacts, the result of inexperience with the sensitive scope. Somehow, it seems highly improbable that the Boston scientists allowed pesky artifacts to creep into their striking photographs made using the temperamental instrument and that the pretty pictures of what's going on within neurons generated by Brainbow will stand the test of time, as have many Rembrandts and the *Mona Lisa*.

Out of Darkness Comes Brilliant Light

On November 6, 2007, just five days after publication of the Brainbow paper in *Nature*, another science writer — Gareth Cook, on the staff of *The Boston Globe*, the hometown paper of Harvard's Brainbow creators — likened Brainbow to another renowned artist, Claude Monet, the late 19th to early 20th century French painter, who created highly recognizable impressionistic paintings — the ones many of us have purchased for home or office wall hangings (Fig. 30.15).

Wrote Cook, reflecting upon a colorful Brainbow image, "The field of splotches across the middle of the image evoke Claude Monet's impressionistic brushwork. The rich colors — from fiery reds to grass greens and sky

Fig. 30.15. *On the Bank of the Seine, Bennecourt* painted by Claude Monet in 1868. An early example of plein-air impressionism, in which a gestural and suggestive use of oil paint was presented as a finished work of art.

(*Source*: Reproduced with permission from the publisher.) Reference 45.

blues — would be at home in his famous painting of a sunny poppy field near Argenteuil, France".[46] Ironically, the bright Brainbow images Cook was referring to came out of a "dimly lit Harvard laboratory" equipped with an "exquisitely sensitive microscope."[46] "Brainbow," he believed, "is part of a renaissance in microscopy that could have far-reaching implications for biology. . . . In just the last few years there have been dramatic improvements in biologists' ability to highlight features in a cell they want to see by using fluorescent colors. This, combined with genetic engineering and advances in the microscopes themselves, is illuminating living worlds never seen before".[46]

Beauty is in the Eye of the Beholder

As you might imagine, if adjacent and nearby neurons within a nerve were colored the same, it would be (virtually) impossible to visually separate individual nerve cells. But if neighboring cells were labeled with different colors, it would become quite possible to follow the trail of individual nerve cells no matter how tangled they became. In a separate commentary on Brainbow by Emily Singer published online by the Massachusetts Institute of Technology in Technology Review on November 1, 2007, Singer reminded readers that "fluorescent proteins [were] derived from jelly fish,"[48] taking us back to Friday

Harbor, *circa* 1961. Then, she said, referring to the bacteria-derived Cre gene which activates color genes haphazardly, "When the mice are fed a [nonhallucinogenic] compound that activates the enzyme, each cell undergoes a random molecular process in which subsets of the color-coding genes are knocked out. The remaining genes produce the three colored fluorescent compounds in different amounts, which combine to form a unique new hue."[48] "We get a wide range of colors — about 100," said Lichtman (Fig. 30.16), an opinion echoed by a third member of the Brainbow team, Josh Sanes (Fig. 30.18). Said Singer, "The ability to paint individual brain cells with such a broad palette will allow neuroscientists to explore neural circuits like never before." Or, as Elly Nedivi, a neuroscientist at M.I.T. who was not a part of the Brainbow research team, put it, "This will be an incredibly powerful tool."[48]

Just how vividly Brainbow can "paint" neural circuits and how vividly a rainbow can "paint" a darkened sky is seen in Fig. 30.17. It is images such as these that led Connecticut College's Marc Zimmer, who posted online the Brainbow image in Fig. 30.16, to say, "Never before has a brain been so beautiful. . . .Using a brainbow of colors, researchers [can] now. . .map the neural circuits of the brain. The individually colored neurons [should] help define the complex tangle of neurons that comprise the brain and nervous system. By creating [such] a wiring diagram of the brain, [it ought to be possible to] identify the defective wiring found in neurodegenerative diseases such as

Fig. 30.16. Jeff Lichtman, neuroscientist in the Department of Cellular and Molecular Biology at Harvard, and coauthor of Brainbow as well as a leader in the new field of connectomics. Lichtman aims to construct a connectome — a "complete map of neural wiring in the human brain" and all of its "100 billion neurons and several trillion synaptic connections."

(*Source:* Modified (cropped) and reproduced with permission from the publisher.) Reference 47.

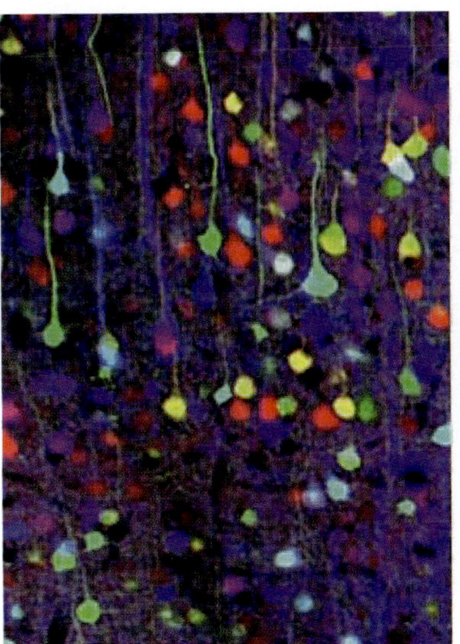 

Fig. 30.17.   I suppose a rainbow is still prettier than a Brainbow. But it's tough to pick the winner, don't you think? Men, typically, are good judges of female pulchritude, but women, I believe, are better judges of beauty in general, be it beauty in clothes, cars, or furniture, or in many other forms. According to *The Boston Globe's*, Gareth Cook, when Tamily Weismann, the sole woman on the Brainbow team, saw the first image of a mouse's brain in Technicolor on her computer screen, she "ran [excitedly] to get [Jeff] Lichtman,"[47] the head of one of the two laboratories that combined their staffs to create Brainbow mice. "It was amazing," she said.[48]

Although there may not be a pot of gold at the end of the above rainbow, there may well — probably will — be one at the end of the Brainbow, in the form of grants and honoraria for the Harvard neurobiologists who figured out how to make the brain glow in color. As Claudia Wiedeman, writing for *Nature Reviews of Neuroscience*, mentioned, "[Brainbow is destined to] profoundly influence neuroscience research in the future"[50] . . .and the lives of the Brainbow team for a long, long time.

The photograph on the left, by the way, shows roundish, multicolored, neuronal cell bodies and their same-colored axons coursing upward in the cerebral cortex of a Brainbow mouse. The cerebral cortex occupies only the outermost 2–4 mm (less than ¼ inch) of the human brain, yet plays a key role in memory. In a dead, unpreserved brain it has a gray color, leading to the term "gray matter." It's hard to believe "gray matter" can be transformed into such colorful splendor.

(*Source*: Reproduced with permission from the publisher.) References 25 and 50.

Alzheimer's and Parkinson's disease," which are Josh Sanes' (Fig. 30.18) research focus. The breathtaking beauty of Brainbow images led Zimmer to muse, "Brainbow will have a similar effect on neuroscience as Google Earth had on cartography," in 2006.[25]

Fig. 30.18.   Josh Sanes, neuroscientist, Director of the Center for Brain Sciences at Harvard, and coauthor of Brainbow. Beyond Brainbow, Sanes' research focuses on the tiny synaptic connections (Figs. 30.4 and 30.7) where electrical impulses are transmitted from one nerve to the next "in animal models of [varied] neurological and psychiatric diseases."

(*Source*: Reproduced with permission from the publisher.) References. 49.

And all this by using just four fluorescent protein colors randomly mixed to "create a palette of ninety distinctive hues and colors."[25] Don't go far away. We'll come back to those "hues and colors" shortly.

References

1. http://www.spineuniverse.com/displayarticle.php/article2524.html   (accessed August 7, 2010).
2. http://en.wikipepia.org.wiki?Sciatic_nerve (accessed August 2, 2010).
3. http://www.ncbi.nlm.nih.gov/pubmed/827F2540 (accessed July 31, 2010).
4. Livet J, Weissman FA, Kang H, Draft RW, Lu J, Bennis RA, Sanes JR, Lichtman JW. (2007) Transgenic strategies for combinational expression of fluorescent proteins in the nervous system. *Nature* **450**: 56–62.
5. http://en.wikipedia.org/wiki/Neuron (accessed August 1, 2010).
6. Editor's Summary. Over the Brainbow. *Nature* 2007 **450**:
7. About the cover. *Nature* 2007; **450**: vii.
8. http://en.wikipedia.org/wiki/Soma_ (biology) (accessed August 1, 2010).
9. http://faculty.washington.edu/chudler/chnt1.html (accessed August 3, 2010).
10. http://www.enchantedlearning.com/subjects/anatomy/brain/Neuron   (accessed July 29, 2010).
11. http://en.wikipedia.org/wiki/Theodor_Schwann (accessed August 2, 2010).

12. http://en.wikipedia.org/wiki/Schwann_cell (accessed August 2, 2010).

13. Woodhoo A, Gilson SJ, Gilson J, Setzu A, Franklin RJM, Blakemore WF, Mirsky R, Jensen KP. (2007) Schwann cell precursors: a favorable cell for myelin repair in the central nervous system. *Brain* **130**: 2175–2185.

14. http://www.merck.com/mmhe/sec06/ch092/ch092a.html (accessed August 1, 2010).

15. http://www.educater.com/images/brain-nerve-axon.jpg (accessed August 1, 2010).

16. http://www.rikenresearch.riken.jp/research/225/image_923.html

17. http://www.google.com/search?hl=en&ei=hEmOSczEE9CCtwfB1-WUCw&sa=X&oi=spell&resnum=0&ct=result&cd=1&q=Britannica+concise+encyclopedia&spell=1 (accessed July 31, 2010).

18. http://en.wikipedia.org/wiki/White_matter (accessed August 2, 2010).

19. http://en.wikipedia.org/wiki/Multiple_sclerosis (accessed August 1, 2010).

20. http://en.wikipedia.org/wiki/Grey_matter (accessed August 2, 2010).

21. Kocsis JD, Waxman SE. (2007) Schwann cells and their precursors for repair of central nervous system myelin. *Brain* **130**: 1978–1980.

22. http://www.hcm.hitachi.com/electronic_ribbon_cable/intermittent-bonded-rainbow-cable.shtml-cable.shtml

23. http://www.nature.com/nature/journal/v450/n7166/suppinfo/nature06293.html

24. http://news.harvard.edu/gazette/section/science-n-health

25. http://www.conncoll.edu/ccacad/zimmer/GFP-ww/cooluses0.html (accessed August 3, 2010).

26. Golgi, C. (1891) La rette nervosa difusa degli organi centrali del sistema nervoso. Suo significato fisiologico, *Rendiconti del R. Istituto Lombardo di Scienze e Lettere* **24**: 594–603, 656–673.

27. http://www.conncad.com/gallery/Golgi%20stained%20neurons%20in%20the%20dentate%20gyrus%20of%20an%20epilepsy%20patient.jpg (accessed August 7, 2010).

28. http://en.wikipedia.org/wiki/Postdoctoral_researcher (accessed August 5, 2010).

29. http://mcb.harvard.edu/NewsEvents/Lichtman3.html (accessed August 3, 2010).

30. Frankel F. (2008) In living color. Am Scientist **96**: 59–61.

31. http://www.molecularstation.com/research/author/livet-jean.html (accessed August 1, 2010).

32. Lichtman JW, Livet J, Sanes JR. (2008) A Technicolor approach to the connectome. *Nat Rev Neurosci* 9:417–422.

33. Than K. Brain cells colored to create "Brainbow." http://www.livescience.com/1977-brain-cells-colored-create-brainbow.html

34. http:// www.epifluorescencemicroscopes.com (accessed August 4, 2010).

35. http://nobelprize.org/educational_games/physics/microscopes/fluorescence/index.html (accessed August 4, 2010).

36. http://en.wikipedia.org/wiki/Brainbow
37. Hampel S, Chung P, Mc Kellar CE, Hall D, Looger LL, Simpson JH. (2011) Drosophila Brainbow: a recombinase-based fluorescence labeling technique to subdivide neural expression patterns. *Nat Meth* **8**: 253–258.
38. http://ww.nature.com/nature/about/editors/
39. http://www.ukscj.org/speakers/helen-pearson-chief-features-editor-nature.html (accessed September 12, 2012).
40. http://uk.linkedin.com/pub/helen-pearson/27/45/963 (accessed October 1, 2012).
41. http://en.wikipedia.org/wiki/Nature_Publishing_Group (accessed August 1, 2010).
42. Pearson A. (2007) The good, the bad and the ugly. *Nature* **447**: 138–140.
43. Willemse M, Janssen E, de Lange E, Wieringer B, Fransen J. (2007) ATP and FRET — a cautionary note. *Nat Biotechnol* **25**: 170–172.
44. http://hubblesite.org/gallery (accessed July 29, 2010).
45. http://en.wikipedia.org.wiki/Claude_Monet (accessed July 28, 2010).
46. http://www.boston.com/news/globe/health_science/articles/2006/11/06/microscope_renaissance/ (accessed July 27, 2010).
47. http://news.harvard.edu/gazette/story/2007/10/researchers-create-colorful-brainbow-images-of-the-nervous-system/ (accessed August 4, 2010).
48. Singer E. (2007) The Technicolor brain: illuminating neurons with nearly 100 different colors could shed light on the brain. *MIT Technol Rev,* Nov. 1.
49. http://www.techtransfer.harvard.edu/crop/investigators/investigator.php?id=109 (accessed August 1, 2010.
50. Wiedman C. (2007) Rainbows in the brain. *Nat Rev Neurosci* **8**: 907.

31

CRE/LOX: A CUT-AND-PASTE METHOD OF GENE SWAPPING WITHOUT A MAC OR PC AND A NEW SCIENTIFIC TERM WITH VAST SIGNIFICANCE

Don't be frightened by the verbiage. It is not as tough to grasp "Cre/*lox*" as you might think upon first seeing such an odd term. An open mind and a dollop of patience will be rewarded.

But, first, the biochemical, enzymatic, and genetic (gene-swapping) stage for such colorful splendor will be set and the cast of characters introduced. The cast includes such foreign actors as Cre and *lox* and some *very* strange-looking bacteriophage. To begin, let's separate "Cre" from "*lox*" to examine, ever so briefly, each term's grammatical origin and biochemical background.

What Is Cre?

Tackling the term "Cre" first simply because it *is* first, "Cre" is an acronym formed by the first letters of three — in effect, two — scientific words: cyclization recombination protein. The term refers to the ability of Cre — an enzyme and, and by virtue of its "birthright," a protein — to promote *cyclization* — the transformation of molecules aligned in a linear manner within a hydrocarbon (a compound composed primarily or solely of carbon and hydrogen atoms[1] into a circular structure, as shown in Fig. 31.1. Cyclization is followed by *recombination*, meaning (in the jargon of genetics) the rejoining of a DNA strand after the excision of damaged nucleotides — the "bricks" of the DNA molecule (see Chap. 15) — and the insertion of fresh

489

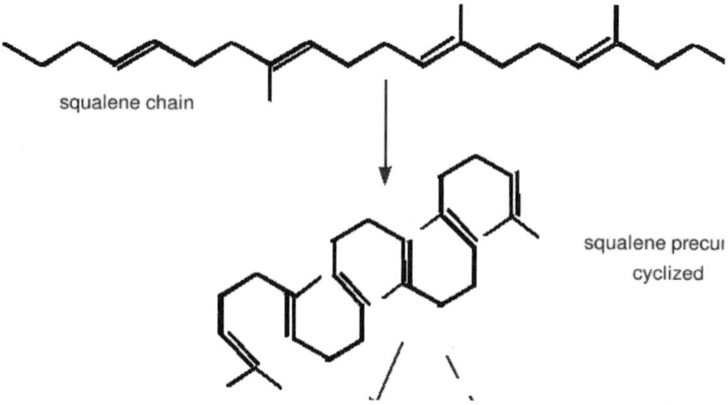

Fig. 31.1.   An example of cyclization. In this case the compound squalene is converted from a linear structure (upper figure) into a ringed or cyclical-shaped structure (lower figure). Squalene is a natural compound derived from shark liver oil and is also synthesized by humans. Its principal claim to fame is as the biochemical precursor of all natural steroids, including cholesterol, testosterone, estrogen, and progesterone — as well as the synthetic steroids used by some athletes to build muscle mass, and vitamin D.[2]

In the case of Cre, the enzyme favors the formation of one or more (benzene) rings (the hexagonal structures above) in an organic compound — an organic compound being one that contains at least one carbon atom — with a linear (linked chain) arrangement before Cre's handiwork (upper panel) into one configured with closed rings (lower panel).

(*Source*: Modified and reproduced with permission from the Massachusetts Institute of Technology). Reference 3.

ones. Following damage — including a complete break in either of the two strands of the sizeable DNA molecule (see Fig. 15.3) — by everyday environmental hazards such as cigarette smoke,[a] radiation, and ultraviolet light.[4] (see Chap. 4) that constantly bombard the molecule, it is "weakened." That is Cre's cue to get busy.

---

[a] I am always puzzled by the number of young women on a university campus who smoke cigarettes, and occasionally pause to speak with some. I say that smoking — and spending a lot of time in the sun or in tanning salons — hastens the development of wrinkles. And since they are bright people, some in the sciences, some in the arts, I may toss in a comment about smoking and sunlight stirring a hornet's nest of free radicals to provide a scientific explanation for the accelerated facial aging *and* as a scare tactic. To an extent that's cheating, because some free radicals do serve a healthful purpose (see Chap. 9). But many coeds are not aware of that tidbit of news. They see free radicals as sinister, *always*. I rationalize my deception by thinking the harmful effects of cigarettes on the heart and lungs far outweigh any beneficial effect a smattering of free radicals may have.

At least that's the origin of the term "Cre" most often given. Using my mind's eye, I view "Cre" as an acronym for "cyclization recombination enzyme," rather than "cyclization recombination protein," because that serves as a reminder that the compound *is* an enzyme, and as such a protein, because (nearly) all enzymes are proteins.[b] As an enzyme, it catalyzes — it causes or accelerates a chemical reaction — without itself being consumed or altered permanently, qualities common to all enzymes. It does what it is "supposed" to do, emerges unscathed, and is ready to catalyze another reaction. It's done, but not spent. Viewing Cre as an enzyme also permits the legitimate use of the term as an acronym, since technically only the initial letters of words in a phrase should be used to form an acronym. [Recall the penchant for acronyms in science and medicine, particularly in clinical trials of therapeutic agents (see Table 20.7)].

An even simpler account of what constitutes cyclization may be that it is a process which converts a compound configured as a straight line into a circular structure (Fig. 31.1). In the case of Cre, the enzyme favors the formation of one or more (benzene) rings (Fig. 31.2) in an organic compound with a linear configuration before Cre's handiwork. (An organic compound contains at least one carbon atom. Examples of organic compounds are the carbohydrates, fats, proteins, and nucleic acids so plentiful in a Western diet, and the steroids cholesterol, testosterone, estrogen, and progesterone.) Cre converts

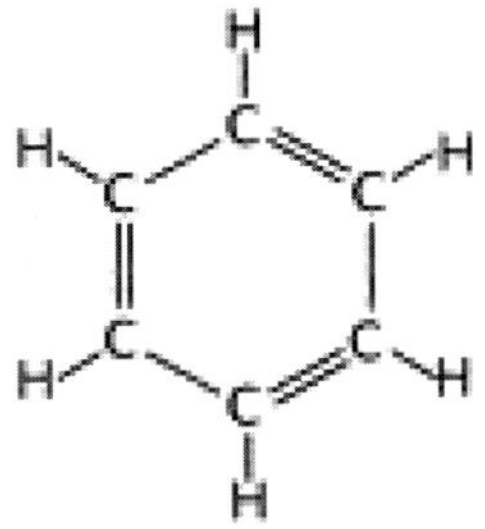

Fig. 31.2. A benzene ring, perhaps the best-known compound in biochemistry, with a circular — actually, hexagonal — arrangement of its carbon and hydrogen atoms.

(*Source*: Reproduced with permission from the publisher.) Reference 5.

------

[b] In 1982, chemist Thomas Cech at the University of Colorado showed that RNA — long known as a messenger of heredity — was also able to cut and splice chemical bonds, just like an enzyme (see Chap. 2). In recognition of this discovery, Cech and Sydney Altman, a physicist and molecular biologist at Yale, who credits much of his inspiration leading to the discovery of the "enzymatic properties of RNA" to George Gamow, a physicist at the University of Colorado in the late 1960s, shared the 1989 Nobel Prize in Chemistry.[6]

such open-ended compounds configured in the shape of a linked chain into a compound configured as a closed-ring structure.

That's the grammatical background of the acronym "Cre." What Cre, the enzyme, does is at least as important as its grammatical basis. First, the enzyme excises the usually short injured segment in either of the two long, helical (twisted) strands (see Fig. 15.3) of the sizeable DNA molecule and then splices the two exposed ends together. So, Cre may be thought of as a repair enzyme. But it is a far more sophisticated fixer of broken parts than a Band-Aid. Cre literally replaces a damaged — even severed — section of DNA with a fresh section, restoring its integrity and leaving it as functional as it was before an environmental villain struck. (Environmentalists are sure to cheer this "recycling" of precious genetic material.) This enzymatic reparative process is known as "homologous recombination," first brought up in Chap. 15.

How About *Lox?*

Now that the "Cre" component of Cre/*lox* has been scrutinized, the "*lox*" component will be dismembered similarly to examine its grammatical origin and biochemical background. Regrettably, the word has zilch to do with the seafood enjoyed by so many folks. While lox — often presented on or with a bagel and a side of cream cheese topped with capers (Fig. 31.3) — is the centerpiece of the "quintessential New York brunch"[7] relished by Yiddish and

Fig. 31.3.   A bagel with lox, cream cheese and onions.

(*Source*: Reproduced with permission from the publisher.) Reference 7.

Gentiles alike, the variety of lox served with the Cre version is far less delectable. It's downright inedible!

A Second Difference Between the Two Types of Lox

A second difference between the two varieties of lox is that the inedible one is typically italicized, because it is the "name of a genetic locus" and, by the "rules" of genetic nomenclature, gene names and loci are italicized.[8] And this particular locus (site) is one important locus, for it is here that homologous recombination occurs. It is here that a section of DNA in a bacteriophage (a bacteria-eating macrophage, first mentioned in Chap. 3, footnote d) "crosses over" (to be explained shortly) and X marks the (cross-over) spot — creating a very different form of lox. That X marks the spot where the phage swaps a section of its DNA for a bacterium's DNA — homologous recombination. But not just *any* bacteriophage! It *must* be a particular type of "phage"[c] (as bacteriophage are commonly called in scientific and medical circles) named "phage P1," soon to be discussed further.

Homologous Recombination and Mother's Little Helper

The process of homologous recombination begins with the enzymatic excision of a short section of nucleotides within either of the two long, helical

---

[c]When there was just one recognized phage, as there was soon after the discovery of the first phage in 1915–1917, no dispute about the plural form of the term "phage" or "bacteriophage" existed. But the grammatical agreement didn't last long. Additional phage were soon discovered and today more than 5000 strains of phage are recognized, causing grammatical warfare. There are two warring camps. One maintains that the term "phage" refers to a "single phage particle or strain," while the plural form — "phages" — "designates a population of phage particles, several phage species or strains, and the sum of all bacterial viruses."[10]

The other side argues that there is only one term — "phage" or "bacteriophage" — which covers a single micro-organism or many. One or more than one organism is inferred by accompanying "pronouns or modifiers [such as] it, this, these, and few and verb forms [such as] these phage, those phage, and a few phage as well as two phage and the sum of all phages, rather than 'phages' " for multiple organisms. "Differentiation [of phage singularity or plurality] is generally impossible in the past tense."[10] In this book, I have chosen to use the word "phage" in its second incarnation — one or many, they are all "phage."

(spiral-shaped) strands of the huge DNA molecule (see Figs. 15.1 and 15.3), and ends when the excised segment is replaced by a similar or identical segment. As an enzyme, Cre does what all enzymes do — they catalyze, they increase the speed of a chemical reaction, often "dramatically"[9]; yet they are unaffected (neither changed nor destroyed) by the reaction they accelerate. But even enzymes — all proteins — are not almighty. Most need the help of a nonprotein "cofactor" to convert the starting compound — the substrate, in this case a short segment of nucleotides within the huge DNA molecule — into the final product [http://en.wikipedia.org/wiki/Cofactor_(biochemistry)] Some nonprotein cofactors — mother's little helpers — are metal ions, such as copper, iron, and zinc ions; others are B-complex vitamins, such as vitamins B_1, B_2, B_6, B_{12}, niacin (vitamin B_3), and folic acid (vitamin B_9). Nucleotide segments within a vitamin serve as "handles" by which Cre can "grasp and lift" the nucleotide to transfer it from one site to another[11]; and all are "bridesmaids" — but never a bride — on hand to support marquee-name enzymes such as amylase and lesser-known enzymes such as zeatin reductase[d]... the lucky brides.

The Nuts and Bolts of Gene Excision and Recombination:
Excising a Nucleotide and Repaining the Hole Left Behind

That's an overview of homologous recombination. The nuts and bolts of the process are a bit more detailed.

Here's how it works. The reparative enzyme Cre is first firmly bound to any of its many receptors aligned along a strand of DNA.[e] [Cre binds to its receptors in the same manner as LDL binds to its receptors (see Chap. 9, Fig. 9.6)]. Next, it excises a short segment of damaged nucleotides in between two *loxP* sites, as illustrated in Fig. 31.4. (A "*loxP*" site is genetic shorthand for a locus of cross-over where X marks the cross-over spot in the bacteriophage P1.) Once the target DNA and half of the two *loxP* sites have been excised, the

---

[d] Zeatin is a not-widely-recognized plant hormone found in corn and coconut milk needed for the maturation of plants. The compound's main claims to fame are that it is a cytokinin,[13] allowing confusion with cytokines (see Chap. 8), and that it has "anti-aging effects on human skin fibroblasts,"...but, alas, so far, only *in vitro*.[14]

[e] Cre binds to *loxP* sites in the DNA molecule at a 75° angle in the manner of a "C-shaped clamp," grasping the DNA in opposite directions.[15]

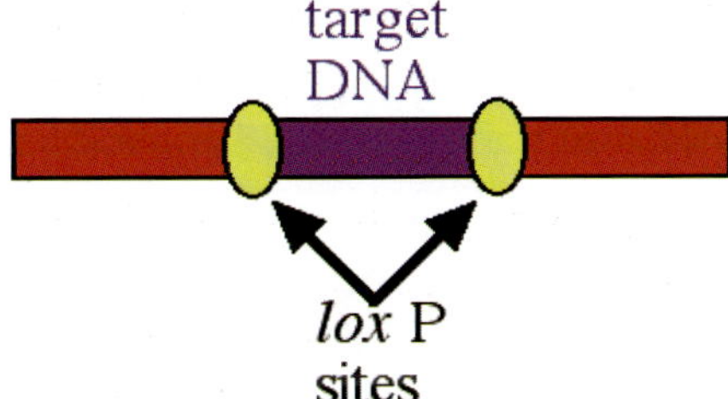

Fig. 31.4.   The enzyme Cre excises a short stretch of nucleotides (labeled "target DNA," in purple) and one-half of each flanking *loxP* site (the yellow ovals), leaving behind the two flanking fragments of DNA.

(*Source*: Reproduced with permission from the publisher.) Reference 12.

remaining *loxP* site and its two flanking fragments of DNA are spliced together, as shown in Fig. 31.5.[f] The target DNA is then degraded and discarded.[g]

Part of the beauty of the Cre/*lox* system is that should a molecular biologist choose to modify some DNA tomorrow morning, all he or she needs to do is to flank the target DNA with a pair of *loxP* sites, stir in some Cre the moment he or she decides to excise the target DNA. . .and, presto, the scientist has the modified DNA *du jour* to study!

---

[f] This excision-and-recombination technique is a lot like the cut and-paste technique that is nearly second nature to children nowadays, at least those beyond age six or so who are comfortable and skilled with the use of Mac and PC computer software. The Cre/*lox* excision–recombination technique may also be more readily understood by persons skilled at activities calling for visual, small-muscle motor coordination — knitting, sewing, calligraphy, and most types of surgery — activities leading to a better end product when both the person and the "object of interest" are stationary. (Women who sew well — who know how to repair a hole leaving an all-but-invisible scar — may be particularly attuned to excising a nucleotide from DNA and repairing precisely the hole left behind.) That statement risks making a sewing technique an avatar of homologous recombination; but sometimes exemplifications — even not very good ones — make points.

[g] So useful is the Cre/*lox* strategy that it is now considered by researchers and their laboratories around the world as "an essential procedure to [determine] gene function in [health] and disease"[16] The technique, introduced in 1988 by chemist Brian Sauer with DuPont Laboratories in Wilmington, Delaware, is credited with "numerous important discoveries that would otherwise have not been possible."[16]

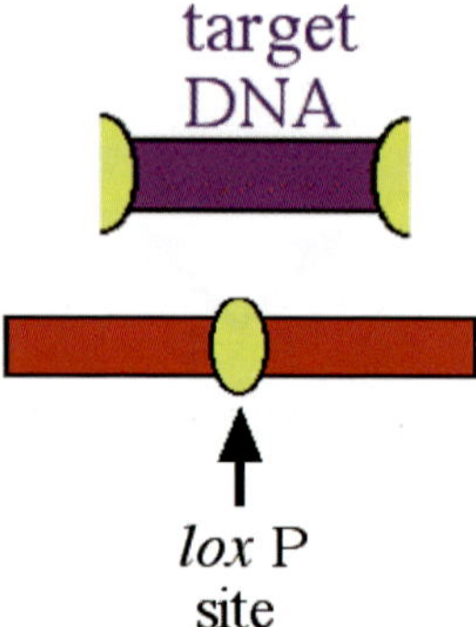

Fig. 31.5. The two flanking fragments are then spliced together to form one *loxP* site. The excised target DNA is degraded and discarded, to tidily complete the process.

(*Source*: Reproduced with permission from the publisher.) Reference 12.

References

1. *Tabor's Cyclopedic Medical Dictionary,* 20th ed. (2005). F.A. Davis, Philadelphia.
2. http://en.wikipedia.org/wiki/Squalene
3. http://www-eaps.mit.edu/geobiology/biomarkers/images/squalene.jpg
4. Nishio T, Morikawa K. (2002) structure and function of nucleases in DNA repair, shape, grip and blade of the DNA scissors. *Oncogene* **21**: 9022–9032.
5. http://en.wikipedia.org/wiki/DNA_repair_and_recombination_protein_RAD54-like
6. http://nobelprize.org/nobel_prizes/chemistry/laureates/1989/press.html
7. http://www.bing.com/images/search?q=photo+of+A+bagel+with+lox%2c+cream+cheese+and+onions.&view=detail&id=A1399DB924DD47A16A70B6A149D520F310BD6542&first=1
8. Personal communication with Peter Kelmenson, Technical Information Scientist, Jackson Laboratories, Bar Harbor, Maine on June 24, 2012.
9. http://en.wikipedia.org/wiki/Enzyme
10. http:www.mansfield.ohio-state.edu/wavy sy`mbolsabedon/bgnws014.htm
11. http://en.wikipedia.org/wiki/Cofactor_(biochemistry)
12. http://www.bio.davidson.edu/courses/genomics/method/CreLoxP.html
13. http://users.rcn.com/jkimball.ma.ultranet/BiologyPages/C/Cytokinins.html
14. http://en.wikipedia.org/wiki/Zeatin
15. Van Duyne GD. (2002) A structural view of tyrosine recombination site-specific recombination. In: Craigie R, Gellert M, Lambowitz AM (eds.), *Mobile DNA II*. Craig NL, ASM, Washington, D.C., Chap. 6, pp. 93–113.
16. http://en.wikipedia.org/wiki/Cre-Lox_Recombination

32

BACTERIOPHAGE, TRANSGENIC MICE AND TRANSGENIC MARMOSETS

Bacteriophage are viruses that infect bacteria.[a] At first, that sentence may be thought a misprint or an oxymoron, maybe? Yet, why not? Why shouldn't, why couldn't a virus infect a bacterium? Yes, they're both from the huge family of infectious micro-organisms, which also includes fungi, parasites, and protozoa. But does that preclude one micro-organism from infecting — invading — another? Did being from the same family prevent North Korea from invading South Korea in 1950 or North Vietnam from infiltrating South Vietnam in the early 1960s? Or, far closer to home, did being from the same country prevent Union troops from attacking a military base in Fort Sumter, South Carolina, a century and a half ago after South Carolina and 10 other Southern states seceded from the United States, triggering the American Civil War?[1] But, you say, that's different. This is biology, not war.

Where Are Phage Found?

"Ubiquitous" is a word commonly linked to the habitat of phage. Phage thrive everywhere that bacteria thrive, which is just about anywhere — "in the atmosphere, in the soil, in rivers, and in the sea."[3] Everywhere! Everywhere that "bacteria breed and prosper," wrote French biologist Alain Dublanchet; "bacteriophage breed and prosper."[3]

---

[a]According to *Stedman's Medical Dictionary*, the verb "infect" refers to "a microorganism that enters, invades or inhabits another organism, causing infection or contamination."[2]

Within the human body (and all other forms of life with a digestive tract), phage are happiest in such haunts as the lower digestive tract and the waste within. And you need not be sick for the intestinal tract to be home sweet home to hungry phage (plenty of them), even those within the GI tract of "individuals in a normal state of health."[3] So constantly are phage on the hunt for bacteria to settle in for their sustenance and survival that it is reckoned there could be "10 times more phages than bacteria. . .in the world."[3] Or, as phage biologist Elizabeth Kutter in Olympia, Washington, pointed out, "Phage are the most abundant life forms on Earth."[4]

Outside the body, phage seem most at home in raw sewage. Such sewage is an "inexhaustible source" of bacteriophage, just as it is of bacteria.[3] (Phage can make a scrumptious meal out of unprocessed sewage.) But because phage are "obligatory parasites," living at the expense of the bacteria they infect, they *must* exist where their host bacteria exist, and that is in the deepest, darkest, dirtiest places. These places include not only raw sewage, but also — and this will not be something ocean/beach lovers from Cape Cod to Malibu want to hear — seawater. Seawater is particularly packed with phage, boasting one of the highest concentrations of phage on Earth. Marine biologists at the University of Maryland estimate there may be as many as "10^7 phages per ml" of ocean water and the nearby Chesapeake Bay.[b,5] *That* is a whopping number!

Phage and Phage Typing

Since phage are *everywhere* in meganumbers, how can one phage be distinguished from all others. The answer is, they can be typed.

Phage typing refers to the laboratory "characterization of cultured bacteria. . . by [the] demonstration of its [*sic*] *susceptibility* to a spectrum of bacteriophage,"[6] much like determining the "*sensitivity*" — the susceptibility — of cultured bacteria to a variety of antibiotics, as done in bacteriology laboratories worldwide. In effect, bacteria get batted around by phage and antibiotics like a tennis ball on a tennis court. The aim of such a 'cellular tennis match' is to kill the bacteria and the type — the brand — of phage, and the

---

[b]The number of phage — and bacteria — in the Bay's waters does fluctuate seasonally, with the lowest count being in the winter months and the highest in the warmer seasons. . .when its beaches and waters are most used.[5]

classification of an antibiotic administered is irrelevant as long as the game is won by the attacking team of 'phage and antibiotics,' just as in a doubles tennis match the brand of the ball batted around doesn't much matter as long as the game is won fairly by one side or the other. *Ergo,* "in order for phage therapy to work, the right 'match' between phage and bacteria must be determined."[7] And, "since some [phage] can only infect a single strain of bacteria," phage typing becomes a simple, key way to distinguish one phage from its brethren, permitting selection of the phage most likely to match and eradicate its target bacteria.

Phage Nicknames, Acronyms, and Abbreviations

Phage are sometimes referred to popularly as "killing machines," which choose nasty bacteria such as *Salmonella, E. coli,* and *Staphylococcus aureus* as their victims. [At least the word "phage" is an abbreviation. The unlucky virus could have been hamstrung by the quasi-acronym "BPG,"[c] which it likely would have been called after 1970, when acronyms became *very* popular in medicine (see Table 21.7 and Fig. 24.4).] Still, abbreviations and acronyms do differ in several ways.

One such difference between the two is that though *both* are shortened forms of words, an acronym forms a *new,* recognizable word (such as "radar", "laser", and "AIDS"), while an abbreviation is the shortened form of an *existing* word or phrase (such as "U.K." for "United Kingdom" and "U.S." for "United States").[8,d] The acronyms "radar," "laser," and "AIDS" stand for "*radio detection*

---

[c]Within medicine, "BPG" is sometimes used as an abbreviation for "bypass graft," referring to veins from either leg being "harvested" to use as replacement vessels for coronary arteries clogged by cholesterol later in life. I have never been a fan of the abbreviation "CABG" — pronounced — for "coronary artery bypass grafts," widely used in medicine instead of "BPG" when discussing the operation to replace such worn-out coronary arteries with healthy leg veins. But tacky abbreviations aren't limited to medicine. In basketball, "BPG" may be an abbreviation for "blocks per game," while the same three letters could mean "bullet-proof glass" in a different setting.

[d]To muddy the grammatical waters, some scholars prefer the term "initialism" to "abbreviation," because both "acronyms" and "abbreviations" are technically abbreviations, both being shortened forms of words in the first place. Hence, the word "initialism" would replace the word "abbreviation," as used in this text, and "BBC" and "U.K." would become "initialisms," while "NATO" would remain an "acronym."[8]

and *r*anging," "*l*ight *a*mplification by *s*timulated *e*mission of *r*adiation," and "*a*cquired *i*mmune *d*eficiency *s*yndrome," respectively.

A second difference between an acronym and an abbreviation is that an acronym is pronounced as a word. For example, the abbreviation for the international peace organization NATO is formed from the words "*N*orth *A*tlantic *T*reaty *O*rganization" but is pronounced as a new, distinct word, whereas an abbreviation is pronounced using individual letters. "BBC," for example, is an abbreviation for the *B*ritish *B*roadcasting *C*orporation and is pronounced "B — B — C," letter by letter.

A third difference between the two is that while an acronym is formed by the first letters of its component words, an abbreviation may be formed by letters other than the first letters of its components. For example, "St.," while an abbreviation for "Street," is also an abbreviation for "Saint," and "Dr." is an abbreviation for "Doctor." Meanwhile, "M.D." is also an abbreviation for "medical doctor," underscoring yet another difference: abbreviations may contain periods, such as "I.D.," "M.D.," and "I.Q." And one need not be a grammaticist — er, grammarian — to know that. (Noah Webster's good book does not include the word "grammaticist.")

Phage Reproduction

That settled, let's focus on how phage reproduce. And, reproduce they do.

The structural simplicity of phage — acellular organisms (meaning organisms or tissues lacking cells) with a wad of DNA tucked away in their micro-millimeter-sized heads — renders the bacteria-scavenger highly reproductive. Though it takes nine months to form a single, anatomically–physiologically complex human baby, it takes just 40 min to spawn acellular phage in astronomical numbers. Thousands of them! Millions of them!! While camping in whichever "restaurant" — whichever bacteria — the parasite[e] has chosen to mooch off for the time being, phage operate what is tantamount to a "phage factory," spewing out enormous numbers of young phage day in, day out. And each and every phage is an amorphous blob[f] of protoplasm able to "spread" its entrails in multiple directions at the same time.

---

[e]Phage *are* parasites. If a parasite is defined as "an organism that lives on or in a host organism in a way that harms or is of no advantage to the host,"[2] then phage qualify as parasites — ubiquitous ones to boot.

[f]There are parallel definitions, with only minor distinctions among dictionaries, of the individual words "amorphous" and "blob," as well as the term "amorphous blob." The online *Free*

Transgenic Mice Upstaged by Transgenic Monkeys

Come May 2009, scientists at Japan's Central Institute for Experimental Animals reported doing something that had not been done before: the passing of a transgene — "a gene that is transferred from an organism of one species to an organism of another species by genetic engineering"[12] — from one generation of a primate [a monkey] to the next.[g] This is key, pointed out an editorial in *Nature*, "to making a useful research model" because to study inherited diseases that affect humans in experimental animals mandates the genetic defect be passed on from one generation to the next.[13]

Five such transgenic monkeys were born, including a set of twins — Kei and Kou ("*keikou*" (恵子 / 敬子) is Japanese for "fluorescence") (Fig. 32.1).[14] The monkeys were created by inserting sperm from a transgenic animal into the egg of a natural animal. Four and one-half to five months later, a transgenic monkey was born.[h] And since the green fluorescent protein gene — the same protein Osamu Shimomura extracted from the jellyfish he caught in Friday Harbor during the summer of 1961 (Chap. 27) — was introduced into their genetic material, the monkeys glowed green under fluorescent light.

The type of monkey bred — a marmoset — is particularly useful, because marmosets "reproduce often and from a young age"[14] — about one year. [And

---

Dictionary defines "blob" as "a soft amorphous mass," while characterizing "amorphous" as "lacking definite form; shapeless" (www.thefreedictionary.com/blob). The Free Merriam-Webster online dictionary calls a blob "something ill-defined or amorphous,"[9] while defining "amorphous" as "having no definite form; shapeless."[10] And, for kids, a "blob" is explained as "a small lump or drop of something thick (like paste or putty)," while "amorphous" is "having no fixed form," like "an amorphous cloud."[9,10] But the *real* question is: Why even use the phrase "amorphous blob"? Isn't if a tautology ("tautology" is defined in Merriam-Webster's good book as "a needless repetition of an idea, statement, or word")? The renowned lexicographer uses as an example of a tautologous statement "A beginner who is just getting started."[11] And, as examples of tautologous words, "new innovation," "close proximity," and "first priority."

[g] The feat, naturally, sparked concern about the possibility of creating transgenic humans. The ethical aspects are sure to spawn debates in scientific journals, as well as the news media.

[h] The actual methodology involved in creating these five transgenic monkeys is relatively straightforward — shrinking the egg "within its outer cover by placing it in a hypertonic [high-concentration] sugar solution, [thereby] freeing up space for the injection of more transgene-containing particles" and then injecting sperm from a transgenic animal into a natural animal — it also took a lot of patience and a meticulous transfer technique by the Japanese team. After the transfer of 80 embryos to 50 surrogate mothers, just seven pregnancies occurred, with five live offspring.[16]

(a)   (b)

(c)   (d)

Fig. 32.1.   Five transgenic marmoset infants, named (a) Hisui, (b) Wakaba, (c) Banko, and (d) Kei and Kou (twins), "*keikou*" meaning "fluorescence" in Japanese. The monkeys were created by splicing a jellyfish gene (see Chap. 27) into an ordinary marmoset. Erika Sasaki and Hideyuki Okano at the Keio University School of Medicine in Japan used a virus to carry the gene for green fluorescent protein into monkey embryos *ex vivo*, which were later implanted into a female monkey. About 135–140 days later, these endearing creatures were born. Subsequent generations can "be produced by natural propagation with the eventual establishment of transgene-specific monkey colonies—a potentially invaluable resource for studying incurable human disorders," noted Gerald Schatten, at the University of Pittsburgh School of Medicine, and Shoukhrat Mitalipov, at the Oregon Health and Sciences University, as they called the event a "milestone" in a commentary accompanying the feature article that appearing in the May 2009 issue of *Nature*.

The insets in the four panels show the fluorescent paws, under UV light, of transgenic marmosets (on the right) and the paws of two natural marmosets, again under UV light (on the left).

(*Source*: Reproduced with permission from the publisher.) Reference 17.

they're cute (Fig. 32.1).] This feature should allow the creation of transgenic monkey lines with such serious neurologic conditions as Parkinson's disease, and the less common Huntington's chorea and amyotrophic lateral sclerosis — Lou Gehrig's disease[15] — to be studied over time.

Drawbacks of transgenic marmosets

Still, marmosets do have drawbacks; they are more removed from humans phylogenetically than rhesus monkeys — until now the standard for nonhuman primate research — and their "brains are too small for positron emission tomography."[15]

Commentaries and editorials prompted by transgenic marmosets

Drawbacks notwithstanding, the publication of the Japanese history-making paper in the May 28, 2009, issue of *Nature* was accompanied by not just one, two, or even three commentaries or editorials — but four! And, like Brainbow two years, earlier, the editor of *Nature* chose to feature the star of the new report — a transgenic marmoset monkey — on its cover (Fig. 32.2). [Two breakthrough papers from two institutions (nearly) polar opposites geographically.] One of the four accompanying commentaries and editorials said this new work "could establish marmosets as a model research organism to rival the more commonly used rhesus (monkey) and usher in a new era of primates as human-disease models."[15] Another raved, "The birth of this transgenic marmoset baby is undoubtedly a milestone," one that is "a potentially invaluable source for studying incurable human diseases."[13] And a third editorialist wrote, "Japanese researchers announce a major accomplishment in this issue of *Nature*: the creation of the first transgenic primates able to pass on a foreign gene to their offspring."[13] Finally, the editor's summary concluded, "A non-human primate model amenable to gene manipulation with transgenic technologies would be invaluable for biomedical research into disease mechanisms and for developing therapies in gene therapy and regenerative medicine."[19] All in all, this is good news for folks with diabetes, Parkinson's disease, and a variety of other inheritable diseases.

The all-important dollar

With regard to the cost of creating and maintaining a transgenic marmoset compared to that of a transgenic mouse (see Chap.15) (Fig. 32.3), the former is far higher than the latter; but the latter isn't low. Ordinary field mice may be a dime a dozen, but converting one into a *transgenic* mouse comes at a cost of about US$10,000, according to the University of Texas Medical Branch in Galveston.[20] That includes an approximate US$5400 fee to knock out a specific gene from a mouse embryonic stem cell and a US$4200 fee for microinjection

Fig. 32.2.   The historic cover of the May 28, 2009, issue of *Nature* depicting a transgenic marmoset, the product of a team of 22 researchers from 7 Japanese institutions. The caption reads, "This new non-human primate model, amenable to gene manipulation with transgenic technologies, should be invaluable for biomedical research into disease mechanisms and for developing therapies in gene therapy and regenerative medicine."

(*Source*: Reproduced with permission From the publisher.) Reference 18.

of the modified embryonic stem cell into a mouse blastocyst, (as shown in Fig. 15.6), followed by its transfer into a foster mother. But once a line is established and the mice begin to procreate, each animal, according to Brian Soper, Ph.D., at the Jackson Laboratories in Bar Harbor, Maine — which makes the mice — costs US$150–300 to create, plus cage costs.[22] Meanwhile, a transgenic marmoset costs approximately US$5000 per animal to create, plus (only) "$10 *per diem* costs," says Gerald Schatten, Ph.D., the Director of Developmental and Regenerative Medicine at the University of Pittsburgh School of Medicine.[23] They're small, they're cute. . .and they don't eat much.

Trying to put the two transgenic species in perspective, Erica Sasaki, the lead author of the transgenic marmoset paper, said, "You must think of the cost–benefit. If you can do it in [transgenic mice] or *in vitro*, we should," implying a transgenic marmoset is more costly to create and maintain than a transgenic mouse.

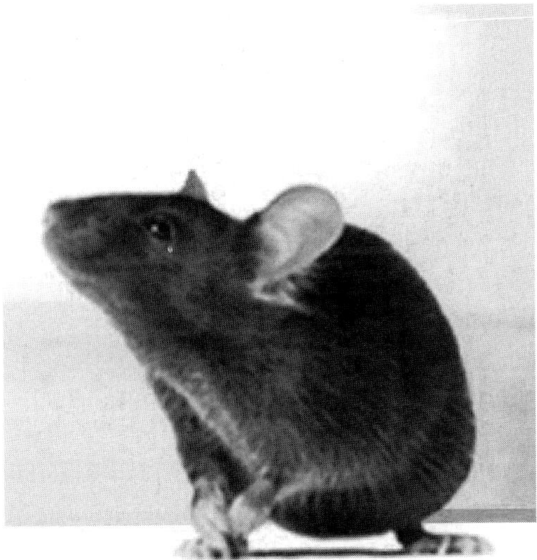

Fig. 32.3. A genetically engineered (transgenic) Brainbow mouse created by the Jackson Laboratories in Bar Harbor, Maine. This particular mouse almost appears to be posing for the photographer. Though genetically engineered mice can be created to simulate a variety of human diseases, they can't (yet) be programmed to pose. This mouse looks "pettable," almost "cuddleable" — but it can still bite!

(*Source*: Reproduced with permission from the Jackson Laboratories.) Reference 21.

(Still, transgenic marmosets have a "relatively short gestational period [about 144 days], reach sexual maturity at 12–18 months, and females have 40–80 offspring during their life."[18]) "But for many diseases, like Parkinson's, there is not a good [mouse] model now."[24] Another advantage of employing nonhuman primates is that they may be used to study the effects of new stem cell therapies or gene therapies — and drug toxicities — before these are tried on people.[25]

References

1. http://en.wikipedia.org/wiki/Causes_of_the_Civil_War
2. *Stedman's Medical Dictionary*, 22nd ed. (1972) Williams & Wilkins, Baltimore.
3. Dublanchet A, Bourne S. (2007) The epic of phage therapy. *Can J Infect Dis Med Microbiol* **18**: 15–18.
4. http://www.intralytix.com/Intral_News_LATimes.htm

5. Wommack KE, Cotwell RC. (2000) Viroplankton: viruses in aquatic ecosystems. *Mol Biol Rev* **64**: 69–114; and Whitman WB *et al.* (1998) Prokaryotes: the unseen majority. *Proc Nat Acad Sci USA* **95**:6578–6583.

6. http://medical-dictionary.the freedictionary.com/phage+typing

7. Thiel K. (2004) Old dogma, new tricks — 21st century phage therapy. *Nat Biotech* **22**: 31–36.

8. http://www.differencebetween.net/language/difference-between-abbreviation-and-acronym

9. http://www.merriam-webster.com/dictionary/blob

10. http://www.merriam-webster.com/dictionary/amorphous

11. http://www.merriam-webster.com/dictionary/tautology

12. http://dictionary.reference.com/browse/transgene

13. http://www.nature.com/nature/journal/v459/n7246/full/459483a.html

14. Dolgin E. (2005) Transgenic primates transmit DNA. *The Scientist,* May 27.

15. Cyranowski D. (2009) Marmoset model takes center stage. Newly created transgenic primate may become an alternative model to rhesus macaques. *Nature* **459**: 492–493.

16. Schatten G, Mitalipov S. (2009) Developmental biology: transgenic primate offspring. *Nature* **459**: 515–516.

17. Sasaki E, Suemizu H, Shimada A, Hanazawa K, Oiwa R, Kamioka M, Tomioka I, Sotomaru Y, Hirakawa R, Eto T, Seiji S, Maeda T, Ito M, Ito R, Kito C, Yagihashi C, Kawai, Miyoshi KH, Tanioka YH, Tamaoki N, Habu S, Okano H, Nomura T. (2009) Generation of transgenic non-human primates with germline transmission. *Nature* **459**: 523–527; doi: 10.1038/nature08090.

18. http://www.nature.com/nature/journal/v459/n7246/index.html

19. About the cover. May 28, 2009, issue of *Nature.*

20. Editor's summary. Biomedical supermodel: germline transmission in a transgenic non-human primate. *Nature* (2009); **459**: XX

21. http://jaxmice.jax.org

22. Soper B. (2012) Personal communication, July 22.

23. Schatten G. (2012) Personal communication, July 24.

24. http://www.nature.com/news/2009/090527/full/459492a.html

25. http://www.smh.com.au/environment/monkeys-glow-with-gene-transfer-breakthrough-20090528-bo2b.html

33

PHAGE GEOMETRY AND SOCCER BALLS

Come now! What can the shape of phage possibly have to do with soccer balls? As it turns out, there really is a connection. This isn't yet another "loose association" — though this one may be disputable.

But let's begin with phage size. It's a bit easier to deal with phage size than with their strange shapes.

Phage Size

While plants and animals are measured in meters and centimeters, phage and other viruses are measured in nanometers — one-billionth of a meter. They may be as long as 24–200 nm and as wide as 100 nm.[1] The large difference in the size of viruses is related to how many genes a virus carries. Phage — like humans — come in a variety of shapes, sizes, and heights, with the word "height" being replaced by "length." Bacteria, in comparison, measure 0.5–5 microns (0.0005–0.005 mm.) in length. The overlap means a large phage can be the size of a small bacterium.

Phage (Head) Shapes

The shapes of Phage and other viruses are more interesting. Fig. 33.1 is a collection of electron micrographs depicting six different viruses — some, such as the influenza virus and the human immune deficiency virus (HIV), are familiar names, and others are not so familiar, yet colorful. There are three fundamental phage forms: the humanoid icosahedral phage, the filamentous (rod-shaped) phage, and the prolate (elongated) icosahedral phage (Figs. 33.2–33.4).

507

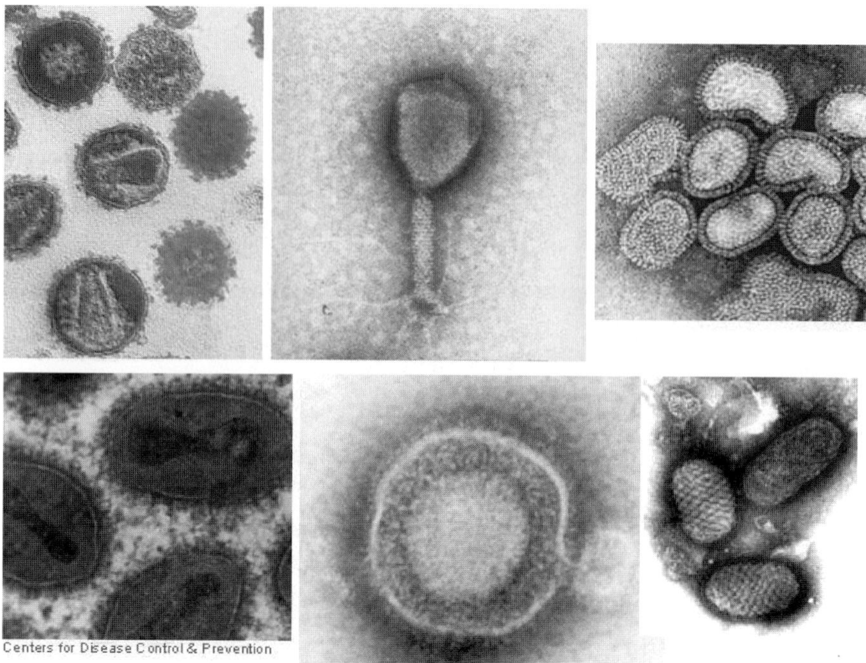

Fig. 33.1. A gallery of electron micrographs of viruses, illustrating their diversity in shape. Going clockwise are the human immunodeficiency virus (HIV), the *Aeromonas* virus, which causes human diarrhea acquired by drinking groundwater from either a municipal system or private wells[2]; the influenza virus; the Orf virus, which causes "sore mouth" disease in sheep and goats; the *Herpes simplex* virus (HSV); and the smallpox virus.

(*Source*: Reproduced with permission from he publisher.) Reference 3.

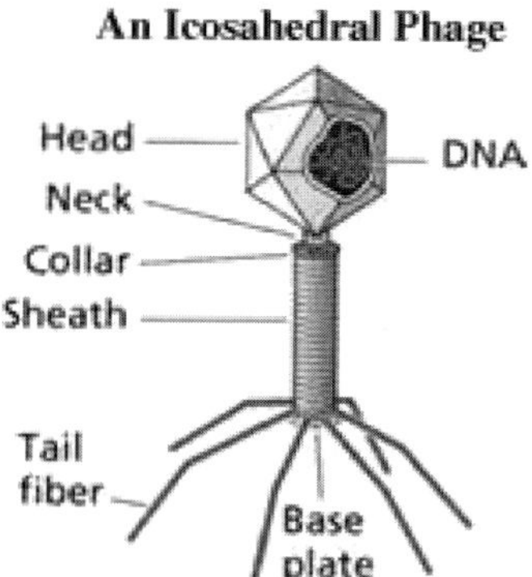

Fig. 33.2. A diagram of a bacteriophage illustrating its humanoid appearance — a head, a neck, a collar, a torso (sheath), a base plate, and tail fibers resembling multiple legs.

(*Source*: Reproduced with permission from the publisher.) Reference 4.

Filamentous Phage

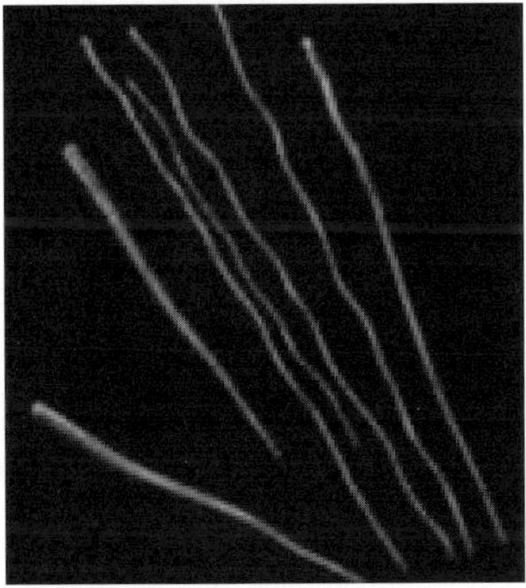

Fig. 33.3. Several filamentous phage.
(*Source*: Reproduced with permission from the publisher.) Reference 5.

Prolate Phage

Fig. 33.4. The prolated (elongated) head and shortened midsection (sheath) of phage T4.
(*Source*: Reproduced with permission from the publisher.) Reference 6.

We'll come back to these funny-looking, fully assembled creatures in a minute, after looking into how they are put together.

Phage Assembly

We generally speak of phage as being "assembled" — rather than built or constructed — from "assorted parts." These assorted parts include a head, a neck or collar, a torso (sheath), a tail, and a base plate equipped with cleats.[3] They're not simply "born" in a polished state. So, how to characterize such a creature? For some folks, if a single word were to describe the appearance of a phage, it might be — to say the least — "strange." Strange, yet put together in an architecturally sound manner.

A base plate, acting as the foundation, is made first. But even it — the foundation — is made from other "parts" — a central "hub surrounded by six wedges,"[7] each made separately and later combined to form the base plate.

Cleats and spikes

Matters only get stranger. Beneath the baseplate lie cleats — to enable an inherently clumsy phage to (more literally than figuratively) dig into nasty bacteria — like a quarterback planting his feet to pass the ball (Fig. 33.5) or bracing himself for a blitz by a defensive linebacker sporting pink cleats! (Fig. 33.6) — but *never, ever* worn by a ballerina, who prefers *pointe* shoes, which enable her to pirouette ever so gracefully (Fig. 33.7).

Next, the head is built. While not every phage has a tail, each and every one has a head — a "capsid," in the language of microbiology.[11] Like humans, phage have distinctive "facial" features.[11] Some sport an *icosahedral*-head (an icosahedron is a geometric object with 20 identical equilateral faces, 30 edges, and 12 vertices) (Fig. 33.9), and others have a more slender

Fig. 33.5. A quarterback planting his cleated right foot, readying himself to pass a football. (*Source*: Reproduced with permission from the publisher.) Reference 8.

Fig. 33.6.   The Cincinnati Bengals' Chad Ochocinco (Chad Johnson) (#85) and a team member (center) wearing pink cleats before a game against the Cleveland Browns in September 2009 to promote Breast Cancer Awareness Month. The next week he added a pink jersey (inset), gloves, a wrist band, and even a pink mouthpiece and chin strap.

(*Source*: Reproduced with permission from the publisher.) Reference 9.

figure — *filamentous,* in the jargon of virology — more of an Abe Lincoln configuration. There's a phage to suit just about everyone's taste!

Then the tail is made…if a particular phage is lucky enough to have a tail. If it does, the tail may be short…or it may be long. Either way, it, too, is made piecemeal and later its several segments are melded together.[13] Though not all phage have a tail, those that do put it to good use. Within the tail is a hollow central channel — a tunnel — through which the phage's nucleic acids — DNA — pass into a bacterium in waiting — waiting for just the "right" phage to come along. The tail, itself, is surrounded by a contractile sheath which constricts to "inject" *its* DNA — in a hypodermic-syringe-type motion (an

Fig. 33.7.   Russian ballerina and actress Tamara Toumanova in *pointe* shoes and Ukrainian *danseur* and choreographer Serge Lifar performing in *Swan Lake*.

(*Source*: Reproduced with permission from the publisher.) Reference 10.

animated version of this DNA transfer is available at http://cronodon.com/ BioTech/Virus_Tech.html) once the phage is securely attached to its bacterium by six or more short "spikes"[a] (Fig. 33.8) at the tail's tip, forcing the virus' DNA into the bacterium...before turning in its cleats.

So, if you ever want to build a phage, that's the recipe!

An explanation for such unorthodoxy

Why such an unorthodox *modus operandus,* you ask? Said geneticist and microbiologist Allan Campbell at Stanford, "The advantage of [using] multiple assembly pathways is that [assembly] mistakes can be rejected for each

---

[a] The nature of phage "spikes" was finally unveiled in 2012. As if to extend the metaphor, a team of five bioengineers from Purdue, Switzerland, and Russia determined that the membrane-penetrating spikes in Fig. 33.9 are in fact composed of a central iron (ion) core surrounded by several layers of protein.[12]

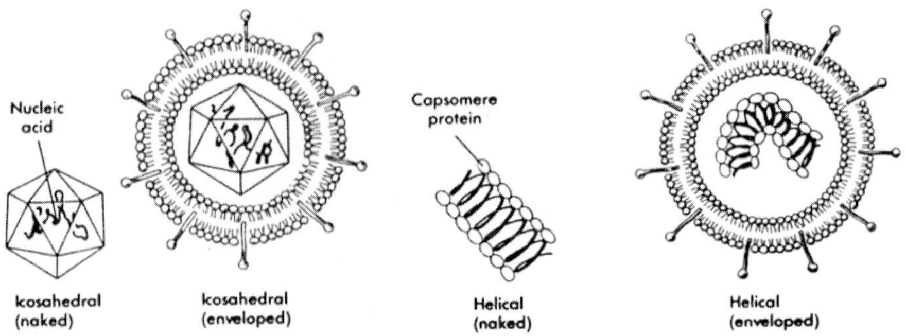

Fig. 33.8. The most common viral morphologies are, (from left to right): A naked — without an outer envelope — icosahedral virus (*e.g.* the poliovirus), an enveloped icosahedral virus (*e.g.* the herpes virus), a naked helical virus (*e.g.* a tobacco mosaic virus) and an enveloped helical virus (*e.g.* the influenza virus). Individual capsomeres — the building blocks of a phage's capsid — are arranged to form the icosahedral or helical capsid housing the phage's DNA. Many animal viruses also have an envelope, which is partly derived from the host cell's membrane but which always contains unique viral *attachment* proteins, drawn here as "spikes." (*Source*: Reproduced with permission from the publisher.) Reference 14.

pathway separately, rather than risking [potentially fatal] damage to the finished product"[b,13] — kind of like how things are done in an assembly line at the Ford Motor Company.

Mission accomplished

Mission accomplished, a well-groomed and well-clad phage goes about courting an equally well-groomed and –well-clad phage for reproductive purposes to ensure propagation of their "species"...and then dies, ending its brief, but busy, life. But what a life phage have! They're "born," reproduce fast and furiously, eat choice, grade-A bacteria to the point of (literally) bursting, and then die — fully

---

[b] And everyone knows damage to the finished product can be "fatal." After all, when a carefree Humpty Dumpty (Fig. 33.10) "had a great fall, all the King's horses and all the King's men couldn't put Humpty together again (Fig. 33.11).[15] The term "humpty dumpty," should you ever have trouble falling asleep wondering where in the world the name came from, describes "a short clumsy person of either sex," according to the *Oxford Dictionary of Nursery Rhymes*,[16] although originally, in the 17th century, it was the name of a drink made from "brandy and ale."[17] (That tidbit of information should cast aside the notion that the wave of unusual mixed-drink names in the 1980s and '90s was a phenomenon spawned by jetsetters lounging around summertime pools.)

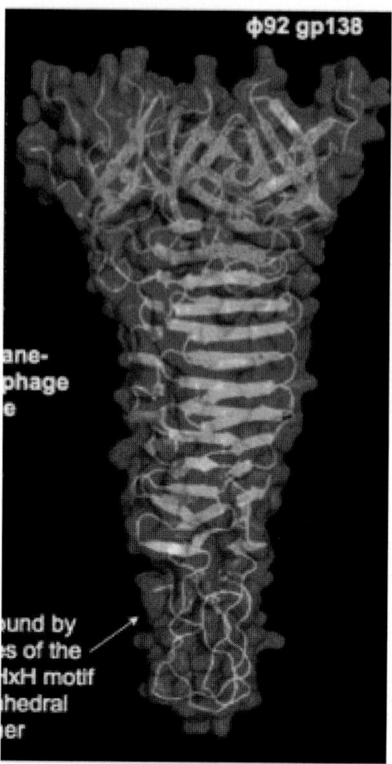

Fig. 33.9.   A phage tail spike consists of an iron (ion) core surrounded by several layers of protein. Together they penetrate the surface of a bacterium, allowing the phage to enter.

(*Source*: Modified (cropped) and reproduced with permission from the editor.) Reference 12.

satisfied — all within a span of 20–40 min. Could this be heaven, nirvana, paradise, and utopia all rolled up into one intense, blissful, decadent, 30-min moment?

Back to the Fully Assembled Phage

All "grown-up" phage, regardless of shape, consist merely of a DNA nucleus wrapped in a protective coat or jacket, a so-called "capsid" in the jargon of genetics. Another common feature of phage is that a given phage can "infect only one" — or a closely related — bacterium in the entire bacterial world. Filamentous phage (Montage 33.1) choose gram-negative bacteria, such as *E. coli* (Fig. 33.12) and *Pseudomonas* — which so plague women (via urethritis and cystitis) because of their shorter urethra[19] — as their victims. Such phage

Fig. 33.10. A carefree, dancing Humpty Dumpty as drawn by cartoonist W.W. Denslow in 1904. (*Source*: Reproduced with permission from the publisher.) Reference 15.

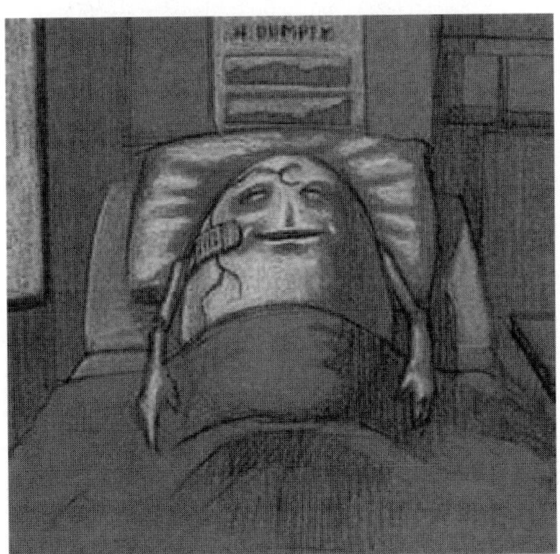

Fig. 33.11. A not-so-care free Humpty Dumpty recuperating after his great fall. (*Source*: Reproduced with permission from the publisher.) Reference 16.

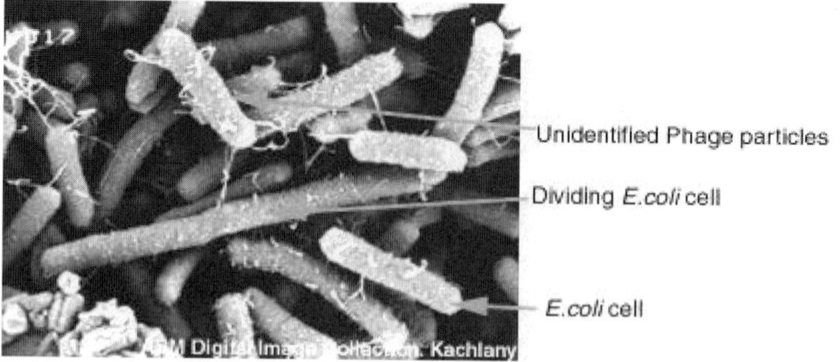

Fig. 33.12.   This scanning electron micrograph shows sausage-shaped *E. coli* bacteria being attacked by phage (which appear as hard-to-see, white dots) attached to the surface of the bacteria. The wrinkled appearance of the bacteria's usually smooth outer membrane is due to lysis (dissolution) of it by the phage. The germ is beginning to crumble. If there is power in numbers, the phage ought to win this contest.

A scanning electron micrograph is one of several types of electron microscopic images made by scanning an object with a high-energy beam of electrons. The resulting high-resolution, three-dimensional images can reveal details less than 1 nm in size. Black-and-white scanning electron micrographs such as this are easy to view, while color images are yet more pleasing to the eye — just as are colorful Brainbow images.

(*Source*: Reproduced with permission fom the publisher.) Reference 18.

carry a single, long, thin strand — a filament — of DNA in their tiny heads, akin to the filament of a light bulb (Fig. 33.16). Icosahedral phage — those with 20 identical, equilateral triangles of an icosahedron (Fig. 33.17) — come with or without an outer protective envelope (capsid) — the former called "enveloped" phage, the latter "naked" phage, as shown in Fig. 33.8. Pretty *risqué*, these phage. Montage 33.2 is a display of icosahedral phage. Prolate (elongated) icosahedral phage have a polar axis of greater length than their equatorial axis — the opposite of Mother Earth. Montage 33.3 is a collection of prolated icosahedral phage equipped with a head, neck, collar, even whiskers, a torso, six legs, and a base plate.

The earth — an oblate ellipsoid

Earth, strictly speaking, is not a sphere, since "it is flattened [a bit] at the poles and bulges [a bit] at the equator."[22] The geometrical form that most accurately describes the Earth's shape is "spheroid" or "oblate ellipsoid."[22]

Problem is — with that explanation of the Earth's configuration — two additional terms demand definition — "spheroid" and "oblate ellipsoid." The

Montage 33.1   Filamentous Phage.

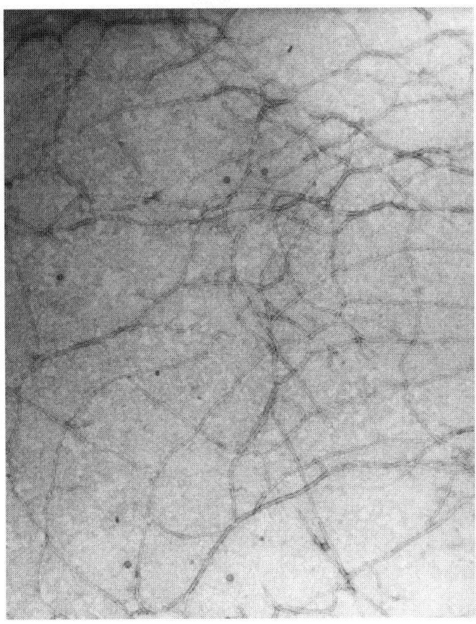

Fig. 33.13.   An electron-microscopic photograph of the flexible filamentous phage which was associated with an epidemic of gastroenteritis caused by a virulent bacterium in Southeast Asia from 1996 to 2000.

(*Source*: Reproduced with permission from the publisher.) Reference 23.

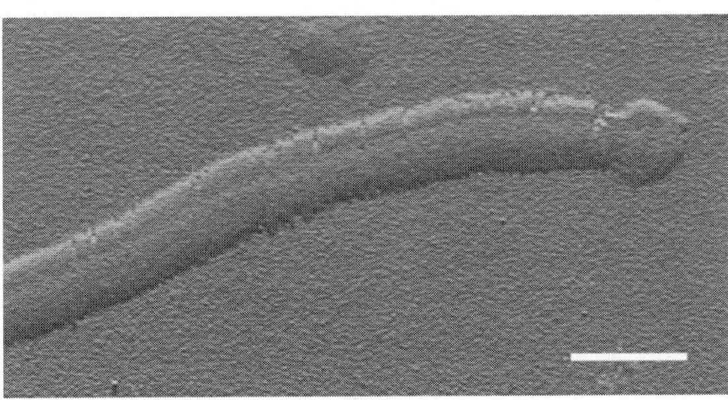

Fig. 33.14.   An electron micrograph of a filamentous phage. The bulblike tip of the phage to the right is likely due to a partially denatured clump of enzyme protein.

(*Source*: Reproduced with permission from the publisher.) Reference 24.

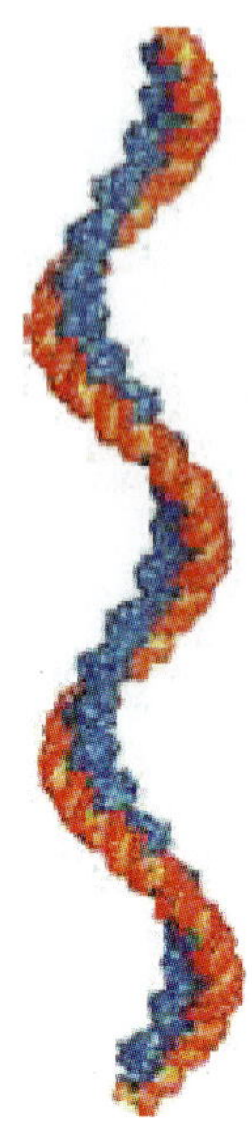

Fig. 33.15.   In this colorful model, a DNA filament from an *E. coli* bacterium has received DNA from a filamentous λ phage producing a hybrid DNA. Proteins in the helical (spiral) strand are colored blue, DNA orange-red.

(*Source*: Modified and reproduced with permission from the publisher.) Reference 25.

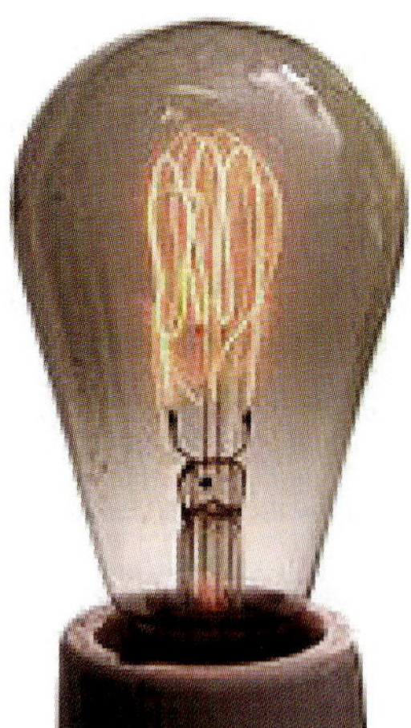

Fig. 33.16.   The glowing filament of an ordinary incandescent light bulb invented by Edison in 1879.

(*Source*: Reproduced with permission from the publisher.) Reference 20.

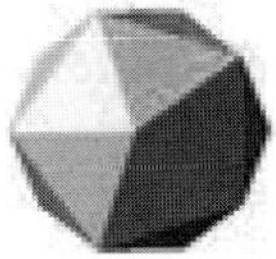

Fig. 33.17.   An icosahedron — a geometric object with 20 identical equilateral faces, 30 edges and 12 vertices.

(*Source*: Reproduced with permission from the publisher.) Reference 21.

Montage 33.2   ICOSAHEDRAL PHAGE.

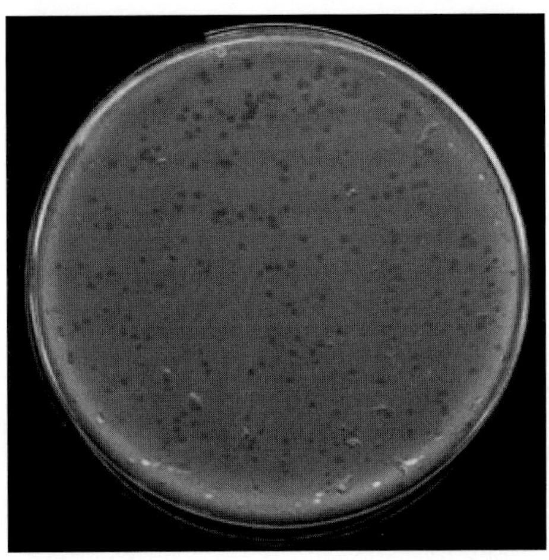

Fig. 33.18.   Holes created by lambda (λ) phage — growing in a laboratory Petri dish — upon lysing (dissolving) a green carpet of *E. coli* bacteria. (Like prolated phage T4 in Fig. 33.4, λ phage eradicate cystitis–urethritis caused by *E. coli*. But T4 phage are more deceitful; they enter *E. coli* flaunting their icosahedral shape. Once within, they trick the bacteria's DNA into not recognizing who is in their house by becoming *circular* and allowing them to stay on and replicate *ad lib*.

These holes are called "plaques" — *but*, in contrast to raised dental and atherosclerotic (cholesterol-filled) plaques (see Fig. X.XX), phage plaques are sunken...yet not empty. In their recesses lie millions of λ phage.

(*Source*: Reproduced with permission from the publisher.) Reference 26.

issue is quickly mitigated, because "spheroid" and "ellipsoid" are essentially interchangeable — and "spheroid" or "ellipsoid" will be used from here on leaving only the terms "oblate spheroid" prolate spheroid" to be accounted for.

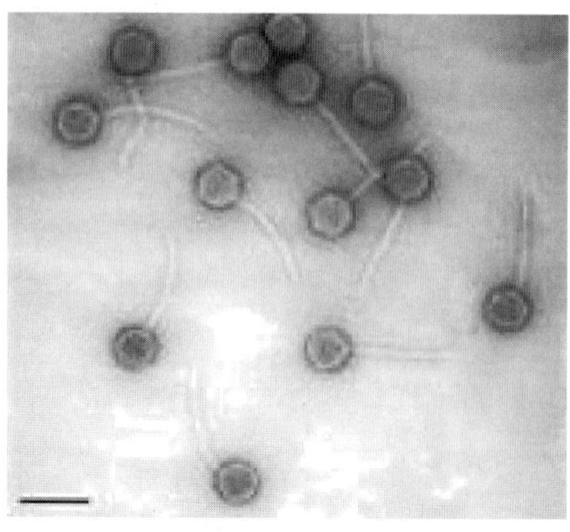

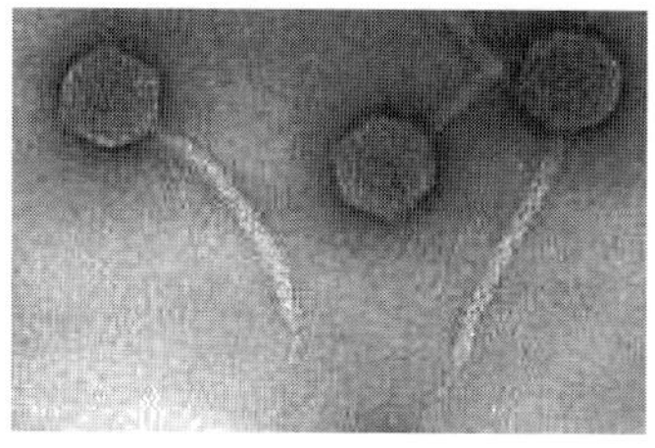

Fig. 33.19. Lambda (λ) phage — the lollipop phage (see enlarged inset) — that infects *E. coli* bacteria. λ phage treat their victims more kindly than *E. coli* treat theirs, which is part of the reason they are called lollipop phage. They are also referred to as "temperate" phage. Temperate in the sense they are self-restrained — as are more temperate imbibers of alcoholic beverages. In microbiology, temperate phage exist within their host bacteria — even inserting their DNA into the host's DNA — *but* rarely, if ever, cause lysis (destruction) of the bacterium. When the bacterium's DNA duplicates, the virus' DNA replicates as well — and its life goes on.

(*Source*: Reproduced with permission from the publisher.) Reference 27.

While an *oblate spheroid* sounds intimidating, the term really isn't so hard to envision. It refers to the object formed by "rotating an ellipse[c] [around] its

---

[c] An ellipse is a "[cone-shaped] structure formed by the intersection of a right circular cone by a plane that cuts the axis and the surface of the cone."[33] For those few who choose to do so, "an ellipse can be drawn on a sheet of paper using two pins, a pencil, and a loop of string. The pins are placed vertically at two separate points (which will become the foci of the ellipse), and the loop of string is placed around the pins and the pencil. The pencil is held vertically such that the string is taut and forms a triangle. If the pencil is moved around so that the string stays taut, it will trace out an ellipse, because the sum of the distances from the pencil to the pins will remain a constant."[34] And if you don't care to take time to collect two pins, a pencil, and a loop of string, just peek at Fig. 33.24. It shows the rings of Saturn, which in fact are circles — but appear to be ellipses when viewed at an angle.

Montage 33.3   Prolated icosahedral phage.

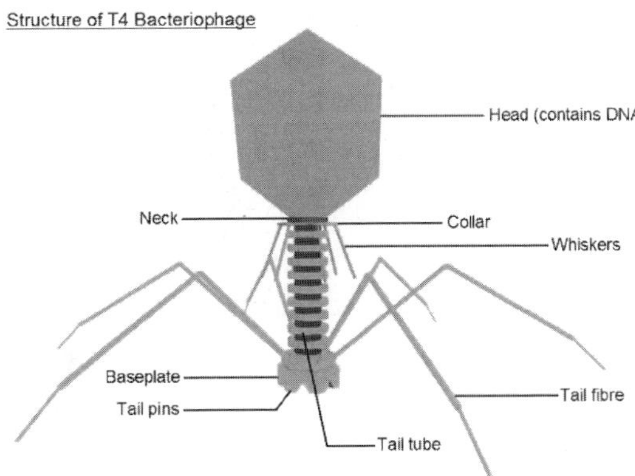

Fig. 33.20.   A prolated T4 phage, this one with six legs (tail fibers) and even six whiskers. The whiskers are actually made of a protein — fibritin — and they serve as a "environmental sensors" that switch the phage into a noninfectious state under "unfavorable conditions."[28]

(*Source*: Reproduced with permission from the publisher.) Reference 28.

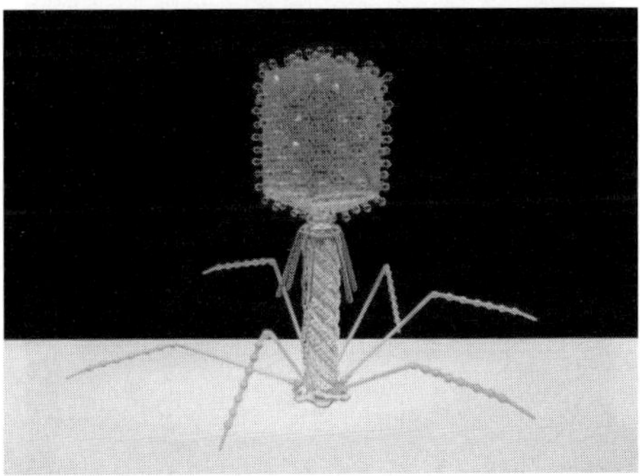

Fig. 33.21.   This is another rendition of a bacteria-eating T4 phage with its six long legs, stretched midriff, short neck, prolated head, and chin whiskers! If you'd like to see him dance, he'll do that — at no cost — if you log on to the website below. He's dancing in delight atop a bacterium he's about to inject his DNA into to make more phage *just like him.*

(*Source*: Reproduced with permission from the publisher.) Reference 28.

Fig. 33.22.   This is his *objet d'amore* — an ordinary *E. coli* bacterium. (It ought to be shaded green for that's the natural hue of *E. coli*. Three colonies of the bacteria are shown growing in a laboratory culture dish in the inset). A phage dances like a spider on a thread (in the animated version at the website just mentioned) just above the bacterium until the moment he decides to puncture its surface membrane to insert his wad of DNA...and then the rhapsody ends.

(*Source*: Reproduced with permission from the publisher.) Reference 28.

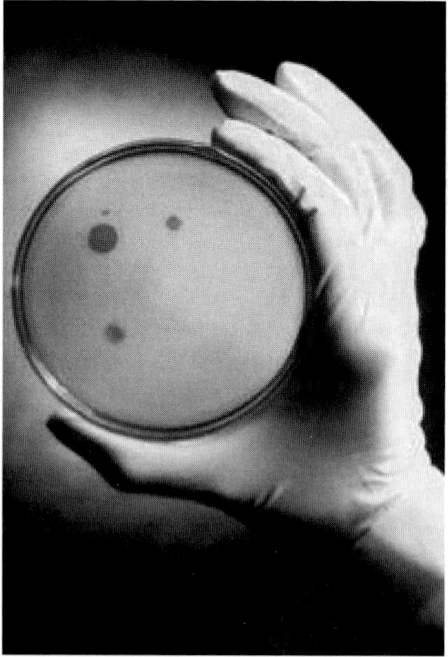

Fig. 33.23.   Source of three *E. coli* colonies. Reference 29.

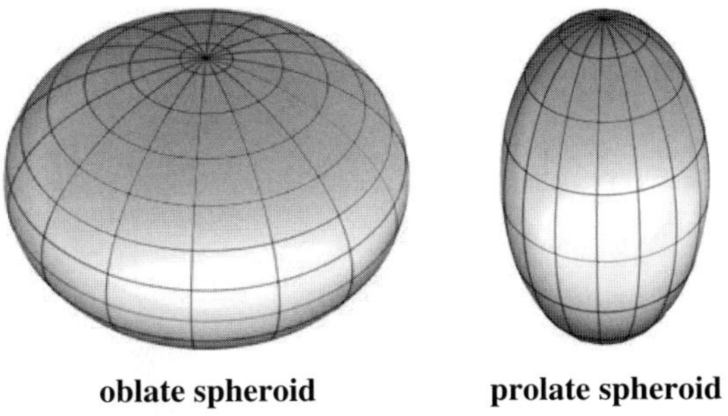

Fig. 33.24.

(*Source*: Reproduced with permission from the publisher.) Reference 30.

Fig. 33.25.   The rings of Saturn are circular, but when seen partially edge on, as in this image, they appear to be ellipses.

(*Source*: Reproduced with permission from the publisher.) Reference 36.

shorter [polar] axis."[30] The resultant object resembles a pumpkin (Fig. 33.24), or the shape of a once-spherical foam pillow (chair) that has been lain on too long, squooshing it a trifle. (In a like manner, the Earth[d] assumes the shape of an oblate spheroid, slightly flattened at its poles — making the term "oblate spheroid" more accurate than "sphere.") In contrast, a more elongated object, similar in shape to an American football or a rugby ball — a prolate

---

[d] The Earth measures 3959 miles (6357 km) between its poles and has a 3963 mile (6378 km.) equatorial diameter,[35] making it ever-so-slightly spheroidal, rather than spherical.

Fig. 33.26.   A soccer ball.

(*Source*: Reproduced with permission from the publisher.) Reference 40.

(elongated) spheroid — is formed when an ellipse is rotated around its *longer* (equatorial) axis (Fig. 33.24.[31] And, lastly, if the "generating ellipse is a circle, the [resultant object] is a sphere."[32]

Icosohedral Phage and Soccer Balls

Nearly everyone has seen a soccer ball and a goodly number of Americans, Mexicans, Europeans, Southeast Asians, and Australians have played or now do play soccer. Yet few recognize that the leather cover of the familiar black-and-white soccer ball has 12 pentagons and 20 hexagons (Fig. 33.26). Still fewer grasp that the 20-sided head of an icosahedral phage is composed of 20 identical equilateral triangles.[37,38] Peter Dragnev is one of those few.

Born and schooled in Bulgaria, Dragnev came to the States in 1992, where he graduated from the University of South Florida (Tampa) in 1997 with a Ph.D. degree in Mathematical Sciences. From the Tamiami Trail and the subtropical Everglades of South Florida, which alligators, snakes, and turtles call home, he shifted his work north to the more hospitable surroundings of the less natural world of Fort Wayne, Indiana, where he joined the faculty of the Department of Mathematical Sciences at Indiana University–Purdue University.[39] But not your everyday mathematical scientist is Dragnev. Evidence of his uniquity soon surfaced, when he put together a website called

"The Chemistry, Physics, and Mathematics Behind the Soccer Ball Design. Or, How to Distribute Dictators on a Planet?". This gives scientific, yet witty, account of just how those 12 black pentagons and 20 white hexagons form the cover of a soccer ball. It's easy, says he.[41]

Here's How

"Those 32 interlocking geometric shapes forming the cover of a soccer ball," announces the whiz-bang mathematician, physicist, ed biochemist "are [merely] a function of its generalized energy [represented mathematically]." And since a soccer ball is a sphere, its "generalized energy"[e] may be calculated according to equations used in the so-called "science of spherical arrangements," with the applicable equation for a sphere being $E_a(T_N) = \Sigma_{ij} |X_i - X_j|$. If then its (generalized) energy is "minimized...over all possible configurations when <0 and [then] maximized when this energy is >0, as well as when $=0$, we minimize the logarithmic energy....[Then] this external energy may be expressed by the equation E (", N)."[41]

The 12 Black Pentagons

Next, Dragnev ponders a question thousands and thousands of kids and professional soccer players have kicked around since time immemorial: "How many pentagons are needed for a soccer ball design? (We shall assume that exactly three edges emanate from each vertex)."

---

[e] Discussions on "generalized energy" invariably, it seems, lead to the following question: Just what is "generalized energy"? Is it energy in general, or the energy generally associated with a nutritious diet? With caffeine? With a budding romance? With the perceived energy boost from amphetamines, cocaine, and the like? Is it the adrenaline-induced energy, the "rush," the endorphin–enkephalin spike that goes with winning Olympic gold? With skydiving? With walking a tight line across the Grand Canyon? Toeing a slack line 30 feet, 60 feet, 90 feet above the ground? The floor? Without a safety net? Just what is "generalized energy"?

Turns out, in science and medicine it is none of the above. The term "generalized energy" is used less often than the term "generalized coordinates," which is "a holdover from a period when Cartesian coordinates were the default coordinate system."[42] In practice, physicists acknowledge, it may be "easier to solve a problem using...Euler-Lagrange equations than Newtons laws...because not only may more appropriate generalized coordinates...be chosen to exploit symmetries in the system, but constraint forces are replaced with simpler relations."[42]

Professor Dragnev answers the question instantly, and then provides the mathematical proof. The answer, he says, is "exactly 12." He then postulates that if one lets one number be such and such and another number be such and such, where

F_p = # of pentagons [and] F_q = # of hexagons,

then

$F = F_p + F_q.$ [Now] there is a new formula:

$5F_p + 6F_q = 2E.$

But still

$3V = 2E.$

Then, if the following equations are substituted in Euler's formula,

$$F + V - E = F_p + F_q + \frac{1}{3}(5F_p + 6F_q) - \frac{1}{2}(5F_p + 6F_q) = 2,$$
$$6(F_p + F_q) + 2(5F_p + 6F_q) - 3(5F_p + 6F_q) = 12,$$
$$(6 + 10 - 15)F_p + (6 + 12 - 18)F_q = 12,$$
$$1F_p = 12.$$
$$\text{Q.E.D.}^{f}$$

The 20 White Hexagons

That settled, he asks the question everyone has been hoping would be asked, the question asked millions and millions of times a day worldwide. "Is it possible to tile [meaning to cover with tiles] a soccer ball with only hexagons?" The answer is a quick, emphatic "No!". That response, naturally, raises an equally emphatic question, "Why not?" And *that* prompts the response of "Euler's formula" — which invariably leads to the following question: "Euler's formula?" Reiterated, "WHOSE formula?" "Euler's formula!" he repeats, thinking on the side, *Don't you get it?* "It's based on the number of faces, vertices, and edges of an object," any object. Sensing a need for clarification, he

---

[f]"Q.E.D". is "an initialism of the Latin phrase *quod erat demonstrandum*," which when placed "at the end of a mathematical proof or philosophical argument means 'What was to be demonstrated has been demonstrated.'"[43]

chooses to use a cube to make his point — an object *everyone* is familiar with — and reminds those still viewing the website that a cube has "6 faces, 8 vertices and 12 edges." Then, he says, perhaps a bit didactically, "Suppose we… cut the edges in half." And then, "each vertex [will have] three-half edges ($3V = 2E$) and each face [will have] six edges ($6F = 2E$). Thus, $3V = 6F$ and $V = 2F$. [As a result], $F + V - E = F + 2F - 3F = 0$ 2, which contradicts Euler's… formula." And *that* is tantamount to blasphemy.

Now, should you be wondering just how all this mathematical gobbledygook assists in the construction of a Voronoi diagram using Dirichlet cells, Professor Dragnev has an explanation for that as well. But that's a topic for another book, another day, another year…another lifetime.

References

1. http://pathmicro.med.sc.edu/mayer/phage.htm
2. http://wwwnc.cdc.gov/eid/article/9/2/02–0031_article.htm
3. http://www.textbookofbacteriology.net/themicrobialworld/Phage.html
4. http://www.bing.com/images/search?q=Bacteriophage+Structure&FORM=RESTAB
5. http://www.bing.com/images/search?q=filamentous+phage+photo&id=2142F6BECAC084CF5ECB50247008B4CDE04E92FA&FORM=IQFRBA
6. http://www.bing.com/images/search?q=figure+of+phage+T4&qpvt=figure+of+
7. Miller ES, Kutter E, Mosig G, Arisaka F, Kunisawa T, Rüger W. (20003) Bacteriophage T4 genome. *Microbiol Mol Biol Rev* **67(1)**: 86–156. doi: 10.1128/MMBR.67.1.86–156.2003
8. http://www.bing.com/images/search?q=photo+quaterback+throwing+a++football&qpvt=photo+quaterback+throwing+a++football&FORM=IGRE
9. http://upload.wikimedia.org/wikipedia/commons/b/ba/Tamara_Toumanova_%26_Serge_Lifar.jpg (accessed June 10, 2011.)
10. *Pointe* shoes.
11. http://en.citizendium.org/wiki/Bacteriophage (accessed June 4, 2011).
12. Browning C, Shneider MM, Bowman VD, Schwarzer D, Leiman PG. (2012) Phage pierces the host cell membrane with iron-loaded spike. *Structure* **20:** 326–329.
13. Campbell AM. Bacteriophages. In: Knipe DM, Griffin DE, Lamb RA, Martin MA, Roizman B, Strauss SE (ed.). *Fields Virology,* 5th ed. Wolters Kluwer Williams & Wilkins, Philadelphia, Baltimore, New York, London, Buenos Aires, Hong Kong, Sydney, Tokyo. Chap. 23, pp. 769–791.
14. http://www.textbookofbacteriology.net/themicrobialworld/Phage.html

15. http://en.wikipedia.org/wiki/Humpty_Dumpty

16. Opie J. and Opie P. (ed.). (1951) Humpty Dumpty, *Oxford Dictionary of Nursery Rhymes.* Oxford at the Clarendon Press, pp 213–216.

17. http://eclipsee.rutgers.edu/goose/rhymes/dump/1.aspex

18. Kaehtany S, Cornell University, Ithaca NY, Microbiology Library: http://pathmicro.med.sc.edu/mayer/coliphage1.jpgFigur

19. http://www.umm.edu/altmed/articles/urethritis-000167.htm

20. http://en.wikipedia.org/wiki/Incandescent_light_bulb

21. http://en.wikipedia.org/wiki/Icosahedron

22. http://en.wikipedia.org/wiki/Figure_of_the_Eart

23. Nasu H, Iida T, Sugahara T, Yamaichi Y, Park K-S, Yokoyama K, Makino K, Shinagawa H, Honda T. (2000) A filamentous phage associated with recent pandemic *Vibrio parahaemolyticus* O3:K6 strains. *J Clin Microbiol* 2000; **38:** 2156–2161.

24. http://ntmf.mf.wau.nl/cor/m13.htm,

25. http://www.pnas.org/content/104/7/2109/F5.expansion

26. http://en.wikipedia.org/wiki/Lambda_phage

27. http://english.turkcebilgi.com/Lambda+phage

28. http://www.google.com/imgres?imgurl=http://cronodon.com/images/T4_labelled_v2.jpg&imgrefurl=http://cronodon.com/BioTech/Virus_Tech.html&usg=__64-UVxhmWUj9RjeFh8ohXjSm7FY=&h=655&w=887&sz=119&hl=en&start=29&sig2=_hiHz7mYdc4K4ojd6ku_aQ&zoom=1&tbnid=cSHDPJ-LhbapIM:&tbnh=108&tbnw=146&ei=z7pkUKOyL4exyQGEqIG4Aw&prev=/search%3Fq%3Dprolate%2Bicosahedral%2Bphage%2Bwith%2Bhelical%2Btail%26start%3D20%26hl%3Den%26sa%3DN%26gbv%3D2%26tbm%3Disch&itbs=1

29. http://www.ehow.com/info_8507841_colony-characteristics-ecoli.htm

30. http://en.wikipedia.org/wiki/Figure_of_the_Earth

31. http://en.wikipedia.org/wiki/Spheroid

32. http://www.britannica.com.EBchecked/topics/559619/sphere

33. http://en.wikipedia.org/wiki/Ellipse

34. http://www.newworldencyclopedia.org/entry/Ellipse#Drawing_an_ellipse

35. *The New Encyclopædia Britannica* (2010). Encyclopædia Britannica, Inc., Chicago, London, New Delhi, Paris, Seoul, Sydney, Taipei, Tokyo, Vol. 4, pp 320–321.

36. http://en.wikipedia.org/wiki/Rings_of_Saturn

37. http://en.wikipedia.org/wiki/Icosahedral

38. http://en.wikipedia.org/wiki/Icosahedral, http://pathmicro.med.sc.edu/mayer/phage.htm

39. http://users.ipfw.edu/dragnevp/vitae.htm

40. http://www.bing.com/search?q=Soccer+Ball&FORM=QSRE1

41. http://www.bing.com/search?q=dragnev+chwmistry+physics+mathmatics+soccer+ball&src=IE-Address. Also, Dragnev P. "The Chemistry, Physics, and Mathematics Behind the Soccer Ball Design. Or, How to Distribute Dictators on a Planet."

42. http://en.wikipedia.org/wiki/Lagrangian_mechanics

43. http://en.wikipedia.org/wiki/QED

34

EULER'S FORMULA

Named in honor of the 18th century Swiss mathematician and physicist Leonhard Euler (1707–1783), Euler's formula is sometimes referred to as "Euler's number," a number akin to perhaps the more widely recognized term "Avogadro's number"[a] — a term many nonscience majors will still recall from high-school biology/chemistry classes. And though Avogadro's number is praiseworthy, the accolades for the Swiss mathematician–scientist and the formula bearing his name are spectacular. Just as the clever math guy Peter Dragnev is not an average mathematician, Euler's formula is not your 'average' formula. The thoughts of other mathematicians — while maybe not quite as humorous as Dragnev's expressions — but giants in the field nonetheless have had to say about it will be traced. But, first, a bit about Leonhard Euler.

---

[a] The designation "Avogadro's number" recognizes the significant scientific contributions of Amedeo Avogadro, the (mostly) 19th century (1776–1856) Italian physicist–mathematician, who first proposed that "the volume of a gas [at a given pressure and temperature] is proportional to the number of atoms or molecules regardless of the nature of the gas."[1] In 1909, French physicist Jean Perrin proposed applying the moniker "Avogadro's number" to both the number of molecules in one mole of oxygen *and* the number of atoms, molecules, ions, or even electrons in 12 g of the oxygen isotope of carbon-12.[2] (A mole, in science, is simply the molecular weight of a substance — any substance — expressed in grams and has nothing to do with either the small, dark skin blemishes that flourish in sunlight or the spies and double agents glorified in the riveting espionage novels, by Londoner Ian Fleming and sought by James Bond — code-named Agent 007 — navels enjoyed by so many Americans and Brits.) That magical number is — $6.02214179 \times 10^{23}$ mol^{-1}. While the number of "particles" is the same irrespective of the substance, the weight of one mole of a given substance differs. For example, one mole of the gas ammonia, NH_3, contains 6.022×10^{23} molecules and weighs about 17 g. while 1 mole of the solid copper also contains 6.022×10^{23} atoms, but weighs 63.54 g, nearly four times as much as NH_3. For those less fastidious about mathematics, yet desirous of scientific accuracy and brevity, Avogadro's number may be mercifully abridged to 6.02, knowing that a slew of digits really follow, says the *Britannica Concise Encyclopedia*.[3]

Fig. 34.1.   Leonhard Euler (1707–83) is considered to be the "preeminent mathematician of the 18th century and one of the greatest" of all time.

(*Source*: Reproduced with permission from the publisher.) Reference 4.

Leonhard Euler and his Famous Formula

The man who figured out this fabulous formula more than 250 years ago was a Swiss-born mathematician–physicist. He shifted his life, at age 20, to St. Petersburg, Russia, to attend the city's famed Academy of Sciences. 'Fore long, he was its Chairman of the Mathematics Department. But Germany's Frederick the Great enticed him to come to Berlin to teach at the celebrated Berlin Academy — which he did for 25 years, before falling out of Frederick's good graces in 1766. But that mattered little, because Catherine II quickly asked Euler (Fig. 34.1) to return to Russia. He did, but soon lost his sight (due to cataracts), but not his mathematical vision, and remained highly productive till his death in 1783.[5]

Now, What is Euler's Famous Formula?

Simply put, Euler's formula is a "mathematical formula in complex analysis that establishes the deep relationship between the trigonometric functions and the complex exponential function."[6] The exact formula is $e^{ix} = \cos x + i \sin$, x, where, for any given number, "e is the base of the natural logarithm, i is the imaginary unit,[b] and cos and sin are the trigonometric functions cosine and sine, respectively, with the argument[c] x given in radians."[8] Its brevity belies its complexity and applicability to a variety of geometric forms. The formula may not even be readily understood by some mathematicians, though Peter Dragnev would certainly apprehend it quickly — and, you can bet, come up with a humorous, clever account of what it means.

Mathematical Beauty

If one were asked to choose the adjective most often associated with Euler's formula, it would have to be "beautiful." Many of us may find it hard to imagine a formula as "beautiful," let alone hear of the phrase "mathematical beauty." Yet, that descriptor and that phrase have cropped up again and again in writings about Euler's formula. Table 34.1 records how giants in mathematics spanning nearly 300 years have felt about this "mysterious," yet legendary, formula.

Let us begin with the noted 19th-century American physicist–mathematician and Harvard professor *Benjamin Peirce,* who also was a "devoutly religious person."[18] Peirce said of the formula one day in class, after having validated it during a lecture, "It is absolutely paradoxical; we cannot understand it, and we don't know what it means, but we have proved it, and therefore we know it must be the truth" (Table 34.1).[19] For his enormous contributions to mathematics, Peirce was later dubbed the "father of American mathematics." (Peirce's mind has been said to be "so complex…that it was difficult for all but the most proficient mathematicians to understand much of his work…[which made] it difficult for him to describe his work"[21] to fellow mathematicians, leaving him isolated, even within the arcane microcosm of mathematicians and physicists.

Far more recently, on America's West Coast, Stanford professor *Keith Devlin* (1948–), the Math Guy on N.P.R.'s Weekend Edition, described Euler's formula as being, "like a Shakespearian sonnet that captures the vast essence

---

[b] An imaginary unit — i.e. number — is a number whose square is less than or equal to zero for example $\sqrt{-25}$.[7] The concept of an "imaginary number" was first put forward by Euler.

[c] In mathematics, the "argument" is an independent variable of a function, with a "function" being "the relation between two sets in which one element of the second set is assigned to each element of the first set, as in the expression $y = x^2$.[9]

Table 34.1　Praise for Euler's formula.

Author	Year	Comment
Devlin[10]	—	"Euler's number is like a Shakespearean sonnet that captures the very essence of love, or a painting that brings out the *beauty* of the human form that is far more than just skin deep, Euler's equation reaches down into the very depths of existence."
Wikipedia[11]	—	"Euler's number is considered by many to be remarkable for its mathematical *beauty*."
The Mathematical Intelligencer Magazine[12]	1990	"Euler's number is "the most *beautiful* theorem in mathematics."
Physics World[13]	2004	Euler's number "is the greatest equation ever."
Reid[14]	—	Euler's formula is "the most famous formula in mathematics."
Nahin[15]	2006	"Euler's formula is the *gold standard* for mathematical *beauty*."
Peirce[16]	—	"Euler's number is absolutely paradoxical, we cannot understand it, we don't know what it means, but we have proved it, and therefore we know it must be the truth."
Feynman[17]	—	Euler's formula is "*our jewel*" and "the most remarkable formula in mathematics. It is our jewel and one of the most remarkable, almost astounding formulas in all of mathematics."

of love, or a painting that brings out the beauty of the human form that is far more than just skin deep... [a formula that] reaches down into the very depths of existence."[11] That rhetoric is matched by the words of Emeritus Professor *Paul Nahin* (1940–) at the University of New Hampshire, who, in a 400-page monograph titled *Dr. Euler's Fabulous Formula* (published in 2006), calls Euler's formula "the gold standard for mathematical beauty."[11] Biographer *Constance Reid* (1918–2010), a woman who had a deep interest in mathematics, though no formal training in the discipline, referred to Euler's formula as "the most famous formula in all mathematics."[14]

In the 20th–21st century' era of (seemingly endless) polls, Euler's formula has continued to fare well. In 1990, readers of the esoteric magazine *The Mathematical Intelligencer* voted it "the most beautiful theorem in

Sin-Itiro Tomonaga

1906–1979

Julian Schwinger

1918–1994

Richard P. Feynman

1918–1988

Fig. 34.2. The 1965 Nobel laureates in Physics. The Prize went to these three men for their "fundamental work in quantum electrodynamics, with deep-ploughing consequences for the physics of elementary particles."

(*Source*: Reproduced with permission from the publisher.) Reference 24.

mathematics,"[22] while a 2004 poll by *Physics World* of its readers named the formula "the greatest equation ever."[22] And yet another American physicist, *Richard Feynman*[d] (1918–88), known best for his work on "the path integral formulation of quantum mechanics, the theory of quantum electrodynamics and...particle physics,"[23] referred to Euler's formula as "the most remarkable formula in mathematics."[17] Feynman's efforts in the field of "quantum electrodynamics" were acknowledged by the Nobel Prize Committee — just as it honored Einstein's work in 1921 — when it jointly awarded Feynman, *Julian Schwinger* (University of California, Berkeley, and Purdue University), and *Sin-Itiro Tomonaga* (Tokyo University of Education, now Tsukuba University) (Fig. 34.2) the 1965 Nobel Prize in Physics for their "fundamental work in quantum electrodynamics, with deep-ploughing consequences for the physics of elementary particles."[24] Still, in spite of his "brains" and such lofty awards, Feynman may have been most appreciated by his students for "taking great care when giving explanations."[23] (It truly takes a special talent for a college or postgraduate professor to make simplicity out of apparent complexity.)

Still, there was one other trait that endeared Feynman to all. It might simply be called his practical jokes, his cleverness. After all, when you're "isolated"

---

[d] Feynman had "a perfect score on the graduate school entrance exams to Princeton University in mathematics and physics — an unprecedented feat."[23]

(because of the top-secret nature of the work) and "bored" in Los Alamos — or, in Feynman's words, "There wasn't anything to *do* there" — you *create* things to do. Being a scientist with ample gray matter, he combated the boredom and "indulged his curiosity by learning to pick the combination locks on cabinets and desks used to secure papers."[23] On one occasion, "he found the combination to a locked filing cabinet by trying the numbers a physicist would use (it proved to be 27–18–28 after the base of [the] natural logarithm $e = 2.71828....$), and found that the three filing cabinets where a colleague kept a set of atomic bomb research notes all had the same combination. He left a series of notes as a prank, which initially spooked his colleague Frederic de Hoffman into thinking a spy or saboteur had gained access to atomic bomb secrets."[23] (On several occasions, Feynman drove to Albuquerque to see his ailing wife in a car borrowed from Klaus Fuchs, who was later discovered to be a *real* spy for the Soviets, transporting nuclear secrets in his car to Santa Fe.[23])

Physics holds that the mass of a body — any body — is directly related to its strength and inversely related to its energy level. And when that body is a human body, the heavier a body the greater its strength — to a point — and the leaner a body, the greater its energy level. It's a truism highly relevant to overweight-to-obese 21st century American children on the fast track to heart attacks and strokes.

Started in 1941, "the Manhattan Project was the [program] conducted by the United States, United Kingdom, and Canada to develop the first atomic bomb. The military aspects of the [endeavor] were under the direction of General J. Leslie Groves (1896–1970),[25] while its scientific aspects were directed by J. Robert Oppenheimer, an American physicist."[26] The project sprang from the fear among American scientists that "Nazi Germany might be developing a nuclear weapon" and was carried out at "more than thirty sites across the United States, United Kingdom, and Canada."[27] "The three major research and production sites were the at [the current] U.S. Hansford site in Washington state, the uranium enrichment [plant] in Oak Ridge, Tennessee, and the research and design laboratory in Los Alamos, New Mexico."[27]

Lise Meitner: The Forgotten Lady

Not so long ago — June 17, 2006 — a Public Broadcasting Service documentary, *Path to Nuclear Fission: The Story of Lise Meitner and Otto Hahn*, retold "the story of the double whammy that for many years prevented nuclear physicist Lise Meitner from being recognized as the person who first explained nuclear fission."[28] I shall retell it here. It typifies the ever-thinking scientific mindset. Hypotheses and scientific revelations have no place or time limitations. And, ideally, personal fame — and financial gain — ought not to compete with sound scientific research meant to benefit all.

Austria-educated, Germany-confined, Sweden-exiled

The story goes like this. The Austria-born, University of Vienna–educated Meitner was a brilliant scientist–physicist who conceived the process of nuclear fission. She had worked with Hahn at the Kaiser Wilhelm Institute in Berlin (Fig. 34.3) from 1907 to 1938,[30] but when Hitler came to power she requested a passport, which was denied because of concern she might speak harshly of the Third Reich. She fled the country illegally, in 1938, going first to Holland and then Sweden, where she found a safe haven at the Nobel Institute for Experimental Physics.

Meanwhile, in Berlin, Hahn was busily irradiating uranium with neutrons. He expected to get radium, a heavy element close to uranium on the periodic table, but instead got barium, a much lighter element. It is said that "Hahn had no idea what accounted for this and asked Meitner during a secret meeting in Copenhagen what [she thought] could account for this."[31] She didn't have a ready answer, but, naturally, told him she would give it some "heavy"

Fig. 34.3. The former Kaiser Wilhelm Institute in Berlin, Germany, where Lise Meitner and Otto Hahn discovered nuclear fission.

(*Source*: Reproduced with permission from the publisher.) Reference 29.

thought and get back in touch with him. At the time, no one had ever thought of "nuclear fission."

A walk in the snow

She did — after discussing Hahn's unanticipated results with Otto Frisch, her nephew in Denmark, who was also a physicist. The documentary cited one snowy day in 1938 when Meitner was walking with Frisch, she on foot, he on skis. Frisch recalled, "At [some] point we both sat down on a tree trunk and started to calculate on scraps of paper [whereupon they reasoned] that the charge of a uranium nucleus is…large enough to destroy the effect of surface tension almost completely, [leaving] the uranium nucleus…a very wobbly, unstable drop ready to divide [split] itself at the slightest provocation [such as the impact of a neutron]."[32] But there was a catch. "When the smaller two drops separated they would be driven apart [by electrical repulsion] and would acquire a very large energy.…Where would that energy come from?" Fortunately, "Lise Meitner remembered how to compute the masses of nuclei…and [then and there] they worked out that the two nuclei [so] formed…would be lighter by about one-fifth the mass of a proton. Now [since] whenever [nuclear] mass disappears energy is created, in accord with $E = mc^2$, one-fifth of a proton mass was equivalent to 200 MeV. So here was the source for the energy. [Suddenly], it all fitted!"[33]

Collaborators Compete

Meitner (Fig. 34.4) relayed her deductions to Hahn, who promptly published a paper in January 1939 regarding nuclear fission without recognizing Meitner as the creator of the idea — perhaps because he was "afraid the Nazi government might punish him for continuing to collaborate with a Jew."[31] Hahn (Fig. 34.5), a Protestant, is said to have "had great compassion for persecuted German Jews."[36] Meitner, who is reported to have been "painfully shy,"[31] still had the spunk to publish a paper — along with her snow-skiing nephew — depicting the nuts and bolts of nuclear fission brought about by irradiation of uranium[e] — a chain reaction with the potential of causing "an

---

[e] Hahn's paper can be found in the January 9, 1939, issue of *Naturwissenschaften,* while Meitner and Frisch's paper can be found in the February 11, 1939, issue of the science journal *Nature.* The term "nuclear fission" was coined and first published in the Meitner–Frisch paper.

Fig. 34.4. Lise Meitner, a very bright, sophisticated lady. She appears to be holding a cigarette, a thin stream of smoke wafting from its tip. At first, that might be thought a surprise. *But,* the photograph was made *circa* 1928, long before the now widely known health risks of smoking cigarettes were recognized. *And,* this was in the Roaring Twenties, when it was stylish for women to smoke cigarettes, perhaps leading her to intentionally pose for this photo with one in her hand.

(*Source*: Reproduced with permission from the publisher.) Reference 34.

enormous explosion."[37] (Fig. 34.6 is a 21st century conceptual diagram of the three-step process). It is said that this report "had an electrifying effect on the scientific community."[39]

Because of this potential, and the possibility that the information could be used by Germany to create a "weapon of mass destruction," as we refer to it today, the Hungarian physicist Leo Szilard and two other scientists "jumped into action, persuading Albert Einstein, who had the celebrity, to write [America's] President Franklin D. Roosevelt a warning letter [reprinted earlier], [which] led directly to the establishment of the Manhattan Project." While Meitner's nephew accepted an offer to work with Manhattan Project engineers in Los Alamos, she refused, declaring, "I will have nothing to do with a bomb."[33]

Fig. 34.5.   Lise Meitner, the scientist, with Otto Hahn in her beloved laboratory. (*Source*: Reproduced with permission from the publisher.) Reference 35.

The Nobel Prize for Nuclear Fission: Who Should Have Received It?

In 1944, Otto Hahn received the Nobel Prize for Chemistry for the discovery of nuclear fission, while, according to the Atomic Archive, "Meitner was ignored."[40] In the opinion of many scientists, Meitner should have at least shared the prize. The oversight was partially rectified in 1966, when Hahn, Fritz, Strassmann, *and* Meitner together received the Enrico Fermi Award.[f]

---

[f] Enrico Fermi was an Italian-born, naturalized American citizen who won the Nobel Prize in Physics in 1938. The award bearing his name is an annual "presidential award and is one of the oldest and most prestigious science and technology honors bestowed by the U.S. Government. The Enrico Fermi Award is given to encourage excellence in research in energy science and technology benefiting mankind; to recognize scientists, engineers, and science policymakers

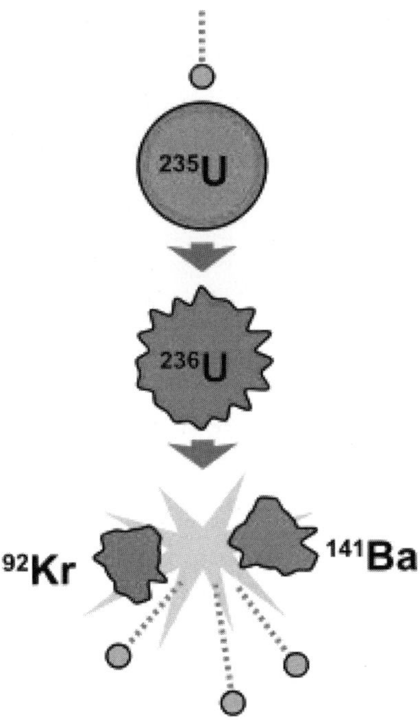

Fig. 34.6. A neutron (light blue circle at top) is absorbed by a urnium-235 nucleus converting it briefly unto a highly excited uranium-236 nucleus. The U-236 nucleus then splits into a krypton-92 fragment and a barium-141 fragment and releases three neutrinos, all packing large amounts of kinetic energy.

(*Source*: Reproduced with permission from the publisher.) Reference 38.

Lise Meitner died in 1968, two weeks shy of age 90. She was buried in Bramley, U.K., where she had moved to from Sweden a few years earlier. Her tombstone (Fig. 34.7) inscription was composed by her nephew, Otto Frisch. It says, "Lise Meitner: A physicist who never lost her humanity."

---

who have given unstintingly over their careers to advance energy science and technology; and to inspire people of all ages through the examples of Enrico Fermi, and the Fermi Award laureates who followed in his footsteps, to explore new scientific and technological horizons."[42]

Fig. 34.7. Lise Meitner's grave in the village of Bramley, U.K. Her tree-stump, calculating nephew, Otto Frisch, composed the inscription on her headstone. It reads: "Lise Meitner: A physicist who never lost her humanity."

(*Source*: Reproduced with permission from the publisher.) Reference 41.

Gone but not forgotten: In memoriam

In 1982, 14 years after Meitner's death, German scientists synthesized a new heavy element — No. 109 in the periodic table. They named it after the shy, brilliant physicist who lived her life in exile — meitnerium.

References

1. Hinshelwood CN, Pauling L. (1956) Amedeo Avogadro. *Science* 124: 708–713.
2. Perrin J. (1909) "*Mouvement brownien et réalité moléculaire.*" *Annales de Chimie et de Physique, 8ᵉ Série* **18:** 1–114. Extract in English, as translated by Frederick Soddy.
3. http://www.answers.com/library/Britannica%20Concise%20Encyclopedia-cid-958435
4. http://sco.wikipedia.org/wiki/File:Leonhard_Euler.jpeg
5. http://www.britannica.com/EBchecked/topic/195201/Leonhard-Euler
6. http://en.wikipedia.org/wiki/Euler%27s_formula
7. http://en.wikipedia.org/wiki/Imaginary_number
8. http://www.bing.com/search?q=euler's+formula&src=IE-Address
9. *Random House Webster's Dictionary,* 2005 ed.

10. http://en.wikiquote.org/wiki/Leonhard_Euler.

11. http://en.wikipedia.org/wiki/Eulers_identity.

12. Brede M. *The Mathematical Intelligencer* 2005; 27: 6–7.

13. Crase RP. The greatest equation ever. Physics Web. October 2004.

14. http://en.wikipedia.org/wiki/Constance_Reid

15. http://www.youtube.com/watch?v+7rrgRIP-KG8.

16. http://en.wikipedia.org/wiki/Benjamin_Peirce.

17. http://en.wikipedia.org/wiki/Mathematical_beauty

18. http://www.answers.com/topic/Benjamin-peirce

19. http://www.answers.com/topic/benjamin-peirce

20. http://en.wikipedia.org/wiki/Benjamin_Peirce

21. http://en.wikipedia.org/wiki/Benjamin_Peirce

22. http://en.wikipedia.org/wiki/Euler%27s_identity

23. http://en.wikipedia.org/wiki/Richard_Feynman

24. http://www.nobelprize.org/nobel_prizes/physics/laureates/1965

25. http://en.wikipedia.org/wiki/Leslie_Groves

26. While its scientific aspects were directed by J. Robert Oppenheimer, an American physicist.

27. ref is at bottom of page in *New World Encyclopedia*).

28. http://www.jewishsightseeing.com/dhh_weblog/2006-blog/2006–06/2006–06–17-pbs-meitner-

29. http://en.wikipedia.org/wiki/Kaiser_Wilhelm_Institute

30. http://www.atomicarchive.com/Bios/Meitner.shtml

31. http://www.jewishsightseeing.com/dhh_weblog/2006-blog/2006–06/2006–06–17-pbs-meitner-hahn.htm

32. Cropper WH. Great Physicists: The life and times of leading physicists. 2004, Chapter 22. p. 340.

33. http://repo-nt.tcc.virginia.edu/book/chap4/chapter4sec1.htm

34. Lise Meitner photo source

35. http://en.wikipedia.org/wiki/Otto_Hahn

36. great compassion for persecuted German Jews

37. an enormous explosion

38. http://en.wikipedia.org/wiki/Nuclear_fission

39. electrifying effect on the scientific community

40. http://www.atomicarchive.com/Bios/Meitn

41. http://en.wikipedia.org/wiki/Lisa_Meitner

42. http://science.energy.gov/fermi)

35

$E = MC^2$ AND EINSTEIN'S BRAIN

Enter Albert...

But one 20th century German luminary (Figs. 35.1–35.3) would have understood him easily. That man earned his Ph.D. in Physics at Switzerland's University of Zurich in 1905 — his so-called *annus mirabilis*[a] ("miracle year) — and in later years honorary doctorates in Physics, as well as in Medicine and even Philosophy, everywhere he went. As the hoof beats of the Third Reich grew louder in 1933, Einstein, a German Jew, elected to exit Germany to begin a new life in a safer place — a place called the United States of America. And, like Osamu Shimomura (Chap. 27) some 30 years later, Einstein chose to base his research at Princeton University, where the powers that be appointed him to the position of Professor of Theoretical Physics,[3] a position, a university, and a city where he remained for the remainder of his career and life.[b]

---

[a]Einstein's "miracle year" was 1905. One hundred years later, National Public Radio (N.P.R) recognized the occasion by reminding listeners that a century earlier a then 25-year-old scientist named Albert Einstein was "working as a patent clerk [in Switzerland to make ends meet]," when he published not one or two, but *four* "groundbreaking papers about space, time, atoms and the strange nature of light."[1] C.B.S radio agreed. In 2005, C.B.S referred to Einstein's 1905 quartet of revolutionary papers as "colossal ideas about time and space which would prove to be groundbreaking."[2] The print media joined the electronic media in recalling Einstein's miracle year when *Time Magazine* featured the scientist [presumptively] one last time on the cover of its date issue (Fig. 30.BB).

[b]Einstein died from a ruptured aortic aneurysm on April 18, 1955, at the Princeton University Hospital. As part of an autopsy, his celebrated brain was "removed and preserved."[4] *But,* whether his famed brain was "removed and preserved" *with his prior permission* or the *postmortem permission of his family* is a matter of dispute to this day.[5] The dispute is highlighted in two biographies.

A 1984 bio by Ronald W. Clark contends that Einstein had "insisted that his brain should be used for research" and then his remains "should be cremated."[6] (The postmortem examination,

545

Fig. 35.1. Albert Einstein as a dapper young man in 1905 upon his graduation from the University of Zurich at age 25.

(*Source*: Reproduced with permission from the publisher.) Reference 10.

Euler's equation is brief, yet powerful — as is the equation E = mc^2 (formulated by Einstein in 1905), where *E* represents energy, *m* mass or matter, and *c* the speed of light in a vacuum squared[c] — a *very* large number

---

by the way, was performed by pathologist Thomas Harvey from Princeton at the University of Pennsylvania's pathology laboratory, some 50 miles away.) Harvey "noticed immediately" that parts of Einstein's renowned brain were, of all things, "missing" or "vacant"!.[4] Once Harvey had "removed, weighed, and dissected" the brain *in toto*, he bestowed "pieces" of it upon himself while giving other sections to "leading pathologists."[6] He also removed Einstein's "+eyes," perhaps on the premise that his vision contributed to his brilliance.[6] The eyes are *not* routinely removed at the time of an autopsy.

Clark's biography contrasts with that by Walter Isaacson, a gifted journalist and biographer; a former Managing Editor of *Time Magazine* (1996–2000); the current President and CEO of the Aspen Institute, a "nonpartisan educational and policy studies organization" based in Washington, D.C.; and President Barack Obama's choice to preside over the Broadcasting Board of Governors, which oversees the Voice of America and Radio Free Europe.[7] Isaacson's 2008 Einstein biography, *Einstein: His life and Universe,* has been welcomed by freelance science writer Andrew Zimmerman Jones as "a warm exploration of the life of one of history's most endearing characters and one of physics' greatest minds."[8]

[c]In 2005, the centennial of Einstein's "miracle year," a team of physicists selected by the American Institute of Physics, headquartered in College Park, Maryland, made the most accurate test yet

Fig. 35.2. The genius Albert Einstein riding a simple transportation device in Santa Barbara in 1933, at age 54.

(*Source*: Reproduced with permission from the publisher.) Reference 10.

(http://www.aip.org/history/einstein/emc1.htm). Einstein's original paper is entitled "*Ist die Trägheit eines Körpers von seinem Energieinhalt abhängig?*" Or, translated into English, "Does the Inertia of a Body Depend upon Its Energy Content?." It was published in *Annalen der Physik* **18**: 639–643, (1905).

The Odd Odyssey of Einstein's Brain

The reaction to Harvey's pirating of Einstein's brain was swift, and damaging to his career. Not impressed by Harvey's qualifications, to study Einstein's brain, but impressed by his (mis)handling of a delicate situation, Harvey's boss — who, as fate would have it, was also the hospital's director — brought Harvey's "tenure

---

of his equation. They measured the tiny change in mass of radioactive atoms before and after the atoms emitted gamma rays. They also measured the energy of the gamma rays. The "missing mass times c^2 equaled the energy of the gamma rays to within 4 hundred-thousandths of one percent." That might well become the scientific definition of "precision" in the 21st century.[9]

Fig. 35.3. A yet older Einstein, who finally takes on the appearance of a "mad scientist"...which he was not.

(*Source*: Reproduced with permission from the publisher.) Reference 10.

as a [Princeton] pathologist...to an end...within [a] few months [of Einstein's death]...for refusing to surrender his precious specimen."[11]

When it rains, it pours

Not only did he lose his job, he lost his wife. Soon after he left Princeton Hospital, he took the brain to Philadelphia, where he had a "technician section it into over two hundred blocks [each] embedded...in collodion[d]...and [then] placed [them] in two formalin-filled jars."[11] Next, he returned to his home in Princeton and placed the brain in "a cider box stashed under a beer cooler" in the basement of his house. That did not sit well with his estranged wife, who "didn't want the specimen in her house" — irrespective of *who* it belonged to — "and threatened to dispose of [it]."[11]

His career and his wife gone, Harvey retrieved the brain and headed west on a journey that would last years and years. He landed first in Wichita, Kansas, next in Weston, Missouri, and sometime substantially later in

---

[d]Colloidion (pyroxylin) is a syrupy substance used to stick unlike surfaces — such as an EKG electrode to skin — together.[12] In this instance, it was used to glue brain tissue to glass slides, later put into a formalin (formaldehyde) solution — a preservative.

Lawrence, Kansas, all the time carting his prize in his car's trunk as he hopscotched across the country, finally returning to Princeton — though not the University — some 40 years later, nearly broke.

All along his journey, Harvey tried to market his trophy, but there were no buyers. Finally, in 1996, he offered the brain — at no cost — to McMaster University neuroscientist Sandra Witelson in Ontario, who carried out a detailed examination of the showpiece. Her prize was the majority of the original roughly $240 \times 10 \text{ cm}^3$ blocks of brain, the rest having been given to other researchers over the years. She wouldn't squander this once-in-a-lifetime opportunity.

The Findings of Sandra Witelson's Thorough Examination of Einstein's Brain

Here's what she discovered. There was *no difference* in the size or weight of Einstein's brain compared to the brains of 91 other persons, or the brains of eight men — like Einstein — aged 65 or more years of age, even though there was moderate atrophy, believe it or not, of Einstein's brain.[e] (The human brain normally diminishes — variably — in size over a lifetime, making an age- and gender-matched comparison group pertinent.) And there was *no difference* between the anatomy of Einstein's frontal lobes (the human brain consists of four lobes[f] — the frontal, parietal, temporal, and occipital lobes

---

[e]Witelson was quick to point out that the extent of atrophy was "common for a person in their eighth decade." Still, Einstein's brain weighed 1230 g, while the average weight of the brains from the control group was 1400 g, a 12% difference.

[f]A lobe is typically a rounded part of the body, such as an earlobe. The four lobes of the brain — also rounded — correspond in nomenclature to the four bones of the overlying skull — the frontal, parietal, temporal, and occipital bones — with the border between the frontal and parietal lobes being shifted posteriorly (backward) slightly with respect to the corresponding bones of the skull.[13] The human brain also includes the cerebellum. The cerebellum is considered by some neuroanatomists to be a separate lobe — a fifth lobe. Its *functions* include motor coordination and cognitive abilities such as attention and language, as well as the sensations of fear and pleasure. *Anatomically*, the cerebellum seems to be a "separate structure" tucked beneath the cerebral hemispheres. (The Latin term "cerebellum" means "little brain" in English.) In fact, "the surface of the cerebellum is covered with finely spaced parallel grooves, in striking contrast to the broad irregular convolutions of the cerebral cortex. These parallel grooves conceal the fact that the cerebellum is actually a continuous thin layer of tissue" — the cerebellar cortex, an extension of the cerebral cortex — "tightly folded in the style of an accordion."[14] For that reason, categorizing the cerebellum as a distinct lobe is contested.

(Fig. 35.4), one on each side, the cerebellum, and the brain stem) and that of the frontal lobes of the comparison group. On the other hand, Einstein's parietal lobes were a bit wider and a bit more spherical than the parietal lobes of the comparison group. (The inferior parietal region of the brain is thought to be related to mathematical thinking and visual–spatial intelligence.[16] This "unusual brain anatomy" led Witelson to speculate that it "may explain why Einstein thought the way he did."[16]

But the principal differences between Einstein's remarkable brain and the brains of a comparison group were the absence of the so-called "parietal operculum"[g] (Fig. 35.4) on both the right and the left side of the genius' brain *and* the fact that the parietal lobes of Einstein's brain were symmetrical when they are normally somewhat — not very much — asymmetrical. Indeed, absence of the parietal operculum was not seen in *any* of 35 brains from males and 56

The *brain stem* does not suffer from any such anatomical–grammatical dispute. Though small, the brain stem controls both the sensory and motor innervations of the face and neck through the 12 cranial nerves that sprout from it, and, because of its position midway between the (central) cerebral cortex and the peripheral nerves innervating the rest of the body, the large nerve tracts (bundles) originating in the cerebral cortex must pass through the brain stem to reach their destinations.[15] Even the heart and lungs are, to some extent, regulated by nerves traversing the brain stem. The brain stem is a short, high-traffic area — like freeway on- and off-ramps.

[g]"Operculum" is a word with a broad application. Meaning "little lid" in English, the Latin-based word crops up in marine, freshwater, and land biology, where the operculum serves "as a sort of 'trapdoor'"[17] to close the shell of a snail when it recoils into its "cave"; in freshwater fish, where the operculum serves as a cover for their gills[18]; in birds, where the operculum — their beaks — covers their mandibles (jawbones) and nares (nose)[18.5]; in some plants that have an operculum covering and protecting their seeds, spores, and flowers, the protective cover falling off as they open[19]; and, in humans, where the *teeth* have an operculum — a flap covering yet-to-erupt teeth; the *uterus* has an operculum in the form of a mucus plug that seals the uterus' canal after conception to hopefully ensure fertilization[20]; and, finally, the *brain* has an operculum covering its precious gray matter, which forms the entire surface of each cerebral hemisphere, the gray matter itself being folded into gyri, much of it hidden in the cerebral fissures, where it works day and night to "integrate higher mental functions, general movement, perception and behavior" by day,[21] sleep and dreaming by night,[22] and "visceral functions" 24 h a day, with "visceral functions" simply being the operation (the functioning) of the organs (the viscera), particularly those within the abdominal cavity used for digestion and urination.[21] Finally, there are humans with a figurative operculum who under minimal provocation, are apt to "blow their lids."

And…it was inevitable, I suppose, that the term would become commercialized. Operculum is now the brand name of a contraceptive device used to block — like a little lid — human Fallopian tubes, avoiding tubal ligation[22] — though its manufacturer points out "this device is not FDA-approved," nor is it "for sale in the U.S."

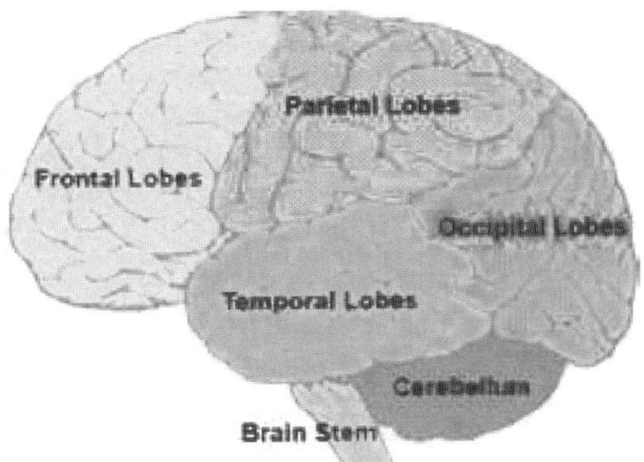

Fig. 35.4.   The four lobes of the human brain — frontal, parietal, temporal and occipital — override the cerebellum (a disputed lobe) and the brain stem. The brain stem is comprises the medulla oblongata, the pons, and the midbrain (not shown). The meandering dark lines over all four lobes represent grooves separating the brain's surface into gyri or gyruses, depending on your tendency to apply Latin or English plural suffixes to singular Latin terms, such as "alumnus". Is more than one college graduate "alumni" or "alumnuses"? (Most of us say "alumni.") are alumni *all* college graduates or are *male* graduates "alumni" and *female* "alumnae"? (The word "alumnus" in Latin is a masculine noun; the word "alumna," a feminine noun.) To avoid sexism, mixed groups may be referred to as the *alumni/alumnae* of a university, or the *alumni* and *alumnae* of a university. And, if you have any patience left, what is the plural of "pleura"? The thin, transparent membrane covering a single lung — so painful during pleurisy — is a *pleura*;, both lungs, *pleurae*. The word "pleural," meaning "of or pertaining to the pleura," has nothing to do with the word "plural," unless there is inflammation of both pleurae and then there would be inflammation of plural pleurae!

(*Sources*: Reproduced with permission from the publisher.) References 24 and 25.

from women (total 91 brains) — or even in the subgroup of brains from the 8 men aged 65 or more. (Witelson, incidentally, published her work, entitled "The Exceptional Brain of Albert Einstein," in the June 19, 1999, issue of *The Lancet*.) The 35 control male brains and the 56 female brains represented "all the. . . specimens available at the time" housed in Witelson's brain museum.[23] Einstein's brain, then, differed from all the control brains in being a bit rounder and a bit more symmetrical than the brains from the control group[h] brains, and, most significantly, its lack of the inferior parietal operculum (Figs. 35.5 and 35.6). The

---

[h]The control group featured the "brains [of] research volunteers with normal neurological and psychiatric examinations" during life and a normal IQ. In fact, no one volunteered to sacrifice his or her life for the sake of scientific research — illustrating the strength or lack, of subpar syntax.

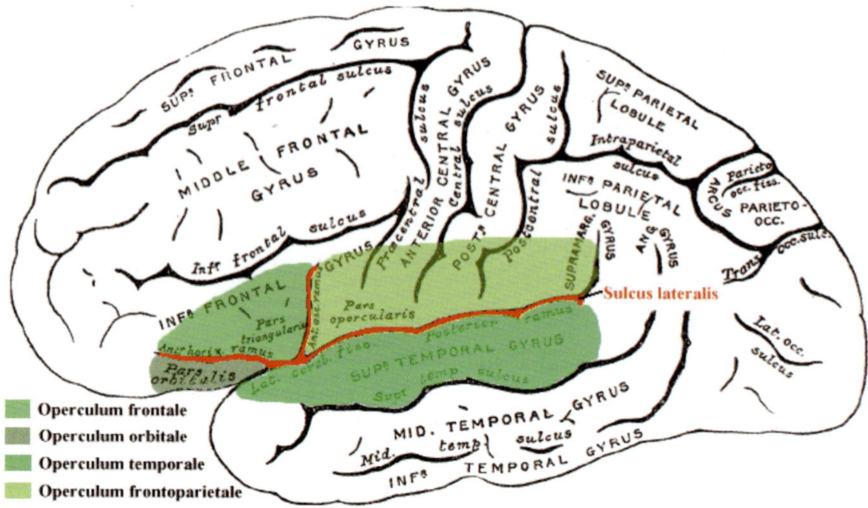

Fig. 35.5.   This rather "complicated" figure of the human brain is being used because it shows what simpler, more colorful figures do not, *i.e.* the specific region of the parietal lobe (shaded a sea green) without its operculum, its "lid." The lack of a lid over the inferior parietal lobe may have allowed that region — the seat of "mathematical thought" — to "bulge" a bit in the "right" place. A bit more "storage space" in the "mathematics department" *might* have translated into Einstein's enhanced conceptual clarity in the fields of physics and mathematics.

(*Source:* Reference 26. This figure is from the 20th U.S. edition of *Gray's Anatomy of the Human Body*, originally published in 1918, which now lies in the public domain, its copyright having expired worldwide. The text is available under the Creative Commons Attribution-ShareAlike License. Reproduced with permission from the publisher.)

absent operculum, it is speculated, may have led to an "increased expansion of the small inferior [portion of the] operculum in the parietal region," which could "provide a functional advantage"[23] — the brain of a genius.

Witelson also pointed out that "the increased expansion of. . .Einstein's parietal region" had been seen in the brains of other physicists and mathematicians — the likes of Swedish physicist Per Siljestrom (1815–92)[27] and German mathematician Carl Gauss (1777–1855).[28] For all that Einstein gave to science and mankind during his 76 years on Earth, he was selected by *Time Magazine* as the "Person of the Century," his image appearing on the December 31, 1999, issue of the publication (Fig. 35.8). Such an honor for a German Jew who fled his homeland as the dark cloud of the Third Reich settled over Germany in 1933.

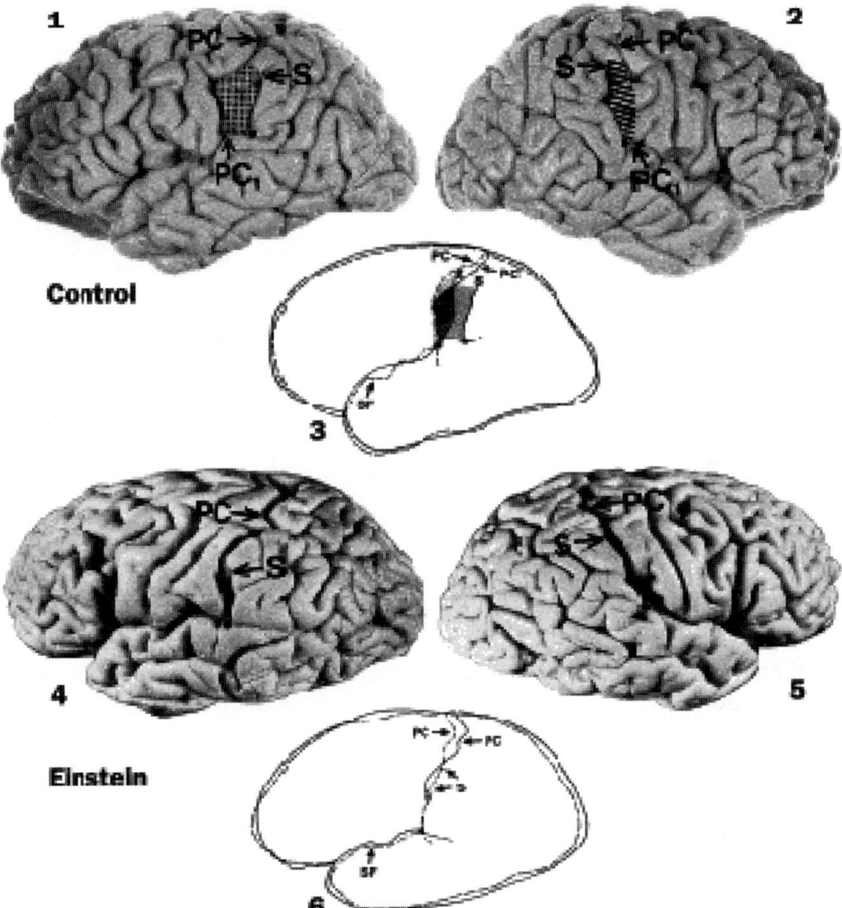

Fig. 35.6. Photographs of Einstein's brain taken at the time of autopsy in 1955 and those of a control brain from Sandra Witelson's brain museum some 40 years later are shown side by side. The lateral photographs and drawings of the left hemisphere (solid line) and right hemisphere (dashed line) of a *control male* of "average intelligence" (1, 2, and 3) are superimposed on *Einstein's* brain (4, 5, and 6). The photographs of the control brain show the parietal operculum in the left (stippled) and right (hatched) hemisphere, situated between the postcentral (PC) sulcus and the posterior ascending branch of the Sylvian fissure (SF), which originates at the point of bifurcation (l) and terminates at S. PC1 is the inferior end of PC at SF. The tracing of the superimposed hemispheres (3) shows the asymmetry in position and size between the parietal opercula. The tracing of Einstein's hemispheres (6) highlights the confluence of PC and the posterior ascending branch of SF in each hemisphere, the *absence of the parietal opercula*, and the *symmetry* of the sulcal morphology between hemispheres. Comparison of the tracings shows the relatively anterior position of the SF bifurcation in Einstein, and the associated greater posterior parietal expanse, particularly in his left hemisphere compared with the control brain.

(*Source*: Reproduced with permission from the publisher.) Reference 23.

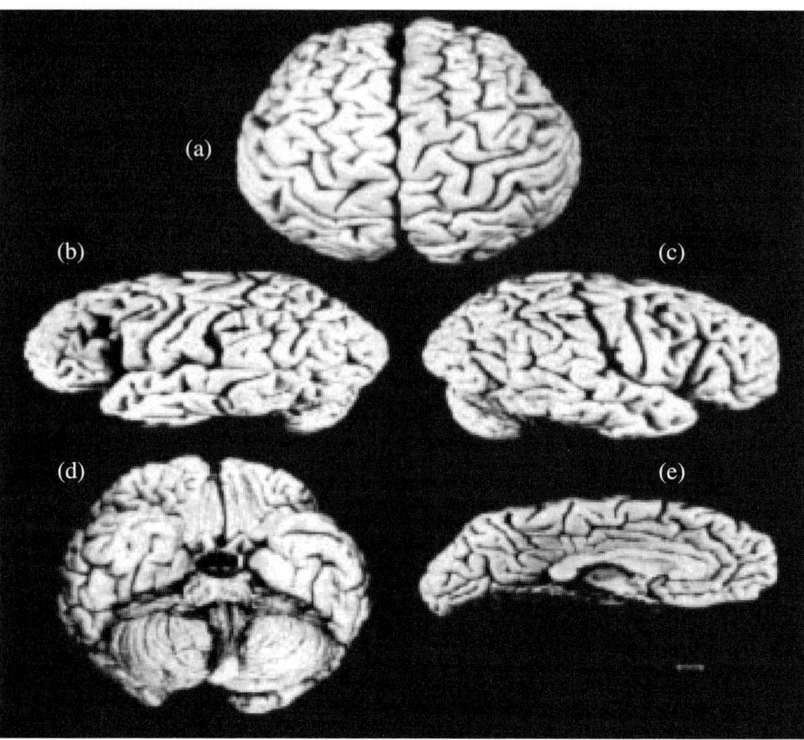

Fig. 35.7. Five photographs of Einstein's brain taken (a) from above, (b) from the left side, (c) from the right side, (d) from below, and (e) of the left hemisphere alone cut halfway though from front (left) to back (right). In the top photo, labeled "A", moderate *cerebral atrophy* is present in the two hemispheres adjacent to the prominent fissure separating the two hemispheres. In the middle photos, labeled "b" and "c", the posterior ascending branch of the so-called Sylvian fissure is confluent with the postcentral sulcus at the point indicated by the two small, black, horizontal, hard-to-see arrows in each hemisphere (*cf.* Fig. 35.6, control). Consequently there is *no operculum* over either hemisphere. In contrast, none of the brains from the 35 male controls or those from the 56 female controls displayed this anatomy.

(*Source*: Reproduced with permission from the publisher.) Ref. 23.

What to Expect Down the Road

In the future, Witelson imagined, we can expect to learn more about "the structure–function relationship of the general population through qualitative and quantitative assessment of brain anatomy and function through computerized imaging techniques such as MRI," not available during Einstein's lifetime.

Fig. 35.8. Albert Einstein — *Time Magazine's* "Person of the (20th Century"). He was chosen because the editors felt he was "the preeminent scientist in a century dominated by science. The touchstones of the era — the Bomb, the Big Bang, quantum physics, and electronics — all bear his imprint." This was Einstein's fifth appearance on the cover of *Time* in the past century. The July 1, 1946 cover of *Time* is displayed in Fig. 35.9.

Einstein was depicted by Fredric Golden, the author of the feature story, as "unfathomably profound — [a] genius among geniuses who discovered merely by thinking about it, that the universe was not as it seemed."

(*Source*: Reproduced with permission from *Time Magazine.*) Reference 29.

But we may have to wait a lifetime for another Einstein-like brain to come along.

Einstein Eulogies Accrue

One of the 20th century's most revered scientific minds, Einstein was eulogized by fellow theoretical physicist J. Robert Oppenheimer (Fig. 35.10) (1904–67)[31] from the University of California, Berkeley, as being "almost wholly without sophistication and wholly without worldliness. . . .There was always with him a wonderful purity at once childlike and profoundly stubborn."[31] Oppenheimer himself was chosen, in 1942, by General Leslie Groves to supervise the all-important scientific aspects of the Manhattan Project. General Groves, in turn, had been designated by President Franklin D.

Fig. 35.9.   The cover of *Time* Magazine on July 1, 1946, illustrating the complex association between Einstein's famed equation — $E = mc^2$ — and the atomic bomb. The artist cleverly superimposed the equation on a mushroom-shaped atomic cloud — the cloud itself being an amalgamation of the Hiroshima and Nagasaki explosions. The small print reads, "Cosmoclast Einstein. All matter is speed and flame."

(*Source*: Reproduced with permission from the publisher.) Reference 30.

Fig. 35.10.   Oppenheimer (right) with Einstein in the 1940s.

(*Source*: Reproduced with permission from the publisher.) Reference 31.

Roosevelt as the Project's Director, an enormous responsibility — the planning and production of the world's first atomic bomb, as well as the secrecy of the far-flung Project. (More about the famed Manhattan Project on tap.)

Opinions of Einstein's Peers About Einstein the Man

As a scientist, Albert Einstein will undoubtedly be remembered as "the most epic among 20th-century thinkers," invoked Einstein biographer Walter Isaacson.[32] Albert Einstein as a man, a husband, and a father, on the other hand, is a much more challenging person to characterize, and what little is known of his personal life, even by those within his innermost circle, is less than "epic" and gleaned largely from Ronald W. Clark's 1984 biography and Isaacson's 2007 account of his life, *Einstein: His life and Universe.* Isaacson (Fig. 35.11) is particularly qualified to record Einstein's life. The New

Fig. 35.11.   Walter Isaacson (1952–), New Orleans native, Harvard graduate (B.A. in History and Literature), Rhodes Scholar, *Time Magazine* political correspondent, national editor, and 14th editor in 1996, Chairman and C.E.O. of C.N.N. in 2001. President and C.E.O. of the Aspen Institute in 2003, as well as being appointed Chairman of the Broadcasters Board of Governors, which oversees the Voice of America and Radio Free Europe, by President Barrack Obama in 2008   — and still has found time to author well-written biographies of four famous Americans— Henry Kissinger (1992), Benjamin Franklin (2003), Albert Einstein (2007), and Apple Inc. cofounder Steve Jobs (released on November 21, 2011 — six weeks after his premature — age 56 — death from pancreatic cancer.[33]

(*Source*: Reproduced with permission from the publisher.) Reference 32.

Orleans native, Harvard graduate, Rhodes Scholar, and *Time Magazine* political-correspondent-turned-national-editor is also Barrack Obama's appointee to the Broadcasters Board of Governors. Isaacson has made a habit of writing addicive biographies of well-known Americans, the likes of Henry Kissinger (1992) and Benjamin Franklin (2003), as well as Einstein (2007), and a 2011 biography of the late cofounder of Apple Inc. — one Steve Jobs (Fig. 35.12).

Isaacson's Einstein bio caught the eye of Anne Bartholomew, a book reviewer for AmazonCrossing (a publishing imprint of Amazon.com[34], who also writes commentaries on behalf of N.P.R. These were her words regarding that biography, *Einstein: His life and Universe*: "This book brings Einstein's experience of life, love, and intellectual discovery into brilliant focus. The book is the first biography to tackle Einstein's enormous volume of personal correspondence that heretofore had been sealed from the public, and it's hard to imagine another book that could do such a richly textured and complicated life as Einstein's the same thoughtful justice. Isaacson is a master of. . .form and this latest opus is at once arresting and wonderfully revelatory."[32]

Fig. 35.12.   Steve Jobs (1955–2011), cofounder of Apple Inc., proudly holding one of his iPhones. Said *New York Times* reporter David Streitfeld of Jobs, "Rarely has a major company and industry been so dominated by a single individual, and so successful[ly]. His influence has gone far beyond the iconic personal computers that were Apple's principal product for its first 20 years. In the last decade, Apple has redefined the music business through the iPod, the cell phone business through the iPhone and the entertainment and media world through the iPad. Again and again, Mr. Jobs has gambled that he knew what the customer would want, and again and again he has been right."

(*Source*: Reproduced with permission from the publisher.) Reference 35.

References

1. http://www.npr.org/series/4645093/1905-science-s-miracle-year
2. http://www.npr.org/series/4645093/1905-science-s-miracle-year
3. http://www.nobelprize.org/nobel_prizes/physics/laureates/1921/einstein-bio.html
4. http://en.wikipedia.org/wiki/Albert_Einstein%27s_brain
5. http://en.wikipedia.org/wiki/Albert_Einstein
6. Clark RW. (1984) *Einstein: The Life and Times.* HarperCollins.
7. http://en.wikipedia.org/wiki/Walter_Isaacson
8. http://physics.about.com/od/physicsbooks/gr/einsteinwibio.htm
9. http://www.aip.org/history/einstein/emc1.htm (American Institute of Physics).
10. http://www.aip.org/history/einstein/emc1.htm (American Institute of Physics).
11. Burrel B. *The long, strange journey of Einstein's Brain, Postcards from the Brain Museum.* Random House. 2005, New York.
12. http://en.wikipedia.org/wiki/Collodion
13. http://en.wikipedia.org/wiki/Human_brain
14. http://en.wikipedia.org/wiki/Cerebellum
15. http://wikipedia.org/wiki/Brain_stem
16. http://www.psych.ualberta.ca/~gcpws/Witelson/Biography/Witelson_bio5.htm
17. http://en.wikipedia.org/wiki/Operculum_(gastropod)
18. http://www.biology-resources.com/fish-01.html
18.5 http://en.wikipedia.org/wiki/Cere#Cere
19. http://www.thefullwiki.org/Operculum_(botany)
20. http://medical-dictionary.thefreedictionary.com/operculum
21. *Mosby's Medical Dictionary of Medicine, Nursing and the Health Profession.* Elsevier, 7th ed. (2006).
22. http://operculum.org
23. Witelson S, Kigar DL, Harvey T. (1999) The exceptional brain of Albert Einstein. *Lancet* **353**:2149–2153.
24. http://www.bing.com/images/search?q=lobes+of+brain&qpvt=lobes+of+brain&FORM=IQFR
25. *Random House Webster's Unabridged Dictionary,* 2nd ed. (2001). Random House, New York.
26. http://en.wikipedia.org/wiki/Operculum_(brain)
27. http://en.wikipedia.org/wiki/Operculum_(brain)
28. http://www.britannica.com/EBchecked/topic/227204/Carl-Friedrich-Gauss
29. http://www.amazon.com/gp/product/images/B00006FCRO/ref=dp_image_0/183-1872432-2647835?ie=UTF8&n=283155&s=books
30. *Time* Magazine cover, July 1, 1946.
31. http://en.wikipedia.org/wiki/J._Robert_Oppenheimer

32. http://wazon.com/Einstein-Life- Universe-Walter-Isaacson/dp/0743264738
33. http://www.nytimes.com/2011/10/06/business/steve-jobs-of-apple-dies-at-56. html?_r=1&pagewanted=all
34. http://en.wikipedia.org/wiki/Amazon.com#Amazon_Publishing
35. http://www.nytimes.com/2011/08/25/technology/jobs-stepping-down-as-chief-of- apple.html?_r=1&ref=stevenpjobs

36

THE MANHATTAN PROJECT
AND ITS CONNECTION
WITH ALBERT EINSTEIN

The Manhattan Project was a joint WWII venture among the United States, Canada, and the United Kingdom — the three countries that collaborated to design and construct the world's first atomic bomb.[1] General Leslie Groves was designated the project's administrative director, as just mentioned, by then-president Franklin D. Roosevelt, while American physicist J. Robert Oppenheimer, also just mentioned, kept a critical eye on its scientific aspects.[a,1]

While Einstein was the person who set forth $E = mc^2$ and expounded the theories of relativity and mass equivalence at the equation's core, his 19th century German roots were enough to negate his winning the top-security clearance he needed to work on the Manhattan Project and the development of an atomic bomb. And that…and that, even though he had fled his native Germany in 1932 fearful of Adolf Hitler, *and* had been appointed to Princeton's newly

---

[a]While American scientists created the destructive bomb, years later many "became crusaders against nuclear weapons,"[1] others became proponents of research to improve nuclear weapons, and still others "applied their…mastery of nuclear fission to the development of nuclear reactors for generating electricity from small quantities of nuclear fuel" in a more environmentally conscious (mostly) peacetime nation.[1] As evidence of America's intent to use nuclear energy for peaceful purposes, the duties of the Manhattan Project were transferred to the United States Energy Commission upon its creation in January 1947. That agency now controls nuclear weapon production within the United States and its territories.

While sharp differences of opinion about how to use nuclear fission existed among nuclear physicists following WWII, the release of the "nuclear genie" paved the road to a "nuclear arms race between the United States and the Soviet Union" — the so-called Cold War — "that led to a massive proliferation of nuclear armaments."[1] And, today, the goal of nuclear scientists and the United States Armed Forces is to never, never let the genie out of the bottle again.

561

created Institute for Advanced Study in 1933, *and* had won U.S. citizenship (in 1940), *and* — possibly most significantly — had been disturbed enough about Hitler and the Third Reich's potential to build an atomic bomb for use against the U.S. to write his famous letter to then-president Roosevelt in August 1939 warning of an unsuspected — an unheard-of — deadly force…a "nuclear chain reaction" that could "lead to the construction of a [nuclear] bomb." That letter is reprinted here:[2]

               Albert Einstein
               Old Grove Rd.
               Nassau Point
               Peconic, Long Island

               August 2nd, 1939

F.D. Roosevelt,
President of the United States,
White House
Washington, D.C.

Sir:

  Some recent work by E.Fermi and L. Szilard, which has been communicated to me in manuscript, leads me to expect that the element uranium may be turned into a new and important source of energy in the immediate future. Certain aspects of the situation which has arisen seem to call for watchfulness and, if necessary, quick action on the part of the Administration. I believe therefore that it is my duty to bring to your attention the following facts and recommendations:

  In the course of the last four months it has been made probable - through the work of Joliot in France as well as Fermi and Szilard in America - that it may become possible to set up a nuclear chain reaction in a large mass of uranium,by which vast amounts of power and large quantities of new radium-like elements would be generated. Now it appears almost certain that this could be achieved in the immediate future.

  This new phenomenon would also lead to the construction of bombs, and it is conceivable - though much less certain - that extremely powerful bombs of a new type may thus be constructed. A single bomb of this type, carried by boat and exploded in a port, might very well destroy the whole port together with some of the surrounding territory. However, such bombs might very well prove to be too heavy for transportation by air.

-2-

The United States has only very poor ores of uranium in moderate quantities. There is some good ore in Canada and the former Czechoslovakia, while the most important source of uranium is Belgian Congo.

In view of this situation you may think it desirable to have some permanent contact maintained between the Administration and the group of physicists working on chain reactions in America. One possible way of achieving this might be for you to entrust with this task a person who has your confidence and who could perhaps serve in an inofficial capacity. His task might comprise the following:

a) to approach Government Departments, keep them informed of the further development, and put forward recommendations for Government action, giving particular attention to the problem of securing a supply of uranium ore for the United States;

b) to speed up the experimental work,which is at present being carried on within the limits of the budgets of University laboratories, by providing funds, if such funds be required, through his contacts with private persons who are willing to make contributions for this cause, and perhaps also by obtaining the co-operation of industrial laboratories which have the necessary equipment.

I understand that Germany has actually stopped the sale of uranium from the Czechoslovakian mines which she has taken over. That she should have taken such early action might perhaps be understood on the ground that the son of the German Under-Secretary of State, von Weizsäcker, is attached to the Kaiser-Wilhelm-Institut in Berlin where some of the American work on uranium is now being repeated.

Yours very truly,

A. Einstein

(Albert Einstein)

The Beginnings of the Manhattan Project

Why build an atomic bomb and where to build such a weapon?

The impetus for development of a nuclear weapon sprang from "*scientists' fears. . .that Nazi Germany might be developing a nuclear weapon of its own.*"[1] Born out of a relatively unheralded research program in 1939, the Manhattan

Project mushroomed into a major wartime effort that "employed more than 130,000 people and cost nearly US$2 billion" (US$24 billion in 2008 dollars when linked to the Consumer Price Index). The Project's three "principal research and production sites were the plutonium production facility at what is today the Hanford Site in Washington state, the uranium enrichment facility in Oak Ridge, Tennessee, and the Weapons Research and Design Laboratory, now known as the Los Alamos National Laboratory," situated atop an isolated mesa in the heart of New Mexico in the town of Los Alamos just 34 miles (55 km) northwest of Santa Fe (Fig. 36.1). Actual construction — not production of the radioactive uranium and plutonium isotopes — of the bomb itself took place at "more than thirty [secret] sites [scattered] across the United States, Canada and the United Kingdom."[1]

The project's leadership, raw materials — uranium and plutonium — and heavy-duty machinery needed to build a nuclear weapon

Come August 1942, the Manhattan Project was launched.[4] Major General Leslie Groves was chosen by Colonel G. Bryan Conrad as its administrative director, while J. Robert Oppenheimer was appointed as its scientific director. Together, they were to build an atomic bomb.[b]

Groves' first goal was to come up with enough precious *uranium* to build the bomb. He didn't have to look far. He found 1250 tons of the stuff in barrels "sitting on Staten Island's docks."[4] But it didn't belong to any company in the U.S. or, for that matter, to the government of the United States. It was the lawful property of Belgium, which, fearful it might fall into the hands of neighboring Nazi Germany, moved it to New York City.[8] Groves, of course, "immediately purchased all of it."[4]

Next, he needed a *nuclear reactor,*[c] which, *à la* Ancel Keyes (Chap. 27), he found buried under the seats of Stagg Field at the University of Chicago.[4] Earlier in 1942, Italian physicist Enrico Fermi had immigrated to the U.S.

---

[b]Groves had directed the construction of the Pentagon from 1941 to 1943, and would now have the honor and onerous responsibility of codirecting the construction of an atomic bomb.

[c]A nuclear reactor is an apparatus in which a nuclear fission chain reaction can be initiated, sustained, *and* controlled during peacetime to generate constructive, useful radiation.

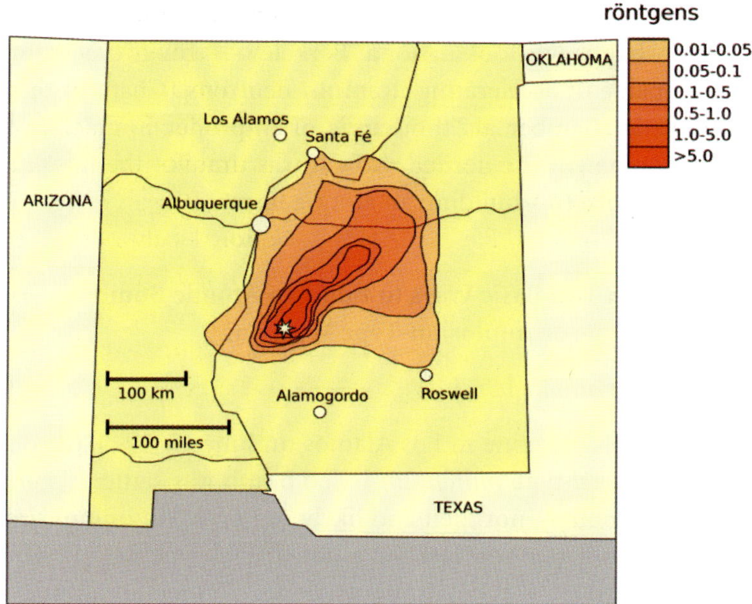

Fig. 36.1. Los Alamos, a town in north-central New Mexico about 36 miles (58 km.) from Santa Fe was the dwelling nearest the site of the Trinity test bomb. The first atomic bomb was exploded at 5:30 am on July 16, 1945, at a site on the Alamogordo air base 120 miles (193 km.) south of Albuquerque, New Mexico. A red arrowhead that blends in with the Röentgen map marks the site The radiation fall out near the Trinity test site, as measured in Röentgens, is shown in the upper right corner as the radioactive cloud moved to the northeast. Gradually decreasing, though still high radiation levels, were measurable within about 100 miles (160 km) northeast of the explosion. Now, more than 65 years after the test, residual radiation at the test site measures about ten times higher than normal. To put that statement in perspective, the amount of radioactivity received during a one-hour visit to the site is about one-half of the amount a U.S. adult receives on an "average" day from natural and medical sources.

(*Source*: Reproduced with permission from the publisher.) Reference 3.

fearful of fascist dictator Benito Mussolini, as had Einstein in 1933 dreading Hitler and his all-but-certain termination of independent scientific research.[d]

---

[d]Fermi (1901–54) was awarded the 1938 Nobel Prize in Physics for his work in radioactivity, including the concept and design of the first nuclear reactor. When he left Stockholm, he headed straight to New York City, rather than returning to Italy, then in the hands of a dictator.[4]

Now, a *cyclotron* and a *particle accelerator* moved to center stage.[e] This time, Groves found what he was looking for at Berkeley — a huge cyclotron, 20 feet tall.[4] It was capable of accelerating uranium neutrons (Chap. 4) through its central, magnetic chamber at "thousands of miles per hour."[4] He had the "heavy-duty equipment" he needed. Now it was time for the *raison d'être* of the Manhattan Project — building an atomic bomb!

Choosing Between Two Basic Ways to Build an Atomic Bomb: A "Gun-Barrel"-Type or Implosion-Type Weapon?

"Gun-barrel"-type atomic bomb

On a fateful day, at a meeting in Los Alamos on June 17, 1944, the destiny of a so-called *gun-barrel*-type *plutonium*-based bomb was settled. There would not be such a weapon — not then and likely not ever. Plutonium simply has a *spontaneous* fission rate too fast for a gun-barrel-type design. The bomb would spontaneously and prematurely explode.[5] Rather, an atomic weapon with a *uranium* core utilizing the gun-barrel-type construction technique (Fig. 36.2) — and an alternate *implosion* method for creating a *plutonium* weapon — were sanctioned.

---

[e]These two devices are similar, in the sense that both of them are huge and expensive — and they both use a magnetic field in a vacuum to accelerate particles. (Hence, a cyclotron is a type of particle accelerator.) A cyclotron initiates a stream of *nonradioactive* particles — ions — moving in a *circular* path through such a magnetic field, and it then "feeds" them into a *linear* particle accelerator, which speeds them up a bit more before shooting the resulting *radioactive* particles at the target[4] — which in modern medicine may be a malignancy. (The transfer of kinetic radioactive energy to its malignant target is remindful of the transfer of the kinetic energy of a moving cue ball to a stationary ball hoping to sink it in a nearby pocket.) Upon exiting, two isotopes exist which are collected in separate containers.[4] An isotope is any of two or more forms of a (chemical) element having the same number of protons in its nucleus. Every currently recognized element has known isotopic forms.

A cyclotron also differs from a *nuclear reactor* by using a stable, nonradioactive starting material, while a nuclear reactor uses a not-so-stable radioactive substance, such as uranium, as its raw material. Nuclear reactors, as we all know, *vis-à-vis* Chernobyl, Three Mile Island and, most recently, the Fukushima Daiichi (Japan) disaster, can emit dangerous quantities of radiation during a nuclear-core meltdown. Cyclotrons are far safer. They use electricity — not atomic fission — as their power source and may be housed near their sites of medical application — the Radioisotope Imaging Laboratory (lung scans, heart scans, brain scans, etc.) and the facility's Magnetic Resonance Imaging (MRI) Laboratory.

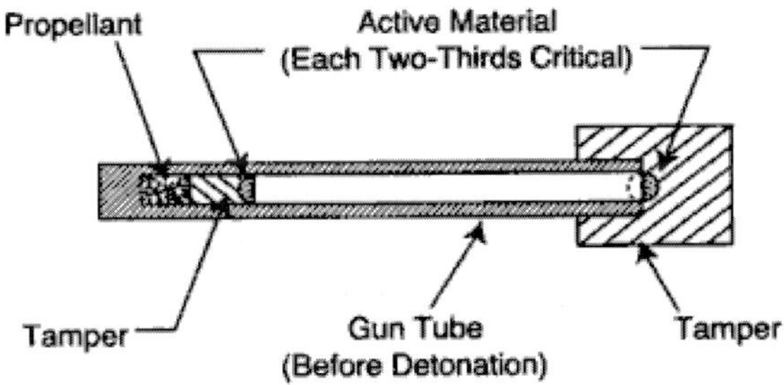

Fig. 36.2.   The assembly mechanism of the gun-barrel design atomic bomb takes two chunks of a fissionable material — uranium — each less than a "critical mass," and fuses them very quickly into a single "supercritical mass." The radioactive material at both ends of the barrel is surrounded by a "tamper," also known as a "reflector," which reflects escaping neutrons back into the fissioning central radioactive core, prolonging its lifetime and raising the yield. The fusion chamber is tubular, as is a gun barrel, allowing a standard explosive such as urea nitrate to propel one subcritical chunk of fissionable material from one end to the other, where another chunk awaits its arrival and then. . .the fireworks begin.

(*Source*: Reproduced with permission from the publisher.) Reference 6.

Implosion-type atomic bomb

In the implosion version of an atomic bomb — AKA the "Fat Man" — its radioactive uranium *or* plutonium core is surrounded by several concentric shells of fissionable material (Fig. 36.3), all wrapped in a jacket — called a "tamper" — of heavy metals, such as lead, beryllium, or even uranium itself in a stable form (there are multiple forms of uranium, some stable, some not.[7]). In effect, "The heart of [an implosion-type] atomic bomb [consists of] a cantaloupe of U-235 or an orange of Pu-239 surrounded by a watermelon of ordinary uranium [acting as a] tamper."[8] The jacket serves to reflect any neutrons thinking about escape straight back into its fissionable layers. These naughty neutrons lead to yet more nuclear fissions — the splitting of a uranium-235 or plutonium-239 nucleus[f] into two fragments, each with nearly half the number of protons and neutrons of its parent — and yet more neutrons. The ongoing fissioning process is a chain reaction, unstoppable until "all the fissionable material is consumed."[10] The three most meaningful words in atomic weaponry may be "neutrons," "fission," and "chain reaction."

---

[f]Because the destructive power of an atomic bomb stems from *nuclear* fission, it is appropriately named a nuclear weapon.

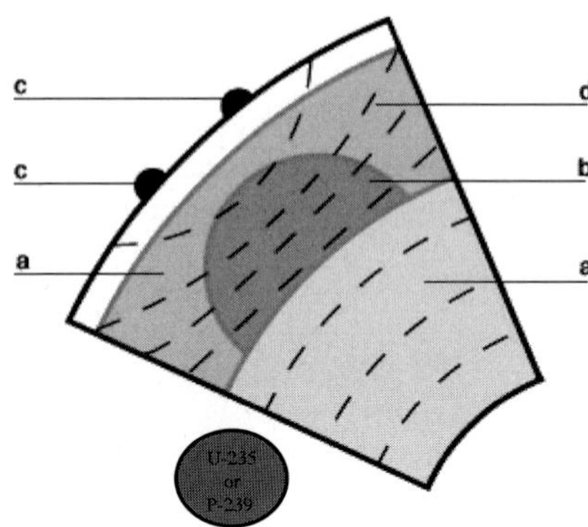

Fig. 36.3.   A core of fissionable material — uranium-235 or plutonim-239 — is surrounded by concentric rings of: (a) a fast explosive, (b) a slow explosive, (c) and a detonator that leads to a detonation (shock) wave (d) the shock wave moves faster than the speed of sound, producing a great (squeezing) pressure increase. A "jacket," not shown, envelops all the inner rings. The shock wave compresses the fissionable uranium or plutonium instantly. As the radioactive core is compressed, it becomes "critical" and then "supercritical," whereupon the fission chain reaction grows exponentially and unstoppably — or at least until the force created by the fragments of the fission process exceeds the implosion pressure caused b the shock wave.
(*Source*: Modified and reproduced with permission from the publisher.) Reference 9.

The urea nitrate jacket is an explosive, as well as serving as a reflector. When exploded, reports the Atomic Archive, it rapidly "compresses [the] sphere of plutonium [forming its nucleus] to a density sufficient to make the weapon become 'critical' and the Fat Man[g] explode" (Fig. 36.4).[12] Both methods of building an atomic bomb, sophisticated as they may be rely on a standard explosive.

---

[g]In reality, the Fat Man's radioactive, deadly core weighed just 13.6 pounds (6.2 kg). The key to the Fat Man's greater "efficiency" was "the inward momentum of the massive U-238 tamper" (which did not undergo fission). Once the chain reaction started in the plutonium, the momentum of the implosion had to be reversed before expansion could stop the fission. By holding everything together for a few hundred nanoseconds more, its efficiency was increased.[11]

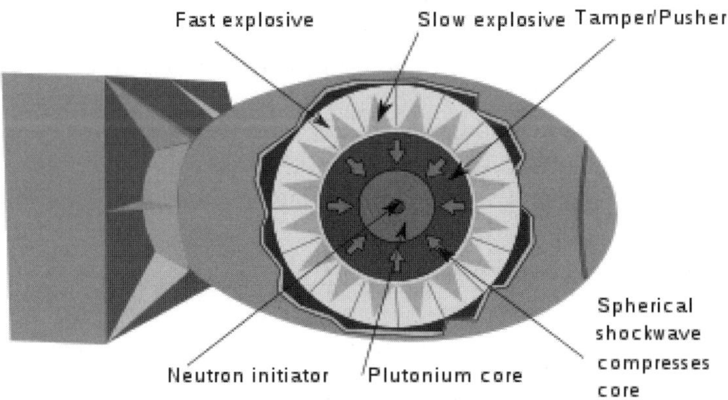

Fig. 36.4. The Fat Man, dropped above Nagasaki, used just 13 pounds (6.2 kilograms) of plutoniuma-239 wrapped in a fast-acting explosive jacket comprised of a heavy metal such as lead, beryllium or uranium in a stable form of uranium. Upon detonation of the explosive shell, the nucleus of plutonium is squeezed uniformly and "goes" critical and the Fat Man explodes.

(*Source*: Reproduced with permission from the publisher.) Reference 13.

The Standard Explosive

An atomic bomb itself, while all-destructive, needs the help of a conventional explosive to trigger the nuclear fissioning process. That conventional explosive can be simply urea nitrate $[CO(NH_2)_2.HNO_3]$ (Fig. 36.5), itself a simple chemical with a whopping kick. So whopping is it that it is "frequently used by terrorists" in the Middle East, say the Israeli Police and the Casali Institute of Applied Chemistry in Jerusalem.[14] It also happened to be the choice of the terrorist group that originally imperiled — using an underground bomb — the North Tower of Manhattan's World Trade Center in 1993. Its popularity among those intent on death and destruction resides in the unfortunate fact that not only is it "exceptionally easy to make," according to the Israeli Police, but also that bomb experts find it "difficult to identify…in post-explosion debris, since only a very small fraction survives the blast."[14] As if *that* wasn't enough, "in the presence of water, [the compound] readily decomposes [into] its original components, urea and nitric acid."[14] So…a little rain after a blast can wash away evidence of the chemical's existence!

An outfit in central Los Angeles named, inappropriately or appropriately, "Outlaw Laboratories" sets forth the urea nitrate recipe — for those who wish

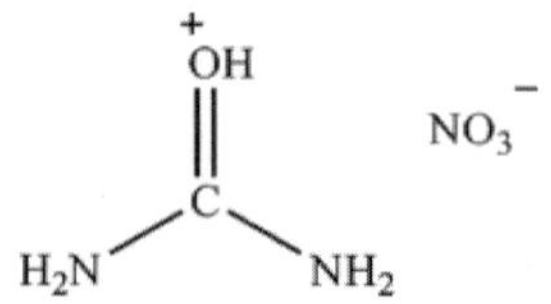

$$\overset{+}{OH} \quad\quad NO_3^{-}$$

Fig. 36.5. The simple structure of the explosive urea nitrate used to abet an atomic explosion.

(*Source*: Reproduced with permission from the publisher.) Reference 5.

Table 36.1 Ingredients to concoct a conventional urea nitrate explosive

[1]	1 cup concentrated solution of uric acid ($C_5 H_4 N_4 O_3$)
[2]	1/3 cup of nitric acid
[3]	4 heat-resistant glass containers
[4]	4 filters (such as coffee filters)*

(*Source*: Reproduced with permission from the publisher.) Reference 16.

to know. All it takes is the ingredients in Table 36.1, available at many chemical supply wholesalers or retailers, followed by about 15 min of active time in a domestic or commercial laboratory.

Once the ingredients are assembled, just pass the concentrated solution of uric acid through a filter to remove any impurities. Next, slowly add a 1/3 cup of nitric acid to the solution and let the mixture stand for 1 h. Filter again. This time the urea nitrate crystals will collect on the filter. Wash the crystals by pouring water over them while they are in the filter. Remove the crystals from the filter and take in a movie or two, followed by a good night's sleep, for it will take 16 h for the crystals to dry. Oh, and don't forget, the explosive will need a blasting cap to detonate.

The Outlaw outfit reminds aspiring bomb makers of two things. First, a caveat: "It may be necessary to make a quantity larger than the aforementioned list [reprinted in Table 36.1) calls for." Second, a disclaimer: "The information conveyed in its atomic bomb web site recipe is strictly for academic use alone. Outlaw Labs and all publishers of this document will bear no responsibility for any use otherwise. It would be wise to note that the personnel who design and construct these devices are skilled physicists and are more knowledgeable in these matters than any layperson can ever hope to be.

Should a layperson attempt to build a device such as this, chances are s/he (*sic*) would probably kill his/herself not by a nuclear detonation, but rather through radiation exposure. We here at Outlaw Labs do not recommend using this document beyond the realm of casual or academic curiosity."[16]

References

1. http://www.newworldencyclopedia.org/entry/Manhattan_Project
2. www.pbs.org/wgbh/americanexperience/features/primary-resources/truman-ein39/
3. http://en.wikipedia.org/wiki/Trinity_site
4. DeGroot G. (2005) *The Bomb*: A Life Harvard Press.
5. http://en.wikipedia.org/wiki/Thin_Man_bomb
6. Gun barrel figure
7. http://web.ead.anl.gov/uranium/guide/ucompound/forms/index.cfm
8. Rhodes R. (1986) *The Making of the Atomic Bomb*. Simon & Schuster, New York, London, Toronto, Singapore, Sydney, Tokyo, Chap. 14, pp. 443–485.
9. http://www.atomicarchive.com/Fission/Fission10.shtml
10. *The New Encyclopedia Britannica*, (2010). Encyclopedia Britannica, Inc., Chicago, London, New Delhi, Paris, Seoul, Sydney, Taipei, Tokyo, Vol. 1, pp. 676–677.
11. http://www.universityessays.com/example-essays/engineering/nuclear
12. http://www.atomicarchive.com/FissionFission9.shtml
13. http://en.wikipedia.org/wiki/Fat_Man
14. Tamiri T, Rozin R, Lemberger N, Almog J. (2009) Urea nitrate, an exceptionally easy-to-make improvised explosive; studies towards trace characterization. *Anal Bioanal Chem* **395**: 421–429.
15. http://auraria.summon.serialssolutions.com/search?s.q=Urea+nitrate%2C+an+exceptionally
16. http://www.serendipity.li/more/atomic.html#explosive_charges

37

A GADGET, A LITTLE BOY AND A FAT MAN

The final chapter of WWII began at four separate, secret bomb construction sites in the U.S. The sites were secret and Project scientists were sworn to secrecy — no discussions whatsoever regarding the Project with *anyone*, including their wives and those few friends with whom they might have wished to maintain an amiable relationship as the war raged on. It's not difficult to imagine that during those deadly serious wartime years (1941–45) of building the world's first atomic bomb, Project engineers — in an effort to create a bit of lightheartedness — named the two nuclear weapons the "Little Boy" and the "Fat Man." And, of course, that allowed them to blithely discuss the bombs in a codelike manner No one knew what they were talking about. A Little Boy and a Fat Man?

Continuing with the code, WWII bomb engineers dubbed the test of a nuclear weapon in Los Alamos, New Mexico, on July 16, 1945, the "Trinity."

Trinity

At 5:30 a.m. local time that morning, the United States Army detonated its "Gadget," the codename given to the bomb itself, on a roughly 36,000-acre desert site atop an isolated mesa in north-central New Mexico[1] (Fig. 37.1). Historic photographs of the explosion were taken from a bunker 10 miles (16 km) from the test site. At the moment of explosion, the weapon released energy equivalent to that of 20 kilotons of TNT leaving a crater 10 feet (3 m) deep and 1100 feet (330 m) wide, while the mushroom–shaped cloud rose to a height of 7.5 miles (12 km) above the test site.[2]

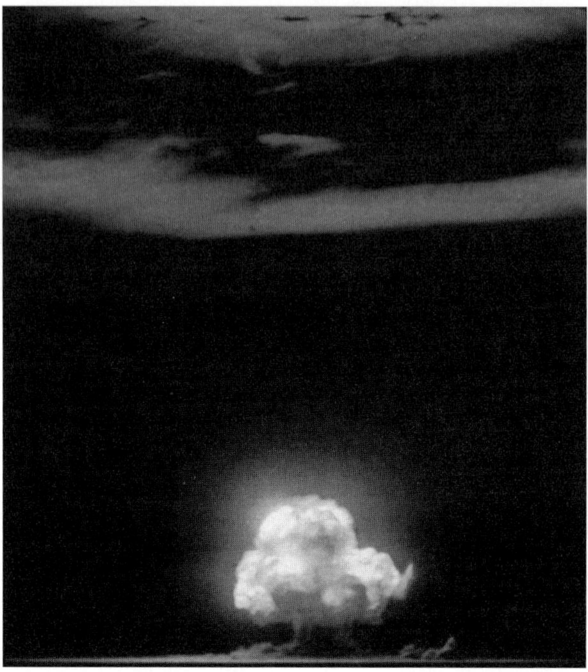

Fig. 37.1. A Color photograph — the only known surviving color photograph — of the famed "Trinity" explosion, the first test of a nuclear weapon conducted by the United states, on July 16, 1945, near Los Alamos, New Mexico, on what is now the White Sands Missile Range, not far from Alamogordo, New Mexico. Trinity tested an: implosion-design plutonium bomb" similar to the bomb dropped three weeks later on August 9, 1945, on Nagasaki, Japan, which, along with the atomic bomb dropped on Hiroshima three days earlier, marked the end of WWII and the beginning of the Atomic Age. Photograph by jack Aeby as a member of the Special Engineering Detachment of the U.S. Army while stationed at the Los Alamos laboratory in 1945.
(*Source*: Reproduced with permission from the publisher) Reference 1.

Origin of "Trinity"

The origin of the term "Trinity" is unclear and will remain so barring the unearthing of a "lost" document. Some ascribe it to J. Robert Oppenheimer, the nuclear physicist with an Eastern religion inclination, as well as a literary bent. When asked years later about the name, Oppenheimer acknowledged, "Why I chose the name is not clear, but I know what thoughts were in my mind. There is a poem [by] John Donne, written just before his death, which I know and love. [In it is] a quotation: 'As West and East/In all flat Maps — and I am one — are one, / So death doth touch the Resurrection." That still

does not form a trinity, but in another, better-known devotional poem,
Donne wrote:

> Batter my heart, three-person'd God; for you
> As yet but knock, breathe, shine, and seek to mend;
> That I may rise and stand o'erthrow me, and bend
> Your force to break, blow, burn and make me new,
> I, like a usurped town, to another due,
> Labor to admit you, but O, to no end;
> Reason, your viceroy in me, me should defend,
> But is captived, and proves weak or untrue.
> Yet dearly I love you, and would be loved fain,
> But am betrothed into your enemy,
> Divorce me, untie or break that knot again;
> Take me to you, imprison me, for I,
> Except you enthrall me, never shall be free,
> Nor chaste, except you ravish me."
> (*Source*: Reference.[3])

Though Oppenheimer may not have recalled why he chose the name
"Trinity" for the first test of a nuclear weapon in July 1945, the author(s) of a
referenced piece posted on Wikipedia say in a matter-of-fact manner, "The
first nuclear weapon test, Trinity was named by J. Robert Oppenheimer after
some extracts of [John Donne's] *Holy Sonnets*."[4] Many Americans might
readily accept that a "three-person'd God" is tantamount to the Christian
Trinity — one God with three distinct beings, the Father, the Son, and the
Holy Spirit; or, put another way, one God — omnipresent omnipotent, and
omniscient. National Defense Research Committee Chairman James Conant[a]
(1893–1978) chose General Leslie Groves to oversee the military aspects of the
Manhattan Project[8] and nuclear scientist Oppenheimer to supervise its scientific side (Fig. 37.2). Conant's selection of Oppenheimer was influenced by his

---

[a] Conant was the President of Harvard from 1933 to 1953. As President, he was "instrumental
in transforming Harvard...into an increasingly diverse, world-class research university. He
introduced aptitude tests into the undergraduate admissions system so that students would be
chosen for their intellectual promise and merit, rather than their social connections." Those
tests eventually evolved into today's (beleaguered) Scholastic Aptitude Test (SAT).[6] When
WWII broke out, the Harvard-trained chemist simultaneously served as the Chairman of the
National Defense Research Committee.[7]

Fig. 37.2.   On the left is the heavy-set General Leslie Groves, the Director of the Manhattan Project, described by his closest aide as "ruthless, egotistical, and confident" (and that might have been the appropriate disposition for a person in such an unusual position)...as well as the "biggest S.O.B. I have ever worked for."[5]

On the right is J. Robert Oppenheimer, described as a "gaunt" person and "a philosophical man, attracted to Eastern mysticism who work[ed] long hours." Oppenheimer was rarely seen without a cigarette between his lips and would later die from throat cancer, a less common consequence of heavy smoking than lung cancer, at age 62.

(*Source*: Reproduced with permission from the publisher.) Reference 5.

"singular grasp of not just physics, but chemistry, metallurgy, ordnance and engineering."[9] The New York City–born, Harvard-educated scientist's appointment also won the imprimatur of then-president Franklin D. Roosevelt.

Oppenheimer and Hitler: Death Dealers of the 1940s — an Unkind and Unwarranted Analogy

While Oppenheimer was responsible for the all-important scientific aspects of the Manhattan Project, he did not savor the prospect of wiping out thousands upon thousands of Japanese, even though Japan had triggered the Pacific chapter of WWII by bombing Pearl Harbor nearly four years earlier.

He was merely doing what had to be done; but he really didn't want to do it. A philosophical man, he would say, years later, of the all-destructive nuclear weapon, "[The bomb] brought to mind words from the [Hindu scripture] *Bhagavad Gītā*," which means "*Song of God.*"[10] "I [became]," recalled Oppenheimer, "Death, the destroyer of worlds."[11] He wasn't alone. A number of WWII veterans would live the rest of their lives tormented by memories of WWII battlefield killings as well as Nazi Germany's POW camps and Hitler's carnage — particularly of German Jews — and his atrocious concentration camps, where captives wasted away (Fig. 37.3).

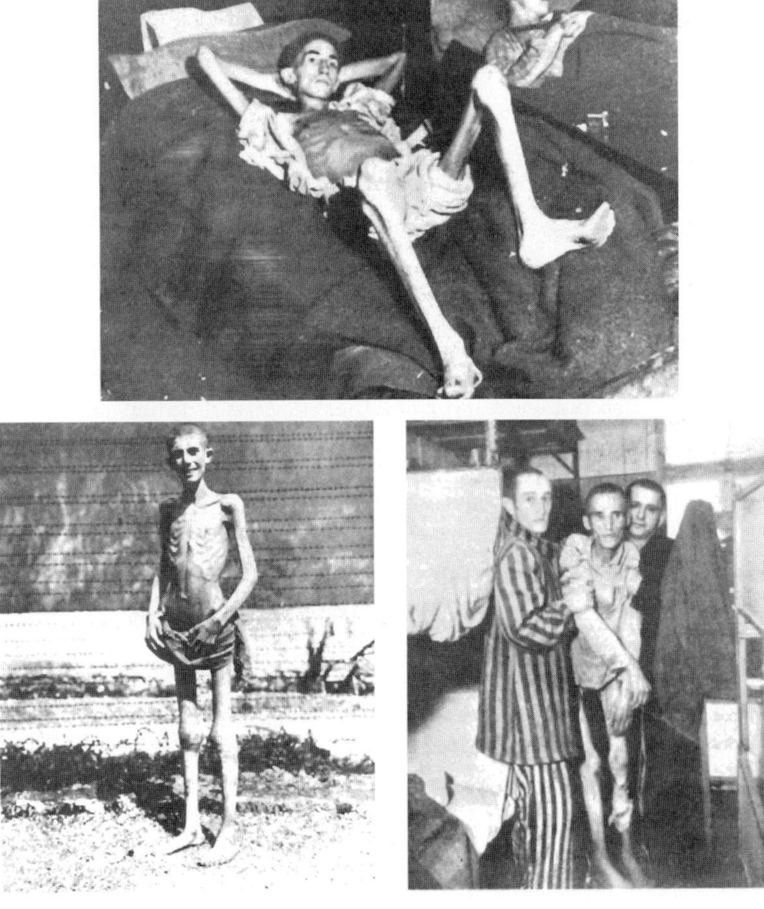

Fig. 37.3.　These photographs needs no words.

(*Source*: Reproduced with permission from the publisher.) Reference 12.

The Island of Tinian

The bomb in the pipeline, out of nowhere the tiny island of Tinian became the focus of the United States' armed forces in the summer of 1944. Seeking an airbase (relatively) near Japan to launch B-29 bombers aimed at Japan, the U.S. Marines captured the remote South Pacific island (Fig. 37.4) — all 41.74 square miles / 108.1 km² of it — from Japan in July 1944.[b] In short order, they built six runways on its north end, and regimentally named them alphabetically — runway A (Able), B (Baker), C (Charlie), and D (Dog) (Fig. 37.5), two

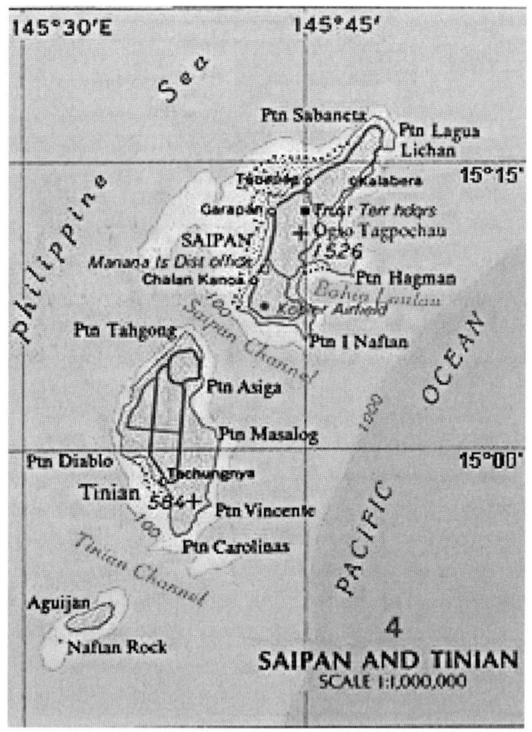

Fig. 37.4. The latitude of the islands of Tinian and Saipan in the South Pacific. The U.S. Marines captured them from Japan in 1944 while the U.S. Air Force used the islands to build runways for their B-29 bombers to leave from with their atomic payloads destined for Japan, some 1500 miles away.

(*Source*: Reproduced with permission from the publisher.) Reference 13.

---

[b] Today, Tinian is a Commonwealth of the United States. It now boasts two gas stations and one casino for American and European tourists to enjoy.[14]

Fig. 37.5. B-29 bombers on runways in Tinian during the summer of 1944, while awaiting deployment to Japan. The island had been "a Japanese stronghold" since the end of WWI. When the U.S. captured Tinian and converted it into a US airbase in 1944, its runways became the longest in the world[14] at that time.

(*Source*: Reproduced with permission from the publisher.) Reference 15.

of them large enough for B-29 bombers to take off and land after dropping their atomic payloads on Japan's mainland some 6 h and 1500 miles away.

But they weren't finished naming things just yet. Driven by nostalgia perhaps, the Marines went on to build a "mini-Manhattan," creating a grid street pattern resembling that of New York City's Manhattan Island — all 22.7 square miles/59 km^2 of it — and named its sections accordingly. They nicknamed the southernmost tip of the now-Marine-controlled island "The Village," reflecting the location of the real Village — Greenwich Village — on the south side of Manhattan Island. A sizeable square area between the west and north airfields was dedicated to the construction of a base hospital...but was otherwise left in its natural, undeveloped state — as is Manhattan's Central Park. And the major north–south street was named — what else? — Broadway.[13] Sentimental and industrious were those Marines so far from home

And Then Came Hiroshima and Nagasaki

On the morning of August 6, 1945, an American B-29 bomber named the *Enola Gay* lumbered onto a runway at an airbase on the tiny, remote South

Fig. 37.6. The *Enola Gay* was named after the mother, Enola Gay Tibbets, of the plane's pilot, Colonel Paul Tibbets — pictured above waving just before takeoff for Hiroshima, who flew the B-29 that carried the Little Boy released over Hiroshima on August 6, 1945.

(*Source*: Reproduced with permission from the publisher.) Reference 16.

Pacific island of Tinian.[c] [The *Enola Gay* (Figs 37.6 and 37.7) was named after the mother — Enola Gay Tibbets — of the plane's copilot, Paul Tibbets.[18] In its belly rested the Little Boy, a uranium-235 atomic bomb, its destination being Hiroshima, Japan. Six hours and 1500 miles (Fig. 37.9) afterward, the bomb was dropped. Forty-three seconds later, at an altitude of 1890 feet, the *Little Boy* exploded.[20] Left in its wake were "80,000 to 140,000 people dead[d] — many literally vaporized (Figs. 37.10 and 37.11). Poof! Nothing to bury, cremate, or freeze for (possible) future reanimation of the frozen

---

[c] Tinian is a member of the Mariana Islands, lying roughly 1500 miles southeast of Hiroshima.
[d] The official U.S. Army website places the Hiroshima toll "between 70,000 and 80,000 people with...a like number [seriously] wounded."[23] The Army estimates that 40,000 more died in Nagasaki, with 60,000 seriously injured. A number of other websites and encyclopedias state that 70,000–160,000 people perished. Who's to ever know the correct toll? After all, how can you count bodies fragmented beyond recognition or vaporized?

Fig. 37.7. The *Enola Gay* and its crew, Little Boy who dropped the atomic bomb on Hiroshima.

(*Source*: Reproduced with permission from the publisher.) Reference 22.

Fig. 37.8. *Bock's Car* was the name of the B-29 that carried the Fat Man, dropped over Nagasaki on August 9, 1945. The plane was usually flown by Captain Frederick C. Bock, but a last minute crew change put the plane in the hands of Major Charles W. Sweeney and Lt. Charles D. Albury. The name *Bock's Car* was a pun on the name of the plane's customary aviator, Captain Bock.

(*Source*: Reproduced with permission from the publisher.) Reference 17.

Fig. 37.9.   The flight paths of the Enola Gay and Bock's Car from Tinian to Hiroshima and Nagasaki, respectively, in August 1945. The planes covered the 1,500 mile span in about six hours.

(*Source*: Reproducd with permission from the publisher.) Reference 19.

cadaver — "and 100,000 more seriously injured."[e,21] As 80% of its buildings crumbled,[20] the city of Hiroshima disappeared under a thick, churning atomic cloud of flames and smoke (Fig. 37.12 and 37.13).

Years later, the B-29's copilot, Robert A. Lewis, would recall in a "faltering" voice words he wrote in his logbook at the time the bomb fell. Gazing down at the destruction and devastation, he was moved to exclaim that early August day, "My God, what have we done?"[f,27]

---

[e]Appalling as the death and serious injury tolls were, they paled in comparison with the "massive fire-bomb raid on Tokyo on March 9, 1945...[when] the Japanese suffered more fatalities than at Hiroshima."[24]

[f]Lewis died in 1983, and copilot Paul Tibbets in 2007.

Fig. 37.10.   The stone steps at the main entrance to the Sumitomo Bank just 820 feet (250 m) from the Hiroshima bomb's hypocenter. It is believed that a person sat down on the steps facing the direction of the hypocenter, possibly waiting for the bank to open. He or she was smitten by a flash of heat "well over 1000 degrees centigrade (1832 degrees Fahrenheit) and was incinerated on the stone steps, leaving behind only a shadow.

Roughly 10 years after the explosion, the shadow remained clearly visible on the stone. With more time and exposure to rain and wind, the shadow became "gradually blurred." The bank was later rebuilt. But its historic stone steps were first removed and remain preserved today at the Hiroshima Peace Memorial Museum . . . thank God, thank Buddha and the pantheon of deities worshipped by Japan's Shinto population.

(*Source*: Reproduced with permission from the publisher.) Reference 25.

August 9, 1945

Three days later, a second American B-29 bomber, dubbed *Bock's Car*, left a Tinian runway carrying another atomic bomb, this one destined to be released over Kokura, Japan. But a cloud cover — plus haze and smoke from the Hiroshima bombing — left Kokura partially obscured and forced a last-minute decision to unload the weapon above Nagasaki. That bit of fate — let's call it "Moirai" (Fig. 37.14), the personification of fate, as we're talking about Japanese history (though "Moirai" is a decidedly Greek term) — led columnist Nicholas Kristof, writing for the *New York Times* on the historical events 50 years earlier, to remark on August 7, 1995, "The smaller cloud that blocked

Fig. 37.11.   In this photo, an asphalt surface not minimally protected by a nearby structure was "scorched," while that surface falling in the shadow of the handrail was protected, a bit.

The physical thermal explanation for these two animate and inanimate protective effects lies in the superintense amount of heat during the "3 seconds" after the explosion. There is minimal thermal loss by conductivity during that brief but intense interval. This site was about 2624 feet (800 m) from the bomb's hypocenter. Within a 0.6-mile (1-km) radius, "roof tiles melted." Casualties, naturally, correlated with the distance from ground zero, as shown in Tables 37.1 and 37.2. The percent mortality followed a similar pattern within a 2 mile radius. The 20 persons killed per square mile in the 6550–9850-foot district in Table 37.2 versus the 12.5–0.0% mortality among those persons in the 6000–10.000-foot range in Table 37.1 is not explained in the accompanying text.

(*Source*: Reproduced with permission from the publisher.) Reference 26.

Mr. Yoshio's view[g] that morning was the best thing that ever happened to the city of Kokura — and the worst that ever happened to nearby Nagasaki."[29]

Bock's Car (Fig. 37.8) was named after its customary pilot, Captain Frederick C. Bock (1918–2000), but a second last-minute change of plans

---

[g] Kenji Yoshio was a 14-year-old boy at the time, "scanning the sky...for the flash that would signal the detonation of a new bomb that America had devised. But the sky was partly cloudy, so he strained to look through the patches of blue...for the American bomber or the telltale flash." And then nothing happened. "Three times, Bock's Car passed over Kokura, bomb bays open, a hum in the cockpit signaling" to the plane's young bombardier, Kermit Beahan, that "the bomb was ready for release.

But though the radar scope was locked on to Kokura, (his) orders were to drop the bomb only on visual identification of the huge arms factory that was the target." Following orders, Beahan did not drop the weapon and the plane veered toward Nagasaki...and Kokura was spared.

Fig. 37.12.   The atomic cloud hovering over Hiroshima on August 6, 1945.

(*Source*: Reproduced with permission fom the publisher.) Reference 40.

placed Major Charles Sweeney (1919–2004) in the cockpit. (Captain Bock piloted a separate aircraft designed to collect scientific data on the bomb's effects.) The payload this time was the Fat Man packing a punch equivalent to that of 21,000 tons of TNT.[30]

The Fat Man, an implosion-type bomb, was released at 11:02 a.m.[36] about 4500 feet above the city of Nagasaki, lying on the floor of Japan's narrow Urakami Valley. Fifty seconds later, it detonated at an altitude of about 1800 feet. Instantly, Nagasaki became a fireball, an inferno, a hell-on-earth "thousands of degrees hotter than the surface of the sun" and the graveyard for 74,000 of the city's 286,000 inhabitants; as another 75,000 suffered nonfatal, severe injuries. The grim totals for the pair of cities: at least 175,000 people dead, representing more than 30% of their combined populations, and 175,000 more *seriously* wounded. Figure 37.13 depicts just how destructive, how grim the bombings were, while Tables 37.1–37.5 record the astounding number of casualties and deaths of the twin bombings. Still, given the ground fires with mass cremation of countless people, the loss of official records due to the bombings and the vaporization of who knows how many people, an accurate count of the dead was not possible.

E=mc²!

Before

After

Fig. 37.13.   Death and destruction.

(*Source*: Reference 35.)

Fig. 37.14.   The three Moirai (Moira, singular) might be *thought* to have been Japanese characters, given the pronunciation of the second syllable of their name, akin to that of *bonsai*.

But they weren't.

The three Greek women were "white-robed incarnations of destiny" in Greek mythology that "controlled the metaphorical thread of life in every mortal from birth to death."

Bonsai is the Japanese art of growing miniature shrubs and trees enjoyed by so many Americans.

"Bonsai" and "banzai" are spelled similarly, pronounced a bit differently, and have decidedly meanings. "Bonsai" is pronounced like the English words "bun" and "sigh" while "banzai" — the suicidal battle cry of WWII Japanese kamikaze (神風) pilots — is pronounced like "bonds" and "eye."
And then there's "*haikai*" — the type of Japanese verse in the 19th century that evolved into haiku and haikus, the 3-line, 17-syllable verses so popular today.

(*Source*: Reproduced with permission from the publisher.) Reference.[28]

Table 37.1

Distance from point of explosion, in feet	Percent mortality at various distances from ground zero
0–1000	93.0
1000–2000	92.0
2000–3000	86.0
3000–4000	69.0
4000–5000	49.0
5000–6000	31.5
6000–7000	12.5
7000–8000	1.3
8000–9000	0.5
9000–10,000	0.0

Table 37.2   The Total Casualties — Those Killed, Injured, and Missing — and the Number Killed per Square Mile from the Point of Explosion above Nagasaki

Distance from Ground Zero (feet)	Killed	Injured	Missing	Total Casualties	Killed per square mile
0–1,640	7,505	960	1,127	9,592	24,700
1,640–3,300	3,688	1,478	1,799	6,965	4,040
3,300–4,900	8,678	17,137	3,597	29,412	5,710
4,900–6,550	221	11,958	28	12,207	125
6,550–9,850	112	9,460	17	9,589	20

(*Source*: Modified (table caption only) and reproduced with permission from the publisher.) Reference 39.

Table 37.3   Estimates of casualties.

	Hiroshima	Nagasaki
Pre-raid population	255,000	195,000
Dead	66,000	39,000
Injured	69,000	25,000
Total Casualties	135,000	64,000

(*Source*: Reproduced with permission from the publisher) Reference BB.

Table 37.4   Causes of immediate death and per-cent of total deaths

City	Cause of death	Percent of total deaths
Hiroshima	Burns	60
	Falling debris	30
	Other	10
Nagasaki	Burns	95
	Falling debris	9
	Flying glass	7
	Other	7

∗ The mathematical error was not noted in the original publication.
(*Source*: Reference 39.)

Table 37.5 Vital wartime statistics concerning size, population and number of fatalities incurred at moment of explosion and within four months (from radiation poisoning) from the time the little boy and the fat man were dropped on hiroshima and nagasaki, respectively, in august 1945.

	Hiroshima	**Nagasaki**
Population (estimated) at moment of explosion	255,000	240,000
Number killed instantly	70,000–140,000	40,000–70,000
Number dying from radiation poisoning within 4 months		
Percentage of population dead from blast (using US Army's numbers).	31%	17%
TNT equivalence	13,000 tons	21,000 tons
Type of bomb	Gun-barrel design assembly Uranium-235 core	Implosion design assembly Plutonium-239 core
Radius of total destruction	1 mile/1.6 km.	0.8 mile/1.3 km.*

* The Nagasaki bomb was a bit stronger than the Hiroshima bomb, but the area of total destruction was a bit less due to the city's hillier terrain limiting the lateral spread of the shock wave from the bomb's epicenter.

And so the war that began on December 7, 1941,[h] when Japan attacked Hawaii's Pearl Harbor, came to an end on August 9, 1945, when the U.S. and its allies dropped the second atomic bomb on Japan. The Little Boy and the Fat Man had done their jobs and, hopefully, there will never be an Act II of the drama.

---

[h] The war, of course, actually began on September 1, 1939, when Nazi Germany invaded Poland. It was Albert Einstein, a German Jew, who fled Germany fearing for his life as Adolph Hitler and the German war machine mustered power in the 1930s, who came to be the man, as history would have it, who (unwittingly) would end WWII through two decimating nuclear explosions. The Japanese armed forces were, in the end, beaten by one genius, three letters, and a superscript (Fig. 37.13), and more than 100 million military personnel. Among those 100 million personnel were 408,306 American troops* who never got to pass the story on to their wives and families. (* This number does not include the 670,846 American servicemen wounded who survived.[37] Nor does it include the estimated 50,000,000 civilian casualties suffered by the 24 Allied nations.[36–39] Those countries were, in order of diminishing casualties: the Soviet Union, Poland, Yugoslavia, Czechoslovakia, the United States, the United Kingdom, the Netherlands, Greece, Belgium, Canada, India, Australia, Albania, Bulgaria, New Zealand, Norway, South Africa, Ethiopia, Luxembourg, Malta, Denmark, and Brazil.[38])

Table 37.6 Size and population of the "Big Three" allied countries and Japan at the time of WWII.

	United States	United Kingdom	Union of Soviet Socialist Republics (Soviet Union, Russia)	Japan
Size (sq. miles/km²)	3,717,813/ 9,629,090	94,060/ 243,610	8,649,538/ 22,402,206	145,925/ 377,944
Population	312,441,000	62,262,000	293,047,571	128,056,028

(*Sources*: References 31, 32, 33–34.)

The Beginning of the End in the Sky High Above Hiroshima…and a Joyous Ending on the Streets of N.Y.C.

Fig. 37.15. The atomic cloud hovering over Hiroshima on August 6, 1945.

(*Source*: Reproduced with permission from the publisher.) Reference 40.

Fig. 37.16.   The death and destruction shown four pages earlier may be offset just a bit by this Times Square photo shot by Alfred Eisenstaedt of *Life* on V-J Day, 1945. The photo legend reads, "A white-clad girl clutches her purse and skirt as an uninhibited sailor plants his lips squarely on the lips of an unknown girl."

(*Source*: Reproduced with permission from the publisher.) Reference 41.

References

1. http://en.wikipedia.org/wiki/Trinity_nuclear_test
2. http://en.wikipedia.org/wiki/Trinity_site
3. Holy sonnets. XIV Chambers. EK(ed.). (1896) *Poems of John Donne.* Lawrence & Bullen, London. Vol. I. P. 165.
4. http://en.wikipedia.org/wiki/Holy_Sonnets
5. http://www.atomicarchive.com/History/mp/p4s26.shtml
6. http://en.wikipedia.org/wiki/Scholastic_Aptitude_Test#2008_changes
7. http://www.geni.com/people/James-Bryant-Conant/6000000002804522196).
8. http://www.newworldencyclopedia.org/entry/Manhattan_Project
9. http://en.wikipedia.org/wiki/J._Robert_Oppenheimer
10. http://en.wikipedia.org/wiki/Bhagavad_Gita

11. From the 1965 television documentary *The Decision to Drop the Bomb.*
12. *Acta Medica Scand Suppl* This was in the 1950s.
13. http://en.wikipedia.org/wiki/Tinian
14. Tinian, in *Encyclopaedia Britannica,* 2010 ed.
15. http://www.scribd.com/doc/279834/Tinian-Island
16. http://en.wikipedia.org/wiki/Enola_Gay
17. http://en.wikipedia.org/wiki/Bockscar
18. http://www.bing.com/search?q=wikipedia+enola+gay&src=IE-Address
19. htpp:///en.wikipedia.org/wiki/Hiroshima_bombing
20. http://www.usaaf.net/ww2/hittinghome/hittinghomepg14.htm.
21. htpp://cni-guide/history/ww2/12/main/12.html
22. http://en.wikipedia.org/wiki/Hiroshima_bombing
23. http://www.usaaf.net/ww2/hittinghome/hittinghomepg14.htm
24. *Encyclopedia Americana,* international edition. Scholastic Library, Danbury, CT 2006, vol. 2, pp. 641–642.
25. http://www.gensuikin.org/english/photo.html
26. http://www.gensuikin.org/english/photo.html
27. http://en.wikipedia.org/wiki/Robert_A_Lewis
28. http://en.wikipedia.org/wiki/Moirai
29. Kristof ND. (1995) *The New York Times,* August 7. Kokura, Japan: Bypassed by A-bomb.
30. http://en.wikipedia.org/wiki/United_States_Army_Air_Forces
31. http://en.wikipedia.org/wiki/United_States
32. http://en.wikipedia.org/wiki/United_Kingdom
33. http://en.wikipedia.org/wiki/Soviet_Union
34. http://en.wikipedia.org/wiki/Japan
35. http://www.gensuikin.org/english/photo.html
http://www.nationalmuseum.af.mil/factsheets/factsheet.asp?id=2549
36. http://www.americanwarlibrary.com/allwars.htm
37. http://warchronicle.com/numbers/WWII/deaths.htm
38. http://www.cddc.vt.edu/host/atomic/hiroshim/hiro_me
39. http://en.wikipedia.org/wiki/Hiroshima_bombing
40. http://en.wikipedia.org/wiki/V-J_day_in_Times_Square

38

CAMILLO GOLGI AND SANTIAGO RAMÓN Y CAJAL: BITTER RIVALS TO THE END

That makes us wonder, How *did* neurobiologists examine nerves, how *did* they track individual neurons densely packed in a nerve or in the brain before November 2007?

Neurobiologists performed their tedious work using a black-and-white staining technique strikingly different from the colorful Brainbow technique. We've all seen old black-and-white movies, even silent B&W ones — and enjoyed them knowing that we could quickly change channels or pop in a color video at any time. And we feel a sense of appreciation and respect for the actors and actresses and producers and directors who made those relics. Now, we need to "sublimate" those feelings, if we can, to apprehend how a neurobiologist felt — from 1872 to the early 21st century — when looking into his microscope day in, day out at B&W images of neurons in the brain or in a nerve as shown in Fig. 38.1 until a full-fledged color image became available (in transgenic mice) in November 2007.

Camillo Golgi and Santiago Ramón y Cajal

The old black-and-white method of staining neurons was introduced by Camillo Golgi (1843–1926) (Fig. 38.2), an Italian pathologist who studied medicine at the University of Pavia, where he received his M.D. degree in 1865.[1] He soon saw that the tissue-staining techniques then available were "inadequate for studying nervous tissue."[1] Moved to improve the available methods, he developed an alternative technique that impregnated (infused, saturated) neurons with silver (silver staining) by simply soaking brain tissue

595

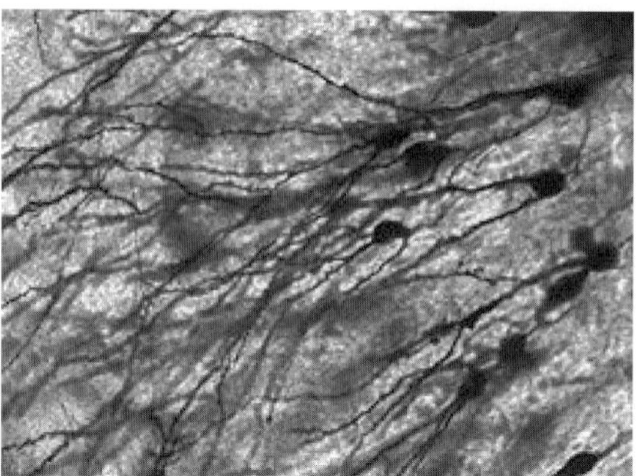

Fig. 38.1.　These are neurons in a human brain stained with the widely used Golgi staining method developed by Camillo Golgi in 1873.[1] The Golgi technique remains in widespread use in the early-21st century. (The dazzling Barinbow technique of coloring neurons within the central nervous system of transgenic mice is considered by one of its principal investigators, Josh Sanes at Harvard, to "unfortunately" not have "any application to human tissue at this point."[2] The black, oval cell bodies (somas) of the neurons give rise to slender, black axons that transmit the soma's electrochemical impulses to other neurons in synapses.

(*Source*: Reproduced with permission from the publisher.) References 1 and 3.

Fig. 38.2.　The white-haired Camillo Golgi at work in his cluttered laboratory at the University of Pavia *circa* 1920.

(*Source*: Reproduced with permission from the publisher.) Reference 6.

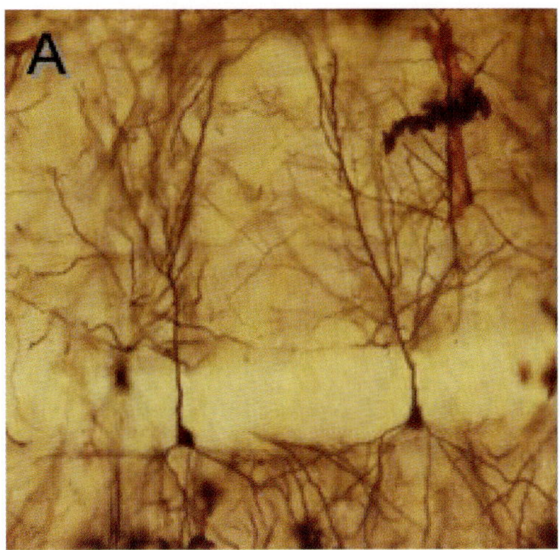

Fig. 38.3.   An example of the Golgi method of staining brain neurons. The filamentous axons and dendrites of the nerve cells are clearly visible when stained dark-brown to black and viewed against a transparent yellow–orange background.

(*Source*: Reproduced with permission from the publisher.) Reference 4.

in a silver (chromate) solution for four days.[4] (Nearly simultaneous work by photographers with an artistic bent led to the introduction of "silver staining" in first black-and-white still photography and later motion pictures.[5] The result was that the cell body and axon of a neuron stained a "stark black" color, generating an "exceedingly clear and well contrasted picture of a neuron [when viewed] against a yellow background,"[1] as portrayed in Fig. 38.3.

As it turned out, Golgi's new tissue-staining technique was discovered "quite unexpectedly,"[7] just as the discovery of penicillin by Alexander Fleming in 1928 and of superoxide dismutase in 1969 (see Chap. 4) by McCord and Fridovich were "by accident." (The quotation marks are to remind us of Louis Pasteur's words more than 150 years ago: "Chance favors the prepared mind."[8]) Fleming (unintentionally) left a culture of staphylococci exposed to laboratory air for two weeks while on vacation. When he returned, he saw that the culture dish had been contaminated by a mold, presumably present in the air. The staph grew luxuriantly everywhere *except* in the zone immediately adjacent to the mold. He put two and two together, and figured that "the mould must have been producing some sort of diffusible substance, which killed or inhibited the growth of the bacteria."[9] A little more work disclosed that the mold was the fungus *Penicillum notatum*.[10] And so penicillin was born.

An accident on purpose

In Golgi's case, the "accident" was that he (intentionally) applied a silver solution developed by someone else to the thin, delicate membrane (pia mater) that covers the surface of the brain. Upon doing so, he "accidentally ["unexpectedly" might be a better word, eh?] saw single nerve cells in the adjacent brain substance which were stained dark-brown by the…caustic silver nitrate solution." That "led him to continue with this method…[of staining] single nerve cells"[1] (Fig. 38.1) The real beauty of the Golgi method,[a] however, was and still is that it "stains a limited number of [nerve] cells at random [yet] in their entirety."[11] Or, restated, a few *individual* nerve cells are stained haphazardly, making staining-by-chance desirable. How this happens remains "largely unknown" to this day.[11]

Cajal refines golgi's staining method and applies it to a host of small animals

Golgi's contribution to science and medicine was matched by that of Santiago Ramón y Cajal (1852–1934), a contemporary Spanish anatomist. (An anatomist is a biological scientist who studies the structure — the anatomy — of a living organism, very often a human being. He or she usually holds a Ph.D. degree in a particular basic science, such as genetics, cellular biology, neuroscience, or histology, among a number of other disciplines.[13] A pathologist, on the other hand, holds an M.D. degree and focuses on human beings with disease — be the persons living or not — although veterinary and plant pathologists certainly exist.) Cajal, however, was both an anatomist and a pathologist.[14] In 1887, he was introduced to the Golgi method. His dual science background expedited his application of the staining method developed

______

[a] Golgi prepared a manuscript in the late 1880s detailing his new method of staining neurons. He entitled it "*Studi sulla fina anatomia degli organi centrali del sistema nervoso*" ("Studies on the fine anatomy of the organs of the central nervous system"). The paper won first place in an 1880s competition among European neuroscientists. It was deemed a "grand and extremely meticulous work illustrated by an atlas so rich, clear, and elegant that in itself would be worthy of the prize."[12]

by Golgi to *individual* neurons. So amazed was he by the "marvelous staining of nerve cells that he wondered why this discovery had not yet formed the basis of a scientific revolution."[11]

Cajal immediately began to apply his adaptation of the Golgi method and created his own revolution by "examining practically the whole nervous system" using the Golgi stain in an array of small animals.[11] A makeshift lab in his kitchen became the hub of his research (Fig 38.4).[15] One year later, he published his "first important article" on the subject.[12] Many years later (in 1917), as he reflected on the Golgi staining method during the twilight of his career, Cajal would recall his feelings the first time he saw brain neurons colored black by the Golgi stain. He wrote in his autobiography, "I expressed… the surprise which I experienced upon seeing with my own eyes the wonderful revelatory powers of the chrome-silver reaction and the absence of any excitement in the scientific world aroused by its discovery." [How surprised might Cajal — shown in his home laboratory in Fig. 38.4 — have been to

Fig. 38.4. Santiago Ramón y Cajal in his home lab (note the sulfuric acid next to the olive oil and vinegar). Using the tools of his trade — a simple light microscope, a pen, and a piece of paper to draw what he saw — he examined specimens of brain tissue and formulated revolutionary concepts.

(*Source*: Reproduced with permission from the publisher.)

see "with his own eyes" the Technicolor Brainbow images reproduced several pages ago and how pleased might he have been to witness the excitement generated in today's scientific world?) "How can one explain such strange indifference? Today, when I am better acquainted with the psychology of scientific men, I find it very natural."[16]

Awe, yet completely opposite conclusions

Despite Cajal's apparent awe of the Golgi silver-staining technique, the two Mediterranean pathologists reached "completely opposite conclusions" regarding the structure of the central nervous system, as studied using the Golgi stain.[17] Golgi held that one nerve is connected to the next and it to another nerve, and so on, forming a continuous network, akin to the uninterrupted network of arteries, capillaries, and veins that form the circulatory system. Cajal, on the other hand, championed the notion that neurons forming circuits in the brain — or long nerves in the periphery — "are not continuous with one another and must communicate by contact, not continuity,"[17] much like runners in a relay race passing on a baton. This scientific difference of opinion led to a personal dislike of each other — even scientists have their faults and foibles, and sometimes egos that need nurturing.

Albert Von Kölliker

Meanwhile, back in 1889, Albert von Kölliker, a Swiss anatomist *and* pathologist,[18] who was largely educated in Germany, was being recognized as one of the greatest histologists of that time (histology is simply the study of anatomy at the microscopic level, as opposed to "gross" anatomy). von Kölliker was so impressed by Cajal's discoveries that he said to him, "The results that you have obtained are so beautiful that I am planning to undertake a series of confirmatory studies immediately, adopting your technique. I have discovered you, and I wish to make my discovery known in Germany."[11] (Germany was the undisputed epicenter of neuropathology at the turn of the 20th century.) The "beautiful" personal discovery von Kölliker referred to was really the sum of an innovation — Cajal's application of the Golgi method (itself a remarkable discovery) of staining nerve cells that enabled him to study individual neurons "without much overlap from other cells in the densely packed brain"[4] — and Cajal's wondrous artwork. (Figures 38.5–38.7)

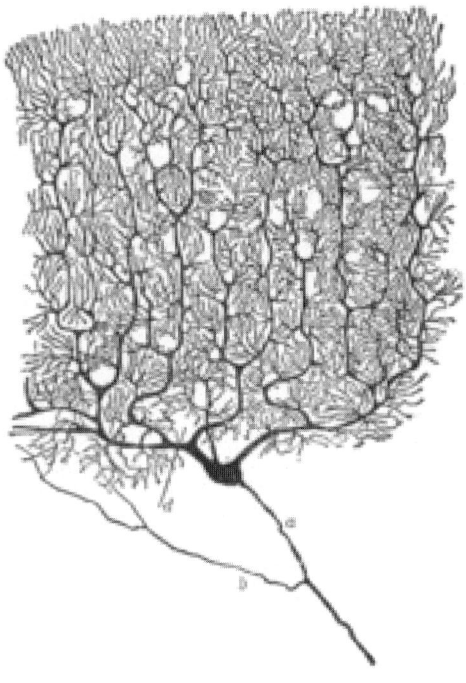

Fig. 38.5.　A freehand drawing of a Prukinje cell in a kitten brain following application of the Golgi stain. It is also an elegant application of artistry to neurobiology. The single, globular cell body gives rise to a massive maze of branching dendrites and a single axon running diagonally in the lower right quadrant of the drawing.

(*Source*: Reproduced with permission from the publisher.) Reference 18.

are masterly drawings by Cajal of a single Purkinje cell[b] (Fig. 38.5) and multiple Purkinje cells (Figs. 38.6 and 38.7) after application of Golgi's de los centros nervosa de las aves. SOURCE Figure "37.17" i.e. Figure 38.5 is Gray's Anatomy of the Human Body 1918 or www.prohealthsys.com/anatomy/grays/neurobiology/structure.php Source Figure "37.18" i.e. Figure 38.6 is

---

[b]A Purkinje cell is a neuron in the brain that transmits electrochemical impulses. The cell is named after its discoverer, Czech anatomist Jan Evangelista Purkyn .[21] The function of Purkinje cells is poignantly illustrated by what happens when a significant number of Purkinje cells are lost, most frequently the result of alcohol abuse or, more specifically, the vitamin B_1 (thiamine) deficiency that goes hand in hand with chronic, excessive alcohol use. A condition known as the Wernicke–Korsakoff syndrome (Aka "wet brain"[22] often develops. It is characterized by a number of neurologic symptoms, perhaps most notably a profound memory loss.[21]

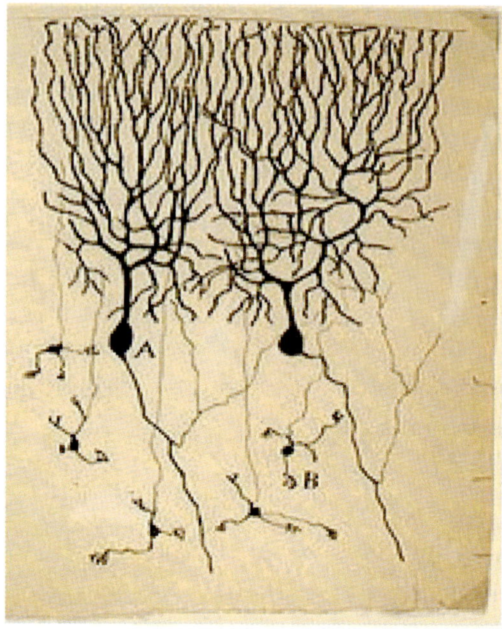

Fig. 38.6.　A drawing of two Purkinje cells in the brain of a pigeon by Santiago Ramón y Cajal. (*Source*: Modified and reproduced with permission from the publisher.) Reference 19.

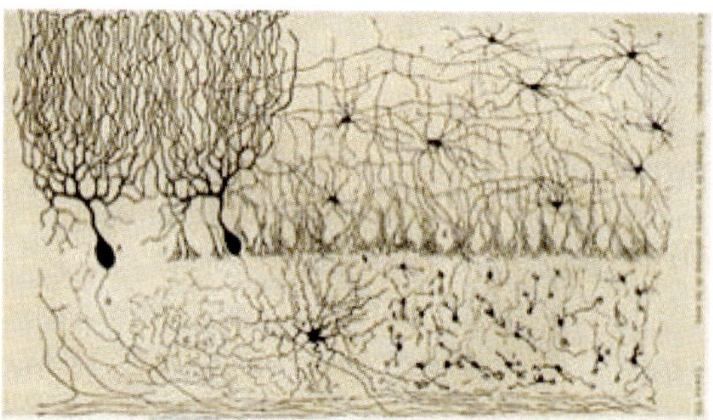

Fig. 38.7.　A drawing of Purkinje cells in the brain of a chicken by Santiago Ramón y Cajal. (*Source*: Reproduced with permission from the publisher.) Reference 19.

Cajal R. Estructura *Rev. Trim. Histol. Norm. Patol.* 1888; 1:1–10. Stain in a kitten, pigeon, and chicken brain, respectively.] In other words, Cajol held that neurons are "discrete cells" *not* connected in a network like telephone wires, possibly a communication system totally unknown or barely known to Cajal, who graduated from medical school in Madrid in 1883,[23] just seven years after Alexander Graham Bell invented the telephone in the U.S.[24]

Cajal's 19th Century Drawings Versus 20th Century Photographs Versus 21st Century Brainbow Images

Just how exquisite Cajal's late 19th century, freehand drawings of Purkinje cells are — and his transfer of what he saw through his microscope onto paper — may be best realized by comparing them with a *photograph* of the brain made in the 20th century after staining cerebral neurons using the Golgi technique. Figure 38.8, for example, is a photograph of a tiny slice of the cerebral cortex from a mouse stained by the Golgi method. Cajal deserves kudos for his artwork which so closely matches the eye of a camera, don't you think? However, drawback of Golgi staining is that it labels all neurons in one color — black. The upshot of single-color labeling is that if "too many cells are labeled, their branching axons and dendrites become impossible to untangle"[26]. A better way was needed to identify easily the individual neurons that make up a given brain circuit at one time.

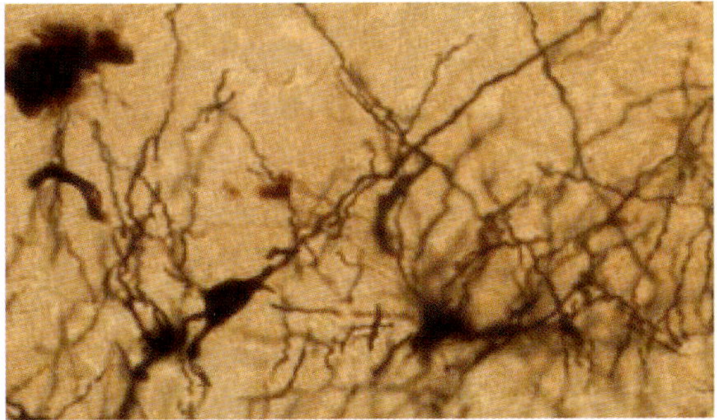

Fig. 38.8. A photograph of Purkinje cells in the cerebral cortex of a mouse stained by the Golgi method. The background hue is similar to that in Fig. 38.6.

(*Source*: Reproduced with permission from the publisher.) Reference 25.

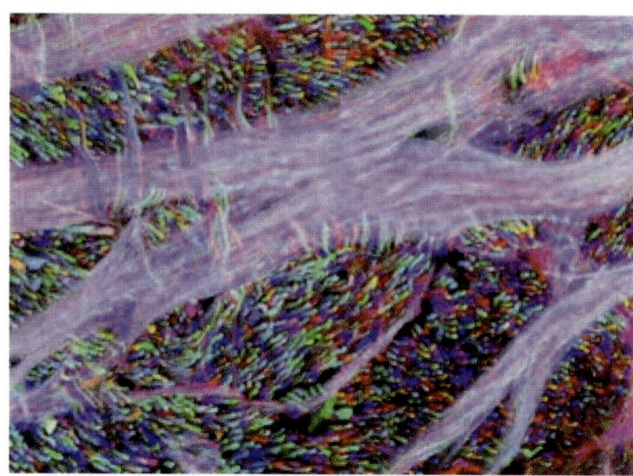

Fig. 38.9. A montage of colorful neurons from the brain stem of a Brainbow transgenic mouse. In Brainbow mice, neurons are randomly stained with an array of fluorescent proteins, allowing each to glow in a different hue. This multicolor labeling system provides a visually appealing way to distinguish even adjacent and nearby neurons and to visualize neural circuits in the brain — appealing enough to capture the 2008 Nikon Small World Prize. The Brainbow scientist who first saw this colorful image was Tammy Weissman.

(*Source*: Reproduced with permission from the publisher.) Reference 27.

But neither Cajal's wondrous drawings or later photographs of Golgi-stained Purkinje fibers are as easy to look at as a multicolored image of the brain stem from a Brainbow transgenic mouse (Fig. 38.9). Entire brain circuits can be visualized in about 90 distinct fluorescent hues. Intertwining nerve fibers can be followed no matter how many twists and turns they make, how tangled they become. The Brainbow technique corrects the major shortcoming of the clever, randomly staining Golgi method — not being able to identify at a glance separate neurons constituting a brain circuit.[27]

Bitter Rivals to the End

It might make intuitive sense that two men less than ten years apart in age, from two Mediterranean countries, *and* two men with a common interest in a (esoteric) field would become the best of friends. Yet, reality didn't follow intuition during the lives of Golgi and Cajal. It was, in fact, just the opposite. Reality mocked intuition.

In real life, alas, they became and remained "bitter rivals to the end."[28] At the root of the rivalry was the fact that the majority of Cajal's papers were published in Spanish scientific journals. That became a concern because leading neuroscientists of the time were German or English and they preferred to publish their works in German and English scientific journals, certainly not Spanish ones.

As a result, Cajal's writings were "largely ignored. Who was to pay attention to the daring theories of an unknown professor from a Spanish university?."[7] Cajal soon decided to somehow get into the international neuroscience circuit. But international travel took money, of course. Universities often reimburse faculty members for education-research-related travel expenses. But, in 1889, Cajal's home university — then the University of Barcelona — turned down his request for a travel stipend.[c] Miffed, he made up his mind to foot the bill of a trip to Germany "out of his own pocket."[29] Once in Berlin, he soon captured "the attention of Professor von Kölliker,[d] the patriarch of German neuroscience, who was so impressed by Cajal's discoveries that he took the brilliant Cajal (Fig. 38.10) under his wing and began to promote the Spaniard's detailed artwork.[33] From that day on, "things began to change for Cajal."[33] Other European scientists soon started to give him "due recognition" for his work.

During the 1890s and the early years of the 20 century, Cajal's research flourished and his fame spread. In 1900, he received the coveted Moscow Prize[33] at the International Congress of Medicine in Paris. Upon his return home, the Spanish government announced that it would subsidize a research laboratory at the University of Madrid, where he had moved in 1892. Five years later, in 1905, he traveled back to Germany — this time on Germany's pocketbook — where he garnered the lauded Helmholtz Medal (Fig. 38.11) from Germany's Royal Academy of Sciences. That award is given every two

---

[c] Even in 1889, scarce money limited scientific research. During the preparation of a manuscript, Cajal adopted a technique that cut down the number of pages displaying his meticulous drawings of Purkinje cells.[31] (still today, medical journal editors stress brevity to conserve space. Manuscripts are often limited to 15, double-spaced pages, including figures and references.) He carefully combined multiple drawings of neurons in a given region of the brain into one image. This amounted to a cut and-paste way to create an illustration — without a computer!

[d] Born Albert Kölliker; the honorary title "von", implying nobility, was added to his name in Germany and Austria.[32] von Kölliker's contributions to histology (the microscopic examination of animal tissues) were widespread."[32] He studied a variety of tissues including "smooth muscle, striated muscle, skin, bone, teeth, blood vessels and other viscera."[32] Sounds reminiscent of the composition of a teratoma. (See Fig. 28.8 of Chapter 28.)

Fig. 38.10. Cajal not only drew delicate Purkinje cells coaxed from the brain and spinal cords of common domestic animals, he sketched a self-portrait while gazing into his beloved instrument.

(*Source*: Reproduced with permission from the publisher.) Reference 30.

Fig. 38.11. The Helmholtz Medal

(*Source*: Reproduced with permission from the publisher.) Reference 34.

years to the person who "has made the most significant discoveries in any field of human knowledge."[34]

But the most acclaimed reward for his years of underpaid research and meticulous drawings would come in 1906.

Stockholm, December 1906

The telegram, written in German, from Sweden's *Carolinische Institut* (Caroline Institute), arrived in Pavia and Madrid in October 1906. It invited both men to be in Stockholm in December to receive the most coveted of all awards — the Nobel Prize.[33]

The award had existed for just five years in 1906.[e] This would be the first time the Prize would be shared, although shared awards would become common in the future. (It would also be the first time Golgi and Cajal met.[45] The person who nominated both Golgi and Cajal (Fig. 38.14) was Albert von Kölliker,[11] the man who had been so greatly impressed back in 1889 by Cajal's detailed drawings of the brain as seen under a microscope after Golgi staining. The Swiss pathologist had become equally enchanted by the new-at-the-time

---

[e] The Nobel Prize sprang from the will of Alfred Nobel,[36] who was born in Stockholm in 1833 and died in San Remo, Italy. He moved to Italy in 1891 from his beloved Sweden for his health, aggravated by Sweden's winter weather.[37] He died in Italy on December 10, 1896. All awards have been made on the anniversary of his death.[37] Nobel was a multilingual — Swedish, Russian, French, German, and English — chemist. He was also a pacifist[38] — a pacifist in spite of his family's wealth coming from the manufacturie of explosives particularly nitroglycerin (NTG), the same medication taken (in tiny amounts) under the tongue by so many for the quick relief of heart pain (angina) due to coronary heart disease. It is the vasodilating properties of NTG that relieve angina by increasing blood flow to oxygen-deprived heart muscle…and can lead to explosive headaches. Nobel did not discover NTG, but he established that when it was mixed with gunpowder, that led (perhaps not surprisingly) to an even more powerful compound — dynamite.[39] Nobel's will called for the establishment of five prizes. (A picture of the Nobel Medal is displayed in Fig. 38.12 and a photograph of the first Nobel Award ceremony in 1901 is shown in Fig. 38.13.) The categories were to be Physics, Chemistry, Physiology or Medicine, Literature and Peace. (A sixth category — Economic Sciences — was added in 1969.[40]) The recipients were to be "those who, during the preceding year, shall have conferred the greatest benefit on mankind."[41] Although Swedish through and through, Nobel specified, with regard to the annual distribution of capital, that "it is my express wish that in…awarding the prizes no consideration whatever shall be given to the nationality of the candidates, so that the most worthy shall receive the prize, whether he be from Scandinavia or not."[41] The pronoun "he" was (rightfully) interpreted to apply to women as well as men in this case. Marie Curie, in fact, won the third Nobel Prize in *Physics* in 1903 and then the 1911 prize in *Chemistry*.[42] The first Nobel Prize was worth 150,000 Swedish kronor, the equivalent of about US$20,000, and the 2008 Prize was about 10 million Swedish kronor, the equivalent of slightly more than US$1,000,000 or 1,000,000 euros.[1,43] Yet, even at US$1,000,000 or 1,000,000 euros, the Prize is not in the same financial stratosphere as many U.S., Canadian, or European professional athletes' annual salaries.

Fig. 38.12.   A photograph of the Nobel Medal.

(*Source*: Reproduced with permission from the publisher.) Reference 35.

Fig. 38.13.   A photograph shot at the first Nobel Award ceremony in 1901 at the Royal Academy of Music in Stockholm.

(*Source*: Reproduced with permission from the publisher.) Reference 44.

silver staining method for studying the central nervous system developed by Golgi.[46] von Kolliker was a natural for proposing both Golgi and Cajal for the 1906 Nobel Prize in Science or Medicine. The award that year was to honor their "meritorious work on the structure of the nervous system."[47,48]

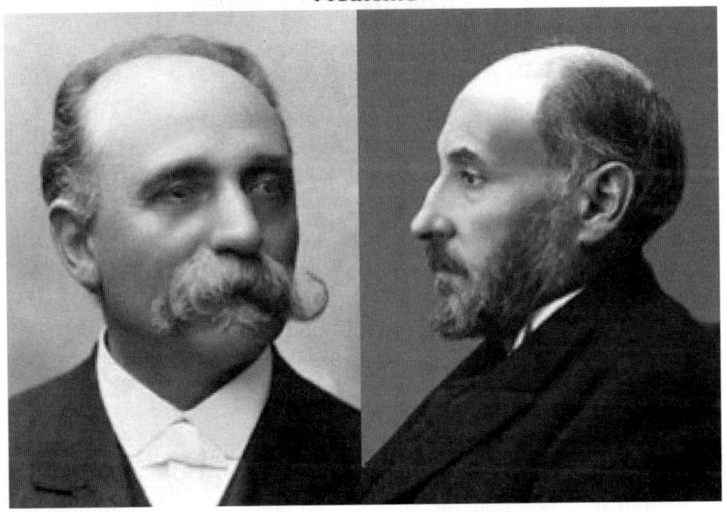

Camillo Golgi	Santiago Ramón y Cajal
(1843–1926)	(1852–1934)
Pavia University, Italy	University of Madrid, Spain
½ of the Prize	½ of the Prize

Fig. 38.14.

(*Source*: Reproduced with permission from the publisher.) Reference 48.

The atmosphere was tense and the Nobel acceptance speeches of both recipients were contentious in Stockholm.[49] One hundred years later (May 2007) at a symposium held at the *Carolinische Institut* honoring the two men, the 1906 Nobel ceremony was described as one of the most dramatic on record.[31,49] So icy was their relationship that the rivals were scheduled to deliver their speeches on separate days, presumably to avoid an in-person confrontation.[50,51]

Golgi spoke first.[f] He emphasized his firm belief that neurons in the central nervous system are "fused together"[49] to form an uninterrupted network similar to arteries and veins in the circulatory system. Cajal followed with a longer

---

[f] The Italian and Spanish rivals agreed to make their presentations in French.[11] Thank goodness for the so-often-multilingual Europeans — exemplified by the Swiss.

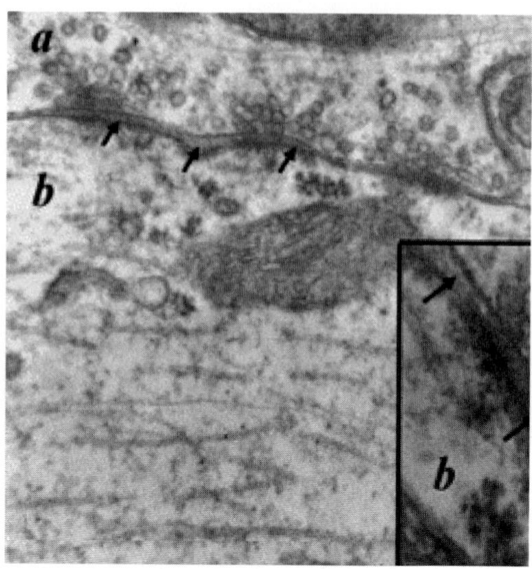

Fig. 38.15. Definitive proof that the nervous system is not a continuous network was obtained in 1954 using the electron microscope (EM). This photograph of a synapse in a human cord as seen through an EM at the standard (a) and a greater magnification (b) shows that there is a tiny space (arrows) — only 20 nm wide — between two neurons. (At 20 nm, it is 20,000,000 times less than a millimeter.) It isn't much, but it establishes unequivocally that although nerve endings attain "great intimacy," the nervous system is not continuous. Transmission of an impulse across the space is accomplished through the release of a neurotransmitter such as dopamine. "At last…there was…physical and tangible evidence of the key structure…the synapse."[52]

(*Source*: Reproduced with permission from the publisher.) Reference 52.

speech, going into considerably "more detail,"[49] reminiscent of the detail that characterized his careful drawings. (In effect, the lectures were a defense of their opposing theories regarding the basic structure of the nervous system.) He argued forcefully that neurons are *not* connected with one another. [Proof that the central nervous system is not a continuous network of nerve cells finally came in the 1950s, upon the invention of the electron microscope (Fig. 38.15),[g] just as Cajal had predicted 50 years earlier. At last, Cajal was vindicated, 20 years

---

[g] The electron microscope was, actually, developed in the 1930s, but the tiny cleft between neurons in a neural circuit awaited refinements of the delicate instrument in the 1950s.[53]

after his death.] Cajal, reflecting on the 1906 Nobel ceremony 11 years later, wrote, "What a cruel irony of fate to pair, like Siamese twins united at the shoulders, scientific adversaries of such contrasting character."[7] Then, referring to Golgi's diametrically opposite view of the fundamental structure of the nervous system, Cajal, noted for his fiery rhetoric, cranked his verbiage up a notch, saying, "The misfortune was that in defending his extravagant lucubration he made such an immoderate display of pride and self-worship that it produced a deplorable effect upon those assembled. Not even incidentally did he allude to the almost innumerable neurological works that had appeared outside Italy. ... Likewise, he considered it unnecessary to correct any of his old theoretical errors, or the lapses in his observations. The noble...Swedish neurologists and histologists looked at the speaker [Golgi] with stupefaction. I was trembling with impatience as I saw that the most elementary respect for the conventions prevented me from offering a suitable and clear correction of so many odious errors and so many deliberate omissions."[7] (This is an English translation by the authors. As such, it may not convey Cajal's emotions and choice of words as faithfully as would the original Spanish version. Yet, how disparate can the two be?[h]) After the award ceremony, the two men returned to their home countries, never to meet again.

In spite of Cajal's lambasting Golgi's Nobel acceptance speech in 1906, he had kinder words for Golgi's staining method, if not Golgi the man. In 1909, he wrote, almost lyrically, how he felt upon seeing for the first time brain tissue

---

[h] A measure of Cajal's temperament emerges from a commentary written by Ronald S. Fishman, now a retired ophthalmologist living in suburban Maryland. Fishman describes Cajal as "by nature, a born polemicist, unafraid of confrontation, a loner by inclination."[49] Rachel Katz-Sidlow, a pediatrician at the Albert Einstein College of Medicine in the Bronx, suggests that Cajal's concept of the central nervous system as being composed of individual, independent — free — neurons is but an extension of his personality.[54] She calls his 1917 autobiography[16] a "profound and brilliantly written memoir" that records his "intellectual achievements" and provides "insight into his upbringing and family life."[54] (For those who prefer an English translation, Ref.[55] will fill the bill.) Like Robinson Crusoe, his role model, Cajal "consistently challenged authority, maintained his individuality, and enjoyed working in solitude." His somewhat eccentric personality made him a bit unapproachable, even to his peers. On the other hand, as Marcus Jacobson, a professor at the University of Utah who reviewed Cajal's many publications in 1998 while writing a historical book covering the first 100 years of neuroscience, put it, "We have no right to expect a genius to be amiable. The majority of them are singularly unamiable characters, and Cajal was no exception."[56]

stained with Golgi's silver stain. Cajal modified the Golgi method — basically converting it into a double-impregnation technique — and it "produced results that he could not possibly have hoped for.…What a fantastic sight! On a yellow, completely transparent background, there appear sparsely scattered black fibres, smooth and small or thick and prickly, as well as black, triangular, star-or rod-shaped bodies! Just like fine India ink drawings on transparent Japanese paper. The scientist gazes upon it in astonishment. He is more accustomed to the chaotic images produced by [tissue stains], which yield one dubious interpretation after another. Here, on the other hand, everything is absolutely clear, without any possibility of confusion. There is nothing more to interpret; one need only observe and note these cells, with their different, ramified extensions, like plants in the morning frost.…The delighted and astonished gaze cannot tear itself away from this fantastic sight. Methodic wishful thinking has become reality. Metal impregnation has produced a magnificent and unexpected slide."[11]

Madrid, September 2006

On the 100th anniversary of the 1906 Nobel Prize, Juan De Carlos and José Borrell pondered the conflicting positions of the two men and their enormous impact on neuroscience in an article entitled "A Historical Reflection on the Contributions of Cajal and Golgi to the Foundations of Neuroscience." They wrote, "100 years after the award of the [1906] Nobel Prize (in Physiology or Medicine) to these [two] scientific revolutionaries, we can consider that if the studies of Camillo Golgi on the structure of the elements of the nervous system marked an era, either due to the method he developed or the results he obtained, the works of Santiago Ramón y Cajal were of such intensity and were so fruitful in terms of the new discoveries they produced, that we could consider him the greatest anatomist of the nervous system in his own right, and essentially the founder of modern neurobiology."[7] That conclusion favors Cajal's contribution over Golgi, yet does not negate their equal sharing of the 1906 Prize.

While the opinion is from two current members of the *Instituto Cajal* in Madrid, it coincides with that of Pedro Andres-Barquin, now a neuroscientist working for the Hoffmann La Roche pharmaceutical company in Switzerland. (While Andres-Barquin was born and educated in Spain, he has no close [i.e. financial] contact with the *Instituto Cajal*.) Reflecting on

Cajal's work nearly 100 years later, Andres-Barquin described Cajal's career as an "extraordinary scientific adventure which led him to…publish the most profound body of work by a single scientist in the history of neuro science."[57] He pointed out that, in all, the Spaniard authored more than 300 scientific papers, monographs, and books.[i] Such a prolific professional life seems like it would leave little time for a personal life. It also seems improbable that someone such as Cajal, who savored his solitude to draw neurons in great detail and write sparkling scientific papers, would be a family man. That turns out not to be the case, as belied by five children with his wife, Silveria (Fig. 38.16). Andres-Barquin summed up referring to Cajal[j] as "an incredible scientist…[and] probably the most prominent neuroscientist of all time."[k,57]

Life came to an end for Golgi on January 21, 1926.[60] He died in Pavia, Italy, just 128 miles (206 km) from Cortina, Italy, his birthplace. Cortina is a small village (population 2000) in the central Italian Alps, quite close to the Swiss border (Fig. 38.17). It lies in a scenic valley in Brescia province, which itself is in the Lombardy Region of northern Italy, encircled by the Dolomites. (The Dolomites are a particularly picturesque range of mountains in the heart of the Italian Alps.) Pavia is a larger municipality (population 71,000) km 22 miles (35 km) south of Milan, also in the Lombardy Region (Pavia province). It is the home of Pavia University, where Golgi created and taught his famed technique.

---

[i] All Cajal's writings were in Spanish, French, or German, none in English, leading to numerous English translations of his works. His clarity of thought in print was translated into his exceptionally clear drawings. And he even had the perspicacity to construct "composite drawing[s] of several [photo] micrographs [to conserve space] and present a clearer, more persuasive, scientific picture."[49]

[j] Like so many Spanish family names, "Ramon y Cajal" is made up of the paternal surname ("Ramon") followed by the maternal surname ("Cajal")[58] to foster unity — akin to the use of hyphenated surnames in the 21st century.

[k] Senior European scientific writer Alison Abbott of the superb science journal *Nature*, noted that when Cajal first gazed through his microscope's lenses at individual nerve cells stained by the Golgi method, he exclaimed, "All was sharp as a sketch with Chinese ink."[59] So hooked was Cajal that he promptly "dedicated [himself] to neuroanatomy," giving up "casino visits, social outings. . .[even] painting, his first passion."[59] Through his "artistic eye," Cajal was able to see brain cells individually, not as a continuous network of neurons, as had Golgi and others. And *that* observation permitted him to infer that "neurons receive information at one end and convey it in [a single direction] to [their] junction[s] with the next [nerve] cell."[59]

Fig. 38.16.   A portrait of the Cajal family, made in 1897. From left to right are Luis, Pilar, Jorge, Paula, Santiago, his wife Silveria, and Fe.

(*Source*: Reproduced with permission from the publisher.) Reference 57.

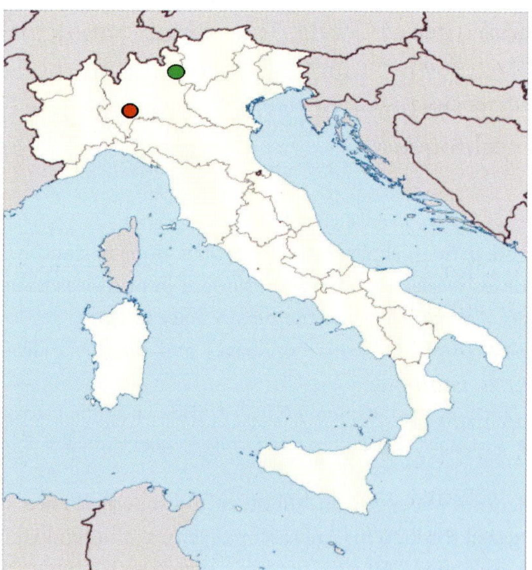

Fig. 38.17.   Cortina (green) and Pavia (red), Italy. Both towns lie in the beautiful Italian Alps, where Golgi was born, worked, and died.

(*Source*: Reproduced with permission from the publisher.) Reference 61.

Fig. 38.18.   This marble statue of Camillo Golgi is situated among the old buildings in the center of the University of Pavia.

(*Source*: Reproduced with permission from the publisher.) Reference 62.

Life After Death

Although life ended for Golgi in 1926, his spirit lives on in Pavia. A marble statue now stands among the old buildings at the University of Pavia (Fig. 38.18). Its base bears the following inscription: "*Camillo Golgi/patologo sommo/della scienza istologica/antesignano e maestro/la segreta struttura/del tessuto nervoso/con intenta vigilia/sorprese e descrisse/qui operò/qui vive/guida e luce ai venturi/MDCCCXLIII — MCMXXVI.*" Translated into English it reads, "Camillo Golgi/outstanding pathologist/of histological science/precursor and master/the secret structure/of the nervous tissue/with strenuous effort/ discovered and described/here he worked/here he lives/here he guides and enlightens future scholars/1843–1926."[63]

Life ended for Cajal in 1934. Cajal, like Golgi, was born in a small, mountain town. His birthplace was Petilla, Spain, a village in the Pyrenees.[64] Yet the

Fig. 38.19.   Known as "*el Lapicero*" ("the Pencil") in Spain, this tall, marble statue of Santiago Ramón y Cajal stands in central Madrid.

(*Source*: Cropped and reproduced with permission from the publisher.) Reference 65.

rivalry[l] was maintained by his champions, who erected a marble statue in his honor in 1931, three years *before* his death, five years following Golgi's.[64] And, Cajal's followers commissioned his statue — later placed in the "Cajal Garden" in the center of Madrid (Fig. 38.19) — to be considerably taller than Golgi's. In Spanish it is called "*el Lapicero*" ("the Pencil" in English). On the day the sculpture was dedicated, the Cajal disciple who orated at the commemoration ceremony told the crowd that Cajal's cherished wish would be for them "to study"…not admire his statue.[65]

---

[l]While the two giants of neurobiology quibbled during their lives and their followers built huge statues of them, the tiny South Asian country of Bhutan issued a tiny stamp posthumously in their honor (Fig. 38.20).

Fig. 38.20. The tiny South Asian country of Bhutan, however, had no difficulty honoring both men — sit issued a stamp bearing their images in 2010.

(*Source*: Reproduced with permission from the publisher.) Reference 66.

One month before his death, Cajal made his will. In the handwritten — of course — document he instructed that his works be preserved in the institution he founded in 1902. Initially, they were, although they were "not well curated."[66] During Spain's Civil War (1936–39), many drawings were "lost," only to be later found amidst "rubble in the [institution's] basement."[67] Some of his drawings and letters were then exhibited at a small Madrid museum from 1945 on, but most remained in museum basement boxes. Finally, in the late 1990s, a decision was made to construct a new, larger museum. It was to have opened in 2002, the 150th anniversary of Cajal's birth. But some disappointed members of Cajal's family who "never accepted that he left his entire scientific and artistic legacy to the institute in Madrid" that he formed stymied all efforts to construct a new building to house all Cajal's works.

In 2007 — more than 70 years following his death — the Spanish government stepped in and built just such a museum to display his more than 300 well-written scientific works, his entire collection of unparalleled hand drawings, his 3700 microscopic slides, his many, many paintings, and his photographic equipment — *Everything!* Today that building is called the *Consejo Superior de Investigaciones Científicas*, or C.S.I.C. (Translated into English, the "Spanish National Research Council," also sometimes referred to as the "Cajal Institute.") For those traveling in Europe, its street address is 37 Avenida del Doctor Arce, 28002 Madrid, Spain.

Shimomura, Chalfie, and Tsien

These three men are the ones whose names, it was promised, would "come up again." They are the three chemists who individually contributed to the use of green fluorescent protein as a biological marker in living cells. This application may one day allow tracking of tumor metastases (spread) — without actually removing any tissue — in human beings. Green fluorescent protein is a safe way to illuminate living cells, as opposed to other fluorescent molecules, which are "phototoxic when used in living cells."[68] Phototoxicity refers to the property of "rendering the skin abnormally sensitive to light,"[69] i.e. rendering the skin easily damaged by sunlight. The prefix "photo," in this context, refers to a "photon" — an "elementary particle, the 'basic' unit of light."[70] (A laser emits a steady stream of photons.) Phototoxicity is caused by "oxygen free radicals" (Chap. 4) when "a fluorescent molecule is introduced into living cells."[71]

For their scientific works, Shimomura, Tsien, and Chalfie and (Fig. 38.21) were also invited to Stockholm, 102 years after Golgi and Cajal. There, on December 10, 2008, the king of Sweden presented each with the Nobel Prize in Chemistry for "the discovery and development of the green fluorescent protein."[72] The jellyfish in Friday Harbor never could have imagined that their luminescence would one day brighten the brain of mice — and maybe one day

Fig. 38.21.   The three chemists who won the 2008 Nobel Prize in Chemistry for "taking the ability of some jellyfish to glow (green) and transforming it into a ubiquitous tool of molecular biology for watching the dance of living cells and the proteins within them." Osamu Shimomura is now an emeritus professor at the Marine Biological Laboratory in Woods Hole Massachusetts and Boston University medical school, Roger Tsien is a professor of pharmacology, chemistry, and biochemistry at the University of California, San Diego, and Martin Chalfie is a professor of biological sciences at Columbia University in New York City.

(*Source*: Reproduced with permission from the publisher.) Reference 73.

in man — and lead to a Nobel Prize. *The New York Times* reported that Shimomura was recognized for identifying the green fluorescent protein in jellyfish and showing that "it glowed green under ultraviolet light," Chalfie for demonstrating that when the "green gene" is inserted into the DNA of a living organism the fluorescent protein functions as a "biological identification tag," and Tsien for taking the biological coloring scheme proteins a step further by creating additional fluorescent proteins allowing scientists to "track" more than one cellular process at the same time.[58] Upon returning to New York, Chalfie said a high-school acquaintance he had not seen in many years wrote him and told him that "several of [her] friends had crushes on you." In a return letter, Chalfie quipped, "Why are you telling me this now? Back then it would have been useful information."[59] And now, it's only a matter of time until the Brainbow team members (Fig. 38.22) travel to Stockholm to receive their Prize!

It shouldn't take long.

Fig. 38.22.   The seven members — all neurobiologists in the Department of Molecular and Cellular Biology at Harvard — of the Brainbow team. They are, from left to right, Ju Lu, Jeff Lichtman, Jean Livet, Joshua Sanes, Tamily Weissman, Ryan Draft, and Hyuno Kang.

(*Source*: Reproduced with permission from the publisher.) Reference 26.

References

1. http://en.wikipedia.org/wiki/Camillo_Golgi
2. Sanes J. (2012) Personal communication, Nov. 5.
3. Creative Commons Attribution-Share Alike 2.5 generic license.
4. Friedland DR, Los JG, Ryugo DK. (2006) A modified Golgi staining protocol for use in the human brain and cerebellum. *J Neurosci Meth* **150**: 90–95.
5. http://en.wikipedia.org/wiki/Photography
6. Muscatelllo U. Golgi's contribution to medicine. (2007). *Brain Res Rev* **55**: 3–7.
7. De Carlos JA, Borrell J. (2007) A historical reflection of the contributions of Cajal and Golgi to the foundations of neuroscience. *Brain Res.* Rev **55**: 8–16.
8. http://en.wikiquote.org/wiki/Louis_Pasteur
9. Horrick D, Butcher L. *Nelson Modular Science*, Nelson Thomas, Cheltenham, UK, p. 50.
10. http://history1900s.about.com/od/medicaladvancesissues/a/penicillin.htm
11. Grant G. (2007) How the 1906 Nobel Prize in Physiology or Medicine was shared between Golgi and Cajal. *Brain Res Rev* **55**: 490–498.
12. Mazzarello P. Buchtel HA, Bandiani A. (2009) *The Hidden Structure: A Scientific Biography of Camillo Golgi*. Translated and edited by Buchtel HA, Badian A. University of Michigan and University of Rome "*La Sapienza*," Oxford University Press, Chap. 10, pp. 118–141.
13. http://careers.stateuniversity.com/pages/383/Anatomist.html
14. De Felipe J. (2002) Sesquicentenary of the birthday of Santiago Ramón y Cajal, the father of modern neuroscience. *Trends Neurosci* **25**: 481–484.
15. http://pathology.wustl.edu/corbolab/Cajal.htm
16. Cajal SR. (1917) *Recuerdos de mi Vida*. Moya, Madrid.
17. Baer MF. Connors BW, Paradiso MA. (2007) *Neuroscience: Exploring the Brain*. Lippincott Williams & Wilkins, p. 27.
18. Gray H. (1918) *Gray's Anatomy of the Human Body*. Lea & Febiger.
19. Cajal S. (1888) Estructure de los centres nervosa de los aves. *Rev Trim Histol Norm Patol* **1**: 1–10.
20. http://es.wikipedia.org/wiki/Doctrina_de_la_neurona
21. http://en.wikipedia.org/wiki/Wernicke-Korsakoff_syndrome
22. http://www.ninds.nih.gov/disorders/wernicke_korsakoff/wernicke-korsakoff.htm
23. http://www.britanica.co/EBchecked/topic/490729/Santiago-Ramon-y-Cajal
24. http://en.wikipedia.org.wiki/Alexander_Graham_Bell
25. http://www.ihcworld.com/imagegallery/special-stain.htm
26. Bennis R, Draft R, Kang H, Livet J, Weissman T, Lichtman J, Sanes J. Technicolor brains: mapping neural circuits in Brainbow mice (http://www.mcbharvard.edu/mcb/news-detail/3409/technicolor-brains-mappinng-neural-circuits-inbrainbow-mice).

27. http://www.newscientist.com/gallery/dn14971-smallworld-gallery/9

28. Simpkins CA, Simpkins AM. (2013) *Neuroscience for Clinicians: Evidence, Models, and Practice.* Springer Science Business Media, New York, Chap. 1. (The birth of a New Science).

29. "Out of pocket."

30. http://www.scholarpedia.org/article/Santiago_Ramon_y_Cajal

31. Swanson LW, Grant G, Grillners S, Hoskell T, Jones G, Morrison JH. (2007) A century of neuroscience discovery: reflecting on the Nobel Prize awarded Golgi and Cajal in 1906. *Brain Res Rev* **55**: 191–192.

32. http//www.biounity.com/en/encyclopedia/Albert_von-K%C3%B6olliker.html

33. http://www.nobelprize.org_nobel_prizes/medicine/laureats/1905/cajal-bio.html

34. http://www.google.com/search?tbm=isch&hl=en&source=hp&q=helmholtz+medal&gbv=2&oq=helmholtz+medal&gs_l=img.3..0i24.12047.17937.0.22250.15.9.0.6.6.0.109.874.4j5.9.0...0.0...1ac.1.o5FiJqlcD3

35. http://en.wikipedia.org/wiki/Nobel_Prize

36. http://nobelprize.org/alfred_nobel/will/will-full.html

37. Bergengran C. (1962) *Alfred Nobel: The Man and His Work.* Thomas Nelson and Sons, London, Edinburgh, Paris, Melbourne, Johannesburg, Toronto, New York, Chap. 21, pp. 130–139.

38. Preston R. (1958) The adventures of a pacifist. *The New Yorker*, Mar. 15, p. 49.

39. Encyclopaedia Britannica (2010). Chicago, London, New Delhi, Paris, Seoul, Sydney, Taipei, Tokyo, Vol. 8, p. 729.

40. *Ibid.,* pp. 738–739.

41. Excerpt from the will of Alfred Nobel http://www.nobelprize.org/alfred_nobel/will/short_testament.html

42. http://en.wikipedia.org/wiki/Female_Nobel_Prize_laurestes

43. http://www.nobelpreis.org/english/index.html

44. http://nobelprize.org/alfred_nobel/biographical/articles/frangsmyr/index.html

45. Parrish M. Santiago Ramon y Cajal: "The Father of Neuroscience" (http://brain-connection.positscience.com/topics/?main=fa/cajal).

46. Golgi C. (1891) *La rette difusa negli organi centrali del sistema nervosa. Sue significato fisiolegico; Reconditi del R Instituto Lombardu di scienze e Lett* **24**: 594–603, 656–673.

47. http://nobelmedicine.co.uk/Santiagoramonycajal.htm

48. http://www.nobelprize.org_nobel_prizes/medicine/laureats/1906/index.html

49. Fishman RS. (2007) The Nobel Prize of 1906. *Arch Ophthalmol* **125**: 690–694.

50. http://www.scientificlearning.com

51. http://brainconnections.positscience.com/topics?main=fa.cajal4

52. López-Muñoz F, Boya J, Alamo C. (2006) Neuron theory, the cornerstone of neuroscience, on the centenary of the Nobel Prize award to Santiago Ramón y Cajal. *Brain Res Bull* **70**: 391–404.

53. *Encyclopaedia Britannica* (2010). Chicago, London, New Delhi, Paris, Seoul, Sydney, Taipei, Tokyo, Vol. 24, pp. 66–67.

54. Katz-Sidlow RJ. (1998), The formulation of the neuron doctrine. The island of Cajal. *Arch Neurol* **55:** 237–240.

55. Cajal S. (1989) *Recollections of My Life.* MIT Press, Cambridge, Massachusetts. Translated by Craigie EH with the assistance of Cano J.

56. Jacobson M. (1993) *Foundations of Neuroscience.* Plenum, New York, pp. 151–290.

57. Andres-Barquin PJ. (2001) Ramón y Cajal: a century after the publication of his masterpiece. *Endeavour* **25**: 13–17.

58. http://www.scholarpedia.org/article/Santigo_Ram%C39%B3n_y_Cajal

59. Abbott A. (2008) Hidden treasures: the Cajal collection in Madrid. The perceptive drawings, paintings, photographs and slides of Spain's neuroanatomy pioneer record a tale of ambition and rivalry. *Nature* **452**: 940.

60. Encyclopaedia Britannica (2010). Chicago, London, New Delhi, Paris, Seoul, Taipei, Tokyo. Vol. 5, p. 349.

61. http://www.google.com/images?q=italy+map&hl=en&gbv=2&rlz=1R2ADFA_enUS405&tbm=isch&ei=iV7wUM29D8njigLE04G4DA&start=60&sa=N

62. http://www.bing.com/image/search?q=golgi+staue+in+pavia&view+detail&id=9ACFE49BE5EDA22817EB0B64330FDC071F671C47&first=1

63. http://en.wikipedia.org/Camillo_Golgi

64. http://www.britannica.com/EBchecked/topic/4901729/Sntiago-Rmon-y-Cajal

65. http://himetop.wikidot.com/santiago-ramon-y-cajal-statue-in-the-colegio-de-medicos

66. http://www.japi.org/march_2010/Article_16.pdf

67. http://www.scribd.com/36087928/NPG-nature-vol-409-Issue-6819-Jan

68. Edwards SN, Donnelly YA, Sayre R. *et al.* Quantitative *in vitro* assessment of photoxicity using a human skin mode, Skin2. Photodermatol. Photoimmunol. CPL Press; 1999: 160–6.

69. www.encyclo.couk/define/photoxicity

70. www.miriam-webster.com/dictionary/photon

71. www.ncbi.nih.gov/books/NAK26893

72. www.nobelprize.org/nobel_prizes/chemistry/laureates/2008

73. Chang K. (2008) Three chemists win Nobel Prize. *The New York Times,* Oct. 9, p. A 25.